Mathematical Physics with Partial Differential Equations

Mathematical Physics with Partial Differential Equations

Second Edition

James Kirkwood
Professor of Mathematical Sciences,
Sweet Briar College, Sweet Briar, VA, USA

Academic Press is an imprint of Elsevier
125 London Wall, London EC2Y 5AS, United Kingdom
525 B Street, Suite 1800, San Diego, CA 92101-4495, United States
50 Hampshire Street, 5th Floor, Cambridge, MA 02139, United States
The Boulevard, Langford Lane, Kidlington, Oxford OX5 1GB, United Kingdom

Library of Congress Cataloging-in-Publication Data
A catalog record for this book is available from the Library of Congress

British Library Cataloguing-in-Publication Data
A catalogue record for this book is available from the British Library

ISBN: 978-0-12-814759-7

For information on all Academic Press publications visit our website at https://www.elsevier.com/books-and-journals

Publisher: Katey Birtcher
Acquisition Editor: Katey Birtcher
Editorial Project Manager: Lindsay Lawrence
Production Project Manager: Bharatwaj Varatharajan
Designer: Victoria Pearson

Typeset by TNQ Books and Journals

To Bessie, Katie, and Elizabeth. The lights of my life.

Contents

Preface

The major purposes of this book are to present partial differential equations (PDEs) and vector analysis at an introductory level. As such, it could be considered a beginning text in mathematical physics. It is also designed to provide a bridge from undergraduate mathematics to the first graduate mathematics course in physics, applied mathematics, or engineering. In these disciplines, it is not unusual for such a graduate course to cover topics from linear algebra, ordinary differential equations and partial differential equations (PDEs), advanced calculus, vector analysis, complex analysis and probability, and statistics at a highly accelerated pace.

In this text we study in detail, but at an introductory level, a reduced list of topics important to the abovementioned disciplines. In PDEs, we consider Green's functions, the Fourier and Laplace transforms, and how these are used to solve PDEs. We also study using separation of variables to solve PDEs in great detail. Our approach is to examine the three prototypical second-order PDEs—Laplace's equation, the heat equation, and the wave equation—and solve each equation with each method. The premise is that in doing so, the reader will become adept at each method and comfortable with each equation.

The other prominent area of the text is vector analysis. While the usual topics are discussed, an emphasis is placed on viewing concepts rather than formulas. For example, we view the curl and gradient as properties of a vector field rather than as simply equations. A significant portion of this area deals with curvilinear coordinates to reinforce the idea of conversion of coordinate systems.

Reasonable prerequisites for the course are a course in multivariable calculus, familiarity with ordinary differential equations to the point of being able to solve a second-order boundary problem with constant coefficients, and some experience with linear algebra.

In dealing with ordinary differential equations, we emphasize the linear operator approach. That is, we consider the problem as being an eigenvalue/eigenvector problem for a self-adjoint operator. In addition to eliminating some tedious computations regarding orthogonality, this serves as a unifying theme and a more mature viewpoint.

The level of the text generally lies between that of the classic encyclopedic texts of Boas and Kreyszig and the newer text by McQuarrie, and the PDE books of Weinberg and Pinsky. Topics such as Fourier series are developed in a mathematically rigorous manner. The section on completeness of eigenfunctions of a Sturm–Liouville problem is considerably more advanced than the rest of the text and can be omitted if one wishes to merely accept the result.

The text is written at a level where it can be used as a self-contained reference as well as an introductory text. There was a concerted effort to avoid situations where filling in details of an argument would be a challenge. One thought in writing the text was that it would serve as a source for students in subsequent courses that felt "I know I'm supposed to know how to derive this, but I don't."

A couple of such examples are the fundamental solution of Laplace's equation and the spectrum of the Laplacian.

The major changes from the first edition are first that a chapter on generating functions has been added. This gives a nice way of considering solutions to LaGuerre equations, Hermite equations, Legendre's equations, and Bessel equations. Second, in depth analyses for the rigid rotor, one-dimensional quantum mechanical oscillator, and the hydrogen atom are presented. Also, more esoteric coordinate systems have been moved to an appendix.

CHAPTER

Preliminaries

1

1.1 SELF-ADJOINT OPERATORS

The purpose of this text is to study some of the important equations and techniques of mathematical physics. It is a fortuitous fact that many of the most important such equations are linear, and we can apply the well-developed theory of linear operators. We assume knowledge of basic linear algebra but review some definitions, theorems, and examples that will be important to us.

Definition:

A *linear operator* (or *linear function*) from a vector space V to a vector space W, is a function $\mathcal{L} : V \to W$ for which

$$\mathcal{L}(a_1 \widehat{v}_1 + a_2 \widehat{v}_2) = a_1 \mathcal{L}(\widehat{v}_1) + a_2 \mathcal{L}(\widehat{v}_2)$$

for all $\widehat{v}_1, \widehat{v}_2 \in V$ and scalars a_1 and a_2.

One of the most important linear operators for us will be

$$\mathcal{L}[y] = a_0(x)y(x) + a_1(x)y'(x) + a_2(x)y''(x)$$

where $a_0(x)$, $a_1(x)$ and $a_2(x)$ are continuous functions.

Definition:

If $\mathcal{L} : V \to V$ is a linear operator, then a nonzero vector $\widehat{v}$ is an *eigenvector* of $\mathcal{L}$ with *eigenvalue* λ if $\mathcal{L}(\widehat{v}) = \lambda \widehat{v}$.

Note that $\widehat{0}$ cannot be an eigenvector, but 0 can be an eigenvalue.

Example:

For $\mathcal{L} = \dfrac{d}{dx}$, we have

$$\mathcal{L}(e^{ax}) = \frac{d}{dx}(e^{ax}) = ae^{ax}$$

so e^{ax} is an eigenvector of $\mathcal{L}$ with eigenvalue a.

An extremely important example is $\mathcal{L} = \dfrac{d^2}{dx^2}$. Among its properties are

$$\mathcal{L}(\sin nx) = \frac{d^2}{dx^2}(\sin nx) = -n^2 \sin nx \quad \text{and}$$

$$\mathcal{L}(\cos nx) = \frac{d^2}{dx^2}(\cos nx) = -n^2 \cos nx.$$

Mathematical Physics with Partial Differential Equations. https://doi.org/10.1016/B978-0-12-814759-7.00001-6

We leave it as Exercise 1 to show that if $\widehat{v}$ is an eigenvector of $\mathcal{L}$ with eigenvalue λ, then $a\widehat{v}$ is also an eigenvector of $\mathcal{L}$ with eigenvalue λ.

Definition:

An *inner product* (also called a *dot product*) on a vector space V with scalar field F (which is the real number $\mathbb{R}$ or the complex number $\mathbb{C}$) is a function $\langle\ ,\ \rangle: V \times V \to F$ such that for all $f, g, h \in V$ and $a \in F$

$$\langle af, g\rangle = a\langle f, g\rangle;$$

$$\langle f, ag\rangle = \overline{a}\langle f, g\rangle, \quad \text{where } \overline{x+iy} = x - iy;$$

$$\langle f+g, h\rangle = \langle f, h\rangle + \langle g, h\rangle;$$

$$\langle f, g\rangle = \overline{\langle g, f\rangle};$$

$\langle f, f\rangle \geq 0$ with equality if and only if $f = 0$.

A vector space with an inner product is called an inner product space.

If $V = \mathbb{R}^n$, the usual inner product for $\widehat{a} = (a_1, \ldots, a_n)$, $\widehat{b} = (b_1, \ldots, b_n)$ is

$$\left\langle \widehat{a}, \widehat{b} \right\rangle = a_1 b_1 + \cdots + a_n b_n.$$

If the vector space is $\mathbb{C}^n$, then we must modify the definition, because, for example, under this definition, if $\widehat{a} = (i, i)$, then

$$\langle \widehat{a}, \widehat{a} \rangle = i^2 + i^2 = -2.$$

Thus, on $\mathbb{C}^n$, for $\widehat{a} = (a_1, \ldots, a_n)$, $\widehat{b} = (b_1, \ldots, b_n)$, we define

$$\left\langle \widehat{a}, \widehat{b} \right\rangle = a_1 \overline{b_1} + \cdots + a_n \overline{b_n}.$$

We use the notation $\langle f, f\rangle = \|f\|^2$, which is interpreted as the square of the length of f, and $\|f - g\|$ is the distance from f to g.

We shall be working primarily with vector spaces consisting of functions that satisfy some property such as continuity or differentiability. In this setting, one usually defines the inner product using an integral. A common inner product is

$$\langle f, g\rangle = \int_a^b f(x)\overline{g(x)}dx,$$

where a or b may be finite or infinite. There might be a problem with some vector spaces in that $\langle f, f\rangle = 0$ with $f \neq 0$. This problem can be overcome by a minor modification of the vector space or by restricting the functions to being continuous and will not affect our work. We leave it as Exercise 4 to show that the function defined above is an inner product.

On some occasions it will be advantageous to modify the inner product above with a weight function $w(x)$. If $w(x) \geq 0$ on $[a,b]$, then

$$\langle f, g\rangle_w = \int_a^b f(x)w(x)\overline{g(x)}dx$$

is also an inner product as we show in Exercise 5.

Definition:

A linear operator A on the inner product space V is *self-adjoint* if $\langle Af, g\rangle = \langle f, Ag\rangle$ for all $f, g \, \varepsilon \, V$.

Self-adjoint operators are prominent in mathematical physics. One example is the Hamiltonian operator. It is a fact (Stone's theorem) that energy is conserved if and only if the Hamiltonian is self-adjoint. Another example is shown below. Part of the significance of this example is due to Newton's law $F = ma$.

Example:

The operator $\frac{d^2}{dx^2}$ is self-adjoint on the inner product space

$V = \{f | f$ has a continuous second derivative and is periodic on $[a, b]\}$,

with inner product

$$\langle f, g\rangle = \int_a^b f(x)\overline{g(x)}dx.$$

We must show

$$\langle f'', g\rangle = \langle f, g''\rangle;$$

that is,

$$\int_a^b f''(x)\overline{g(x)}dx = \int_a^b f(x)\overline{g''(x)}dx.$$

To do this, we integrate by parts twice. Let

$$u = \overline{g(x)} \quad du = \overline{g'(x)}$$
$$dv = f''(x) \quad v = f'(x)$$

so

$$uv - \int v\, du = \overline{g(x)}f'(x)\Big|_a^b - \int_a^b f'(x)\overline{g'(x)}dx.$$

The periodicity of f and g forces $\overline{g(x)}f'(x)\Big|_a^b = 0$. Thus,

$$\int_a^b f''(x)\overline{g(x)}dx = -\int_a^b f'(x)\overline{g'(x)}dx.$$

Integrating the integral on the right by parts with

$$u = \overline{g'(x)} \quad du = \overline{g''(x)}$$
$$dv = f'(x) \quad v = f(x)$$

we have

$$-\int_a^b f'(x)\overline{g'(x)}dx = -\left[f(x)\overline{g'(x)}\Big|_a^b - \int_a^b f(x)\overline{g''(x)}dx \right] = \int_a^b f(x)\overline{g''(x)}dx.$$

Notice that if $[a,b]$ is of length 2π, then $\{\sin(nx)\ \cos(nx)|n \in \mathbb{Z}\}$ is a subset of V. We next prove two important facts about self-adjoint operators.

Theorem:

If $\mathcal{L} : V \rightarrow V$ is a self-adjoint operator, then

1. The eigenvalues of $\mathcal{L}$ are real;
2. Eigenvectors of $\mathcal{L}$ with different eigenvalues are orthogonal; that is, their inner product is 0.

Proof:

1. Suppose that f is an eigenvector of $\mathcal{L}$ with eigenvalue λ. Then

$$\langle \mathcal{L}f,f\rangle = \langle \lambda f,f\rangle = \lambda\langle f,f\rangle$$

and

$$\langle f,\mathcal{L}f\rangle = \langle f,\lambda f\rangle = \overline{\lambda}\langle f,f\rangle.$$

Since $\mathcal{L}$ is self-adjoint,

$$\langle \mathcal{L}f,f\rangle = \langle f,\mathcal{L}f\rangle \text{ so } \lambda\langle f,f\rangle = \overline{\lambda}\langle f,f\rangle$$

and since $\langle f,f\rangle \neq 0$, we have $\lambda = \overline{\lambda}$, so λ is real.

2. Suppose

$$\mathcal{L}f = \lambda_1 f \text{ and } \mathcal{L}g = \lambda_2 g \text{ with } \lambda_1 \neq \lambda_2.$$

Then

$$\langle \mathcal{L}f,g\rangle = \langle \lambda_1 f,g\rangle = \lambda_1\langle f,g\rangle$$

and

$$\langle \mathcal{L}f,g\rangle = \langle f,\mathcal{L}g\rangle = \langle f,\lambda_2 g\rangle = \lambda_2\langle f,g\rangle.$$

So

$$\lambda_1\langle f,g\rangle = \lambda_2\langle f,g\rangle$$

and thus $\langle f,g\rangle = 0$ because $\lambda_1 \neq \lambda_2$.

Example:

We have

$$\int_0^{2\pi} \sin(nx)\cos(mx)\, dx = 0,$$

$$\int_0^{2\pi} \cos(nx)\cos(mx)\,dx = 0, \quad \text{if } m \neq n;$$

$$\int_0^{2\pi} \sin(nx)\sin(mx)dx = 0, \quad \text{if } m \neq n;$$

for m and n integers. This is because sin (nx) and cos (mx) are eigenfunctions of the self-adjoint operator d^2/dx^2 with the inner product defined above with different eigenvalues.

We shall use the technique of the example above to prove the orthogonality of functions such as Bessel functions and Legendre polynomials without having to resort to tedious calculations.

FOURIER COEFFICIENTS

We now describe how to determine the representation of a given vector with respect to a given basis. That is, if $\left\{\widehat{b}_1, \widehat{b}_2, \ldots\right\}$ is a basis for the vector space V, and if $\widehat{v} \,\varepsilon\, V$, we want to find scalars $a_1, a_2, \ldots$ for which

$$\widehat{v} = a_1\widehat{b}_1 + a_2\widehat{b}_2 + \cdots.$$

If the basis satisfies the characteristic below, then this is easy.

Definition:

If $\left\{\widehat{b}_1, \widehat{b}_2, \ldots\right\}$ is a set of vectors from an inner product space for which

$$\left\langle \widehat{b}_i, \widehat{b}_j \right\rangle = 0 \quad \text{if } i \neq j,$$

then $\left\{\widehat{b}_1, \widehat{b}_2, \ldots\right\}$ is called an *orthogonal set*. If, in addition,

$$\left\langle \widehat{b}_i, \widehat{b}_i \right\rangle = 1 \quad \text{for all } i,$$

then $\left\{\widehat{b}_1, \widehat{b}_2, \ldots\right\}$ is called an *orthonormal* set. A basis that is an orthogonal (orthonormal) set is called an orthogonal (orthonormal) basis.

Theorem:

If $\left\{\widehat{b}_1, \widehat{b}_2, \ldots\right\}$ is an orthogonal basis for the inner product space V, and if

$$\widehat{v} = a_1\widehat{b}_1 + a_2\widehat{b}_2 + \cdots,$$

then

$$a_k = \frac{\left\langle \widehat{v}, \widehat{b}_k \right\rangle}{\left\langle \widehat{b}_k, \widehat{b}_k \right\rangle} = \frac{\left\langle \widehat{v}, \widehat{b}_k \right\rangle}{\left\| \widehat{b}_k \right\|^2}.$$

Proof:

We have

$$\left\langle \widehat{v}, \widehat{b}_k \right\rangle = \left\langle a_1\widehat{b}_1 + a_2\widehat{b}_2 + \cdots, \widehat{b}_k \right\rangle$$
$$= a_1\left\langle \widehat{b}_1, \widehat{b}_k \right\rangle + \cdots + a_k\left\langle \widehat{b}_k, \widehat{b}_k \right\rangle + \cdots = a_k\left\langle \widehat{b}_k, \widehat{b}_k \right\rangle.$$

Thus

$$a_k = \frac{\left\langle \widehat{v}, \widehat{b}_k \right\rangle}{\left\langle \widehat{b}_k, \widehat{b}_k \right\rangle} = \frac{\left\langle \widehat{v}, \widehat{b}_k \right\rangle}{\left\| \widehat{b}_k \right\|^2}.$$

Note that if $\left\{\widehat{b}_1, \widehat{b}_2, \ldots\right\}$ is an orthonormal basis, then $a_k = \left\langle \widehat{v}, \widehat{b}_k \right\rangle$.

Definition:

The constants $\{a_1, a_2, \ldots\}$ in the theorem above are called the *Fourier coefficients* of $\widehat{v}$ with respect to the basis $\left\{\widehat{b}_1, \widehat{b}_2, \ldots\right\}$.

Fourier coefficients are important because they provide the best approximation to a vector by a subset of an orthogonal basis in the sense of the following theorem.

Theorem:

Suppose $\widehat{v}$ is a vector in an inner product space V, and $\mathcal{B} = \left\{\widehat{b}_1, \widehat{b}_2, \ldots\right\}$ is an orthogonal basis for V. Let $\{c_1, c_2, \ldots\}$ be the Fourier coefficients of $\widehat{v}$ with respect to $\mathcal{B}$. Then

$$\left\| \widehat{v} - \sum_{i=1}^{n} c_i\widehat{b}_i \right\| \le \left\| \widehat{v} - \sum_{i=1}^{n} d_i\widehat{b}_i \right\|$$

for any numbers d_i. Equality holds if and only if $c_i = d_i$ for every $i = 1, \ldots, n$.

Proof:

We assume the constants are real, and the basis is orthonormal to simplify the notation. We have

$$\left\| \widehat{v} - \sum_{i=1}^{n} d_i\widehat{b}_i \right\|^2 = \left\langle \widehat{v} - \sum_{i=1}^{n} d_i\widehat{b}_i, \widehat{v} - \sum_{i=1}^{n} d_i\widehat{b}_i \right\rangle$$
$$= \langle \widehat{v}, \widehat{v} \rangle - 2\sum_{i=1}^{n}\left\langle \widehat{v}, d_i\widehat{b}_i \right\rangle + \left\langle \sum_{i=1}^{n} d_i\widehat{b}_i, \sum_{i=1}^{n} d_i\widehat{b}_i \right\rangle. \tag{1}$$

Now,

$$\left\langle \sum_{i=1}^{n} d_i\widehat{b}_i, \sum_{i=1}^{n} d_i\widehat{b}_i \right\rangle = \sum_{i=1}^{n} d_i^2,$$

since $\left\{\widehat{b}_1, \widehat{b}_2, \ldots\right\}$ is an orthonormal basis, as we verify in Exercise 10. Also, $\left\langle \widehat{v}, d_i\widehat{b}_i \right\rangle = c_i d_i$.

Thus, the right-hand side of Eq. (1) is

$$\langle \widehat{v}, \widehat{v}\rangle - 2\sum_{i=1}^{n}\left\langle \widehat{v}, d_i\widehat{b}_i\right\rangle + \left\langle \sum_{i=1}^{n} d_i\widehat{b}_i, \sum_{i=1}^{n} d_i\widehat{b}_i \right\rangle = \langle \widehat{v}, \widehat{v}\rangle - 2\sum_{i=1}^{n} c_i d_i + \sum_{i=1}^{n} d_i^2$$

$$= \langle \widehat{v}, \widehat{v}\rangle - \sum_{i=1}^{n} c_i^2 + \left(\sum_{i=1}^{n} c_i^2 - 2\sum_{i=1}^{n} c_i d_i + \sum_{i=1}^{n} d_i^2\right)$$

$$= \langle \widehat{v}, \widehat{v}\rangle - \sum_{i=1}^{n} c_i^2 + \sum_{i=1}^{n} (c_i - d_i)^2.$$

Following the first steps in the argument above, we get

$$\left\| \widehat{v} - \sum_{i=1}^{n} c_i\widehat{b}_i \right\|^2 = \langle \widehat{v}, \widehat{v}\rangle - 2\sum_{i=1}^{n} c_i c_i + \sum_{i=1}^{n} c_i^2 = \langle \widehat{v}, \widehat{v}\rangle - \sum_{i=1}^{n} c_i^2. \tag{2}$$

Finally,

$$\langle \widehat{v}, \widehat{v}\rangle - \sum_{i=1}^{n} c_i^2 \le \langle \widehat{v}, \widehat{v}\rangle - \sum_{i=1}^{n} c_i^2 + \sum_{i=1}^{n} (c_i - d_i)^2,$$

with equality if and only if $c_i = d_i$ for all $i = 1,\ldots,n$.

Note that from Eq. (2), we have *Bessel's inequality*

$$\sum_{i=1}^{n} c_i^2 \le \langle \widehat{v}, \widehat{v}\rangle = \|\widehat{v}\|^2.$$

Example:

In this example, we demonstrate an application of eigenvalues and eigenfunctions (eigenvectors) to solve a problem in mechanics.

Suppose that we have a body of mass m_1 attached to a spring whose spring constant is k_1. See Fig. 1.1.1. We assume that the surface is frictionless. If x_1 is the

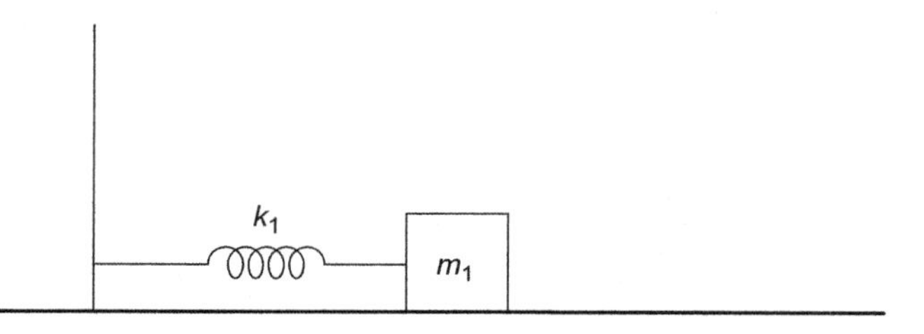

FIGURE 1.1.1

displacement of the spring from equilibrium, then, according to Hooke's law, the spring creates a force $\widehat{F} = -\, x_1\, k_1$.

Then

$$\widehat{F} = m_1 \frac{d^2\, x_1}{dt^2} = -\, x_1\, k_1.$$

Now consider the coupled system shown in Fig. 1.1.2.

We use the convention that force is positive if it pushes a body to the right.

We suppose that both springs are under no tension if the masses are at points a and b.

Suppose that the masses are at points x_1 and x_2.

Force on mass m_1:

1. Force due to spring 1: If $x_1 > a$, then spring 1 is stretched an amount $x_1 - a$ and pulls m_1 to the left. If the spring constant of spring 1 is k_1, then the force on m_1 due to spring 1 is

$$F_{1,1} = -k_1(x_1 - a\ ).$$

2. Force due to spring 3: If $x_2 - x_1 < b - a$, then spring 3 is compressed an amount $(b - a) - (x_2 - x_1)$. If the spring constant of spring 3 is k_3 then spring 3 pushes the body m_1 to the left with force

$$F_{1,3} = -k_3[(b - a) - (x_2 - x_1)] = -k_3[(b - x_2) + (x_1 - a)].$$

Thus, the total force on m_1 is

$$F_1 = F_{1,1} + F_{1,3} = -k_1(x_1 - a) - k_3[(b - x_2) + (x_1 - a)]. \tag{3}$$

Force on mass m_2:

1. Force due to spring 2: If $x_2 < b$, then spring 2 is stretched an amount $b - x_2$ and pulls m_2 to the right. If the spring constant of spring 2 is k_2, then the force on m_2 due to spring 2 is

$$F_{2,2} = k_2(b - x_2).$$

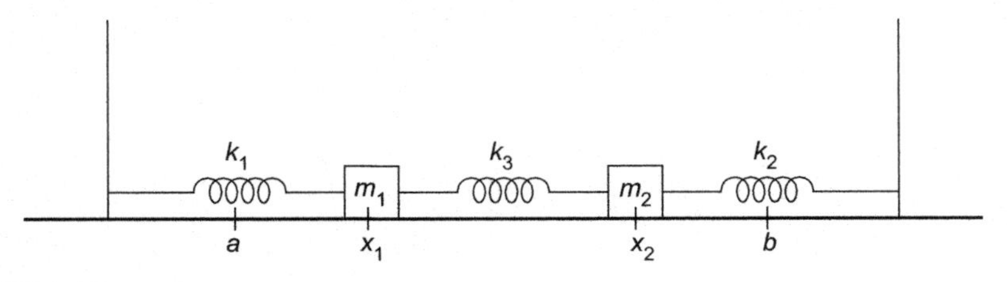

FIGURE 1.1.2

2. Force due to spring 3: If $x_2 - x_1 < b - a$, then spring 3 is compressed an amount $(b - a) - (x_2 - x_1)$. If the spring constant of spring 3 is k_3 then spring 3 pushes the body m_2 to the right with force

$$F_{2,3} = k_3[(b - a) - (x_2 - x_1)] = k_3[(b - x_2) + (x_1 - a)].$$

Thus, the total force on m_{21} is

$$F_2 = F_{2,2} + F_{2,3} = k_2(b - x_2) + k_3[(b - x_2) + (x_1 - a)]. \tag{4}$$

If we let $z_1 = x_1 - a$ and $z_2 = x_2 - b$, we get from Eq. (3)

$$F_1 = -k_1(x_1 - a) - k_3[(b - x_2) + (x_1 - a)] = -(k_1 + k_3)z_1 + k_3 z_2$$

and we get from Eq. (4)

$$F_2 = k_2(b - x_2) + k_3[(b - x_2) + (x_1 - a)] = k_3 z_1 - (k_1 + k_3)z_2.$$

Using $F = ma = -kx$, we get

$$m_1 \frac{d^2 z_1}{dt^2} = -(k_1 + k_3)z_1 + k_3 z_2$$

$$m_2 \frac{d^2 z_2}{dt^2} = k_3 z_1 - (k_1 + k_3)z_2$$

or

$$\frac{d^2 z_1}{dt^2} = -\left(\frac{k_1 + k_3}{m_1}\right) z_1 + z_2 \frac{k_3}{m_1} \tag{5}$$

$$\frac{d^2 z_2}{dt^2} = \frac{k_3}{m_2} z_1 - \left(\frac{k_2}{m_2} + \frac{k_3}{m_2}\right) z_2. \tag{6}$$

Eqs. (5) and (6) can be written as the matrix equation

$$\frac{d^2}{dt^2}\begin{pmatrix} z_1 \\ z_2 \end{pmatrix} = \begin{pmatrix} -\dfrac{k_1 + k_3}{m_1} & \dfrac{k_3}{m_1} \\ \dfrac{k_3}{m_2} & -\dfrac{k_2 + k_3}{m_2} \end{pmatrix}\begin{pmatrix} z_1 \\ z_2 \end{pmatrix} = A\begin{pmatrix} z_1 \\ z_2 \end{pmatrix}$$

where

$$A = \begin{pmatrix} -\dfrac{k_1 + k_3}{m_1} & \dfrac{k_3}{m_1} \\ \dfrac{k_3}{m_2} & -\dfrac{k_2 + k_3}{m_2} \end{pmatrix}.$$

Suppose that λ is an eigenvalue for A, so that

$$A\begin{pmatrix} z_1 \\ z_2 \end{pmatrix} = \lambda\begin{pmatrix} z_1 \\ z_2 \end{pmatrix}.$$

Then we would have

$$\frac{d^2}{dt^2}\begin{pmatrix} z_1 \\ z_2 \end{pmatrix} = \lambda\begin{pmatrix} z_1 \\ z_2 \end{pmatrix} \text{ so } \frac{d^2 z_i}{dt^2} = \lambda z_i$$

and

$$z_i(t) = B_i e^{\sqrt{\lambda} t}.$$

The eigenvalues of A are those values of λ for which $\det(A - \lambda I) = 0$. With the values we have, this would best be done with a computer algebra system (CAS); however, if we set the value of each mass to be m, and the value of each spring constant to be k, then the matrix A is

$$\begin{pmatrix} -\frac{2k}{m} & \frac{k}{m} \\ \frac{k}{m} & -\frac{2k}{m} \end{pmatrix}$$

and

$$\det(A - \lambda I) = \begin{vmatrix} -\frac{2k}{m} - \lambda & \frac{k}{m} \\ \frac{k}{m} & -\frac{2k}{m} - \lambda \end{vmatrix} = \lambda^2 + \frac{4k}{m}\lambda + \frac{3k^2}{m^2} = \left(\lambda + \frac{k}{m}\right)\left(\lambda + \frac{3k}{m}\right).$$

Thus, the eigenvalues for A are $\lambda = -\frac{k}{m}$ and $\lambda = -\frac{3k}{m}$.

Before continuing, we note there is an alternate method to calculate the equations of motion. If V is the potential energy of the system, then

$$m_i \frac{d^2 z_1}{dt^2} = -\frac{\partial V}{\partial z_i}.$$

In the spring setting,

$$V = \sum_i \frac{k_i}{2} d_i^2$$

where d_i is the distortion of the *ith* spring from equilibrium. In our problem

$$V = \frac{k_1}{2} {z_1}^2 + \frac{k_2}{2} {z_2}^2 + \frac{k_3}{2}(z_1 - z_2)^2.$$

We now find the eigenvectors for the eigenvalues. For the eigenvalue $\lambda = -\frac{k}{m}$. Suppose that

$$\begin{pmatrix} -\frac{2k}{m} & \frac{k}{m} \\ \frac{k}{m} & -\frac{2k}{m} \end{pmatrix} \begin{pmatrix} z_1 \\ z_2 \end{pmatrix} = -\frac{k}{m} \begin{pmatrix} z_1 \\ z_2 \end{pmatrix}.$$

Then

$$-\frac{2k}{m} z_1 + \frac{k}{m} z_2 = -\frac{k}{m} z_1$$
$$\frac{k}{m} z_1 - \frac{2k}{m} z_2 = -\frac{k}{m} z_2$$

so $z_1 = z_2$ and $\begin{pmatrix} 1 \\ 1 \end{pmatrix}$ is an eigenvector for $\lambda = \frac{k}{m}$.

For the eigenvalue $\lambda = -\frac{3k}{m}$. Suppose that

$$\begin{pmatrix} -\frac{2k}{m} & \frac{k}{m} \\ \frac{k}{m} & -\frac{2k}{m} \end{pmatrix} \begin{pmatrix} z_1 \\ z_2 \end{pmatrix} = -\frac{3k}{m} \begin{pmatrix} z_1 \\ z_2 \end{pmatrix}.$$

Then

$$-\frac{2k}{m} z_1 + \frac{k}{m} z_2 = -\frac{3\,k}{m} z_1$$
$$\frac{k}{m} z_1 - \frac{2k}{m} z_2 = -\frac{3k}{m} z_2$$

so $z_1 = -z_2$ and $\begin{pmatrix} 1 \\ -1 \end{pmatrix}$ is an eigenvector for $\lambda = \frac{3k}{m}$.

Now the motion of the two masses is given by

$$z(t) = C_1 \begin{pmatrix} 1 \\ 1 \end{pmatrix} \exp\left(i\sqrt{\frac{k}{n}} t \right) + C_2 \begin{pmatrix} 1 \\ -1 \end{pmatrix} \exp\left(i\sqrt{\frac{3k}{n}} t \right)$$

where $z(t) = \begin{pmatrix} z_1(t) \\ z_2(t) \end{pmatrix}$.

The linear operators that we shall use throughout the text will be differential operators. A typical example of which is

$$L[y] = y''(x) + p(x)y'(x) + q(x).$$

We shall often use the *Principle of Superposition*, which states that if $y_1(x)$ and $y_2(x)$ are solutions to

$$L[y] = y''(x) + p(x)y'(x) + q(x) = 0$$

then

$$c_1 y_1(x) + c_2 y_2(x)$$

is also a solution to

$$L[y] = y''(x) + p(x)y'(x) + q(x) = 0$$

for any constants c_1 and c_2.

EXERCISES

1. Show that if $\hat{v}$ is an eigenvector for A with eigenvalue λ, then for any scalar a, the vector $a\hat{v}$ is an eigenvector of A with eigenvalue λ.
2. For, $\langle\ ,\ \rangle$ an inner product, show that $\langle f, g+h\rangle = \langle f, g\rangle + \langle f, h\rangle$.
3. Show that

$$\mathcal{L}[y] = a_0(x)y(x) + a_1(x)y'(x) + a_2(x)y''(x)$$

where $a_0(x)$, $a_1(x)$ and $a_2(x)$ are continuous functions, is a linear operator.

4. Show that the function

$$\langle f, g\rangle = \int_a^b f(x)\overline{g(x)}dx$$

where $f(x)$ and $g(x)$ are continuous functions on $[a,b]$, is an inner product.

5. Show that the function

$$\langle f, g\rangle_w = \int_a^b f(x)w(x)\overline{g(x)}dx$$

where $f(x)$, $w(x)$ and $g(x)$ are continuous functions on $[a,b]$ and $w(x) > 0$, is an inner product. What if $w(x) < 0$?

6. In $\mathbb{R}^2$ or $\mathbb{R}^3$ the angle θ between the vectors $\hat{u}$ and $\hat{v}$ is determined by

$$\cos\theta = \frac{\hat{u}\cdot\hat{v}}{\|\hat{u}\|\|\hat{v}\|}.$$

   a. Verify that $\left(\frac{1}{2}, \frac{\sqrt{3}}{2}, 5\right)$ and $\left(\frac{\sqrt{3}}{2}, \frac{\sqrt{5}}{2}, \sqrt{24}\right)$ are points on the sphere of radius $\sqrt{26}$ feet.
   b. Find the angle between $\hat{u}$ and $\hat{v}$.
   c. What is the distance between $\hat{u}$ and $\hat{v}$ traveling along the surface of the sphere?

7. Let $\mathcal{L}[y] = (1-x^2)y''(x) - 2xy'(x)$. Show that $\mathcal{L}$ is self-adjoint with the inner product

$$\langle f, g\rangle = \int_{-1}^{1} f(x)g(x)dx.$$

8. Find the eigenvalue(s) and eigenfunction(s) for the following boundary value problems:

a. $-\dfrac{d^2f(x)}{dx^2} = \lambda f(x), \quad f(0) = f(\pi) = 0.$

b. $-\dfrac{d^2f(x)}{dx^2} = \lambda f(x), \quad f(0) = f(L) = 0.$

c. $-\dfrac{d^2f(x)}{dx^2} = \lambda f(x), \quad f(-\pi/2) = f(\pi/2) = 0.$

d. $-\dfrac{d^2f(x)}{dx^2} = \lambda f(x), \quad f(-L) = f(L) = 0.$

9. Show that if $\{\widehat{x}_1, \ldots, \widehat{x}_n\}$ is a basis for the vector space V, then every vector in V can be written as a linear combination of $\widehat{x}_1, \ldots, \widehat{x}_n$ in exactly one way.

10. Show that if $\{\widehat{x}_1, \ldots, \widehat{x}_n\}$ is an orthogonal basis for the vector space V, and

$$\widehat{v} = \sum_{i=1}^{n} a_i \widehat{x}_i \text{ and } \widehat{w} = \sum_{i=1}^{n} b_i \widehat{x}_i$$

then

$$\langle \widehat{v}, \widehat{w} \rangle = \sum_{i=1}^{n} \sum_{j=1}^{n} a_i b_j \langle \widehat{x}_i, \widehat{x}_j \rangle = \sum_{i=1}^{n} a_i b_i \langle \widehat{x}_i, \widehat{x}_i \rangle.$$

What if $\{\widehat{x}_1, \ldots, \widehat{x}_n\}$ is an orthonormal basis?

11. Suppose that $\{\widehat{x}_1, \ldots, \widehat{x}_n\}$ is a basis for the vector space V and $T{:}V \to V$ is a linear transformation for which $T(\widehat{x}) = \widehat{0}$ only if $\widehat{x} = \widehat{0}$.

a. Show that T is a one-to-one function.

b. Show that $\{T(\widehat{x}_1), \ldots, T(\widehat{x}_n)\}$ is a basis for V.

12. For a function $f(t)$, determine which of the following are linear transformations:

$$T(f) = af''(t) + bf'(t)$$

$$T(f) = af''(t) + bf'(t) + 1$$

$$T(f) = e^t f'(t)$$

$$T(f) = (f(t))^2.$$

13. Suppose that $\{\widehat{x}_1, \widehat{x}_2\}$ is an orthonormal basis for V and $T{:}V \to V$ is a linear transformation for which $\{T(\widehat{x}_1), T(\widehat{x}_2)\}$ is also an orthonormal basis for V. Show that for any vector $\widehat{x}$, $\|T(\widehat{x})\| = \|\widehat{x}\|$. (This is also true for the case of any finite basis.)

14. Recall that if $T{:}V \to V$ is a linear transformation and V is an n-dimensional vector space then T can be represented as multiplication by an $n \times n$ matrix. The matrix depends on the choice of the basis. In particular, if $\{\widehat{x}_1, \ldots, \widehat{x}_n\}$ is a basis for the vector space V and if

$$T(\widehat{x}_i) = a_{1i}\widehat{x}_1 + \cdots + a_{ni}\widehat{x}_n$$

then

$$T(b_1\hat{x}_1 + \cdots b_n\hat{x}_n) = \begin{pmatrix} a_{11} & \cdots & a_{1n} \\ \vdots & \ddots & \vdots \\ a_{n1} & \cdots & a_{nn} \end{pmatrix} \begin{pmatrix} b_1\hat{x}_1 \\ \vdots \\ b_n\hat{x}_n \end{pmatrix} = A\hat{x}$$

where

$$A = \begin{pmatrix} a_{11} & \cdots & a_{1n} \\ \vdots & \ddots & \vdots \\ a_{n1} & \cdots & a_{nn} \end{pmatrix} \text{ and } \hat{x} = b_1\hat{x}_1 + \cdots b_n\hat{x}_n.$$

Show that if $T{:}V \to V$ is a linear transformation and there is a basis of V, $\{\hat{x}_1, \ldots, \hat{x}_n\}$, consisting of eigenvectors of T, so that $T(\hat{x}_i) = \lambda_i\hat{x}_i$, then the matrix of T with respect to this basis is the diagonal matrix

$$\begin{pmatrix} \lambda_1 & \cdots & 0 \\ \vdots & \ddots & \vdots \\ 0 & \cdots & \lambda_n \end{pmatrix}.$$

This is important because computations with diagonal matrices are particularly simple.

15. A linear transformation $U{:}V \to V$ is called an orthogonal transformation (or a unitary transformation if the field is the complex numbers rather than the real numbers) if $\|U\hat{x}\| = \|\hat{x}\|$ for every $\hat{x} \in V$.

a. Show that if U is an orthogonal transformation then $\langle U\hat{x}, U\hat{y}\rangle = \langle \hat{x}, \hat{y}\rangle$ for every pair of vectors $\hat{x}, \hat{y} \in V$. Hint: Use $\langle U(\hat{x}+\hat{y}), U(\hat{x}+\hat{y})\rangle = \langle \hat{x}+\hat{y}, \hat{x}+\hat{y}\rangle$ and expand both sides of the equation.

b. What is the physical interpretation of part a in the case that V is $\mathbb{R}^2$ or $\mathbb{R}^3$.

16. Let V be the $n+1$ dimensional space of real polynomials in x.

a. Show that $\{1, x, x^2, \ldots, x^n\}$ is a basis for V.

b. Let $T{:}V \to V$ be defined by $T = \frac{d}{dx}$. Find the matrix of T with respect to this basis.

c. What is the dimension of $T(V)$? Find a basis for $T(V)$.

d. The *kernel* of a linear transformation $T{:}V \to V$ is $\{x \in V \mid T(x) = \hat{0}\}$. Show that the kernel of T is a vector space. Find the kernel of T in the case V is the $n+1$ dimensional space of real polynomials in x of degree less than or equal to n and $T = \frac{d}{dx}$.

17. a. Show that $\{1, \sin(nx), \cos(nx)\ \mathrm{n} = 1, 2, \ldots;\}$ is an orthogonal set of functions with respect to the inner product

$$\langle f, g\rangle = \int_{-\pi}^{\pi} f(x)g(x)dx.$$

b. Find a_0, a_1, a_2, b_1, b_2 for

i. $e^x = a_0 + \sum_{n=1}^{\infty} a_n \cos(nx) + \sum_{n=1}^{\infty} b_n \sin(nx).$

ii. $x + x^2 = a_0 + \sum_{n=1}^{\infty} a_n \cos(nx) + \sum_{n=1}^{\infty} b_n \sin(nx).$

1.2 CURVILINEAR COORDINATES

Many problems have a symmetry associated with them and finding the solutions to such problems—as well as interpreting the solution—can often be simplified if we work in a coordinate system that takes advantage of the symmetry. In this section we describe the methods of transforming some important functions to other coordinate systems. The most common coordinate systems besides Cartesian coordinates are cylindrical and spherical coordinates, but the methods we develop are applicable to other systems as well. In Appendix 3, we include some of the less common systems. The less common systems will not be used in later sections but are included to reinforce the techniques of the transformations.

In Fig. 1.2.1A we give a diagram of how cylindrical coordinates are defined and in Fig. 1.2.1B we do the same for spherical coordinates. **We note that while the**

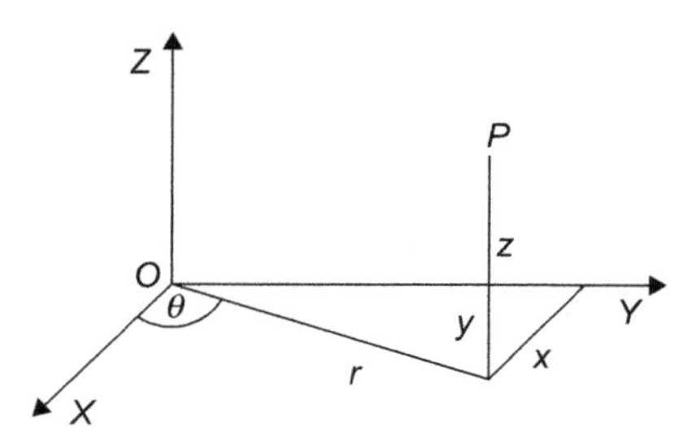

FIGURE 1.2.1A

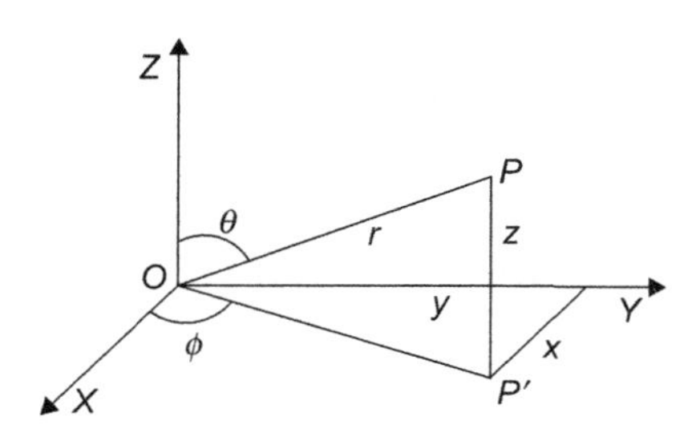

FIGURE 1.2.1B

convention we use for spherical coordinates is common, it is not universal. Some sources reverse the roles of θ and φ.

Our approach will be to describe the general case of converting from Cartesian coordinates (x, y, z) to a system of coordinates (u_1, u_2, u_3). After making a statement that holds in the general case, to visualize that statement, we demonstrate how the statement applies to cylindrical coordinates.

General case: We start with Cartesian coordinates (x,y,z) and select group of variables u_1,u_2,u_3 so that each of x,y,z is expressible in terms of u_1,u_2,u_3; that is, we have

$$x = x(u_1, u_2, u_3), \quad y = y(u_1, u_2, u_3), \quad z = z(u_1, u_2, u_3).$$

Cylindrical case: The variables in cylindrical coordinates are r,θ and z. The relations are

$$x = r\cos\theta, \;\; y = r\sin\theta, \;\; z = z; \quad 0 \le r < \infty, \;\; 0 \le \theta < 2\pi, \;\; -\infty < z < \infty.$$

We write the vector $\widehat{r} = x\widehat{i} + y\widehat{j} + z\widehat{k}$ in terms of u_1,u_2,u_3; that is,

$$\widehat{r} = x(u_1, u_2, u_3)\widehat{i} + y(u_1, u_2, u_3)\widehat{j} + z(u_1, u_2, u_3)\widehat{k}.$$

In cylindrical coordinates, this is

$$\widehat{r} = r\cos\theta\widehat{i} + r\sin\theta\widehat{j} + z\widehat{k}.$$

For some of our relations to be viable, the coordinates (u_1, u_2, u_3) must be orthogonal. This means that the pairs of surfaces $u_i = constant$ and $u_j = constant$ must meet at right angles. In the case of cylindrical coordinates, the surfaces $r = constant$ and $\theta = constant$ are shown in Fig. 1.2.2A, the surfaces $r = constant$ and $z = constant$ are shown in Fig. 1.2.2B, and the surfaces $\theta = constant$ and $z = constant$ are shown in Fig. 1.2.2C. Each pair does indeed meet at right angles. It is also possible to determine that the coordinates are orthogonal by analytical methods, as we now describe.

In the general case the vector $\dfrac{\partial \hat{r}}{\partial u_1}$ will be tangent to the u_1 curve, which is the intersection of the $u_2 = constant$ and $u_3 = constant$ surfaces. Similar relations hold for $\dfrac{\partial \hat{r}}{\partial u_2}$ and $\dfrac{\partial \hat{r}}{\partial u_3}$.

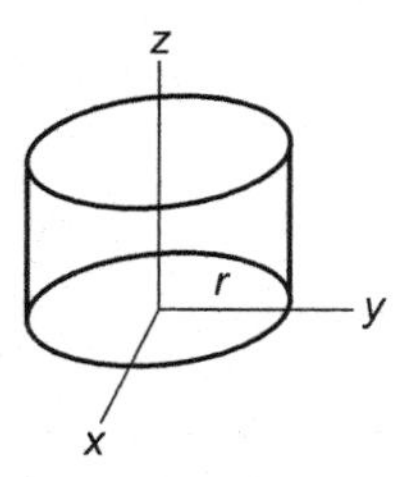

Surface of constant r

FIGURE 1.2.2A

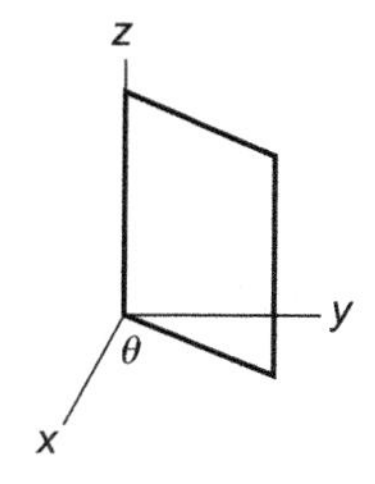

Surface of constant θ

FIGURE 1.2.2B

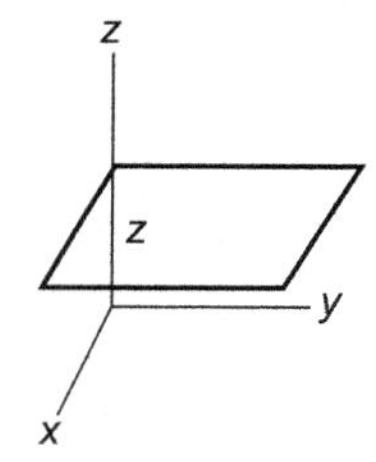

Surface of constant z

FIGURE 1.2.2C

In cylindrical coordinates

$$\frac{\partial \widehat{r}}{\partial u_1} = \frac{\partial \widehat{r}}{\partial r} = \frac{\partial}{\partial r}\left(r \cos \theta \widehat{i} + r \sin \theta \widehat{j} + z\widehat{k}\right) = \cos \theta \widehat{i} + \sin \theta \widehat{j}.$$

$$\frac{\partial \widehat{r}}{\partial u_2} = \frac{\partial \widehat{r}}{\partial \theta} = \frac{\partial}{\partial \theta}\left(r \cos \theta \widehat{i} + r \sin \theta \widehat{j} + z\widehat{k}\right) = -r \sin \theta \widehat{i} + r \cos \theta \widehat{j}.$$

$$\frac{\partial \widehat{r}}{\partial u_3} = \frac{\partial \widehat{r}}{\partial z} = \frac{\partial}{\partial z}\left(r \cos \theta \widehat{i} + r \sin \theta \widehat{j} + z\widehat{k}\right) = \widehat{k}.$$

We can show that a system of coordinates forms an orthogonal coordinate system by showing that the vectors $\frac{\partial \widehat{r}}{\partial u_i}$ are orthogonal; that is, by showing their inner product is zero. In the cylindrical case

$$\left\langle \frac{\partial \widehat{r}}{\partial r}, \frac{\partial \widehat{r}}{\partial \theta} \right\rangle = \left\langle \cos \theta \widehat{i} + \sin \theta \widehat{j}, -r \sin \theta \widehat{i} + r \cos \theta \widehat{j} \right\rangle$$

$$= -r \cos \theta \sin \theta + r \cos \theta \sin \theta = 0$$

$$\left\langle \frac{\partial \widehat{r}}{\partial r}, \frac{\partial \widehat{r}}{\partial z} \right\rangle = \left\langle \cos \theta \widehat{i} + \sin \theta \widehat{j}, \widehat{k} \right\rangle = 0$$

$$\left\langle \frac{\partial \widehat{r}}{\partial z}, \frac{\partial \widehat{r}}{\partial \theta} \right\rangle = \left\langle \widehat{k}, -r \sin \theta \widehat{i} + r \cos \theta \widehat{j} \right\rangle = 0.$$

SCALING FACTORS

We know that in an orthogonal coordinate system, the vectors $\frac{\partial \hat{r}}{\partial u_i}$ are mutually orthogonal. We create an orthonormal system of vectors $\{\widehat{e}_1, \widehat{e}_2, \widehat{e}_3\}$ by setting

$$\widehat{e}_i = \frac{\dfrac{\partial \widehat{r}}{\partial u_i}}{\left\| \dfrac{\partial \widehat{r}}{\partial u_i} \right\|}.$$

We define the scaling factors h_i by $h_i = \left\| \frac{\partial \hat{r}}{\partial u_i} \right\|$, so that

$$\frac{\partial \widehat{r}}{\partial u_i} = h_i \widehat{e}_i.$$

In the case of cylindrical coordinates,

$$\frac{\partial \widehat{r}}{\partial u_1} = \frac{\partial \widehat{r}}{\partial r} = \cos\theta \widehat{i} + \sin\theta \widehat{j}, \quad \frac{\partial \widehat{r}}{\partial u_2} = \frac{\partial \widehat{r}}{\partial \theta} = -r\sin\theta \widehat{i} + r\cos\theta \widehat{j}, \quad \frac{\partial \widehat{r}}{\partial u_3} = \frac{\partial \widehat{r}}{\partial z} = \widehat{k}$$

so

$$h_1 = h_r = \left\| \frac{\partial \widehat{r}}{\partial u_1} \right\| = 1, \quad h_2 = h_\theta = \left\| \frac{\partial \widehat{r}}{\partial u_2} \right\| = r, \quad h_3 = h_z = \left\| \frac{\partial \widehat{r}}{\partial u_3} \right\| = 1.$$

Also,

$$\widehat{e}_1 = \widehat{e}_r = \frac{\dfrac{\partial \widehat{r}}{\partial u_1}}{h_1} = \cos\theta \widehat{i} + \sin\theta \widehat{j},$$

$$\widehat{e}_2 = \widehat{e}_\theta = \frac{\dfrac{\partial \widehat{r}}{\partial u_2}}{h_2} = \frac{-r\sin\theta \widehat{i} + r\cos\theta \widehat{j}}{r} = -\sin\theta \widehat{i} + \cos\theta \widehat{j},$$

$$\widehat{e}_3 = \widehat{e}_z = \frac{\dfrac{\partial \widehat{r}}{\partial u_3}}{h_3} = \widehat{k}.$$

Back to the general case, we have

$$d\widehat{r} = \frac{\partial \widehat{r}}{\partial u_1} du_1 + \frac{\partial \widehat{r}}{\partial u_2} du_2 + \frac{\partial \widehat{r}}{\partial u_3} du_3 = h_1 \widehat{e}_1 du_1 + h_2 \widehat{e}_2 du_2 + h_3 \widehat{e}_3 du_3.$$

For cylindrical coordinates, this is

$$d\widehat{r} = h_1 \widehat{e}_1 du_1 + h_2 \widehat{e}_2 du_2 + h_3 \widehat{e}_3 du_3 = \widehat{e}_r dr + r\widehat{e}_\theta d\theta + \widehat{e}_z dz.$$

VOLUME INTEGRALS

We now describe how to convert volume integrals to other coordinate systems.

General case: Our aim is to determine an expression for an incremental volume element dV in a general coordinate system. The volume of the parallelepiped formed by three noncoplanar vectors $\widehat{A}$, $\widehat{B}$ and $\widehat{C}$ is $\left|\widehat{A}\cdot\left(\widehat{B}\times\widehat{C}\right)\right|$ (See Exercise 1). For the Cartesian case, we compute an incremental volume element dV using

$$\widehat{r} = x\widehat{i} + y\widehat{j} + z\widehat{k}, \quad d\widehat{r} = dx\widehat{i} + dy\widehat{j} + dz\widehat{k}.$$

Then

$$dV = \left|dx\widehat{i}\cdot\left(dy\widehat{j}\times dz\widehat{k}\right)\right| = dxdydz.$$

For the case

$$d\widehat{r} = h_1\widehat{e}_1du_1 + h_2\widehat{e}_2du_2 + h_3\widehat{e}_3du_3.$$

$$\begin{aligned} dV &= |h_1du_1\widehat{e}_1\cdot(h_2du_2\widehat{e}_2\times h_3du_3\widehat{e}_3)| = h_1du_1h_2du_2h_3du_3|\widehat{e}_1\cdot(\widehat{e}_2\times\widehat{e}_3)| \\ &= h_1h_2h_3du_1du_2du_3 \end{aligned}$$

since $|\widehat{e}_1\cdot(\widehat{e}_2\times\widehat{e}_3)| = 1$ because $\{\widehat{e}_1,\ \widehat{e}_2,\ \widehat{e}_3\}$ is an orthonormal system.

Another way to do this computation is to use

$$dV = \left|\frac{\partial\widehat{r}}{\partial u_1}\cdot\left(\frac{\partial\widehat{r}}{\partial u_2}\times\frac{\partial\widehat{r}}{\partial u_3}\right)\right|du_1du_2du_3$$

and that

$$\frac{\partial\widehat{r}}{\partial u_1}\cdot\left(\frac{\partial\widehat{r}}{\partial u_2}\times\frac{\partial\widehat{r}}{\partial u_3}\right) = \begin{vmatrix} \dfrac{\partial x}{\partial u_1} & \dfrac{\partial y}{\partial u_1} & \dfrac{\partial z}{\partial u_1} \\ \dfrac{\partial x}{\partial u_2} & \dfrac{\partial y}{\partial u_2} & \dfrac{\partial z}{\partial u_2} \\ \dfrac{\partial x}{\partial u_3} & \dfrac{\partial y}{\partial u_3} & \dfrac{\partial z}{\partial u_3} \end{vmatrix}. \tag{1}$$

The determinant in Eq. (1) is called the Jacobian of x,y,z with respect to u_1, u_2, u_3 and is denoted $\dfrac{\partial(x,y,z)}{\partial(u_1,u_2,u_3)}$. So we have

$$dV = \left|\frac{\partial(x,y,z)}{\partial(u_1,u_2,u_3)}\right|du_1du_2du_3.$$

We now compute dV for cylindrical coordinates. We demonstrate two methods. First, we use

$$dV = h_1h_2h_3du_1du_2du_3$$

where $u_1 = r$, $u_2 = \theta$, $u_3 = z$ so that $du_1 = dr$, $du_2 = d\theta$, $u_3 = dz$. We have previously found that $h_1 = 1$, $h_2 = \mathrm{r}$, $h_3 = 1$ so

$$dV = h_1 h_2 h_3 \, du_1 \, du_2 \, du_3 = r \, dr \, d\theta \, dz.$$

For the second method we compute the Jacobian. We have

$$x = r\cos\theta, \quad \text{so} \quad \frac{\partial x}{\partial u_1} = \frac{\partial x}{\partial r} = \cos\theta, \quad \frac{\partial x}{\partial u_2} = \frac{\partial x}{\partial \theta} = -r\sin\theta, \quad \frac{\partial x}{\partial u_3} = \frac{\partial x}{\partial z} = 0$$

$$y = r\sin\theta, \quad \text{so} \quad \frac{\partial y}{\partial u_1} = \frac{\partial y}{\partial r} = \sin\theta, \quad \frac{\partial y}{\partial u_2} = \frac{\partial y}{\partial \theta} = r\cos\theta, \quad \frac{\partial y}{\partial u_3} = \frac{\partial y}{\partial z} = 0$$

$$z = z, \quad \text{so} \quad \frac{\partial z}{\partial u_1} = \frac{\partial z}{\partial r} = 0, \quad \frac{\partial z}{\partial u_2} = \frac{\partial z}{\partial \theta} = 0, \quad \frac{\partial z}{\partial u_3} = \frac{\partial z}{\partial z} = 1.$$

Thus

$$\begin{vmatrix} \frac{\partial x}{\partial u_1} & \frac{\partial y}{\partial u_1} & \frac{\partial z}{\partial u_1} \\ \frac{\partial x}{\partial u_2} & \frac{\partial y}{\partial u_2} & \frac{\partial z}{\partial u_2} \\ \frac{\partial x}{\partial u_3} & \frac{\partial y}{\partial u_3} & \frac{\partial z}{\partial u_3} \end{vmatrix} = \begin{vmatrix} \cos\theta & \sin\theta & 0 \\ -r\sin\theta & r\cos\theta & 0 \\ 0 & 0 & 1 \end{vmatrix} = r$$

and $dV = r dr d\theta dz$.

In multivariable calculus, one shows that if $\hat{y} = f(\hat{x})$, then

$$\iiint_V g(\hat{y}) dy_1 dy_2 dy_3 = \iiint_{V_0} g(f(\hat{x})) |\det J(\hat{x})| dx_1 dx_2 dx_3 \tag{2}$$

where $J(\hat{x})$ is the matrix from which the Jacobian is formed.

There are different ways that Eq. (2) is expressed in other sources. One other way is

$$\iiint_V f(x, y, z) dxdydz = \iiint_{V_0} f(x(u, v, w), x(u, v, w), x(u, v, w)) \left| \frac{\partial(x, y, z)}{\partial(u, v, w)} \right| dudvdw.$$

Example:
Evaluate

$$\iint_A e^{x^2+y^2} dxdy$$

where A is the circle $x^2+y^2 \le 9$, by changing to polar coordinates.

In Cartesian coordinates,

$$\iint_A e^{x^2+y^2} dxdy = \int_{x=-3}^{3} \int_{y=-\sqrt{9-x^2}}^{y=\sqrt{9-x^2}} e^{x^2+y^2} dxdy.$$

To convert to polar coordinates, we compute

$$|\det J(\widehat{x})| = \left|\frac{\partial(x,y)}{\partial(r,\theta)}\right| = \begin{vmatrix} \frac{\partial x}{\partial r} & \frac{\partial x}{\partial \theta} \\ \frac{\partial y}{\partial r} & \frac{\partial y}{\partial \theta} \end{vmatrix} = \begin{vmatrix} \cos\theta & -r\sin\theta \\ \sin\theta & r\cos\theta \end{vmatrix} = r.$$

The region A in polar coordinates is $0 \leq r \leq 3$, $0 \leq \theta \leq 2\pi$. So, in this case, Eq. (2) says

$$\iint_A e^{x^2+y^2} dxdy$$
$$= \int_{x=-3}^{3} \int_{y=-\sqrt{9-x^2}}^{y=\sqrt{9-x^2}} e^{x^2+y^2} dxdy = \int_{r=0}^{3} \int_{\theta=0}^{2\pi} e^{r^2} rd\theta dr$$
$$= 2\pi \int_0^3 e^{r^2} rdr = \pi\left(e^9 - 1\right).$$

Example:
We compute

$$\iint_A (x^2 + y^2) dxdy$$

where A is the region bounded by $1 \leq xy \leq 9$, and the lines $y = x$ and $y = 4x$.

We seek a coordinate system (u,v) so that the transformed region of integration will be a rectangle $a \leq u \leq b$, $c \leq v \leq d$. The graph of the region A is shown in Fig. 1.2.3.

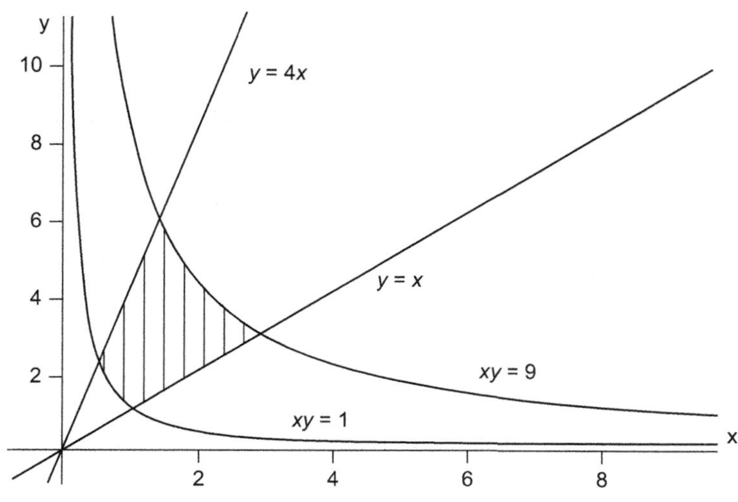

FIGURE 1.2.3

If we let $u = xy$ then $1 \le u \le 9$. If we let $v = \frac{y}{x}$, then $1 \le v \le 4$. Now

$$uv = (xy)\left(\frac{y}{x}\right) = y^2, \quad \text{so } y = \sqrt{uv}$$

$$\frac{u}{v} = \frac{xy}{\frac{y}{x}} = x^2, \quad \text{so } x = \sqrt{\frac{u}{v}}.$$

We compute the Jacobian. We have

$$\frac{\partial x}{\partial u} = \frac{1}{2\sqrt{uv}}, \quad \frac{\partial x}{\partial v} = -\frac{1}{2v}\sqrt{\frac{u}{v}}, \quad \frac{\partial y}{\partial u} = \frac{1}{2}\sqrt{\frac{v}{u}}, \quad \frac{\partial y}{\partial v} = \frac{1}{2}\sqrt{\frac{u}{v}}.$$

Then

$$\frac{\partial(x,y)}{\partial(u,v)} = \begin{vmatrix} \frac{\partial x}{\partial u} & \frac{\partial x}{\partial v} \\ \frac{\partial y}{\partial u} & \frac{\partial y}{\partial v} \end{vmatrix} = \begin{vmatrix} \frac{1}{2\sqrt{uv}} & -\frac{1}{2v}\sqrt{\frac{u}{v}} \\ \frac{1}{2}\sqrt{\frac{v}{u}} & \frac{1}{2}\sqrt{\frac{u}{v}} \end{vmatrix} = \frac{1}{2v}$$

and

$$\iint_A (x^2 + y^2)\,dxdy = \int_{v=1}^{4}\left(\int_{u=1}^{9}\left(\frac{u}{v} + uv\right)\left|\frac{\partial(x,y)}{\partial(u,v)}\right| du\right) dv$$

$$= \int_{v=1}^{4}\left(\int_{u=1}^{9}\left(\frac{u}{v} + vu\right)\frac{1}{2v}du\right) dv = \frac{1}{2}\int_{v=1}^{4}\left(\frac{u^2}{2v^2} + \frac{u^2}{2}\right)\Big|_{u=1}^{9} dv$$

$$= \int_{1}^{4} 20\left(\frac{1}{v^2} + 1\right) dv = 75.$$

In the next section, we shall use the following forms of the change of variables equation:

1. If φ^{-1} exists and is differentiable for $\varphi(a) \le x \le \varphi(b)$, then

$$\int_a^b f(t)h[\varphi(t)]dt = \int_{\varphi(a)}^{\varphi(b)} f\left[\varphi^{-1}(x)\right]h(x)\left[\varphi^{-1}(x)\right]' dx.$$

2. $\iiint_V f(x,y,z)dxdydz = \iiint_{V_0} f(\xi_1,\xi_2,\xi_3)\, h_1h_2h_3 d\xi_1 d\xi_2 d\xi_3$.

THE GRADIENT

Next we determine the gradient of a function f, denoted ∇f. In Cartesian coordinates

$$\nabla f = \frac{\partial f}{\partial x}\hat{i} + \frac{\partial f}{\partial y}\hat{j} + \frac{\partial f}{\partial z}\hat{k}.$$

To compute ∇f in the general case, we set $\nabla f = A\hat{e}_1 + B\hat{e}_2 + C\hat{e}_3$ and write df in two different ways.

First, $df = \nabla f \cdot d\hat{r}$ and using that $d\hat{r} = h_1\hat{e}_1 du_1 + h_2\hat{e}_2 du_2 + h_3\hat{e}_3 du_3$ we get

$$\begin{aligned} df = \nabla f \cdot d\hat{r} &= (A\hat{e}_1 + B\hat{e}_2 + C\hat{e}_3)\cdot(h_1\hat{e}_1 du_1 + h_2\hat{e}_2 du_2 + h_3\hat{e}_3 du_3) \\ &= Ah_1 du_1 + Bh_2 du_2 + Ch_3 du_3. \end{aligned} \tag{3}$$

Second,

$$df = \frac{\partial f}{\partial u_1} du_1 + \frac{\partial f}{\partial u_2} du_2 + \frac{\partial f}{\partial u_3} du_3. \tag{4}$$

From Eqs. (3) and (4), we get

$$Ah_1 du_1 + Bh_2 du_2 + Ch_3 du_3 = \frac{\partial f}{\partial u_1} du_1 + \frac{\partial f}{\partial u_2} du_2 + \frac{\partial f}{\partial u_3} du_3$$

so

$$A = \frac{1}{h_1}\frac{\partial f}{\partial u_1}, \quad B = \frac{1}{h_2}\frac{\partial f}{\partial u_2}, \quad C = \frac{1}{h_3}\frac{\partial f}{\partial u_3}.$$

Thus

$$\nabla f = A\hat{e}_1 + B\hat{e}_2 + C\hat{e}_3 = \frac{1}{h_1}\frac{\partial f}{\partial u_1}\hat{e}_1 + \frac{1}{h_2}\frac{\partial f}{\partial u_2}\hat{e}_2 + \frac{1}{h_3}\frac{\partial f}{\partial u_3}\hat{e}_3.$$

Accordingly, we write ∇ as the operator

$$\nabla = \frac{1}{h_1}\hat{e}_1\frac{\partial}{\partial u_1} + \frac{1}{h_2}\hat{e}_2\frac{\partial}{\partial u_2} + \frac{1}{h_3}\hat{e}_3\frac{\partial}{\partial u_3}.$$

For the cylindrical coordinate case, we again have

$$h_1 = 1, \quad h_2 = r, \quad h_3 = 1; \quad \hat{e}_1 = \hat{e}_r, \quad \hat{e}_2 = \hat{e}_\theta, \quad \hat{e}_3 = \hat{e}_z$$

so in cylindrical coordinates,

$$\nabla = \hat{e}_r\frac{\partial}{\partial r} + \frac{1}{r}\hat{e}_\theta\frac{\partial}{\partial \theta} + \hat{e}_z\frac{\partial}{\partial z} \text{ and}$$

$$\nabla f = \hat{e}_r\frac{\partial f}{\partial r} + \frac{1}{r}\hat{e}_\theta\frac{\partial f}{\partial \theta} + \hat{e}_z\frac{\partial f}{\partial z} = \frac{\partial f}{\partial r}\hat{e}_r + \frac{1}{r}\frac{\partial f}{\partial \theta}\hat{e}_\theta + \frac{\partial f}{\partial z}\hat{e}_z.$$

THE LAPLACIAN

The final function that we consider in this section is the Laplacian, one of the most important operators in mathematics and physics. The Laplacian of the function f, denoted Δf (some authors use $\nabla^2 f$) in Cartesian coordinates is defined by

$$\Delta f = \nabla^2 f = \frac{\partial^2 f}{\partial x^2} + \frac{\partial^2 f}{\partial y^2} + \frac{\partial^2 f}{\partial z^2}.$$

The notation ∇^2 for the Laplacian is suggestive because $\nabla \cdot \nabla f$ gives the Laplacian of f. We have

$$\nabla = \frac{1}{h_1}\widehat{e}_1\frac{\partial}{\partial u_1} + \frac{1}{h_2}\widehat{e}_2\frac{\partial}{\partial u_2} + \frac{1}{h_3}\widehat{e}_3\frac{\partial}{\partial u_3}, \text{ and}$$

$$\nabla f = \frac{1}{h_1}\widehat{e}_1\frac{\partial f}{\partial u_1} + \frac{1}{h_2}\widehat{e}_2\frac{\partial f}{\partial u_2} + \frac{1}{h_3}\widehat{e}_3\frac{\partial f}{\partial u_3}$$

but computing $\nabla \cdot \nabla f$ is not as straightforward as it might seem. This is because $\frac{\partial \hat{e}_i}{\partial u_j}$ is not a simple expression. In fact,

$$\frac{\partial \widehat{e}_i}{\partial u_j} = \frac{\widehat{e}_j}{h_i}\frac{\partial h_j}{\partial u_i} \quad \text{if } i \neq j \text{ and } \frac{\partial \widehat{e}_i}{\partial u_i} = -\sum_{k \neq i}\frac{1}{h_k}\frac{\partial h_i}{\partial u_k}\widehat{e}_k.$$

We demonstrate the validity of

$$\frac{\partial \widehat{e}_i}{\partial u_i} = -\sum_{k \neq i}\frac{1}{h_k}\frac{\partial h_i}{\partial u_k}\widehat{e}_k$$

for cylindrical coordinates with $i = 2$. We have

$$\widehat{e}_2 = \widehat{e}_\theta = \frac{\partial}{\partial \theta}\left(r\cos\theta\widehat{i} + r\sin\theta\widehat{j} + z\widehat{k}\right) = -r\sin\theta\widehat{i} + r\cos\theta\widehat{j}$$

so

$$\frac{\partial \widehat{e}_\theta}{\partial \theta} = -r\cos\theta\widehat{i} - r\sin\theta\widehat{j}.$$

Also

$$\widehat{e}_1 = \widehat{e}_r = \frac{\partial}{\partial r}\left(r\cos\theta\widehat{i} + r\sin\theta\widehat{j} + z\widehat{k}\right) = \cos\theta\widehat{i} + \sin\theta\widehat{j}.$$

Now

$$-\sum_{k \neq 2}\frac{1}{h_k}\frac{\partial h_2}{\partial u_k}\widehat{e}_k = -\left[\frac{1}{h_1}\left(\frac{\partial h_2}{\partial u_1}\right)\widehat{e}_1 + \frac{1}{h_3}\left(\frac{\partial h_2}{\partial u_3}\right)\widehat{e}_3\right]$$

$$= -\left[\left(\frac{\partial h_2}{\partial u_1}\right)\widehat{e}_1 + \left(\frac{\partial h_2}{\partial u_3}\right)\widehat{e}_3\right]$$

since $h_1 = h_3 = 1$. Now $h_2 = r$ and $u_1 = r$ so $\left(\frac{\partial h_2}{\partial u_1}\right) = 1$. Also $u_3 = z$ so $\left(\frac{\partial h_2}{\partial u_3}\right) = 0$. Thus

$$-\left[\left(\frac{\partial h_2}{\partial u_1}\right)\widehat{e}_1 + \left(\frac{\partial h_2}{\partial u_3}\right)\widehat{e}_3\right] = -\widehat{e}_1 = -\left(r\cos\theta\widehat{i} + r\sin\theta\widehat{j}\right)$$

so the formula holds in this case.

In fact, it will be simpler to derive expressions for $\nabla^2 f$ and $\nabla \times \vec{F}$ after we have studied the divergence theorem and Stokes' theorem in Section 2.3. For now, we simply state the results:

$$\nabla^2 f = \frac{1}{h_1 h_2 h_3}\left[\frac{\partial}{\partial u_1}\left(\frac{h_2 h_3}{h_1}\frac{\partial f}{\partial u_1}\right) + \frac{\partial}{\partial u_2}\left(\frac{h_1 h_3}{h_2}\frac{\partial f}{\partial u_2}\right) + \frac{\partial}{\partial u_3}\left(\frac{h_1 h_2}{h_3}\frac{\partial f}{\partial u_3}\right)\right] \tag{4}$$

$$\nabla \times \vec{F} = \frac{1}{h_1 h_2 h_3}\begin{vmatrix} h_1\widehat{e}_1 & h_2\widehat{e}_2 & h_3\widehat{e}_3 \\ \dfrac{\partial}{\partial u_1} & \dfrac{\partial}{\partial u_2} & \dfrac{\partial}{\partial u_3} \\ h_1 F_1 & h_2 F_2 & h_3 F_3 \end{vmatrix} \tag{5}$$

where $\vec{F} = F_1\widehat{e}_1 + F_2\widehat{e}_2 + F_3\widehat{e}_3$.

For the case of cylindrical coordinates, we have

$$h_1 = h_3 = 1,\ h_2 = r; \quad \widehat{e}_1 = \widehat{e}_r,\ \widehat{e}_2 = \widehat{e}_\theta,\ \widehat{e}_3 = \widehat{e}_z; \quad u_1 = r,\ u_2 = \theta,\ u_3 = z,$$

so

$$\begin{aligned}\nabla^2 f &= \frac{1}{r}\left[\frac{\partial}{\partial r}\left(r\frac{\partial f}{\partial r}\right) + \frac{\partial}{\partial \theta}\left(\frac{1}{r}\frac{\partial f}{\partial \theta}\right) + \frac{\partial}{\partial z}\left(r\frac{\partial f}{\partial z}\right)\right] \\ &= \frac{1}{r}\left[r\frac{\partial^2 f}{\partial r^2} + \frac{\partial f}{\partial r} + \frac{1}{r}\frac{\partial^2 f}{\partial \theta^2} + r\frac{\partial^2 f}{\partial z^2}\right] = \frac{\partial^2 f}{\partial r^2} + \frac{1}{r}\frac{\partial f}{\partial r} + \frac{1}{r^2}\frac{\partial^2 f}{\partial \theta^2} + \frac{\partial^2 f}{\partial z^2}.\end{aligned}$$

We also have

$$\begin{aligned}\nabla \times \vec{F} &= \frac{1}{r}\begin{vmatrix} \widehat{e}_r & r\widehat{e}_\theta & \widehat{e}_z \\ \dfrac{\partial}{\partial r} & \dfrac{\partial}{\partial \theta} & \dfrac{\partial}{\partial z} \\ F_1 & rF_2 & F_3 \end{vmatrix} \\ &= \frac{1}{r}\left[\left(\frac{\partial F_3}{\partial \theta} - \frac{\partial}{\partial z}(rF_2)\right)\widehat{e}_r - \left(\frac{\partial F_3}{\partial r} - \frac{\partial F_1}{\partial z}\right)r\widehat{e}_\theta + \left(\frac{\partial}{\partial r}(rF_2) - \frac{\partial F_1}{\partial \theta}\right)\widehat{e}_z\right] \\ &= \left(\frac{1}{r}\frac{\partial F_3}{\partial \theta} - \frac{\partial F_2}{\partial z}\right)\widehat{e}_r + \left(\frac{\partial F_1}{\partial z} - \frac{\partial F_3}{\partial r}\right)\widehat{e}_\theta + \frac{1}{r}\left(r\frac{\partial F_2}{\partial r} + F_2 - \frac{\partial F_1}{\partial \theta}\right)\widehat{e}_z.\end{aligned}$$

SPHERICAL COORDINATES

One of the most commonly used three-dimensional coordinate systems is spherical coordinates. (In the examples above, we used cylindrical coordinates because the computations are simpler.)

The transformations are

$$x = r \sin\theta \cos\varphi$$
$$y = r \sin\theta \sin\varphi$$
$$z = r \cos\theta.$$

The ranges of the variables are $0 \le r < \infty$, $0 \le \theta \le \pi$, $0 \le \varphi < 2\pi$. In Exercise 2, we show

$$h_r = 1, \;\; h_\theta = r, \;\; h_\varphi = r \sin\theta$$

$$dV = r^2 \sin\theta \, dr \, d\theta \, d\varphi$$

$$\nabla f = \frac{\partial f}{\partial r}\widehat{e}_r + \frac{1}{r}\frac{\partial f}{\partial \theta}\widehat{e}_\theta + \frac{1}{r\sin\theta}\frac{\partial f}{\partial \varphi}\widehat{e}_\varphi$$

$$\nabla\cdot\vec{F} = \frac{1}{r^2}\frac{\partial}{\partial r}(r^2 F_r) + \frac{1}{r\sin\theta}\frac{\partial}{\partial\theta}(\sin\theta F_\theta) + \frac{1}{r\sin\theta}\frac{\partial F_\varphi}{\partial\varphi}$$

$$\nabla^2 f = \frac{1}{r^2}\frac{\partial}{\partial r}\left(r^2\frac{\partial f}{\partial r}\right) + \frac{1}{r^2\sin\theta}\frac{\partial}{\partial\theta}\left(\sin\theta\frac{\partial f}{\partial\theta}\right) + \frac{1}{r^2\sin\theta}\frac{\partial^2 f}{\partial\varphi^2}.$$

OTHER CURVILINEAR SYSTEMS

The most common curvilinear systems are cylindrical and spherical coordinates, but there are several lesser known systems. We shall not use these systems in the sequel, but several are given in Appendix 3. Other examples are given in Morse and Feshbach, *Methods of Theoretical Physics* and Arfken, *Mathematical Methods for Physicists*, Second Edition.

APPLICATIONS

As we mentioned at the beginning of the section, the reason for considering different coordinate systems is that many problems can be simplified if the appropriate coordinate system is used. We shall see, for example, that the most important partial differential equations in physics and mathematics—Laplace's equation, the heat equation and the wave equation—can be often be solved by separation of variables if the problem is analyzed using Cartesian, cylindrical, or spherical coordinates. (Separation of variables is a method by which a partial differential equation is solved by solving two or more ordinary differential equations. We investigate this extensively in subsequent chapters.) Another use is that symmetries that exist can change the partial differential equation to an ordinary differential equation. We consider two such examples.

Example:

We solve Laplace's equation $\Delta f = 0$ in spherical coordinates where f is a function that depends only on the distance from the origin.

In spherical coordinates,

$$\Delta f = \frac{1}{r^2}\frac{\partial}{\partial r}\left(r^2\frac{\partial f}{\partial r}\right) + \frac{1}{r^2 \sin\theta}\frac{\partial}{\partial \theta}\left(\sin\theta\frac{\partial f}{\partial \theta}\right) + \frac{1}{r^2 \sin\theta}\frac{\partial^2 f}{\partial \varphi^2} = 0,$$

but if f is independent of θ and φ, this reduces to

$$\Delta f = \frac{1}{r^2}\frac{d}{dr}\left(r^2\frac{df}{dr}\right) = \frac{d^2 f}{dr^2} + \frac{2}{r}\frac{df}{dr} = 0. \tag{6}$$

Letting $g = \frac{df}{dr}$, Eq. (6) is

$$\frac{dg}{dr} = -\frac{2}{r}g \text{ or } \frac{dg}{g} = -2\frac{dr}{r}.$$

Integrating gives

$$\int \frac{dg}{g} = -2\int \frac{dr}{r}$$

so

$$\ln g = -2\ln r + C = \ln\left(\frac{1}{r^2}\right) + \ln C_1 = \ln\frac{C_1}{r^2}.$$

Thus,

$$g = \frac{df}{dr} = \frac{C_1}{r^2}.$$

Integrating again, we get

$$\int df = C_1\int r^{-2}dr, \quad \text{so } f(r) = -\frac{C_1}{r} + C_2.$$

In many situations, it will be the case that $\lim_{r\to\infty} f(r) = 0$, and if that is the case, then $C_2 = 0$.

AN ALTERNATE APPROACH (OPTIONAL)

In the appendix we give a method of computing the Laplacian in cylindrical and spherical coordinates using just the chain rule for derivatives. The computations are tedious, and we include them only because this is the approach that many texts use.

EXERCISES

1. Show that in $\mathbb{R}^3$ the volume of the parallelepiped formed by the noncoplanar vectors $\widehat{A}$, $\widehat{B}$ and $\widehat{C}$ is $\left|\widehat{A}\cdot\left(\widehat{B}\times\widehat{C}\right)\right|$.

2. Let A be the parallelogram shown in Fig. 1.2.4.

a. Show that the equations of the lines that form the parallelogram are

$$y = x,\ \ y = x - 3,\ \ y = \frac{1}{4}x,\ \ y = \frac{1}{4}x + \frac{9}{4}.$$

b. Use the change of variables

$$u = x - y,\ \ v = y - \frac{1}{4}x$$

and find the range of variables for u and v.

c. Find the Jacobian $\dfrac{\partial(x,y)}{\partial(u,v)}$.

d. Use the transformed variables to find

$$\iint_A e^{2x+4y}dxdy.$$

3. Show that if $\widehat{f} = f_1(x,y,z)\widehat{i} + f_2(x,y,z)\widehat{j} + f_3(x,y,z)\widehat{k}$, then $\nabla\times\widehat{f} = \nabla f_1\times\widehat{i} + \nabla f_2\times\widehat{j} + \nabla f_3\times\widehat{k}$.

4. a. Compute ∇r^n where $r = |\widehat{r}| = \left|x\widehat{i} + y\widehat{j} + z\widehat{k}\right|$.

5. Show that for $f(x,y,z,t)$ $df = \nabla f\cdot d\widehat{r} + \dfrac{\partial f}{\partial t}dt$.

6. Use a change of variables so that you are integrating the function $\dfrac{x-y}{x+y}$ over a rectangle to compute the area of the region bounded by the lines $x - y = 0$, $x - y = 2$, $x + y = 1$, $x + y = 4$.

7. Use a change of variables so that you are integrating over a rectangle to compute the area of the region bounded by $ax + by = 0$, $ax + by = 5$, $cx + dy = -6$ and $cx + dy = 12$ where $a/b \neq c/d$.

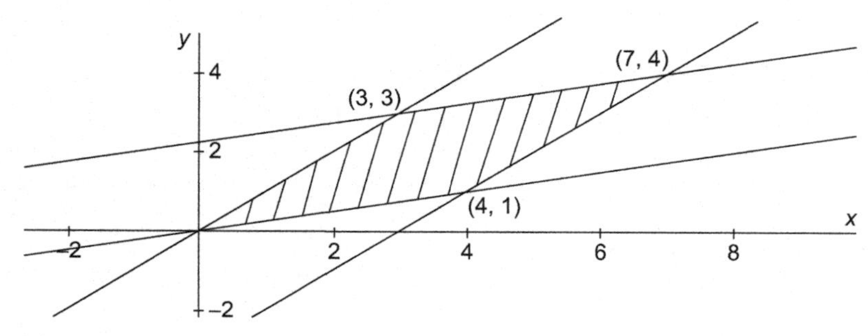

FIGURE 1.2.4

8. Use a change of variables so that you are integrating over a rectangle to compute the area of the region bounded by $y = 2x$, $y = 5x$, $y = x^2$ and $y = 6x^2$.

1.3 APPROXIMATE IDENTITIES AND THE DIRAC-δ FUNCTION

There are many examples in physics and engineering in which factors are extremely concentrated or localized. For example, the force of a hammer blow is nearly instantaneous in time. Point sources arise in electrostatic and gravitation problems, and point sources of heat are often considered in heat conduction.

In this section we establish a mathematical model that addresses these situations. We consider two major topics: approximate identities and the Dirac-δ function. Their link is how they affect the integral of a function at a point where the function is continuous.

APPROXIMATE IDENTITIES

Definition:

A sequence of functions $\{f_n(x)\}$ is an *approximate identity at* $x = 0$ provided

1. $f_n(x) \geq 0$ for $n = 1,2,\ldots$ and $-\infty < x < \infty$;
2. $\int_{-\infty}^{\infty} f_n(x)dx = 1$ for $n = 1,2,\ldots$;
3. Given $\varepsilon > 0$ and $\delta > 0$, there is a number $N(\varepsilon,\delta)$ so that if $n \geq N(\varepsilon,\delta)$, then

$$\int_{-\delta}^{\delta} f_n(x)dx > 1 - \varepsilon.$$

Conditions 1 and 2 say that each $f_n(x)$ is a probability density function, and Condition 3 says that as n becomes large the area under the graph of $f_n(x)$ becomes concentrated near $x = 0$. Fig. 1.3.1 illustrates the idea of an approximate identity.

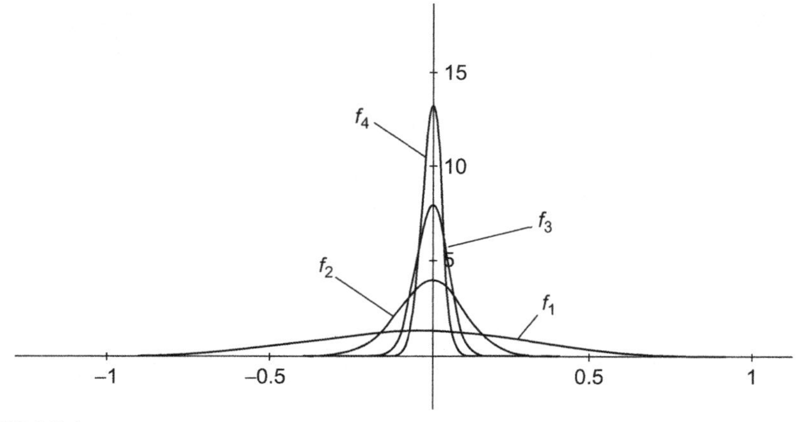

FIGURE 1.3.1

In Exercise 1 we give several examples of approximate identities.

The next theorem gives a crucial feature of approximate identities.

Theorem:

Suppose that $\{f_n(x)\}$ is an approximate identity at $x = 0$ and $g(x)$ is a bounded function that is continuous at $x = 0$. Then

$$\lim_{n\to\infty} \int_{-\infty}^{\infty} f_n(x)g(x)dx = g(0).$$

Proof:

Let $\varepsilon > 0$ be given. We show that if $g(x)$ is a bounded function that is continuous at $x = 0$, then there is a positive integer $N(g,\varepsilon)$ (emphasizing that N depends on both $g(x)$ and ε) so that if $n > N(g,\varepsilon)$, then

$$\left| \int_{-\infty}^{\infty} f_n(x)g(x)dx - g(0) \right| < \varepsilon.$$

The intuition of the proof is simple: For x very close to 0, $g(x)$ is approximately $g(0)$ because $g(x)$ is continuous. Also, because $\{f_n(x)\}$ is an approximate identity at $x = 0$, for δ small

$$\int_{-\infty}^{\infty} f_n(x)g(x)dx \approx \int_{-\delta}^{\delta} f_n(x)g(x)dx \approx g(0) \int_{-\delta}^{\delta} f_n(x)dx \approx g(0)\cdot 1 = g(0).$$

To make the argument rigorous, note that

$$\left| \int_{-\infty}^{\infty} f_n(x)g(x)dx - g(0) \right| = \left| \int_{-\infty}^{\infty} f_n(x)g(x)dx - g(0) \int_{-\infty}^{\infty} f_n(x)dx \right|$$

$$= \left| \int_{-\infty}^{\infty} f_n(x)[g(x) - g(0)]dx \right|.$$

Now

$$\left| \int_{-\infty}^{\infty} f_n(x)[g(x) - g(0)]dx \right|$$

$$= \left| \int_{-\infty}^{-\delta} f_n(x)[g(x) - g(0)]dx + \int_{-\delta}^{\delta} f_n(x)[g(x) - g(0)]dx \right.$$

$$\left. + \int_{\delta}^{\infty} f_n(x)[g(x) - g(0)]dx \right|,$$

for any $\delta > 0$.

Since $g(x)$ is bounded, there is a number $M > 0$ so that $|g(x)| \leq M$ for all x, so

$$\left|\int_{-\infty}^{-\delta} f_n(x)[g(x) - g(0)]dx\right| \leq 2M \int_{-\infty}^{-\delta} f_n(x)dx,$$

and likewise, for $\left|\int_{\delta}^{\infty} f_n(x)[g(x) - g(0)]dx\right|$.

We can now choose $\delta > 0$ so that if $|x - 0| < \delta$, then $|g(x) - g(0)| < \frac{\varepsilon}{2}$. After δ is chosen, we can find an $N(\delta)$ so that if $n > N(\delta)$, then

$$\int_{-\infty}^{-\delta} f_n(x)dx + \int_{\delta}^{\infty} f_n(x)dx < \frac{\varepsilon}{8M}.$$

Thus, we have

$$\left|\int_{-\infty}^{-\delta} f_n(x)[g(x) - g(0)]dx + \int_{-\delta}^{\delta} f_n(x)[g(x) - g(0)]dx + \int_{\delta}^{\infty} f_n(x)[g(x) - g(0)]dx\right|$$

$$\leq \int_{-\infty}^{-\delta} |f_n(x)||g(x) - g(0)|dx + \int_{\delta}^{\infty} |f_n(x)||g(x) - g(0)|dx$$

$$+ \int_{-\delta}^{\delta} |f_n(x)||g(x) - g(0)|dx < 2 \cdot 2M \cdot \frac{\varepsilon}{8M} + 1 \cdot \frac{\varepsilon}{2} = \varepsilon.$$

We now heuristically describe the limit of an approximate identity. If for each n, $f_n(x)$ is a continuous function then, under some rather reasonable conditions, for each $x \neq 0$, $\lim_{n\to\infty} f_n(x) = 0$. With these conditions, it must be that $\lim_{n\to\infty} f_n(0) = \infty$. This is the idea behind the Dirac-δ function *at* $x = 0$, which we denote $\delta_0(x)$.

Definition:

The *Dirac-δ function at* $x = 0$, denoted $\delta_0(x)$, is defined by the condition

$$\int_{-\infty}^{\infty} g(x)\delta_0(x)dx = g(0)$$

for any function $g(x)$ that is continuous at $x = 0$.

We define $\delta_{x_0}(x) = \delta_0(x - x_0)$, so that

$$\int_{-\infty}^{\infty} g(x)\delta_{x_0}(x)dx = g(x_0)$$

for any function $g(x)$ that is continuous at x_0.

The Dirac-δ function is not actually a function, but is what is known as a "generalized function" or a "distribution."

THE DIRAC-δ FUNCTION IN PHYSICS

The Dirac-δ function was used by physicists before mathematicians created a structure that would make it mathematically rigorous. (This structure is called distribution theory.) We explain the ideas of the Dirac-δ function by a physical example.

Suppose we want to model an impulse I as being a large constant force being applied for a very brief period of time, such as a hammer blow. If $f(t)$ is the force at time t, then in general

$$I = \int_0^\infty f(t)dt.$$

We want $f(t)$ to be a positive constant during a small period of time, and 0 otherwise. If $[0,h]$ is the period for which $f(t)$ is nonzero, then

$$I = \int_0^h k dt = k\text{h},$$

so

$$f(t) = \begin{cases} \dfrac{I}{h} & 0 \le t \le h; \\ 0 & \text{otherwise.} \end{cases} \tag{1}$$

The Heaviside function, $H(t)$, is defined by

$$H(t) = \begin{cases} 0 & t < 0 \\ 1 & t > 0 \end{cases}.$$

Some sources define $H(0) = 1/2$, but the difference will not affect our work.

Suppose we take $I = 1$ in Eq. (1). We want to know what happens as $h \to 0$. As $h \to 0, f(t)$ must change as h changes if it is to be true that

$$\int_0^h f(t)dt = 1.$$

So, for a given h, we denote the force function by f_h. With these modifications we have

$$f_h(t) = \begin{cases} \dfrac{1}{h} & 0 < t < h \\ 0 & \text{otherwise} \end{cases}.$$

Note that

$$H(t) - H(t-h) = \begin{cases} 1 & \text{if } t > 0 \text{ and } t - h < 0 \text{ i.e. } 0 < t < h \\ 0 & \text{otherwise} \end{cases}.$$

Thus

$$f_h(t) = \frac{H(t) - H(t-h)}{h}. \tag{2}$$

Now we could attempt to take the limit in Eq. (2) as $h \to 0$. While this is not appropriate mathematically, if we examine the right-hand side of Eq. (2) formally, it appears that we are looking for the derivative of the Heaviside function. We would have to establish a context in which this makes sense, but one could also call

$$\delta(t) = \lim_{h \to 0} \frac{H(t) - H(t-h)}{h}$$

the Dirac-δ function. Thus, in some contexts, the Dirac-δ function is said to be the derivative of the Heaviside function.

We now establish the connection between the Dirac-δ function and approximate identities.

The graphs of $\dfrac{H(t) - H(t-h)}{h}$ for $h = 1, \frac{1}{2}, \frac{1}{4}$ and $\frac{1}{8}$ are shown in Fig. 1.3.2. Note that the area under each graph is 1, the function is positive, and the area becomes more concentrated about $t = 0$ as $h \to 0$. Also

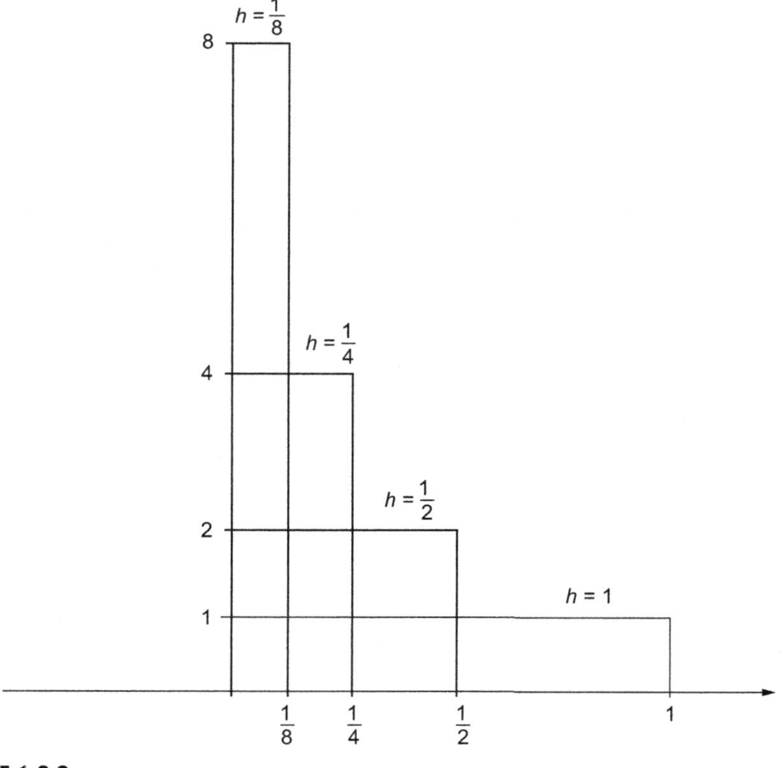

FIGURE 1.3.2

$$\int_a^b \left(\frac{H(t) - H(t-h)}{h} \right) dt = 1$$

if $[a,b]$ is any interval that contains $[0,h]$.

Proceeding heuristically, if we take the limit as $h \to 0$, we have

$$\int_a^b \delta(t) dt = 1$$

whenever $[a,b]$ contains 0.

Obviously, we have done some things that are not mathematically rigorous. Notice, however, that

$$\lim_{h \to 0} \int_a^b \left[\frac{H(t) - H(t-h)}{h} \right] g(t) dt = g(0)$$

if $g(t)$ is a function that is continuous at $t = 0$ and $0 \in (a,b)$. Thus, we could also define $\delta(t)$ by the condition

$$\int_a^b \delta(t) g(t) dt = g(0) \tag{3}$$

for every function $g(t)$ that is continuous at $t = 0$ and every interval (a,b) that contains 0.

Recapping what we have done, $\delta(t)$ or $\delta_0(t)$ is the function (actually the distribution) described by the condition of Eq. (3).

SOME CALCULUS FOR THE DIRAC-δ FUNCTION

We now demonstrate some formulas involving the Dirac-δ function. One such formula is

$$x\delta'(x) = -\delta(x).$$

In this and other formulas for the Dirac-δ function, we mean equality in the weak sense. That is, in the sense that if $g(x)$ is any suitably well-behaved function, then

$$\int_a^b x\delta'(x) g(x) dx = -\int_a^b \delta(x) g(x) dx.$$

We shall proceed formally rather than rigorously.

We give a heuristic interpretation of

$$\int_{-\infty}^{\infty} \delta'(x) g(x) dx.$$

Suppose we consider

$$\int_{-\infty}^{\infty} \lim_{h \to 0} \int_{-\infty}^{\infty} \frac{\delta(x+h) - \delta(x)}{h} g(x) dx.$$

Now

$$\frac{1}{h}\int_{-\infty}^{\infty}\delta(x+h)g(x)dx = \frac{1}{h}g(-h) \text{ and } \frac{1}{h}\int_{-\infty}^{\infty}\delta(x)g(x)dx = \frac{1}{h}g(0),$$

so

$$\int_{-\infty}^{\infty}\frac{\delta(x+h)-\delta(x)}{h}g(x)dx = \frac{g(-h)-g(0)}{h}.$$

Also

$$\lim_{h\to 0}\frac{g(-h)-g(0)}{h} = -g'(0).$$

Thus, we define

$$\int_{-\infty}^{\infty}\delta'(x)g(x)dx = -g'(0).$$

With this convention, one also writes

$$\int_{-\infty}^{\infty}\delta'(x)g(x)dx = -\int_{-\infty}^{\infty}g'(x)\delta(x)dx.$$

Example:
We show

$$x\delta'(x) = -\delta(x)$$

by showing that if $g(x)$ is a differentiable function that is continuous at $x = 0$ then

$$\int_{-\infty}^{\infty}g(x)\, x\delta'(x)dx = -\frac{d}{dx}g(x)x|_{x=0} = -[g(x)+g'(x)x]|_{x=0} = -g(x)$$
$$= -\int_{-\infty}^{\infty}g(x)\,\delta(x)dx.$$

Example:
Problems involving a change of variables.
We want to compute $\delta(f(x))$ where $f(x)$ is some suitably well-behaved function.
To begin, we recall a change of variables formula of integration.
If φ^{-1} exists and is differentiable for $\varphi(a) \le x \le \varphi(b)$, then

$$\int_a^b f(t)h[\varphi(t)]dt = \int_{\varphi(a)}^{\varphi(b)} f\left[\varphi^{-1}(x)\right]h(x)\left[\varphi^{-1}(x)\right]'dx. \tag{4}$$

Taking $h(x) = \delta(x)$ in Eq. (4), we have

$$\int_a^b f(t)\delta[\varphi(t)]dt = \int_{\varphi(a)}^{\varphi(b)} f\left[\varphi^{-1}(x)\right]\delta(x)\left[\varphi^{-1}(x)\right]'dx.$$

Suppose $\varphi(t)$ is monotone increasing for $a \le t \le b$. Now

$$(\varphi^{-1})'(t) = \frac{1}{\varphi'(\varphi^{-1}(t))}.$$

Suppose also that $\varphi(c) = 0$ for some c between a and b. Then

$$\int_{\varphi(a)}^{\varphi(b)} f[\varphi^{-1}(x)]\delta(x)[\varphi^{-1}(x)]'dx = f[\varphi^{-1}(\varphi(c))][\varphi^{-1}(x)]'\Big|_{x=\varphi(c)} = \frac{f(c)}{\varphi'(c)}.$$

If $\varphi(t)$ is monotone decreasing for $a \le t \le b$, we get the same result except for a change in sign (Exercise 7). This yields the following result:

If $\varphi(t)$ is monotone and $\varphi(c) = 0$ and $\varphi'(c) \neq 0$, then

$$\delta(\varphi(t)) = \frac{\delta(t-c)}{|\varphi'(c)|}. \tag{5}$$

Example:

By (5), to compute $\delta(3t-6)$, we have $\varphi(t) = 3t - 6$, so $\varphi(t) = 0$ if $t = 1/2$. Then $c = \frac{1}{2}$ and $|\varphi'(1/2)| = 3$ so $\delta(3t-6) = \frac{1}{3}\delta\left(t - \frac{1}{2}\right)$.

Example:

We next compute $\delta[(t-a)\ (t-b)]$ where $a < b$. In this case, the function $\varphi(t) = (t-a)\ (t-b)$ is not monotone, but is decreasing for $t < (a+b)/2$ and increasing for $t > (a+b)/2$. Also $\varphi(a) = 0$ and $\varphi(b) = 0$. Now

$$\varphi(t) = t^2 - (a+b)t + ab, \quad \text{so } \varphi'(t) = 2t - (a+b).$$

Thus

$$|\varphi'(a)| = |a-b| \quad \text{and } |\varphi'(b)| = |b-a|$$

so

$$\int_{-\infty}^{(a+b)/2} f(t)\delta(\varphi(t))dt = \frac{f(a)}{|a-b|} \quad \text{and} \quad \int_{(a+b)/2}^{\infty} f(t)\delta(\varphi(t))dt = \frac{f(b)}{|b-a|}$$

and thus

$$\int_{-\infty}^{\infty} f(t)\delta(\varphi(t))dt = \frac{1}{|a-b|}[f(a) + f(b)].$$

Hence,

$$\delta[(t-a)(t-b)] = \frac{1}{|a-b|}(\delta(t-a)) + (\delta(t-b)).$$

This example can be generalized to show that if

$$f(x) = \prod_i (x - x_i)$$

and $f'(x_i) \neq 0$ then

$$\delta[f(x)] = \sum_i \frac{\delta(x - x_i)}{|f'(x_i)|}.$$

THE DIRAC-δ FUNCTION IN CURVILINEAR COORDINATES

The representation of the Dirac-δ function in higher dimensions for Cartesian coordinates is fairly obvious. In three dimensions, if

$$\widehat{x} = (x, y, z) \quad \widehat{x}_1 = (x_1, y_1, z_1)$$

then

$$\delta(\widehat{x} - \widehat{x}_1) = \delta(x - x_1)\,\delta(y - y_1)\delta(z - z_1)$$

so that

$$\iiint_V f(\widehat{x})\delta(\widehat{x} - \widehat{x}_1)d\widehat{x} = \begin{cases} f(\widehat{x}_1) & \text{if } \widehat{x}_1 \in V \\ 0 & \text{if } \widehat{x}_1 \notin V \end{cases}.$$

The case of orthogonal curvilinear coordinates is not as straight forward. In spherical coordinates, one might expect that

$$\delta(\widehat{r} - \widehat{r}_1) = \delta(r - r_1)\,\delta(\theta - \theta_1)\delta(\varphi - \varphi_1),$$

but this is not the case. What we are doing in 3–space in evaluating. $\iiint_V f(\widehat{x})\delta(\widehat{x} - \widehat{x}_1)d\widehat{x}$ is evaluating over an increment of volume dV. The reason for the simplicity in Cartesian coordinates, is that in that case $dV = dxdydz$.

In curvilinear coordinates, recall from Section 1.2 that converting from Cartesian coordinates to an orthogonal curvilinear coordinate system (ξ_1,ξ_2,ξ_3) we must introduce the scaling factors

$$h_i = \left[\left(\frac{\partial x}{\partial \xi_i}\right)^2 + \left(\frac{\partial y}{\partial \xi_i}\right)^2 + \left(\frac{\partial z}{\partial \xi_i}\right)^2\right]^{\frac{1}{2}}.$$

For example, in spherical coordinates

$$x = r\sin\theta\cos\varphi, \quad y = r\sin\theta\sin\varphi, \quad z = \cos\theta$$

and $h_1 = 1$, $h_2 = r$, $h_3 = r\,sin\,\theta$. We have

$$d\xi_1 d\xi_2 d\xi_3 = \frac{1}{h_1\,h_2\,h_3}dxdydz$$

so

$$\iiint_V f(x, y, z)dxdydz = \iiint_{V_0} f(\xi_1, \xi_2, \xi_3)\,h_1h_2h_3d\xi_1d\xi_2d\xi_3,$$

so that in spherical coordinates

$$\iiint_V f(x,y,z)dxdydz = \iiint_{V_0} f(r,\theta,\varphi)r^2 \sin\theta \, drd\theta d\varphi.$$

Thus,

$$\iiint_V \delta(\widehat{x}_0)dxdydz = \begin{cases} 1 & \text{if } \widehat{x}_1 \in V \\ 0 & \text{if } \widehat{x}_1 \notin V \end{cases}$$

$$\iiint_{V_0} \delta(\widehat{r}_0)r^2 \sin\theta \, drd\theta d\varphi = \begin{cases} 1 & \text{if } \widehat{r}_0 \in V \\ 0 & \text{if } \widehat{r}_0 \notin V \end{cases}.$$

We then have the correspondence in spherical coordinates,

$$\delta(\widehat{r} - \widehat{r}_1) = \frac{1}{r^2 \sin\theta}\delta(r - r_1)\,\delta(\theta - \theta_1)\delta(\varphi - \varphi_1)$$

and, in the general case,

$$\delta(\widehat{r} - \widehat{r}_1) = \delta((\xi_1, \xi_2, \xi_3) - (\xi_{11}, \xi_{21}, \xi_{31})) = \frac{\delta(\xi_1 - \xi_{11})}{h_1}\frac{\delta(\xi_2 - \xi_{21})}{h_2}\frac{\delta(\xi_3 - \xi_{31})}{h_3}.$$

If there is symmetry in a problem, the integral with respect to the symmetrical variable must be projected (integrated) out. For example, in spherical coordinates, if there is symmetry with respect to φ, we have

$$\int_{\varphi=0}^{2\pi} r^2 \sin\theta d\varphi = 2\pi r^2 \sin\theta$$

so we would have

$$\delta(\widehat{r} - \widehat{r}_1) = \frac{1}{2\pi r^2 \sin\theta}\delta(r - r_1)\,\delta(\theta - \theta_1).$$

We note that in spherical coordinates, it is often useful to replace the variable θ with $\cos\theta$, in which case

$$\delta(\widehat{r} - \widehat{r}_1) = \frac{1}{r^2}\delta(r - r_1)\,\delta(\theta - \theta_1)\delta(\varphi - \varphi_1)$$

as we show in Exercise 6.

EXERCISES

1. Show that the following sequences of functions define an approximate identity at $x = 0$.

   a. $f_n(x) = \begin{cases} \dfrac{n}{2} & -\dfrac{1}{n} < x < \dfrac{1}{n}. \\ 0 & \text{otherwise} \end{cases}$

   b. $f_n(x) = \dfrac{n}{[\pi(1 + n^2x^2)]}.$

c. $f_n(x) = \left(\frac{n}{2\pi}\right)\exp\left(-\frac{n^2x^2}{2}\right).$

d. $f_n(x) = \begin{cases} c_n\left(1-x^2\right)^n & |x|<1 \\ 0 & \text{otherwise} \end{cases}$

where c_n is chosen so that $\int_{-1}^{1} c_n\left(1-x^2\right)^n dx = 1.$

2. Suppose that $\{f_n(x)\}$ is an approximate identity. Show that for any positive integer k, $\{kf_n(kx)\}$ is also an approximate identity.

3. Prove the following properties for the Dirac-δ function:

a. $x^n\delta^{(n)}(x) = (-1)^n n!\delta(x).$

b. $\delta(ax) = \frac{1}{|a|}\delta(x).$

c. $\sin x\, \delta'(x) = -\delta(x).$

d. $\cos x\, \delta'(x) = \delta'(x).$

4. a. Show that in cylindrical coordinates,

$$\delta(\hat{r}-\hat{r}_1) = \frac{1}{\rho}\delta(\rho-\rho_1)\delta(\theta-\theta_1)\delta(z-z_1).$$

b. Find the Dirac-δ function in the case there is symmetry about θ.

5. Find the Dirac-δ function in spherical coordinates when there is symmetry with respect to both the θ and φ variables.

6. Find $\delta(x^2 - x - 12)$.

7. Evaluate

$$\int_{-\infty}^{\infty}(3x-4)\delta(x^2-4)dx.$$

8. Show that in spherical coordinates, if we replace the variable θ with $\cos\theta$, then $\delta(\hat{r}-\hat{r}_1) = \frac{1}{r^2}\delta(r-r_1)\,\delta(\theta-\theta_1)\delta(\varphi-\varphi_1).$

9. Show that if $f(x), f'(x), \ldots, f^{(n)}(x)$ are continuous at $x = 0$, then

$$\int_{-\infty}^{\infty} f(x)\delta^{(n)}(x)dx = (-1)^n f^{(n)}(0).$$

10. Let $\hat{x} = (x_1, x_2, x_3)$ denote an arbitrary point in $\mathbb{R}^3$ and $\hat{\xi} = (\xi_1, \xi_2, \xi_3)$denote an arbitrary point in an orthogonal curvilinear coordinate system. Show that

$$\delta(\hat{x}-\hat{x}_0) = \frac{\delta\left(\hat{\xi}-\hat{\xi}_0\right)}{\left|\frac{\partial(x_1,x_2,x_3)}{\partial(\xi_1,\xi_2,\xi_3)}\right|}$$

where $\frac{\partial(x_1,x_2,x_3)}{\partial(\xi_1,\xi_2,\xi_3)}$ is the Jacobian.

11. The convolution of the functions f and g, denoted $f * g$, is defined by

$$(f * g)(t) = \int_{-\infty}^{\infty} f(x)g(x - t)dx.$$

What is $\delta * g$?

12. We can show

$$\int_{-\infty}^{\infty} \exp\left(-\frac{x^2}{b^2}\right) dx = b\sqrt{\pi}.$$

What is

$$\lim_{b \downarrow 0} \frac{1}{b\sqrt{\pi}} \exp\left(-\frac{x^2}{b^2}\right)?$$

1.4 THE ISSUE OF CONVERGENCE

Many functions in mathematics and physics are expressed as a series of functions, often because the desired function is the superposition of several components. One example is that the sound of a note on a musical instrument is the result of a fundamental frequency and many harmonics. (The reason middle C on a trumpet sounds different from the same note on a violin is that the relative amplitudes of the harmonics are different.)

When dealing with series, we must be careful that processes such as differentiation and integration behave with infinite sums as they do with finite sums. Whether they do depend on the way the series converges.

In this section, we discuss different types of convergence of series of functions, how this impacts mathematical operations on such series and interchanging the order of limits. We also discuss the representation of a function by its Taylor series.

SERIES OF REAL NUMBERS

Series are inextricably linked with sequences, and we begin with some basic ideas of sequences of numbers. This section is intended as a review and will primarily consist of a list of facts. A more complete discussion is available in many sources, including Kirkwood, *An Introduction to Analysis*, Second Edition.

Intuitively, a sequence is an infinite list of numbers. A sequence converges to the number L provided the numbers in the list can be made arbitrarily close to L by going sufficiently far out in the list. More precisely:

Definition:

The sequence $\{a_n\}$ *converges to* L provided given $\varepsilon > 0$, there is a number $N(\varepsilon)$ such that if $n > N(\varepsilon)$, then $|a_n - L| < \varepsilon$. If a sequence does not converge, then it is said to *diverge*.

Definition:

The sequence $\{a_n\}$ is a *Cauchy sequence* provided given $\varepsilon > 0$, there is a number $N(\varepsilon)$ such that if $n,m > N(\varepsilon)$, then $|a_n - a_m| < \varepsilon$.

Theorem:

A sequence $\{a_n\}$ converges if and only if it is a Cauchy sequence.

Intuitively, a series is the sum of an infinite number of numbers. A series is usually represented by

$$\sum_{n=1}^{\infty} a_n = a_1 + a_2 + \cdots,$$

or simply $\sum a_n$.

Associated with each series $\sum a_n$ is a sequence of partial sums $\{S_n\}$, defined by

$$S_k = a_1 + a_2 + \cdots + a_k.$$

Definition:

The series $\sum a_n$ *converges to L* if the associated sequence of partial sums $\{S_n\}$ converges to L. The series $\sum a_n$ *diverges* if $\{S_n\}$ diverges.

As was mentioned above, we shall want to know if a series of functions converges in a manner that permits certain manipulations. A frequently used test (the Weierstrass M-test) relies on knowing that a related series of numbers converges. We now list some of the most common results pertaining to convergence of series of numbers.

Theorem:

If $\sum a_n$ converges, then $\lim\limits_{n\to\infty} a_n = 0$.

It is the contrapositive of this theorem that is most useful; i.e., if $\lim a_n \neq 0$ then $\sum a_n$ diverges. There are many examples of series for which $\lim a_n = 0$ and $\sum a_n$ diverges. (One example: We shall see that $\sum \frac{1}{n}$ diverges.)

Theorem:

If the series $\sum a_n$ converges to L and the series $\sum b_n$ converges to M, then the series $\sum(\alpha a_n + \beta b_n)$ converges to $\alpha L + \beta M$.

A series of the form

$$a + ar + ar^2 + ar^3 + \cdots$$

is called a *geometric series.*

Theorem:

The geometric series

$$a + ar + ar^2 + ar^3 + \cdots \quad (a \neq 0)$$

converges to $\dfrac{a}{1-r}$ if $|r| < 1$ and diverges if $|r| \geq 1$.

Theorem (Comparison tests)

Suppose $0 \leq a_n \leq b_n$.

1. If $\sum b_n$ converges, then $\sum a_n$ converges.
2. If $\sum a_n$ diverges, then $\sum b_n$ diverges.

Theorem:

1. Ratio test: If $\sum a_n$ is a series of positive terms, and

$$r = \lim_{n \to \infty} \frac{a_{n+1}}{a_n},$$

then $\sum a_n$ converges if $r < 1$ and diverges if $r > 1$. The test is inconclusive if $r = 1$.
2. Root test: If $\sum a_n$ is a series of positive terms, and

$$r = \lim_{n \to \infty} (a_n)^{\frac{1}{n}},$$

then $\sum a_n$ converges if $r < 1$ and diverges if $r > 1$. The test is inconclusive if $r = 1$.

Theorem (Integral test):

Suppose $a_1 \geq a_2 \geq a_3 \geq \cdots \geq 0$, and $f(x)$ is a monotone decreasing function with $f(n) = a_n$ for each positive integer n. Then $\sum a_n$ converges if and only if $\int_1^\infty f(x)dx$ converges.

Theorem (p-series test):

The series $\sum \frac{1}{n^p}$ converges if $p > 1$ and diverges if $p \leq 1$.

Theorem:

The series

$$\sum \left(\frac{a_k n^k + a_{k-1} n^{k-1} + \cdots + a_0}{b_j n^j + b_{j-1} n^{j-1} + \cdots + b_0} \right)$$

converges if and only if $\sum \left(\frac{n^k}{n^j} \right)$ converges. (This assumes that $b_j n^j + b_{j-1} n^{j-1} + \cdots + b_0 \neq 0$ for any n, and the coefficients are bounded.)

CONVERGENCE VERSUS ABSOLUTE CONVERGENCE

The series we have considered so far, with the exception of geometric series, have been made up of positive terms. Another type of series that often appears is alternating series. An alternating series is the one in which each term differs in sign from its predecessor. Such series are typically represented as

$$\sum (-1)^n a_n \text{ or } \sum (-1)^{n+1} a_n$$

where $a_n > 0$.

It may be that a series where not all terms have the same sign converges, but if we change the series so that all terms do have the same sign, then the altered series diverges. Such series give rise to different notions of convergence called absolute and conditional convergence.

Definition:

A series $\sum a_n$ is said to *converge absolutely* if $\sum |a_n|$ converges. If $\sum a_n$ converges, but $\sum |a_n|$ diverges, then $\sum a_n$ is *conditionally convergent.*

Theorem:

If $\sum |a_n|$ converges, then $\sum a_n$ converges.

Theorem (Alternating series test)

Suppose

$$a_1 \geq a_2 \geq a_3 \geq \cdots \geq 0;$$

$$\lim_{n \to \infty} a_n = 0,$$

then

$$\sum (-1)^n a_n \text{ and } \sum (-1)^{n+1} a_n$$

converge.

SERIES OF FUNCTIONS

The idea with series of functions is to add infinitely many functions. The approach to developing a theory for series of functions follows that of series of numbers, in that we begin with sequences of functions.

Our setting is for each positive integer n we have a function $f_n(x)$ defined on a set E. We want to know if there is a "limit function" to which the sequence converges. The first question we need to address is what is meant by a limit function. For the limit function $f(x)$ that we consider, we assume that for each $x \in E$, the sequence of numbers $\{f_n(x)\}$ converges, and we define the limit function $f(x)$ by

$$f(x) = \lim_{n \to \infty} f_n(x).$$

We shall see that this idea requires more precision and, as a consequence, we distinguish between pointwise convergence and uniform convergence. So suppose that for each $x \in E$, the sequence of numbers $\{f_n(x)\}$ converges, and we define $f(x)$ as above.

Definition:

We say that the sequence of functions $\{f_n(x)\}$ *converges pointwise* to $f(x)$ on E if given $\epsilon > 0$ and $x \in E$, there is a number $N(x,\epsilon)$, so that if $n > N(x,\epsilon)$, then

$$|f_n(x) - f(x)| < \epsilon.$$

Note that $N(x,\varepsilon)$ depends on both ϵ and x in pointwise convergence.

Definition:

We say that the sequence of functions $\{f_n(x)\}$ *converges uniformly* to $f(x)$ on E if given $\epsilon > 0$ there is a number $N(\epsilon)$, so that if $n > N(\epsilon)$, then

$$|f_n(x) - f(x)| < \epsilon \quad \text{for every } x \in E.$$

In uniform convergence, $N(\varepsilon)$ depends only on ε. That is, with ε fixed, there is a number $N(\varepsilon)$ that works for every $x \in E$.

Example:

The sequence of functions $\{f_n(x)\}$ defined by $f_n(x) = x^n$ on $E = [0,1]$ converges pointwise, but not uniformly, to

$$f(x) = \begin{cases} 0 & 0 \le x < 1 \\ 1 & x = 1 \end{cases}.$$

Table 1.4.1 may help clarify the situation. Notice that for ε fixed, as x gets closer to 1 the value of $N(x,\varepsilon)$ needed to achieve the desired tolerance between x^n and 0 grows without bound.

While the difference between pointwise and uniform convergence might seem subtle, the implications are significant. The examples below give sequences of functions that converge pointwise but not uniformly. In each example, the limit function does not behave as one might expect.

Example:

We continue to consider $f_n(x) = x^n$ on $[0,1]$. Each function $f_n(x)$ is continuous, but the limit function

$$f(x) = \begin{cases} 0 & 0 \le x < 1 \\ 1 & x = 1 \end{cases}$$

is not continuous. Furthermore,

$$\lim_{n\to\infty}\left(\lim_{x\to 1} f_n(x)\right) = 1 \text{ and } \lim_{x\to 1}\left(\lim_{n\to\infty} f_n(x)\right) = 0$$

so we cannot interchange the order of the limits.

Table 1.4.1

		x values				
		.5	.8	.9	.99	.999
	.1	4	11	22	230	2302
ϵ values	.01	7	21	44	459	4603
	.001	10	31	66	688	6904
	.0001	14	42	88	917	9206

Example:
Consider

$$f_n(x) = \begin{cases} n & 0 < x < \dfrac{1}{n} \\ 0 & \text{otherwise} \end{cases}.$$

Then $\lim_{n\to\infty} f_n(x) = 0$ for every x, so $f(x) = 0$. Note that for each n, $\int_0^1 f_n(x)dx = 1$, so

$$\lim_{n\to\infty}\left(\int_0^1 f_n(x)dx\right) = 1, \quad \text{but} \quad \int_0^1 \left(\lim_{n\to\infty} f_n(x)\right)dx \equiv \int_0^1 f(x)dx = 0.$$

Thus, in this case,

$$\lim_{n\to\infty}\left(\int_0^1 f_n(x)dx\right) \neq \int_0^1 \left(\lim_{n\to\infty} f_n(x)\right)dx.$$

Such examples prompt caution when asserting

$$\sum_{n=1}^{\infty}\left(\int_a^b f_n(x)dx\right) = \int_a^b \left(\sum_{n=1}^{\infty} f_n(x)\right)dx.$$

Recalling that a definite integral is the limit of Riemann sums, we again see that we must be careful in interchanging limits.

One can visualize uniform convergence of the sequence of functions $\{f_n(x)\}$ to $f(x)$ by drawing a "belt" of radius ε about $f(x)$ (See Fig. 1.4.1). The sequence $\{f_n(x)\}$ converges uniformly to $f(x)$ provided that if $n > N(\varepsilon)$ then $f_n(x)$ is totally within the belt.

In the two previous examples, the sequence of functions converged pointwise but not uniformly. If we have uniform convergence, then we have some useful implications.

Theorem:

Suppose that each of the functions $f_n(x)$ is continuous, and $\{f_n(x)\}$ converges uniformly to $f(x)$. Then $f(x)$ is continuous.

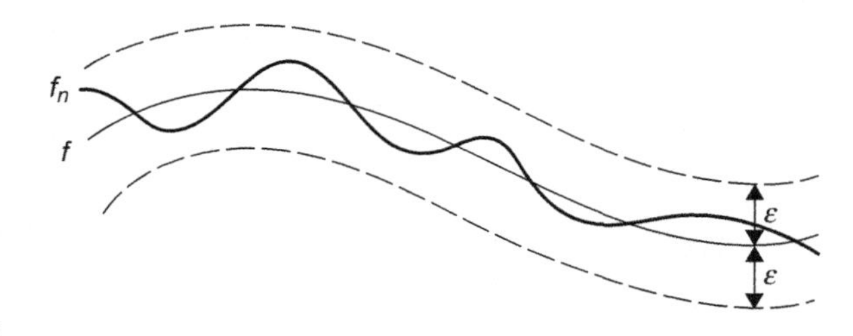

FIGURE 1.4.1

Proof:

Choose c in the domain of each $f_n(x)$. We show that if each function $f_n(x)$ is continuous at $x = c$ and $\{f_n(x)\}$ converges uniformly to $f(x)$. Then $f(x)$ is continuous at at $x = c$.

Let $\varepsilon > 0$ be given. We show that there is a $\delta(\varepsilon) > 0$ so that if $|x - c| < \delta(\varepsilon)$, then $|f(x) - f(c)| < \varepsilon$. We have

$$\begin{aligned}|f(x) - f(c)| &= |f(x) - f_n(x) + f_n(x) - f_n(c) + f_n(c) - f(c)| \\ &\leq |f(x) - f_n(x)| + |f_n(x) - f_n(c)| + |f_n(c) - f(c)|.\end{aligned}$$

Since $\{f_n(x)\}$ converges uniformly to $f(x)$, there is a number $N(\varepsilon)$ so that if $n > N(\varepsilon)$, then $|f_n(x) - f(x)| < \frac{\varepsilon}{3}$ for every x. So we have

$$|f_n(x) - f(x)| < \frac{\varepsilon}{3} \text{ and } |f_n(c) - f(c)| < \frac{\varepsilon}{3},$$

whenever $n > N(\varepsilon)$. Now choose a fixed $N_1 > N(\varepsilon)$. Since $f_{N_1}(x)$ is continuous at $x = c$, there is a number $\delta(\varepsilon)>0$ so that if $|x - c| < \delta(\varepsilon)$, then $|f_{N_1}(x) - f_{N_1}(c)| < \frac{\varepsilon}{3}$. Thus if $|x - c| < \delta(\varepsilon)$, then $|f(x) - f(c)| < \varepsilon$ so $f(x)$ is continuous at $x = c$.

Theorem:

Suppose that each of the functions $f_n(x)$ is integrable on $[a,b]$ and $\{f_n(x)\}$ converges uniformly to $f(x)$ on $[a,b]$. Then $f(x)$ is integrable on $[a,b]$ and

$$\lim_{n\to\infty}\left(\int_a^b f_n(x)dx\right) = \int_a^b \left(\lim_{n\to\infty} f_n(x)\right)dx \equiv \int_a^b f(x)dx.$$

We do not give a proof, which may be found in Kirkwood (1995). The difficult part of the proof is showing that $f(x)$ is integrable on $[a,b]$. We note that uniform convergence does not ensure convergence of improper integrals as the example

$$f_n(x) = \begin{cases} \frac{1}{n} & 0 < x < n \\ 0 & \text{otherwise} \end{cases}$$

shows.

The next example shows that uniform convergence does not ensure that derivatives behave as we might hope.

Example:

Let

$$f_n(x) = \frac{\sin nx}{\sqrt{n}}.$$

Then $|f_n(x)| \leq \frac{1}{\sqrt{n}}$, so $\{f_n(x)\}$ converges uniformly to 0. But

$$f_n'(x) = \sqrt{n}\cos nx,$$

and so $\{f_n'(x)\}$ does not converge.

This provides an example where

$$\lim_{n\to\infty}\left(\lim_{x\to c}\frac{f_n(x)-f_n(c)}{x-c}\right)\neq\lim_{x\to c}\left(\lim_{n\to\infty}\frac{f_n(x)-f_n(c)}{x-c}\right).$$

The situation for derivatives is addressed by the next theorem.

Theorem:

Let $\{f_n(x)\}$ be a sequence of functions that are differentiable on an open interval containing $[a,b]$. Suppose that

1. there is a point $x_0 \in [a,b]$ where $\{f_n(x_0)\}$ converges, and
2. $\{f_n'(x)\}$ converges uniformly on $[a,b]$.

Then

1. $\{f_n(x)\}$ converges uniformly to a function $f(x)$ on $[a,b]$ and
2. $f'(x)=\lim_{n\to\infty} f_n'(x)$ on (a,b).

The proof of this theorem is omitted, and we refer the reader to Kirkwood, *An Introduction to Analysis*, Second Edition.

The next two theorems give criteria for when a sequence of functions converge uniformly.

Theorem:

The sequence of functions $\{f_n(x)\}$ converges uniformly to $f(x)$ on E if and only if $\lim_{n\to\infty} M_n=0$, where

$$M_n=\sup_{x\in E}|f_n(x)-f(x)|.$$

The proof follows directly from the definition as is left as Exercise 19.

Example:

We show $\{xe^{-nx}\}$ converges uniformly to 0 on $[0,\infty)$.

We have

$$\frac{d}{dx}xe^{-nx}=e^{-nx}-xne^{-nx}=e^{-nx}(1-nx).$$

So $\frac{d}{dx}xe^{-nx}=0$ if $1-nx=0$ or $x=\frac{1}{n}$. One can check that $f_n(x)=xe^{-nx}$ attains its maximum on $[0,\infty)$ at $x=\frac{1}{n}$. Thus

$$M_n=\frac{1}{n}e^{-n\left(\frac{1}{n}\right)}=\frac{1}{n}e^{-1} \text{ and } \lim_{n\to\infty} M_n=0.$$

Theorem (Cauchy criterion for uniform convergence of sequences of functions):

A sequence of functions $\{f_n(x)\}$ converges uniformly on E if and only if it is uniformly Cauchy on E; that is, if and only if given $\varepsilon>0$, there is a number $N(\varepsilon)$ so that if $m,n>N(\varepsilon)$, then

$$|f_n(x)-f_m(x)|<\varepsilon$$

for every $x\in E$.

Proof:

Suppose $\{f_n(x)\}$ is uniformly Cauchy on E. We show that $\{f_n(x)\}$ is uniformly convergent. For any $x_0 \in E$, $\{f_n(x_0)\}$ is a Cauchy sequence of real numbers, and therefore converges. For each $x \in E$ define

$$f(x) = \lim_{n\to\infty} f_n(x).$$

Now $\{f_n(x)\}$ converges pointwise to $f(x)$, but we must show that the convergence is uniform. Let $\varepsilon > 0$ be given. There is a number $N(\varepsilon)$ so that if $m,n > N(\varepsilon)$, then

$$|f_n(x) - f_m(x)| < \frac{\varepsilon}{2}$$

for every $x \in E$. Now suppose $n > N(\varepsilon)$, and choose $x_0 \in E$. We shall show that $|f_n(x_0) - f(x_0)| < \varepsilon$. We have

$$|f_n(x_0) - f(x_0)| \leq |f_n(x_0) - f_m(x_0)| + |f_m(x_0) - f(x_0)|$$

for any m. Since $\{f_m(x_0)\} \to f(x_0)$, there is a number $M(\varepsilon, x_0)$ (that depends on ε and x_0) so that if $m > M(\varepsilon, x_0)$, then $|f_m(x_0) - f(x_0)| < \frac{\varepsilon}{2}$. Choose $m > \max\{N(\varepsilon), M(\varepsilon, x_0)\}$. Then

$$|f_n(x_0) - f(x_0)| \leq |f_n(x_0) - f_m(x_0)| + |f_m(x_0) - f(x_0)| < \frac{\varepsilon}{2} + \frac{\varepsilon}{2} = \varepsilon$$

if $n > N(\varepsilon)$.

The proof that a uniformly convergent sequence of functions is uniformly Cauchy is left as Exercise 20.

The work in this section that will be most important for us later deals with series of functions. The connection between series and sequences of functions is the same as with numbers—partial sums. In particular, we say the series of functions $\sum f_n(x)$ converges pointwise (uniformly) to $f(x)$ on E if and only if the sequence of functions $\{S_n(x)\}$ defined by

$$S_n(x) = f_1(x) + f_2(x) + \cdots + f_n(x)$$

converges pointwise (uniformly) to $f(x)$ on E. We give two criteria to determine uniform convergence of series of functions.

Theorem (Cauchy criterion for uniform convergence of series of functions): The series of functions $\sum f_n(x)$ converges uniformly on E if and only if given $\varepsilon > 0$, there is a number $N(\varepsilon)$ so that if $m > n > N(\varepsilon)$, then

$$\left|\sum_{i=n+1}^{m} f_i(x)\right| < \varepsilon \quad \text{for every } x \in E.$$

Proof:

The theorem follows by noting that

$$|S_m(x) - S_n(x)| = \left|\sum_{i=n+1}^{m} f_i(x)\right|,$$

where $\{S_n(x)\}$ is the sequence of partial sums associated with the series $\sum f_n(x)$.

Perhaps the most frequently used test to show uniform convergence of series of functions is the Weierstrass M-test.

Theorem (Weierstrass M-test):

Suppose $\sum f_n(x)$ is a series of functions defined on a set E. Let

$$M_n = \sup_{x \in E} |f_n(x)|.$$

If the series of numbers $\sum M_n$ converges, then $\sum f_n(x)$ converges uniformly on E.

Proof:

We have, for any m and n with $m > n$

$$\left|\sum_{i=n+1}^{m} f_i(x)\right| \le \sum_{i=n+1}^{m} |f_i(x)| \le \sum_{i=n+1}^{m} M_i \quad \text{for every } x \in E.$$

By the Cauchy criterion for convergence of series real numbers, given$\varepsilon > 0$ there is a number $N(\varepsilon)$ so that if $m > n > N(\varepsilon)$, then

$$\sum_{i=n+1}^{m} M_i < \varepsilon.$$

Thus,

$$\left|\sum_{i=n+1}^{m} f_i(x)\right| < \varepsilon \quad \text{for every } x \in E$$

if $m > n > N(\varepsilon)$, so $\sum f_n(x)$ converges uniformly on E.

We note that there are series of functions that converge uniformly for which the Weierstrass M-test fails. One example is developed in Exercise 22.

Uniform convergence of $\sum f_n(x)$ on $[a,b]$ allows us to conclude

$$\sum_{n=1}^{\infty}\left(\int_a^b f_n(x)dx\right) = \int_a^b\left(\sum_{n=1}^{\infty} f_n(x)\right)dx.$$

POWER SERIES

A series of the form

$$a_0 + a_1(x-c) + a_2(x-c)^2 + a_3(x-c)^3 + \cdots = \sum_{n=0}^{\infty} a_n(x-c)^n$$

is called a power series. If we let $z = x - c$, we can express the series as $\sum a_n z^n$.

A major result about power series is the following theorem.

Theorem:

For the power series $\sum a_n z^n$ there is an extended real number R, $0 \leq R \leq \infty$ for which the series

1. converges absolutely if $|z| < R$
2. diverges if $|z| > R$.
3. If $0 < R < \infty$, then the power series converges uniformly on $[-R+\varepsilon, R-\varepsilon]$ for any $\varepsilon > 0$.

If $R = 0$, then the series $\sum a_n z^n$ converges only for $x = 0$. If $R = \infty$, then the series $\sum a_n z^n$ converges absolutely for all values of z, and uniformly on any bounded interval.

The number R in the theorem is called the radius of convergence of the power series. It can be calculated in either of the following ways: For the power series $\sum a_n z^n$, let

$$\lambda = \lim_{n\to\infty} \left(|a_n|^{1/n} \right)$$

or

$$\lambda = \lim_{n\to\infty} \frac{|a_{n+1}|}{|a_n|}$$

(if both limits exist, they are equal) then

$$R = \frac{1}{\lambda}.$$

Examples:

Consider the series $\sum \frac{x^n}{n!}$. Here $a_n = 1/n!$, so

$$\lambda = \lim_{n\to\infty} \frac{|a_{n+1}|}{|a_n|} = \lim_{n\to\infty} \frac{\frac{1}{(n+1)!}}{\frac{1}{n!}} = \lim_{n\to\infty} \frac{1}{n+1} = 0 \text{ and } R = \frac{1}{\lambda} = \infty.$$

Thus, this series converges absolutely for all values of x.

Consider the series

$$\sum \frac{n^3 x^n}{5^n}.$$

Here, $a_n = n^3/5^n$ so

$$\lambda = \lim_{n\to\infty} \left(|a_n|^{1/n} \right) = \lim_{n\to\infty} \frac{\left(n^3\right)^{1/n}}{\left(5^n\right)^{1/n}} = \lim_{n\to\infty} \frac{\left(n^3\right)^{1/n}}{5} = \frac{1}{5}$$

since

$$\lim_{n\to\infty} (n\ )^{1/n} = 1.$$

Thus, $R = \frac{1}{\lambda} = 5$.

TAYLOR SERIES

In Taylor series, we begin with a function $f(x)$ that has derivatives of all orders at $x = c$, and then construct a polynomial of degree n, called the Taylor polynomial of degree n, that provides "the best approximation" to $f(x)$ at $x = c$. To determine this polynomial, we must define what we mean by "the best approximation." In the case of Taylor polynomials, if $P_n(x)$ is the Taylor polynomial of degree n for $f(x)$, we require that

$$P_n(c) = f(c),\ \ P_n'(c) = f'(c),\ \ P_n''(c) = f''(c),\ \ldots, P_n{}^{(n)}(c) = f^{(n)}(c).$$

We now determine a formula for the constants a_k in the Taylor polynomial for $f(x)$ at $x = c$.

Suppose that $f(x)$ that has derivatives of all orders at at $x = c$, and let

$$P_n(x) = a_0 + a_1(x-c) + a_2(x-c)^2 + \cdots + a_n(x-c)^n.$$

The first condition we impose is $f(c) = P_n(c)$. Now

$$P_n(c) = a_0 = f(c), \quad \text{so } a_0 = f(c).$$

Next, we require $f'(c) = P_n'(c)$ We have

$$P_n'(x) = a_1 + 2a_2(x-c) + 3a_3(x-c)^2 + 4a_4(x-c)^3 + \cdots + na_n(x-c)^{n-1}$$

so

$$P_n'(c) = a_1 = f'(c), \quad \text{so } a_1 = f'(c).$$

We seek to determine a formula for a_k. Continuing, we have

$$P_n''(x) = 2a_2 + 2\cdot 3a_3(x-c) + 4\cdot 3a_4(x-c)^2 + \cdots + n(n-1)a_n(x-c)^{n-2}$$

so

$$P_n''(c) = 2a_2 = f''(c), \quad \text{and } a_2 = \frac{f''(c)}{2}.$$

Next,

$$P_n'''(x) = 2\cdot 3a_3 + 4\cdot 3\cdot 2a_4(x-c) + \cdots + n(n-1)(n-2)a_n(x-c)^{n-3}$$

so

$$P_n'''(c) = 2\cdot 3a_3 = f'''(c), \quad \text{and } a_3 = \frac{f'''(c)}{2\cdot 3} = \frac{f'''(c)}{3!}.$$

A pattern begins to emerge (that can be proved using induction) that

$$a_k = \frac{f^{(k)}(c)}{k!}.$$

Definition:

Let $f(x)$ be a function that has derivatives of all orders at $x = c$. The *Taylor series for* $f(x)$ *at* $x = c$ is the power series

$$\sum_{n=0}^{\infty} a_n(x-c)^n \quad \text{where} \quad a_n = \frac{f^{(n)}(c)}{n!}.$$

If $c = 0$, the series is called the Maclaurin series. If there is an open interval about $x = c$ for which

$$f(x) = \sum_{n=0}^{\infty} \frac{f^{(n)}(c)}{n!}(x-c)^n$$

then $f(x)$ is said to be *analytic* at $x = c$.

The most frequently used Taylor series (Maclaurin series) are those for e^x, $\sin x$ and $\cos x$. We demonstrate the computations for e^x and $\sin x$.

Example:

We compute the Maclaurin series for e^x. We have

$$\begin{aligned} f(x) &= e^x & f(0) &= e^0 = 1 & a_0 &= f(0) = 1 \\ f'(x) &= e^x & f'(0) &= e^0 = 1 & a_1 &= \frac{f'(0)}{1!} = 1 \\ f''(x) &= e^x & f''(0) &= e^0 = 1 & a_2 &= \frac{f''(0)}{2!} = \frac{1}{2!} \\ f'''(x) &= e^x & f'''(0) &= e^0 = 1 & a_3 &= \frac{f'''(0)}{3!} = \frac{1}{3!} \end{aligned}$$

and

$$a_k = \frac{f^{(k)}(0)}{k!} = \frac{1}{k!}.$$

Thus, the Maclaurin series for e^x is given by

$$\sum_{n=0}^{\infty} \frac{1}{k!} x^k. \tag{1}$$

The series in (1) converges for all values of x as we showed above. We would like to say that

$$e^x = \sum_{n=0}^{\infty} \frac{1}{k!} x^k = 1 + x + \frac{x^2}{2!} + \cdots$$

for all values of x, and, in fact, this is the case but we do not prove it. (For a discussion of how this can be done, see Kirkwood, 1995.) For the problems that we shall consider, if the Taylor series of a function converges, then it converges to the function.

One use of Taylor series is to replace a function by the first few terms of its Taylor series. This often gives a sufficiently accurate approximation of the function by a simple expression. For example, some applications replace e^x when x is close to 0 by $1+x$ or $1+x+\frac{x^2}{2!}$. To get a feel for the accuracy of the approximation, we let $x=0.02$, and observe that

$$1+0.02=1.02, \text{ and } 1+0.02+\frac{(0.02)^2}{2!}=1.0202$$

and note that to 6 decimal places of accuracy, $e^{0.02}=1.020201$.

Example:

We compute the Maclaurin series for sin x. We have

$$f(x)=\sin x \quad f(0)=\sin 0=0 \quad a_0=f(0)=0$$

$$f'(x)=\cos x \quad f'(0)=\cos 0=1 \quad a_1=\frac{f'(0)}{1!}=1$$

$$f''(x)=-\sin x \quad f''(0)=-\sin 0=0 \quad a_2=\frac{f''(0)}{2!}=0$$

$$f'''(x)=-\cos x \quad f'''(0)=-\cos 0=-1 \quad a_3=\frac{f'(0)}{3!}=-\frac{1}{3!}.$$

By taking enough terms of the series, it is not hard to convince oneself that the Maclaurin series for sin x is

$$x-\frac{x^3}{3!}+\frac{x^5}{5!}-\frac{x^7}{7!}+\cdots.$$

In Exercise 2 we show that the Maclaurin series for cos x is

$$1-\frac{x^2}{2!}+\frac{x^4}{4!}-\frac{x^6}{6!}+\cdots.$$

Example:

Bessel functions are fundamental in solving differential equations where a particular type of symmetry occurs. One way to define the Bessel function $J_0(x)$ is

$$J_0(x)=\frac{1}{2\pi}\int_0^{2\pi} e^{ix\sin\theta}d\theta.$$

We shall find a power series representation for $J_0(x)$.

We use the Maclaurin series expansion for $e^{ix\sin\theta}$, which is

$$e^{ix\sin\theta}=\sum_{n=0}^{\infty}\frac{(ix\sin\theta)^n}{n!}.$$

Now

$$\left|\frac{(ix\sin\theta)^n}{n!}\right| \le \frac{x^n}{n!}$$

and $\sum_{n=0}^{\infty}\frac{x^n}{n!}$ converges for every x. So

$$J_0(x) = \frac{1}{2\pi}\int_0^{2\pi} e^{ix\sin\theta}d\theta = \frac{1}{2\pi}\int_0^{2\pi}\sum_{n=0}^{\infty}\frac{(ix\sin\theta)^n}{n!}d\theta = \frac{1}{2\pi}\sum_{n=0}^{\infty}\int_0^{2\pi}\frac{(ix\sin\theta)^n}{n!}d\theta$$

$$= \frac{1}{2\pi}\sum_{n=0}^{\infty}\frac{(ix)^n}{n!}\int_0^{2\pi}(\sin\theta)^n d\theta.$$

If n is odd, then $\int_0^{2\pi}(\sin\theta)^n d\theta = 0$, so

$$\frac{1}{2\pi}\sum_{n=0}^{\infty}\int_0^{2\pi}\frac{(ix\sin\theta)^n}{n!}d\theta = \frac{1}{2\pi}\sum_{n=0}^{\infty}\frac{(ix)^{2n}}{(2n)!}\int_0^{2\pi}(\sin\theta)^{2n}d\theta.$$

We show in Exercise 17 that

$$\int_0^{2\pi}(\sin\theta)^{2n}d\theta = \frac{(2n-1)!2\pi}{2^{2n-1}(n-1)!n!}.$$

Thus

$$J_0(x) = \frac{1}{2\pi}\sum_{n=0}^{\infty}\frac{(ix)^{2n}}{(2n)!}\frac{(2n-1)!2\pi}{2^{2n-1}(n-1)!n!} = \sum_{n=0}^{\infty}\frac{(-1)^n}{(n!)^2}\left(\frac{x}{2}\right)^{2n}.$$

Another notion of convergence that will be important to us is $L^2[a,b]$ convergence. The setting for $L^2[a,b]$ convergence is a vector space with inner product typically defined by an integral. We say that $\{f_n\}$ converges to f in the L^2 sense provided

$$\lim_{n\to\infty}\|f_n - f\| = 0$$

where

$$(\|f_n - f_2\|)^2 = \langle f_n - f,\ f_n - f\rangle = \int_a^b |f_n(x) - f(x)|^2 dx.$$

This sense of convergence is important in many areas of mathematics, including Fourier analysis.

There are some connections among the notions of convergence that we have seen. These include the following:

(1) Uniform convergence implies pointwise convergence.

(2) Uniform convergence implies L^2 convergence if the interval of integration is finite.

(3) L^2 convergence does not imply pointwise nor uniform convergence.

(4) Pointwise convergence does not imply L^2 convergence.

There will be many instances in which we want to "differentiate under the integral." More precisely, we want to claim

$$\frac{d}{dt}\left(\int_a^b f(t,x)dx\right) = \int_a^b \left[\frac{\partial}{\partial t}f(t,x)\right]dx.$$

This is not always legitimate as the example

$$f(t,x) = \begin{cases} \frac{t^3}{x^2}e^{-t^2/x} & x > 0; \\ 0 & x = 0 \end{cases}$$

shows. (See Gelbaum and Olmsted, *Counterexamples in Analysis*.)

In most of the cases that we shall encounter, the function $f(t,x)$ will be sufficiently well behaved so that the equation

$$\frac{d}{dt}\left(\int_a^b f(t,x)dx\right) = \int_a^b \left[\frac{\partial}{\partial t}f(t,x)\right]dx$$

is valid. The next theorem gives a condition that enables us to move the derivative under the integral.

Theorem:

Suppose that for each $t \in [c,d]$,

$$F(t) = \int_a^b f(t,x)dx$$

exists. If $\frac{\partial f}{\partial t}$ is continuous on $\{(x,t)|a \le x \le b, c \le t \le d\}$, then $\frac{dF}{dt}$ exists for each $t \in (c,d)$, and

$$\frac{dF(t)}{dt} = \int_a^b \left[\frac{\partial}{\partial t}f(t,x)\right]dx.$$

That is,

$$\frac{d}{dt}\left(\int_a^b f(t,x)dx\right) = \int_a^b \left[\frac{\partial}{\partial t}f(t,x)\right]dx.$$

EXERCISES

1. Suppose a sequence $\{a_n\}$ has the property that $|a_{n+1} - a_n| < \frac{1}{n}$. Does this mean that $\{a_n\}$ is a Cauchy sequence? What if $|a_{n+1} - a_n| < \frac{1}{n^e}$?
2. Find the Maclaurin series for $f(x) = \cos x$.

3. Find the Maclaurin series for $f(x) = \ln(1-x)$. What is the radius of convergence of the power series?
4. Use the Maclaurin series for $\sin x$, $\cos x$ and e^x to find the Maclaurin expression for the following functions:
   a. $\frac{\sin x}{x}$
   b. $x^2 \cos(3x)$
   c. $\dfrac{e^x - 1 - x}{x^2}$.
5. If $\sum(-1)^n a_n$ is an alternating series for which $|a_n| \geq |a_{n+1}|$ and $\lim_{n\to\infty} a_n = 0$ and L is the number to which $\sum(-1)^n a_n$ converges, then

$$\left|\sum_{n=0}^{k}(-1)^n a_n - L\right| < a_{k+1}.$$

   Use this fact and a CAS to calculate the following expressions to three decimal place accuracy:
   a. $\int_0^1 \dfrac{\sin x}{x}$
   b. $\int_0^2 e^{-x^2} dx$.
6. Find the Maclaurin series for
   a. $\sinh x$
   b. $\cosh x$.
7. Show that the following series of functions converge uniformly on the given interval
   a. $\sum \dfrac{[\sin(nx)]}{n^2}$ on $(-\infty, \infty)$
   b. $\sum \dfrac{n^k}{x^n}$ on $\left[2, \infty\right)$
   c. $\sum \dfrac{1}{n^{1+\alpha}}$ for $\alpha > 0$ on $[-M, M]$.
8. Show that if the power series $\sum a_n x^n$ has radius of convergence R, then for any positive integer k the series $\sum a_n x^{kn}$ has radius of convergence $\sqrt[k]{R}$.
9. Find the radius of convergence for the series
   a. $\sum n^k x^n$ for k an integer
   b. $\sum \dfrac{(x+2)^n}{\ln n}$
   c. $\dfrac{1\cdot 2\cdot 3\cdots n\cdot x^{2n}}{1\cdot 3\cdot 5\cdots(2n-1)}$.
10. Use the first four terms of the Taylor series for $\sin x$ about $c = \frac{\pi}{6}$ to estimate sin 32°. Note that 32° will have to be converted to radians.
11. Find the first four terms of the Taylor series for the following functions about the given point.
    a. $\cos x$ about $x = \frac{\pi}{4}$
    b. $\ln x$ about $x = 2$

c. $x^{-1/2}$ about $x = 4$

d. x^2 about $x = 3$.

12. The integral

$$\int_0^{\varepsilon} \frac{x^3}{e^x - 1} dx$$

where ε is small, arises in heat capacity in solids.

a. Use Maclaurin series to show

$$\frac{x^3}{e^x - 1} = \frac{x^2}{1 + \frac{x}{2} + \frac{x^2}{6} + \cdots}.$$

b. Write

$$\frac{x^2}{1 + \frac{x}{2} + \frac{x^2}{6} + \cdots} = x^2(1+y)^{-1} \quad \text{where } y = 1 + \frac{x}{2} + \frac{x^2}{6} + \cdots.$$

Use geometric series to show that for $|y|<1$, $(1+y)^{-1} = 1 - y + y^2 - y^3 + \cdots$ so that for powers of x up to x^2

$$\left[1 + \left(\frac{x}{2} + \frac{x^2}{6}\right)\right]^{-1} = 1 - \frac{1}{2}x + \frac{1}{12}x^2 + \cdots.$$

c. Evaluate

$$\int_0^{\varepsilon} x^2\left(1 - \frac{1}{2}x + \frac{1}{12}x^2\right) dx.$$

Note how the higher order terms diminish if $\varepsilon \ll 1$.

13. For the sequence of functions $\{x^n | \quad 0 \le x \le 1\}$, discuss why

$$\lim_{n\to\infty}\left(\lim_{x\to 1} f_n(x)\right) = 1 \text{ and } \lim_{x\to 1}\left(\lim_{n\to\infty} f_n(x)\right) = 0.$$

14. For the sequence of functions, $\left\{\frac{\sin nx}{\sqrt{n}}\right\}$ discuss the difference between

$$\lim_{n\to\infty}\left(\lim_{x\to 1} f_n(x)\right) \text{ and } \lim_{x\to 1}\left(\lim_{n\to\infty} f_n(x)\right).$$

15. In the theory of relativity

$$E = \frac{m_0 c^2}{\sqrt{1 - \frac{v^2}{c^2}}}$$

where m_0 is the rest mass of an object, v is the velocity of the object, E is the energy and c is the speed of light.

a. Determine the first two nonzero terms of the Maclaurin series for $(1-x^2)^{-1/2}$, and use this to expand E in powers of $\frac{v^2}{c^2}$.

b. Subtract the rest mass energy m_0c^2 from the approximation to E obtained in part a to get the kinetic energy K and show

$$K \approx \frac{1}{2}m_0v^2.$$

16. Another formula in the theory of relativity is the displacement x of an object with rest mass m_0 due to a force m_0g is

$$x = \frac{c^2}{g}\left\{\left[1+\left(\frac{gt}{c}\right)^2\right]^{\frac{1}{2}} - 1\right\}.$$

Find an approximation for the right-hand side by using the first two nonzero terms of the Maclaurin expansion of $\left[1+\left(\frac{gt}{c}\right)^2\right]^{\frac{1}{2}}$ and compare with the classical result

$$x = \frac{1}{2}gt^2.$$

17. Show that

$$\int_0^{2\pi} (\sin\theta)^{2n} d\theta = \frac{(2n-1)!2\pi}{2^{2n-1}(n-1)!n!}$$

by using the reduction formula

$$\int \sin^n x\, dx = -\frac{1}{n}\sin^{n-1} x \cos x + \frac{n-1}{n}\int \sin^{n-2} x\, dx.$$

18. Use the integral test to show that the series $\sum \frac{1}{n^p}$ converges if $p > 1$ and diverges if $p \le 1$.

19. Prove that the sequence of functions $\{f_n(x)\}$ converges uniformly to $f(x)$ on E if and only if $\lim_{n\to\infty} M_n = 0$, where

$$M_n = \sup_{x\in E}|f_n(x) - f(x)|.$$

20. Prove that a uniformly convergent sequence of functions is uniformly Cauchy.

21. The L^p–norm of a function $f(x)$ on [0,1] is defined as

$$\|f\|_p = \left(\int_0^1 |f(x)|^p \right)^{1/p} \qquad 1 \le p < \infty.$$

Find a function $f(x)$ for which $\|f\|_1 < \frac{1}{10}$ and $\|f\|_2 < 10$.

22. Show that uniform convergence implies L^2 convergence if the interval of integration is finite.

23. Use the fact that $\int (1+x^2)^{-1} dx = \tan^{-1} x$ to show

a. $\tan^{-1} x = x - (x^3/3) + (x^5/5) - (x^7/7) + \cdots \quad$ for $x \in (-1, 1)$.

b. Find a series expression for $\pi/4$ assuming that part a also holds when $x = 1$.

c. Use this expression to find $\pi/4$ accurate to 0.2.

24. Let

$$f_n(x) = \begin{cases} \dfrac{1}{n} & 2^{n-1} < x \le 2^n \\ 0 & \text{otherwise} \end{cases}.$$

Show that the series $\sum f_n(x)$ converges uniformly, but the Weierstrass M-test fails.

25. Use Maclaurin series to solve

$$\int \frac{e^x - 1}{x} dx.$$

26. The energy E of certain ionic solutions can be described by

$$E(x) = \rho\left[x(1+2x)^{1/2} - x - x^2\right].$$

Expand $(1+2x)^{\frac{1}{2}}$ in a Maclaurin series to show that for small values of x,

$$E \approx \alpha x^3.$$

1.5 SOME IMPORTANT INTEGRATION FORMULAS

In later chapters, we shall solve the heat equation using the Fourier transform. In doing so (and in other applications), we shall need to know the Fourier transform of a Gaussian distribution (a function of the form e^{-ax^2}, $a > 0$). This is an important integral, and we present two ways of doing the integration here. One can study the methods now or delay them until later.

The specific integral that we evaluate is

$$\frac{1}{2\pi}\int_{-\infty}^{\infty} e^{-ikx}e^{-\alpha t k^2}\,dk.$$

Method 1 (Power Series Expansion)

We first compute $\int_{-\infty}^{\infty} e^{-x^2}dx$: We have

$$\left(\int_{-\infty}^{\infty} e^{-x^2}dx\right)^2 = \int_{-\infty}^{\infty} e^{-x^2}dx\int_{-\infty}^{\infty} e^{-y^2}dy$$

$$= \int_{-\infty}^{\infty}\int_{-\infty}^{\infty} e^{-(x^2+y^2)}dxdy$$

$$= \int_{r=0}^{\infty}\int_{\theta=0}^{2\pi} e^{-r^2}rd\theta dr = \pi\int_0^{\infty} e^{-r^2}2rdr = \pi\int_0^{\infty} e^{-u}du = -\pi e^{-u}\Big|_0^{\infty} = \pi.$$

Thus,

$$\int_{-\infty}^{\infty} e^{-x^2}dx = \sqrt{\pi}.$$

Next we compute $\int_{-\infty}^{\infty} e^{-i\alpha x}e^{-\beta x^2}dx$. Now

$$e^{-i\alpha x} = \cos(-\alpha x) + i\sin(-\alpha x) = \cos(\alpha x) - i\sin(\alpha x),$$

so

$$\int_{-\infty}^{\infty} e^{-i\alpha x}e^{-\beta x^2}dx = \int_{-\infty}^{\infty}\cos(\alpha x)e^{-\beta x^2}dx - i\int_{-\infty}^{\infty}\sin(\alpha x)e^{-\beta x^2}dx.$$

Since $\sin(\alpha x)e^{-\beta x^2}$ is an odd function, the second integral is 0. In the first integral, we expand $\cos(\alpha x)$ in a Maclaurin series. We have

$$\int_{-\infty}^{\infty}\cos(\alpha x)e^{-\beta x^2}dx = \int_{-\infty}^{\infty}\left(1 - \frac{(\alpha x)^2}{2!} + \frac{(\alpha x)^4}{4!}\cdots\right)e^{-\beta x^2}dx.$$

We compute $\int_{-\infty}^{\infty} x^{2n}e^{-\beta x^2}dx$. Let

$$u = e^{-\beta x^2} \quad dv = x^{2n}dx$$

$$du = -2\beta x e^{-\beta x^2}dx \quad v = \frac{x^{2n+1}}{2n+1}.$$

So

$$\int_{-\infty}^{\infty} x^{2n} e^{-\beta x^2}\,dx = \int_{-\infty}^{\infty} u\,dv = uv\Big|_{-\infty}^{\infty} - \int_{-\infty}^{\infty} v\,du$$

$$= \frac{x^{2n+1}}{2n+1} e^{-\beta x^2}\Big|_{-\infty}^{\infty} + \frac{2\beta}{2n+1}\int_{-\infty}^{\infty} x^{2n+2} e^{-\beta x^2}\,dx$$

$$= 0 + \frac{2\beta}{2n+1}\int_{-\infty}^{\infty} x^{2n+2} e^{-\beta x^2}\,dx.$$

If we let $I_{2n} = \int_{-\infty}^{\infty} x^{2n} e^{-\beta x^2}\,dx$, we have shown $I_{2n} = \frac{2\beta}{2n+1} I_{2n+2}$, or $I_{2n+2} = \frac{2n+1}{2\beta} I_{2n}$. So

$$I_2 = \frac{1}{2\beta} I_0;$$

$$I_4 = \frac{3}{2\beta} I_2 = \frac{3}{2\beta}\frac{1}{2\beta} I_0 = \frac{1\cdot 3}{(2\beta)^2} I_0;$$

$$I_6 = \frac{5}{2\beta} I_4 = \frac{1\cdot 3\cdot 5}{(2\beta)^3} I_0;$$

$$\vdots$$

$$I_{2n} = \frac{1\cdot 3\cdot 5\cdots(2n-1)}{(2\beta)^n} I_0,$$

and thus

$$\int_{-\infty}^{\infty} \cos(\alpha x) e^{-\beta x^2}\,dx = \int_{-\infty}^{\infty} e^{-\beta x^2}\,dx - \frac{\alpha^2}{2!}\int_{-\infty}^{\infty} x^2 e^{-\beta x^2}\,dx$$

$$+ \frac{\alpha^4}{4!}\int_{-\infty}^{\infty} x^4 e^{-\beta x^2}\,dx \cdots$$

$$= I_0 - \frac{\alpha^2}{2!} I_2 + \frac{\alpha^4}{4!} I_4 - \frac{\alpha^6}{6!} I_6 \cdots = I_0 - \frac{\alpha^2}{2!}\frac{1}{(2\beta)} I_0 + \frac{\alpha^4}{4!}\frac{1\cdot 3}{(2\beta)^2} I_0 - \frac{\alpha^6}{6!}\frac{1\cdot 3\cdot 5}{(2\beta)^3} I_0 \cdots.$$

Consider

$$\frac{1\cdot 3\cdot 5\cdots(2n-1)}{1\cdot 2\cdot 3\cdots(2n)}\cdot\frac{1}{2^n} = \frac{\dfrac{(2n)!}{2\cdot 4\cdot 6\cdots(2n)}}{(2n)!}\cdot\frac{1}{2^n} = \frac{\dfrac{(2n)!}{2^n(n!)}}{(2n)!}\cdot\frac{1}{2^n} = \frac{1}{n!}\cdot\frac{1}{4^n}.$$

Thus

$$\int_{-\infty}^{\infty} \cos(\alpha x)e^{-\beta x^2}\,dx = I_0 - \frac{\alpha^2}{2!}\frac{1}{(2\beta)}I_0 + \frac{\alpha^4}{4!}\frac{1\cdot 3}{(2\beta)^2}I_0 - \frac{\alpha^6}{6!}\frac{1\cdot 3\cdot 5}{(2\beta)^3}I_0\cdots$$

$$= I_0\left(\sum_{n=0}^{\infty}(-1)^n\left(\frac{\alpha^2}{\beta}\right)^n\cdot\frac{1}{n!}\cdot\frac{1}{4^n}\right) = I_0\sum_{n=0}^{\infty}\left(-\frac{\alpha^2}{4\beta}\right)^n\cdot\frac{1}{n!} = I_0e^{-\alpha^2/4\beta}.$$

Now

$$I_0 = \int_{-\infty}^{\infty} e^{-\beta x^2}\,dx = \frac{1}{\sqrt{\beta}}\int_{-\infty}^{\infty} e^{-(\sqrt{\beta}x)^2}\sqrt{\beta}\,dx = \frac{1}{\sqrt{\beta}}\sqrt{\pi} = \sqrt{\frac{\pi}{\beta}}.$$

Thus

$$\int_{-\infty}^{\infty} \cos(\alpha x)e^{-\beta x^2}\,dx = \sqrt{\frac{\pi}{\beta}}e^{-\alpha^2/4\beta},$$

so

$$\frac{1}{2\pi}\int_{-\infty}^{\infty} e^{-i\alpha x}e^{-\beta x^2}\,dx = \frac{1}{2\pi}\sqrt{\frac{\pi}{\beta}}e^{-\alpha^2/4\beta} = \frac{1}{\sqrt{4\pi\beta}}e^{-\alpha^2/4\beta}.$$

For Fourier analysis, we shall use the formula

$$\frac{1}{2\pi}\int_{-\infty}^{\infty} e^{-ikx}e^{-\alpha t k^2}\,dk = \frac{1}{\sqrt{4\pi\alpha t}}e^{-x^2/4\alpha t}$$

which is obtained by replacing α by x, β by αt and x by k.

Method 2 (Cauchy Integral Formula)

To compute

$$\int_{-\infty}^{\infty} \exp\left[-\left(\frac{t^2}{2\sigma^2} + i\mu t\right)\right]dt \tag{1}$$

note that

$$\frac{t^2}{2\sigma^2} + i\mu t = \frac{1}{2}\left(\frac{t^2}{\sigma^2} + 2i\mu t\right) = \frac{1}{2}\mu^2\sigma^2 + \frac{1}{2}\left(\frac{t^2}{\sigma^2} + 2i\mu t - \mu^2\sigma^2\right)$$

$$= \frac{1}{2}\mu^2\sigma^2 + \frac{1}{2}\left(\frac{t}{\sigma} + i\sigma\mu\right)^2.$$

Thus (1) is equal to

$$\exp\left(-\frac{1}{2}\mu^2\sigma^2\right)\int_{-\infty}^{\infty} \exp\left[-\frac{1}{2}\left(\frac{t}{\sigma} + i\sigma\mu\right)^2\right]dt. \tag{2}$$

Let

$$\omega = t + i\sigma^2\mu, \quad \text{so} \quad \frac{\omega}{\sigma} = \frac{t}{\sigma} + i\sigma\mu, \; \omega - i\sigma^2\mu = t \text{ and } d\omega = dt.$$

Then (2) becomes

$$\exp\left(-\frac{1}{2}\mu^2\sigma^2\right)\int_{-\infty-i\sigma^2\mu}^{\infty-i\sigma^2\mu}\exp\left(-\frac{1}{2}\frac{\omega^2}{\sigma^2}\right)d\omega. \tag{3}$$

We evaluate the integral in (3) using the Cauchy integral formula. The function

$$f(z) = \exp\left(-\frac{1}{2}\frac{z^2}{\sigma^2}\right)$$

is analytic, so that

$$\oint_\gamma f(z)dz = 0,$$

for γ any simple closed curve in the complex plane. Consider the simple closed curve γ shown in Fig. 1.5.1.

We have

$$\oint_\gamma f(z)dz = \int_{-x}^{x} f(z)dz + \int_{x}^{x+iy} f(z)dz + \int_{x+iy}^{-x+iy} f(z)dz + \int_{-x+iy}^{-x} f(z)dz = 0,$$

so that

$$\int_{-x}^{x} f(z)dz + \int_{x}^{x+iy} f(z)dz + \int_{-x+iy}^{-x} f(z)dz = -\int_{x+iy}^{-x+iy} f(z)dz = \int_{-x+iy}^{x+iy} f(z)dz.$$

This can be written

$$\int_{-x+iy}^{x+iy} f(z)dz = \int_{-x}^{x} f(z)dz + \int_{x}^{x+iy} f(z)dz + \int_{-x+iy}^{-x} f(z)dz.$$

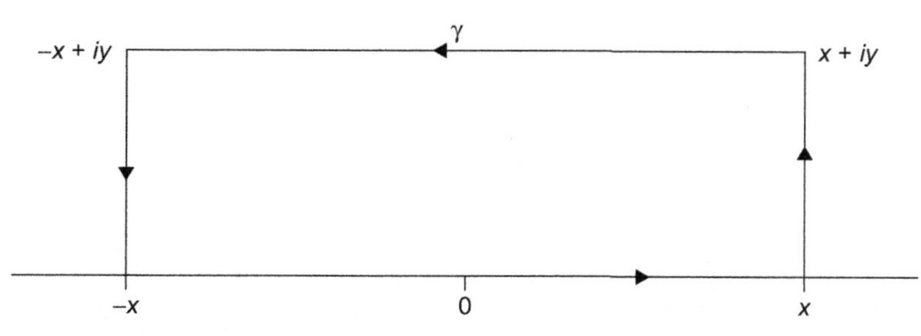

FIGURE 1.5.1

Now take the limit as $x \to \infty$. Note that for $z = x + iy$, $z^2 = x^2 - y^2 + 2ixy$, and

$$\left|e^{-z^2}\right| = \left|e^{-(x^2-y^2+2ixy)}\right| = \left|e^{-x^2+y^2}\right|\left|e^{-2ixy}\right| = \left|e^{-x^2+y^2}\right|$$

since $\left|e^{-2ixy}\right| = 1$. As $x \to \infty$, $\left|e^{-x^2+y^2}\right| \to 0$, so

$$\lim_{x\to\infty}\int_x^{x+iy} f(z)dz = 0 \text{ and } \lim_{x\to\infty}\int_{-x+iy}^{-x} f(z)dz = 0.$$

Thus, we have

$$\lim_{x\to\infty}\int_{-x+iy}^{x+iy} f(z)dz = \lim_{x\to\infty}\int_{-x}^{x} f(z)dz = \lim_{x\to\infty}\int_{-x}^{x}\exp\left(-\frac{1}{2}\frac{x^2}{\sigma^2}\right).$$

Now let

$$s^2 = \frac{x^2}{2\sigma^2}, \quad \text{so } s = \frac{x}{\sigma\sqrt{2}}, \quad dx = \sigma\sqrt{2}ds$$

then

$$\lim_{x\to\infty}\int_{-x}^{x}\exp\left(-\frac{1}{2}\frac{x^2}{\sigma^2}\right) = \sigma\sqrt{2}\int_{-\infty}^{\infty} e^{-s^2}ds = \sigma\sqrt{2\pi}.$$

So

$$\exp\left(-\frac{1}{2}\mu^2\sigma^2\right)\int_{-\infty-i\sigma^2\mu}^{\infty-i\sigma^2\mu}\exp\left(-\frac{1}{2}\frac{\omega^2}{\sigma^2}\right)d\omega = \exp\left(-\frac{1}{2}\mu^2\sigma^2\right)\sigma\sqrt{2\pi}$$

and thus

$$\begin{aligned}\frac{1}{2\pi}\int_{-\infty}^{\infty}\exp\left[-\left(\frac{t^2}{2\sigma^2} + i\mu t\right)\right]dt &= \frac{1}{2\pi}\exp\left(-\frac{1}{2}\mu^2\sigma^2\right)\sigma\sqrt{2\pi} \\ &= \frac{\sigma}{\sqrt{2\pi}}\exp\left(-\frac{\mu^2\sigma^2}{2}\right),\end{aligned}$$

which agrees with Method 1 with $\mu = \alpha$ and $\beta = \frac{1}{2\sigma^2}$.

OTHER FACTS WE SHALL USE LATER

1. $\int_{-\infty}^{\infty} e^{i\mu(x-\xi)}e^{-\mu^2\kappa t}d\mu = \left(\frac{2\pi}{\sqrt{4\pi\kappa t}}\right)\exp\left[\frac{(x-\xi)^2}{4\kappa t}\right]$;
2. if $F(\mu) = \int_{-\infty}^{\infty} f(x)e^{i\mu x}dx$ and $\mu(x,t) = \int_{-\infty}^{\infty} F(\mu)e^{i\mu x}e^{-\mu^2\kappa t}d\mu$, then

$$\mu(x,t) = \int_{-\infty}^{\infty} f(\xi)\frac{e^{-(x-\xi)^2/4\kappa t}}{\sqrt{4\pi\kappa t}}d\xi.$$

Fact (2) follows from

$$\begin{aligned}
&\int_{\mu=-\infty}^{\infty}\left(\frac{1}{2\pi}\int_{\xi=-\infty}^{\infty} f(\xi)e^{-i\mu\xi}\,d\xi\right)e^{i\mu x}e^{-\mu^2\kappa t}\,d\mu\\
&=\int_{\xi=-\infty}^{\infty}\left(\int_{\mu=-\infty}^{\infty}\frac{1}{2\pi}f(\xi)e^{-i\mu\xi}e^{i\mu x}e^{-\mu^2\kappa t}\,d\mu\right)d\xi\\
&=\int_{\xi=-\infty}^{\infty} f(\xi)\left[\frac{1}{2\pi}\int_{\mu=-\infty}^{\infty} e^{-i\mu\xi}e^{i\mu x}e^{-\mu^2\kappa t}\,d\mu\right]d\xi\\
&=\int_{\xi=-\infty}^{\infty} f(\xi)2\pi\frac{1}{2\pi}\frac{e^{-(x-\xi)^2/4\kappa t}}{\sqrt{4\pi\kappa t}}\,d\xi\\
&=\int_{-\infty}^{\infty} f(\xi)\frac{e^{-(x-\xi)^2/4\kappa t}}{\sqrt{4\pi\kappa t}}\,d\xi.
\end{aligned}$$

ANOTHER IMPORTANT INTEGRAL

The integral

$$\int_0^{\infty} e^{-(a^2x^2+b^2/x^2)}\,dx$$

arises in computing the Laplace transform of

$$\frac{2}{\sqrt{\pi}}\int_x^{\infty} e^{-u^2}\,du,$$

and this is used in the analysis of the heat equation. We compute the integral by first considering the integral as a function of b, and showing that it satisfies a differential equation. We let

$$I(b)=\int_0^{\infty} e^{-(a^2x^2+b^2/x^2)}\,dx.$$

Then

$$\begin{aligned}
\frac{d}{db}I(b)=\frac{d}{db}\left(\int_0^{\infty} e^{-(a^2x^2+b^2/x^2)}\,dx\right)&=\int_0^{\infty}\left[\frac{d}{db}\left(e^{-(a^2x^2+b^2/x^2)}\right)\right]dx\\
&=\int_0^{\infty} e^{-(a^2x^2+b^2/x^2)}\left(-\frac{2b}{x^2}\right)dx.
\end{aligned}$$

We make the change of variables

$$z = \frac{b}{ax}.$$

This gives

$$ax = \frac{b}{z}; \quad a^2x^2 = \frac{b^2}{z^2}; \quad \frac{b^2}{x^2} = a^2z^2; \quad -\frac{2b}{x^2} = -\frac{2a^2z^2}{b}; \quad dx = -\frac{b}{a}\cdot\frac{1}{z^2}dz.$$

Also, if $x = 0$, then $z = \infty$, and if $x = \infty$, then $z = 0$. Thus

$$\begin{aligned}
\int_0^\infty e^{-(a^2x^2+b^2/x^2)}\left(-\frac{2b}{x^2}\right)dx &= \int_\infty^0 e^{-\left(\frac{b^2}{z^2}+a^2z^2\right)}\cdot\left(-\frac{2a^2z^2}{b}\right)\left(-\frac{b}{a}\cdot\frac{1}{z^2}\right)dz \\
&= \int_\infty^0 e^{-\left(\frac{b^2}{z^2}+a^2z^2\right)} 2adz \\
&= -2a\int_0^\infty e^{-\left(\frac{b^2}{z^2}+a^2z^2\right)} dz \\
&= -2aI(b).
\end{aligned}$$

Then

$$\frac{d}{db}I(b) = -2aI(b)$$

so

$$\int \frac{dI(b)}{I(b)}db = \int (-2a)db,$$

so

$$\ln I(b) = -2ab + C_0$$

and thus

$$I(b) = e^{C_0}e^{-2ab}.$$

Taking $b = 0$ gives $I(0) = e^{C_0}$, but

$$I(0) = \int_0^\infty e^{-a^2x^2}dx = \frac{\sqrt{\pi}}{2a}.$$

Finally, we have

$$I(b) = \int_0^\infty e^{-(a^2x^2+b^2/x^2)}dx = \frac{\sqrt{\pi}}{2a}e^{-2ab}.$$

EXERCISES

1. Show that $\int_{-\infty}^{\infty} e^{i\mu(x-\xi)}e^{-\mu^2\kappa t}d\mu = 2\pi\frac{e^{(x-\xi)^2/4\kappa t}}{\sqrt{4\pi\kappa t}}$.
2. Evaluate

$$\int_0^{\infty} e^{-a^2x^2}\cos bxdx.$$

   Hint: Let $I(b) = \int_0^{\infty} e^{-a^2x^2}\cos bxdx$. Differentiate with respect to b and integrate by parts.
3. The gamma function, $\Gamma(x)$, is defined by

$$\Gamma(x) = \int_0^{\infty} e^{-t}t^{x-1}dt.$$

   a. Show that the integral that defines $\Gamma(x)$ converges for $x > 0$.
   b. Use integration by parts to show that $\Gamma(x+1) = x\Gamma(x)$ $x > 0$.
   c. Show that $\Gamma(1) = 1$.
   d. Use induction to show that $\Gamma(n) = (n-1)!$ for every positive integer n. (Recall that $0! = 1$.)
   e. Show that $\lim_{x\downarrow 0}\Gamma(x) = \infty$.

   The gamma function is used to extend the idea of factorial to positive non-integer values of x. This is used in several applications, including quantum mechanics.
4. Determine whether $\int_0^{\pi}\frac{\sin x}{x}dx$ converges.

CHAPTER

Vector Calculus

2

2.1 VECTOR INTEGRATION

In this section we present results from vector analysis that pertains to integration. The presentation is somewhat brief, and for a more complete explanation we recommend a standard text in vector analysis such as Marsden and Tromba, *Vector Calculus,* Third Edition. A superb reference for much of what we do in Section 2.2 is *Div, Grad, Curl, and All That* by H. M. Schey.

Definition:

If $f : U \subseteq \mathbb{R}^n \to \mathbb{R}$, the *gradient of f*, denoted ∇f, in Cartesian coordinates is

$$\nabla f = \left(\frac{\partial f}{\partial x_1}, \ldots, \frac{\partial f}{\partial x_n}\right).$$

If $\mathbb{R}^n = \mathbb{R}^3$, as is often the case, then

$$\nabla f = \left(\frac{\partial f}{\partial x}, \frac{\partial f}{\partial y}, \frac{\partial f}{\partial z}\right) = \frac{\partial f}{\partial x}\hat{i} + \frac{\partial f}{\partial y}\hat{j} + \frac{\partial f}{\partial z}\hat{k}.$$

We showed in Section 1.3 that in cylindrical coordinates

$$\nabla f = \frac{\partial f}{\partial r}\hat{e}_r + \frac{1}{r}\frac{\partial f}{\partial \theta}\hat{e}_\theta + \frac{\partial f}{\partial z}\hat{e}_z$$

and in spherical coordinates

$$\nabla f = \frac{\partial f}{\partial r}\hat{e}_r + \frac{1}{r}\frac{\partial f}{\partial \theta}\hat{e}_\theta + \frac{1}{r\sin\theta}\frac{\partial f}{\partial \varphi}\hat{e}_\varphi.$$

Example:

Let $f(x, y, z) = e^x y^2 + 4xy\sin z$. Then

$$\nabla f(x,y,z) = \left(e^x y^2 + 4y\sin z\right)\hat{i} + (2e^x y + 4x\sin z)\hat{j} + (4xy\cos z)\hat{k}.$$

If $f(x, y, z) = constant$, then the graph of the equation is a three-dimensional surface. If (x_0, y_0, z_0) is a point on the surface, and the surface is smooth at that point, then a normal (perpendicular) vector to the surface at that point is $\nabla f(x_0, y_0, z_0)$.

Mathematical Physics with Partial Differential Equations. https://doi.org/10.1016/B978-0-12-814759-7.00002-8

Example:

Let $z = x^2y + 3y$ (or $x^2y + 3y - z = 0$). Then

$$\nabla f(x, y, z) = 2xy\hat{\imath} + (x^2 + 3)\hat{\jmath} - \hat{k}.$$

So a vector normal to the surface at (1, 2, 8) is $4\hat{\imath} + 4\hat{\jmath} - \hat{k}$.

Definition:

If $f : U \subseteq \mathbb{R}^n \to \mathbb{R}$, the *directional derivative* of f at $\hat{x}$ in the direction of the unit vector $\hat{v}$ is $\nabla f(\hat{x}) \cdot \hat{v}$. If $\hat{v} = (v_1, \ldots, v_n)$ then

$$\nabla f(\hat{x}) \cdot \hat{v} = \frac{\partial f}{\partial x_1} v_1 + \cdots + \frac{\partial f}{\partial x_n} v_n.$$

We now describe some different kinds of integrals that are important in vector analysis. These are (in the order in which we discuss them) as follows:

1. Path integrals, where we integrate a real-valued function along a curve.
2. Line integrals, where we integrate a vector-valued function along a curve.
3. Integration of a real-valued function over a surface.
4. Integration of a vector-valued function over a surface.

PATH INTEGRALS

A path integral is very similar to a Riemann integral on $[a, b]$. In constructing the Riemann integral of $f(x)$ on $[a, b]$ we divide $[a, b]$ into n subintervals by choosing $x_0, x_1, \ldots, x_n$, with $a = x_0 < x_1 < \ldots < x_n = b$. This creates subintervals $[x_0, x_1]$, $[x_1, x_2], \ldots, [x_{n-1}, x_n]$ whose union is $[a, b]$. In each subinterval we choose a value of x. Let $\overline{x}_i$ denote the value chosen in the subinterval $[x_{i-1}, x_i]$. We then form the Riemann sum, $\sum_{i=1}^{n} f(\overline{x}_i)\Delta x_i$, where $\Delta x_i = x_i - x_{i-1}$ is the width of the ith subinterval. If $f(x)$ is sufficiently well-behaved, then

$$\lim_{\max \Delta x_i \to 0} \sum_{i=1}^{n} f(\overline{x}_i)\Delta x_i$$

exists, and the limit is denoted $\int_a^b f(x)dx$.

In a path integral and a line integral, a path is used as the quantity over which we integrate instead of an interval on the x-axis. Intuitively, a path is a curve in space that has a direction associated with it. The mathematical definition is as follows.

Definition:

A *path* in $\mathbb{R}^n$ is a continuous function $\sigma : [a, b] \to \mathbb{R}^n$. The *end points* of the path are $\sigma(a)$ and $\sigma(b)$. The image of σ is the *curve* of σ. The path σ is differentiable if each of the components of σ is differentiable.

We often have t as the variable, and think of $\sigma(t)$ as being the position of a particle at time t.

Example:

The function $\sigma(t) = (\cos t, \sin t)$, $0 \le t \le 2\pi$ is a path whose curve is a circle of radius 1 centered at the origin. The motion of the particle that describes the curve begins at (0,1) and travels counterclockwise around the circle.

Example:

The function $\sigma(t) = (t, t^2)$, $-1 \le t \le 2$, is a path whose curve is a parabola that begins at $(-1, 1)$ and ends at $(2, 4)$.

If $\sigma(t)$ is a path, then the velocity vector is $\sigma'(t)$. If $\sigma(t) = (x(t), y(t), z(t))$, then $\sigma'(t) = (x'(t), y'(t), z'(t))$. The speed of $\sigma(t)$ is

$$\|\sigma'(t)\| = \sqrt{(x'(t))^2 + (y'(t))^2 + (z'(t))^2}.$$

We note that $\sigma'(t)$ is tangent to $\sigma(t)$, and the length of $\sigma(t)$ is

$$\int_{t=a}^{b} \|\sigma'(t)\| dt.$$

We often use ds to denote an infinitesimal element of arc length. If this is the case and $s(t) = (x(t), y(t), z(t))$ then

$$ds = \sqrt{\left(\frac{dx}{dt}\right)^2 + \left(\frac{dy}{dt}\right)^2 + \left(\frac{dz}{dt}\right)^2}\, dt.$$

We shall construct Riemann sums for path integrals (where we integrate a real-valued function) and line integrals (where we integrate a vector-valued function). In both cases, the construction involves splitting the path into small pieces and forming a Riemann sum, with the length of the pieces of the path taking the role of Δx in the Riemann integral. We now investigate how these lengths are computed.

Let $\sigma : [a, b] \to \mathbb{R}^n$. Choose $t_0, t_1, t_2, \ldots, t_n$ with $a = t_0 < t_1 < t_2 < \ldots < t_n = b$. Then $\sigma(t_i)$ is a point on the path σ, and the role of $\Delta x_i = x_{i+1} - x_i$ in the Riemann sum is now played by the length between $\sigma(t_{i+1})$ and $\sigma(t_i)$. See Fig. 2.1.1.

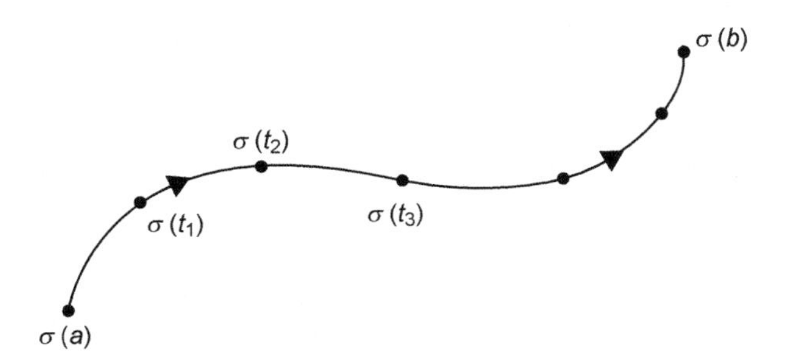

FIGURE 2.1.1

If σ is a path in $\mathbb{R}^3$, and $\sigma(t) = (x(t), y(t), z(t))$ then

$$\sigma(t_{i+1}) - \sigma(t_i) = (x(t_{i+1}) - x(t_i),\ y(t_{i+1}) - y(t_i), z(t_{i+1}) - z(t_i))$$

and

$$\|\sigma(t_{i+1}) - \sigma(t_i)\| = \sqrt{(x(t_{i+1}) - x(t_i))^2 + (y(t_{i+1}) - y(t_i))^2 + (z(t_{i+1}) - z(t_i))^2}.$$

If $\sigma(t)$ is differentiable, then so are the component functions, and by the mean value theorem

$$\frac{x(t_{i+1}) - x(t_i)}{t_{i+1} - t_i} = x'(\bar{t}_i) \text{ or } x(t_{i+1}) - x(t_i) = x'(\bar{t}_i)(t_{i+1} - t_i) \equiv x'(\bar{t}_i)\Delta t_i$$

where $\bar{t}_i$ is between t_i and t_{i+1}. We have the analogous relations for the other two components. Thus

$$\|\sigma(t_{i+1}) - \sigma(t_i)\| \approx \sqrt{[x'(t_i)]^2 + [y'(t_i)]^2 + [z'(t_i)]^2}\Delta t_i = \|\sigma'((t_i))\|\Delta t_i.$$

This method is applicable to curves in dimensions other than three.

We continue to work in three dimensions. Suppose $f : \mathbb{R}^3 \to \mathbb{R}$ and $\sigma : [a, b] \to \mathbb{R}^3$ is a path. To construct a Riemann sum for the path integral we choose $t_0, t_1, t_2, \ldots, t_n$ with $a = t_0 < t_1 < t_2 < \ldots < t_n = b$ as above and from each subinterval $[t_{i-1}, t_i]$ choose a point t_i^*. Form the Riemann sum

$$\sum_{i=1}^{n} f(\sigma(t_i^*))\|\sigma(t_{i+1}) - \sigma(t_i)\|.$$

Notice how this compares with an ordinary Riemann sum. Since in the limit as max $(t_{i+1} - t_i) \to 0$, we have $\|\sigma(t_{i+1}) - \sigma(t_i)\| \to \|\sigma'(t_i)\|dt$, and we have

$$\lim_{\max(t_{i+1}-t^i)\to 0} \sum_{i=1}^{n} f(\sigma(t_i^*))\|\sigma(t_{i+1}) - \sigma(t_i)\| \equiv \int_\sigma f\, ds = \int_{t=a}^{b} f(\sigma(t))\|\sigma'(t)\|dt.$$

Recapping, we have the following:

Definition:

Suppose that $f : \mathbb{R}^n \to \mathbb{R}$ and $\sigma : [a, b] \to \mathbb{R}^n$ is differentiable. The *path integral* of f along σ is

$$\int_\sigma f\, ds = \int_{t=a}^{b} f(\sigma(t))\|\sigma'(t)\|dt.$$

If $f : \mathbb{R}^3 \to \mathbb{R}$, and $\sigma(t) = (x(t), y(t), z(t))$ then

$$\int_\sigma f\, ds = \int_a^b f(x(t), y(t), z(t))\sqrt{\left(\frac{dx}{dt}\right)^2 + \left(\frac{dy}{dt}\right)^2 + \left(\frac{dz}{dt}\right)^2}\, dt.$$

Example:

Let $f(x,\ y,\ z) = x^2 + y^2 + 4z$ and $\sigma(t) = (\cos t,\ \sin t,\ t)\ 0 \le t \le 2$. Then $\sigma'(t) = (-\sin t,\ \cos t,\ 1)$ so $\|\sigma'(t)\| = \sqrt{(-\sin t)^2 + (\cos t)^2 + 1} = \sqrt{2}$ and $f(\sigma(t)) = (\cos t)^2 + (\sin t)^2 + 4t = 1 + 4t$. Thus,

$$\int_\sigma f ds = \int_a^b f(x(t), y(t), z(t))\sqrt{\left(\frac{dx}{dt}\right)^2 + \left(\frac{dy}{dt}\right)^2 + \left(\frac{dz}{dt}\right)^2} dt$$

$$= \int_0^2 (1+4t)\sqrt{2}dt = \sqrt{2}(t + 2t^2)\Big|_0^2 = 10\sqrt{2}.$$

LINE INTEGRALS

In line integrals, we use the idea of a vector field.

Definition:

A *vector field* on $\mathbb{R}^n$ is a function $\vec{F} : \mathbb{R}^n \to \mathbb{R}^n$.

For most of our work, $\vec{F} : \mathbb{R}^3 \to \mathbb{R}^3$. An intuitive way to visualize a vector field is a flowing fluid. At each point, the fluid has a particular velocity. This velocity we can represent as an arrow. Thus each point in space has an arrow associated with it. It may also be visualized as that at each point there is an associated force, as shown in Fig. 2.1.2.

In the case, $\vec{F} : \mathbb{R}^3 \to \mathbb{R}^3$ we often write $\vec{F}(x,y,z) = F_1(x,y,z)\hat{i} + F_2(x,y,z)\hat{j} + F_3(x,y,z)\hat{k}$ where $F_i(x,y,z) : \mathbb{R}^3 \to \mathbb{R}$.

We say that the vector field $\vec{F}$ is continuous (differentiable) if each component function is continuous (differentiable).

The idea of a line integral is well-illustrated by computing work. The work done in moving an object along the x-axis from $x = a$ to $x = b$ by a force $F(x)$ that acts only in the horizontal direction is $\int_a^b F(x)dx$. Suppose that we are in higher

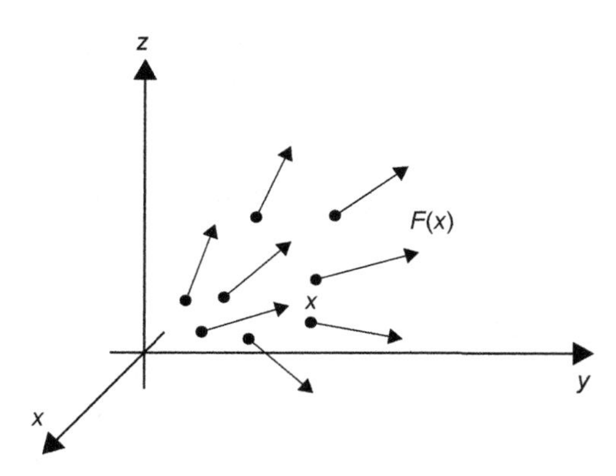

FIGURE 2.1.2

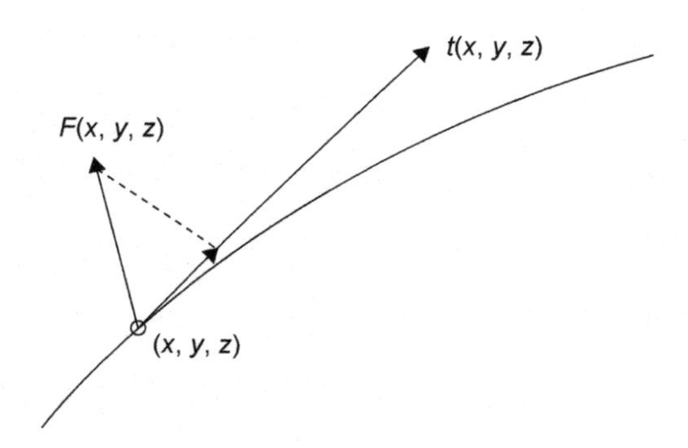

FIGURE 2.1.3

dimensions and we move an object along a path σ from $\sigma(a)$ to $\sigma(b)$, and the object is acted on by a vector field $\vec{F}(x,y,z)$. Consider what happens along an infinitesimal part of the path. See Fig. 2.1.3.

The only portion of $F(x,y,z)$ that contributes to the motion of the body is that part of $\vec{F}(x,y,z)$ that is tangent to the path. If $\hat{t}(x,y,z)$ is a unit vector that is tangent to the path at (x, y, z), then the force that contributes to work at that point is $\vec{F}(x,y,z)\cdot\hat{t}(x,y,z)$. The body moves in almost a straight line along the path from (x, y, z) to $(x + \Delta x, y + \Delta y, z + \Delta z)$. If we denote the distance between those points as Δs, and ΔW is the work done in moving between those points, then

$$\Delta W \approx \vec{F}(x,y,z)\cdot\hat{t}(x,y,z)\Delta s$$

so that in the limit as $\Delta s \to 0$ we get

$$W = \int \vec{F}(x,y,z)\cdot\hat{t}(x,y,z)ds.$$

We now express $\vec{F}(x,y,z)\cdot\hat{t}(x,y,z)\Delta s$ as a quantity that we can integrate. We can express $\vec{F}(x,y,z)$ in Cartesian coordinates as

$$\vec{F}(x,y,z) = F_1(x,y,z)\hat{i} + F_2(x,y,z)\hat{j} + F_3(x,y,z)\hat{k}$$

or

$$\vec{F}(\sigma(t)) = F_1(\sigma(t))\hat{i} + F_2(\sigma(t))\hat{j} + F_3(\sigma(t))\hat{k}.$$

We noted earlier in the section on path integrals that $\sigma'(t)$ is tangent to $\sigma(t)$, so

$$\frac{\sigma'(t)}{\|\sigma'(t)\|}$$

is a unit vector tangent to $\sigma(t)$. We also noted that $\Delta s \approx \|\sigma'(t)\|\Delta t$. Thus, we have

$$W = \int_{t=a}^{b} \vec{F}(\sigma(t)) \cdot \frac{\sigma'(t)}{\|\sigma'(t)\|} \|\sigma'(t)\| \, dt = \int_{t=a}^{b} \vec{F}(\sigma(t)) \cdot \sigma'(t) dt.$$

Definition:

Let $\vec{F}$ be a vector field on $\mathbb{R}^n$ and $\sigma : [a, b] : \to \mathbb{R}^n$. The *line integral* of $\vec{F}$ along σ, denoted $\int_\sigma \vec{F} \cdot d\widehat{s}$, is defined by

$$\int_\sigma \vec{F} \cdot d\widehat{s} = \int_a^b \vec{F}(\sigma(t)) \cdot \sigma'(t) dt.$$

We often have $\vec{F}(x, y, z) = F_1(x, y, z)\widehat{i} + F_2(x, y, z)\widehat{j} + F_3(x, y, z)\widehat{k}$ and $\sigma(t) = (x(t), y(t), z(t))$ and write

$$\int_\sigma \vec{F} \cdot d\widehat{s} = \int_\sigma F_1 dx + F_2 dy + F_3 dz.$$

Example:

Let $\vec{F}(x, y, z) = x^2 y\widehat{i} + xz\widehat{j} + xyz\widehat{k}$ and $\sigma(t) = (t,t^2,4t)$ $0 \le t \le 2$. Then

$$x = t, \quad y = t^2, \quad z = 4t$$

$$dx = dt, \quad dy = 2tdt, \quad dz = 4dt$$

so

$$\int_\sigma \vec{F} \cdot d\widehat{s} = \int_0^2 \left[t^2 t^2 1 + t(4t)2t + t\left(t^2\right)(4t)4\right] dt$$

$$= \frac{t^5}{5} + \frac{8t^4}{4} + \frac{16t^5}{5}\Big|_0^2 = \frac{17\left(2^5\right)}{5} + 2\left(2^4\right).$$

Example:

Compute

$$\int_\sigma xy^2 dx + x^2 dy$$

along the path $y = x^2$, $0 \le x \le 3$. We have $dy = 2x\, dx$ so

$$\int_\sigma xy^2 dx + x^2 dy = \int_0^3 \left[x\left(x^2\right)^2 + x^2(2x)\right] dx = \frac{x^6}{6} + 2\frac{x^4}{4}\Big|_0^3 = \frac{3^6}{6} + \frac{3^4}{2}.$$

Example:

Compute

$$\int_\sigma xy^2 dx + x^2 dy$$

from (0, 0) to (3, 9) along the path σ from (0, 0) to (3, 0) and then from (3, 0) to (3, 9).

Let $\sigma_1$be the path from (0, 0) to (3, 0) and σ_2 be the path from (3, 0) to (3, 9). Then

$$\int_{\sigma} xy^2 dx + x^2 dy = \int_{\sigma_1} xy^2 dx + x^2 dy + \int_{\sigma_2} xy^2 dx + x^2 dy$$

and

$$\int_{\sigma_1} xy^2 dx + x^2 dy = \int_0^3 0 dx = 0$$

since $y = 0$ and $dy = 0$ on σ_1, and

$$\int_{\sigma_2} xy^2 dx + x^2 dy = \int_3^9 3^2 dy = 81 - 27 = 54.$$

Thus,

$$\int_{\sigma} xy^2 dx + x^2 dy = 54.$$

Now compute

$$\int_{\sigma} xy^2 dx + x^2 dy$$

from (0, 0) to (3, 9) along the path $y = 3x$. Then

$$\int_{\sigma} xy^2 dx + x^2 dy = \int_0^3 \left[x(3x)^2 + x^3 3 \right] dx = 12 \left. \frac{x^4}{4} \right|_0^3 = 243.$$

Notice that the answers to the last two examples are different, even though we are integrating the same function between the same end points.

SURFACES

The next two integrals we discuss involve integrating over a surface rather than a curve. Like line integrals that extended integrating a function over an interval to integrating over a curve, surface integrals extend the idea of integrating over a planar region to integrating over a surface.

In developing line integrals it was fundamental that we develop an approximation for an increment of the path—a quantity we denoted Δs. Our first task with creating surface integrals is to develop an approximation for an increment of the surface. We denote this incremental element ΔS.

The simplest situation—the one that we now consider—is when the surface can be expressed $z = f(x, y)$. The cases where $y = g(x, z)$ and $x = h(y, z)$ are conceptually identical.

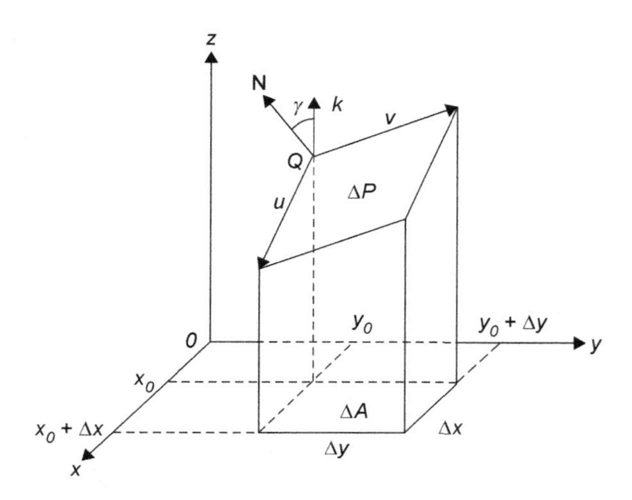

FIGURE 2.1.4

Suppose that D is a region in the x,y plane and $f(x, y)$ has continuous partial derivatives. Divide D into small rectangles whose dimensions are Δx and Δy. We consider the particular rectangle whose corners are (x_0, y_0), $(x_0 + \Delta x, y_0)$, $(x_0 + \Delta x, y_0 + \Delta y)$ and $(x_0, y_0 + \Delta y)$. See Fig. 2.1.4.

Denote this rectangle ΔA. If we project ΔA onto the surface $z = f(x, y)$, we get a portion of the surface that we denote ΔS. To estimate the area of ΔS, choose a point p on ΔS and construct the plane tangent to the surface at that point. It is notationally convenient to choose $p = (x_0, y_0, f(x_0, y_0))$. We project ΔA onto this plane and get a planar region we denote ΔP. We compute the area of ΔP, and this will be our estimate for the area of ΔS. The sides of ΔP are the vectors

$$\widehat{u} = \Delta x\widehat{i} + \frac{\partial f(x_0, y_0)}{\partial x}\Delta x\widehat{k}$$

$$\widehat{v} = \Delta y\widehat{j} + \frac{\partial f(x_0, y_0)}{\partial y}\Delta y\widehat{k}.$$

The area of $\Delta P = \|\widehat{u} \times \widehat{v}\|$. Now

$$\widehat{u} \times \widehat{v} = \begin{vmatrix} \widehat{i} & \widehat{j} & \widehat{k} \\ \Delta x & 0 & \dfrac{\partial f(x_0, y_0)}{\partial x}\Delta x \\ 0 & \Delta y & \dfrac{\partial f(x_0, y_0)}{\partial y}\Delta y \end{vmatrix} = -\Delta x\Delta y\left[\frac{\partial f(x_0, y_0)}{\partial x}\widehat{i} + \frac{\partial f(x_0, y_0)}{\partial y}\widehat{j} - \widehat{k}\right]$$

so

$$\Delta P = \|\widehat{u} \times \widehat{v}\| = \Delta x \Delta y \sqrt{\left(\frac{\partial f(x_0, y_0)}{\partial x}\right)^2 + \left(\frac{\partial f(x_0, y_0)}{\partial y}\right)^2 + 1}.$$

When we develop integrals over surfaces, this will be our incremental surface element if we can write the surface as $z = f(x, y)$.

PARAMETERIZED SURFACES

In situations where the surface is not conveniently expressed as $z = f(x, y)$, such as when cylindrical or spherical coordinates are advantageous, we parameterize the surface with different variables.

Definition:

A *parameterized surface* is a function $\Phi : D \subset \mathbb{R}^2 \to \mathbb{R}^3$, where D is a connected set. The *surface* corresponding to Φ is the image of Φ, $\Phi(D)$. If $\Phi(u, v) = (x(u, v), y(u, v), z(u, v))$ then $S = \Phi(D)$ is continuous (differentiable) if each of the coordinate functions is continuous (differentiable).

Example:

The function $\Phi(\theta, \varphi) = (r\sin\theta\cos\varphi, r\sin\theta\sin\varphi, r\cos\theta)$ $0 \le \theta \le \pi$, $0 \le \varphi < 2\pi$ is a parameterization of the surface of a sphere of radius r.

To compute the element ΔS in parameterized surfaces, we follow the same idea of finding the tangent plane at a point, and estimating ΔS by the area of the increment of the tangent plane. As before, we need two nonparallel vectors that are tangent to the surface to create the tangent plane. We now describe how to do this.

If $\Phi(u,v) = (x(u, v), y(u, v), z(u, v))$ is a surface and $\Phi(u_0, v_0)$ is a point on the surface, then

$$\widehat{T}_u(u_0, v_0) = \frac{\partial x}{\partial u}(u_0, v_0)\widehat{i} + \frac{\partial y}{\partial u}(u_0, v_0)\widehat{j} + \frac{\partial z}{\partial u}(u_0, v_0)\widehat{k}$$

is a vector tangent to the surface at the point $\Phi(u_0,v_0)$ parallel to the curve $\Phi(t,v_0)$ on the surface. Likewise,

$$\widehat{T}_v(u_0, v_0) = \frac{\partial x}{\partial v}(u_0, v_0)\widehat{i} + \frac{\partial y}{\partial v}(u_0, v_0)\widehat{j} + \frac{\partial z}{\partial v}(u_0, v_0)\widehat{k}$$

is a vector tangent to the surface at the point $\Phi(u_0, v_0)$ parallel to the curve $\Phi(u_0, t)$ on the surface. The vector $\widehat{T}_u(u_0, v_0) \times \widehat{T}_v(u_0, v_0)$ is normal to the surface $\Phi(u, v)$ at the point $\Phi(u_0, v_0)$. The surface is said to be smooth at $\Phi(u_0, v_0)$ if $\widehat{T}_u(u_0, v_0) \times \widehat{T}_v(u_0, v_0) \neq \widehat{0}$.

Following what we did in the case where the surface is described by $z = f(x, y)$, our estimate for ΔS is $\|\widehat{T}_u \times \widehat{T}_v\|\Delta u \Delta v$.

Example:

Consider the parameterization of the sphere of radius r, $\Phi(\theta, \varphi) = (r\sin\theta\cos\varphi, r\sin\theta\sin\varphi, r\cos\theta)$ $0 \le \theta \le \pi$, $0 \le \varphi < 2\pi$. Then

$$x = r\sin\theta\cos\varphi, \quad y = r\sin\theta\sin\varphi, \quad z = r\cos\theta$$

so

$$\widehat{T}_\theta = r\cos\theta\cos\varphi\widehat{i} + r\cos\theta\sin\varphi\widehat{j} - r\sin\theta\widehat{k}$$
$$\widehat{T}_\varphi = -r\sin\theta\sin\varphi\widehat{i} + r\sin\theta\cos\varphi\widehat{j}$$

and

$$\widehat{T}_\theta \times \widehat{T}_\varphi = \begin{vmatrix} \widehat{i} & \widehat{j} & \widehat{k} \\ r\cos\theta\cos\varphi & r\cos\theta\sin\varphi & -r\sin\theta \\ -r\sin\theta\sin\varphi & r\sin\theta\cos\varphi & 0 \end{vmatrix}$$
$$= r^2\left[(\sin\theta)^2\cos\varphi\widehat{i} + (\sin\theta)^2\sin\varphi\widehat{j} + \sin\theta\cos\theta\widehat{k}\right].$$

Note that

$$\widehat{T}_\theta \times \widehat{T}_\varphi = r^2\sin\theta\left[\sin\theta\cos\varphi\widehat{i} + \sin\theta\sin\varphi\widehat{j} + \cos\theta\widehat{k}\right]$$
$$= r^2\sin\theta\left(x\widehat{i} + y\widehat{j} + z\widehat{k}\right) = r^2\sin\theta\widehat{r}$$

so that the normal to the tangent plane to a sphere is radially directed, as we would expect.

Also

$$\left\|\widehat{T}_\theta \times \widehat{T}_\varphi\right\| = r^2\sin\theta.$$

Thus the estimate for ΔS is $r^2\sin\theta\Delta\theta\Delta\varphi$.

Fact:

If $\Phi(D)$ is a surface S, then the area of S is

$$\iint_D \left\|\widehat{T}_u \times \widehat{T}_v\right\| dudv.$$

Example:

We compute the surface area S of a sphere. We have

$$S = \iint_D \left\|\widehat{T}_u \times \widehat{T}_v\right\| dudv = \int_{\theta=0}^{\pi}\int_{\varphi=0}^{2\pi}\left(r^2\sin\theta d\varphi\right)d\theta = 2\pi r^2\int_0^{\pi}\sin\theta d\theta$$
$$= -2\pi r^2\cos\theta|_0^{\pi} = 4\pi r^2.$$

INTEGRALS OF SCALAR FUNCTIONS OVER SURFACES

Suppose that $f: \mathbb{R}^3 \to \mathbb{R}$ and S is a surface in $\mathbb{R}^3$. We begin by giving an intuitive description of what $\int_S f\,dS$ should mean. The idea is simpler than it may appear. All we are doing is picking a point on the surface, evaluating the function at that point and then multiplying that number by a small amount of the surface area close to that point. We then sum all the pieces.

As with most integrals, we can go back to Riemann sums. We approximate the surface S by covering S with nonoverlapping planar segments, ΔP_i, each of which is tangent to S at one point. We described how these planar segments are determined above. Let

$$\|\Delta P\|_i = \text{diameter of } \Delta P_i = \sup\{|x_i - y_i| \; |x_i,\ y_i \in \Delta P_i\}$$

so that if $\|P_i\| \to 0$, then every dimension of P_i must go to 0.

For each planar segment P_i, let (x_i, y_i, z_i) denote the point on the surface S at which P_i is tangent to S. Form the product $f(x_i, y_i, z_i)\Delta P_i$ where ΔP_i is the area of P_i. In the work above, we showed how to calculate ΔP_i in the special instances that we consider. Again referring to our work above, $f(x_i, y_i, z_i)\Delta P_i$ is an estimate for $f(x_i, y_i, z_i)\Delta S$. Depending on how we describe the surface, the Riemann sum that we form is

$$\sum f(x_i, y_i, z_i)\left(\sqrt{\left(\frac{\partial f(x_0, y_0)}{\partial x}\right)^2 + \left(\frac{\partial f(x_0, y_0)}{\partial y}\right)^2 + 1}\right)\Delta x \Delta y$$

or

$$\sum f(x_i, y_i, z_i)\|\widehat{T}_u \times \widehat{T}_v\|\Delta u \Delta v = \sum f(\Phi((u_i, v_i)))\|\widehat{T}_u \times \widehat{T}_v\|\Delta u \Delta v.$$

If the function f is reasonably well behaved, then the limit of these Riemann sums exists as $\|\Delta P_i\| \to 0$, and we define

$$\iint_S f dS = \lim_{\Delta x \to 0, \Delta y \to 0} \sum f(x_i, y_i, z_i) \times \left(\sqrt{\left(\frac{\partial f(x_0, y_0)}{\partial x}\right)^2 + \left(\frac{\partial f(x_0, y_0)}{\partial y}\right)^2 + 1}\right)\Delta x \Delta y.$$

(In this case, z is a function of x and y so the integrand is actually a function of x and y.)

Likewise,

$$\iint_S f dS = \lim_{\Delta u \to 0, \Delta v \to 0} \sum f(\Phi((\mathrm{u_i}, \mathrm{v_i})))\|\widehat{T}_u \times \widehat{T}_v\|\Delta u \Delta v.$$

Example:

We compute the surface area of a cone, which is determined by

$$x^2 + y^2 = z^2, \quad 0 \leq z \leq 3$$

by computing $\iint_S 1 dS$. We parameterize the surface by the variables u,v with

$$z = u, \quad x = u \cos v, \quad y = u \sin v, \quad 0 \leq u \leq 3, \quad 0 \leq v < 2\pi.$$

(A more natural choice for the names of the parameterizing variables is z and θ, but we are following the notation we have established.) Then

$$\Phi(u,v) = (u\cos v,\ u\sin v, u)$$

$$\widehat{T}_u = \frac{\partial\Phi}{\partial u} = \cos v\widehat{i} + \sin v\widehat{j} + \widehat{k}$$

$$\widehat{T}_v = \frac{\partial\Phi}{\partial v} = -u\sin v\widehat{i} + u\cos v\widehat{j}$$

$$\widehat{T}_u \times \widehat{T}_v = \begin{vmatrix} \widehat{i} & \widehat{j} & \widehat{k} \\ \cos v & \sin v & 1 \\ -u\sin v & u\cos v & 0 \end{vmatrix}$$

$$= -u\cos v\widehat{i} - u\sin v\widehat{j} + \left(u\cos^2 v + u\sin^2 v\right)\widehat{k}$$

$$\|\widehat{T}_u \times \widehat{T}_v\| = \sqrt{(-u\cos v)^2 + (-u\sin v)^2 + u^2} = \sqrt{2}u.$$

So

$$S = \iint_S 1dS = \int_{u=0}^{3}\left(\int_{v=0}^{2\pi}\sqrt{2}udv\right)du = 2\sqrt{2}\pi\int_0^3 u\,du = \sqrt{2}\pi u^2\Big|_0^3 = 9\sqrt{2}\pi.$$

Example:
Compute the surface integral $\iint_S (5xy^3 + 4xy)dS$ where S is the surface determined by $z = x^2+3y$, $0 \le x \le y \le 1$.

We have

$$z = f(x,y) = x^2 + 3y, \quad \text{so } dS = \sqrt{\left(\frac{\partial f}{\partial x}\right)^2 + \left(\frac{\partial f}{\partial y}\right)^2 + 1} = \sqrt{4x^2 + 10}$$

and

$$\iint_S (5xy^3 + 4xy)dS = \int_{y=0}^{1}\left(\int_{x=0}^{y}(5xy^3 + 4xy)\sqrt{4x^2 + 10}dx\right)dy.$$

We leave it to show

$$\int_{y=0}^{1}\left(\int_{x=0}^{y}(5xy^3 + 4xy)\sqrt{4x^2 + 10}dx\right)dy = \frac{49\sqrt{14}}{15} - \frac{485\sqrt{10}}{168}$$

as an exercise using a computer algebra system (CAS).

SURFACE INTEGRALS OF VECTOR FUNCTIONS

A major reason why we study vector functions over a surface is to measure flux—an idea that we now describe.

Flux is an important concept in electricity and magnetism. In fact, much of what we shall study in the later part of this chapter was developed as part of the effort to give a mathematical formulation to the principles of electricity and magnetism. The culmination of this effort was Maxwell's equations.

An intuitive way to visualize flux is the passage of a fluid through a membrane. We want to measure the amount of fluid that passes through one unit of area of the membrane in one unit of time. Fig. 2.1.5 illustrates how this depends on two factors: (1) the velocity of the fluid and (2) the orientation of the surface with respect to the direction of flow of the fluid. The more closely the surface is to being perpendicular to the flow of the fluid, the more fluid will pass through the surface. Thus, if $\vec{F}$ is the vector field that describes the flow of the fluid, and $\hat{n}$ is the unit vector normal to the surface, then $\vec{F} \cdot \hat{n}$ would be the appropriate mathematical quantity to combine these two conditions.

If S is a given two-sided surface and $\vec{F}$ is the vector field describing the motion of the fluid, we determine how much fluid passes through, dS, a small amount of the surface in one unit of time. (All of our surfaces are two sided. An example of a surface that is one sided is a Mobius strip.)

To compute this volume of fluid, suppose that Φ is a parameterization of S, and $\Phi{:}D \to S$. Partition D into small rectangles, and consider one of the rectangles, D_{ij}. Suppose the length of D_{ij} is Δu and the width is Δv. Let $S_{ij} = \Phi(D_{ij})$ and suppose that (u_0, v_0) is a point in D_{ij}. Then $\hat{T}_u(u_0, v_0) \times \hat{T}_v(u_0, v_0)$ is a vector that is normal to S_{ij} at $\Phi(u_0, v_0)$, and

$$\hat{n} = \frac{\hat{T}_u(u_0, v_0) \times \hat{T}_v(u_0, v_0)}{\left\| \hat{T}_u(u_0, v_0) \times \hat{T}_v(u_0, v_0) \right\|}$$

is a unit vector that is normal to S_{ij} at $\Phi(u_0, v_0)$. In the case that the surface can be described by $z = f(x, y)$, then

$$\hat{n} = \frac{\dfrac{\partial f(x_0, y_0)}{\partial x}\hat{i} + \dfrac{\partial f(x_0, y_0)}{\partial x}\hat{j} - \hat{k}}{\sqrt{\left(\dfrac{\partial f(x_0, y_0)}{\partial x}\right)^2 + \left(\dfrac{\partial f(x_0, y_0)}{\partial y}\right)^2 + 1}}.$$

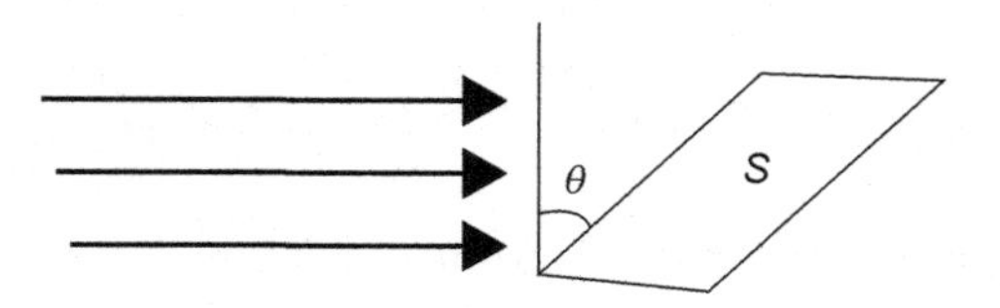

FIGURE 2.1.5

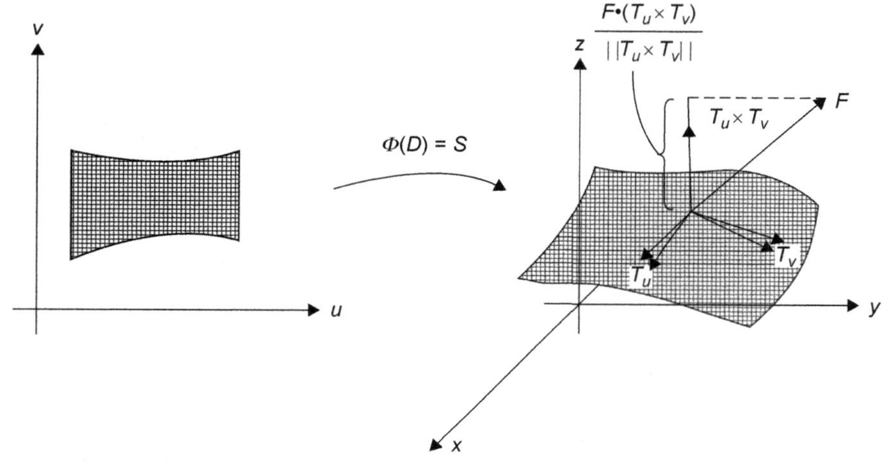

FIGURE 2.1.6

Let θ denote the angle that $\widehat{n}$ makes with $\vec{F}(\Phi(u_0, v_0))$. See Fig. 2.1.6. The parallelogram with sides $\widehat{T}_u(u_0, v_0)\Delta u$ and $\widehat{T}_v(u_0, v_0)\Delta v$ has approximately the same area as S_{ij}. The volume of fluid that passes through S_{ij} in one unit of time is approximately

$$\left\|\vec{F}(\Phi(u_0, v_0))\right\| \cos\theta \left\|\widehat{T}_u(u_0, v_0) \times \widehat{T}_v(u_0, v_0)\right\| \Delta u \Delta v.$$

But

$$\left\|\vec{F}(\Phi(u_0, v_0))\right\| \cos\theta \left\|\widehat{T}_u(u_0, v_0) \times \widehat{T}_v(u_0, v_0)\right\|$$

$$= \vec{F}(\Phi(u_0, v_0)) \cdot \widehat{n} \left\|\widehat{T}_u(u_0, v_0) \times \widehat{T}_v(u_0, v_0)\right\| \Delta u \Delta v$$

$$= \vec{F}(\Phi(u_0, v_0)) \cdot \frac{\widehat{T}_u(u_0, v_0) \times \widehat{T}_v(u_0, v_0)}{\left\|\widehat{T}_u(u_0, v_0) \times \widehat{T}_v(u_0, v_0)\right\|} \left\|\widehat{T}_u(u_0, v_0) \times \widehat{T}_v(u_0, v_0)\right\| \Delta u \Delta v$$

$$= \vec{F}(\Phi(u_0, v_0)) \cdot \left(\widehat{T}_u(u_0, v_0) \times \widehat{T}_v(u_0, v_0)\right) \Delta u \Delta v.$$

Also, this approximation becomes exact as $\Delta u \to 0$, $\Delta v \to 0$. Thus

$$\sum_{i,j} \vec{F}(\Phi(u_0, v_0)) \cdot \left(\widehat{T}_u(u_0, v_0) \times \widehat{T}_v(u_0, v_0)\right) \Delta u \Delta v$$

is a Riemann sum that approximates the amount of fluid that flows through the surface S in one unit of time, and the exact value is given by

$$\iint_D \vec{F}(\Phi(u, v)) \cdot \left(\widehat{T}_u(u_0, v_0) \times \widehat{T}_v(u_0, v_0)\right) du dv. \tag{2}$$

It is common to write expression (2) as $\int_s \vec{F} \cdot d\widehat{S}$ or $\int_S \vec{F} \cdot \widehat{n}\, dS$.

In the case the equation of the surface is $z = f(x, y)$ the integral is

$$\iint_D \vec{F}(x,y,z) \cdot \frac{\left(\frac{\partial f(x,y)}{\partial x}\widehat{i} + \frac{\partial f(x,y)}{\partial y}\widehat{j} - \widehat{k}\right)}{\sqrt{\left(\frac{\partial f(x,y)}{\partial x}\right)^2 + \left(\frac{\partial f(x,y)}{\partial x}\right)^2 + 1}} dxdy$$

keeping in mind that $\vec{F}(x,y,z)$ can be expressed in terms of x and y.

The process of evaluating the integral of a vector function over a surface can be broken down into steps.

Step 1. If the surface is described in parametric form, compute the tangent vectors T_u and T_v as before; that is,

$$\widehat{T}_u = \frac{\partial x}{\partial u}\widehat{i} + \frac{\partial y}{\partial u}\widehat{j} + \frac{\partial z}{\partial u}\widehat{k}$$

$$\widehat{T}_v = \frac{\partial x}{\partial v}\widehat{i} + \frac{\partial y}{\partial v}\widehat{j} + \frac{\partial z}{\partial v}\widehat{k}.$$

Step 2. Form a normal vector. If the surface is described in parametric form, a normal vector is

$$\widehat{T}_u \times \widehat{T}_v.$$

If the surface is described $z = f(x, y)$, a normal vector is

$$\frac{\partial f(x,y)}{\partial x}\widehat{i} + \frac{\partial f(x,y)}{\partial y}\widehat{j} - \widehat{k}.$$

Step 3. Normalize the vector found in Step 2.

If surface is defined parametrically, the vector is $\widehat{T}_u \times \widehat{T}_v$ and then a unit normal vector is

$$\widehat{n} = \frac{\widehat{T}_u \times \widehat{T}_v}{\left\|\widehat{T}_u \times \widehat{T}_v\right\|}.$$

If the surface is defined by $z = f(x, y)$, normalize the vector in Step 2 to get

$$\widehat{n} = \frac{\left(\frac{\partial f(x,y)}{\partial x}\widehat{i} + \frac{\partial f(x,y)}{\partial y}\widehat{j} - \widehat{k}\right)}{\sqrt{\left(\frac{\partial f(x,y)}{\partial x}\right)^2 + \left(\frac{\partial f(x,y)}{\partial x}\right)^2 + 1}}.$$

Step 4. Take the dot product of the unit normal vector with the vector field. This is either

$$\vec{F}(\Phi(u,v))\cdot\frac{\widehat{T}_u\times\widehat{T}_v}{\|\widehat{T}_u\times\widehat{T}_v\|}$$

in the parametric case or

$$\vec{F}(x,y,z)\cdot\frac{\left(\frac{\partial f(x,y)}{\partial x}\widehat{i}+\frac{\partial f(x,y)}{\partial y}\widehat{j}-\widehat{k}\right)}{\sqrt{\left(\frac{\partial f(x,y)}{\partial x}\right)^2+\left(\frac{\partial f(x,y)}{\partial x}\right)^2+1}}$$

in the case $z=f(x,y)$.

Step 5. Integrate the quantity in Step 4 as a scalar function over a surface using dS as the infinitesimal element. For example, integrating over x and y, $dS=dxdy$; integrating over the surface of a sphere of radius R, $dS=\mathrm{R}^2\sin\theta d\theta\, d\varphi$.

Example:

Let $\widehat{F}(x,y,z)=x^2y\widehat{i}+2yz\widehat{j}+4x\widehat{k}$ and S be the surface defined by $z=1-x-2y$ in the first octet. We compute $\iint_S\vec{F}\cdot dS$.

We have

$$\widehat{F}(x,y,z)=x^2y\widehat{i}+2yz\widehat{j}+4x\widehat{k}=x^2y\widehat{i}+2y(1-x-2y)\widehat{j}+4x\widehat{k}.$$

Also,

$$z=f(x,y)=1-x-2y,$$

so

$$\frac{\left(\frac{\partial f(x,y)}{\partial x}\hat{i}+\frac{\partial f(x,y)}{\partial y}\hat{j}-\hat{k}\right)}{\sqrt{\left(\frac{\partial f(x,y)}{\partial x}\right)^2+\left(\frac{\partial f(x,y)}{\partial x}\right)^2+1}}=\frac{-\hat{i}-2\hat{j}-\hat{k}}{\sqrt{(-1)^2+(-2)^2+1}}=\frac{-1}{\sqrt{6}}\left(\widehat{i}+2\widehat{j}+\widehat{k}\right).$$

Then

$$\vec{F}(x,y,z)\cdot\frac{\left(\frac{\partial f(x,y)}{\partial x}\widehat{i}+\frac{\partial f(x,y)}{\partial y}\widehat{j}-\widehat{k}\right)}{\sqrt{\left(\frac{\partial f(x,y)}{\partial x}\right)^2+\left(\frac{\partial f(x,y)}{\partial x}\right)^2+1}}$$

$$=\left[x^2y\widehat{i}+2y(1-x-2y)\widehat{j}+4x\widehat{k}\right]\cdot\left[\frac{-1}{\sqrt{6}}\left(\widehat{i}+2\widehat{j}+\widehat{k}\right)\right]$$

$$=\frac{-1}{\sqrt{6}}\left(x^2y+4y-4xy-8y^2+4x\right).$$

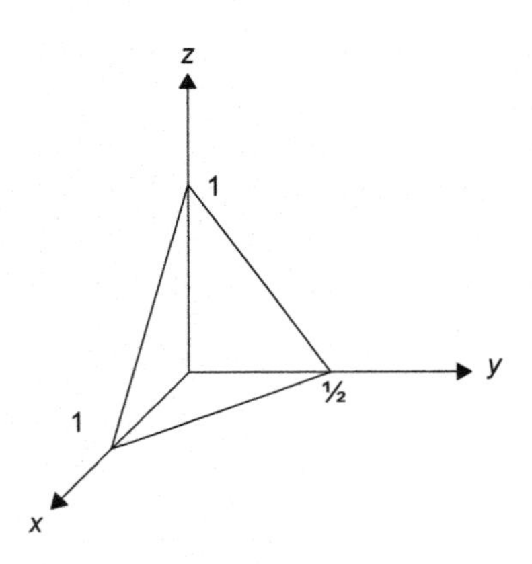

FIGURE 2.1.7A

The surface is shown in Fig. 2.1.7A, and the region of integration for x and y is shown in Fig. 2.1.7B.

We have

$$\iint_S \vec{F} \cdot dS = \int_{y=0}^{\frac{1}{2}} \left(\int_{x=0}^{1-2y} \frac{-1}{\sqrt{6}} (x^2y + 4y - 4xy - 8y^2 + 4x) dx \right) dy.$$

Using a CAS

$$\int_{y=0}^{\frac{1}{2}} \left(\int_{x=0}^{1-2y} \frac{-1}{\sqrt{6}} (x^2y + 4y - 4xy - 8y^2 + 4x) dx \right) dy = \frac{-1}{\sqrt{6}} \left(\frac{91}{240} \right).$$

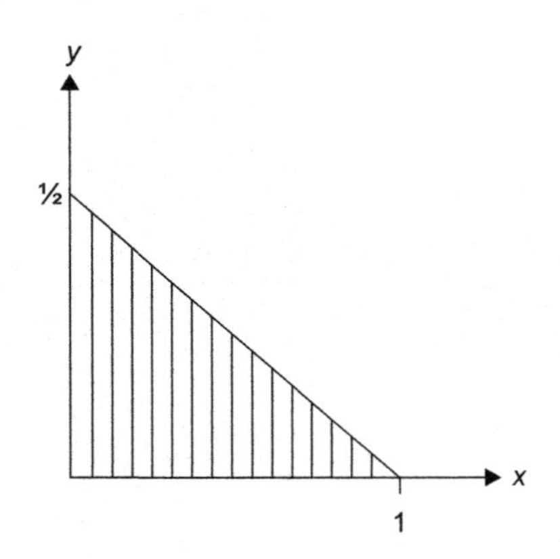

FIGURE 2.1.7B

Example:

We compute the flux of $\vec{\Phi}(x,y,z) = x\hat{i} + y\hat{j} + z\hat{k}$ out of the upper half of the unit sphere.

Parameterizing with spherical coordinates with $r = 1$, we have

$$x = \cos\varphi \sin\theta, \quad y = \sin\varphi \sin\theta, \quad z = \cos\theta.$$

Now,

$$\widehat{T}_\theta = \frac{\partial x}{\partial \theta}\hat{i} + \frac{\partial y}{\partial \theta}\hat{j} + \frac{\partial z}{\partial \theta}\hat{k} = \cos\varphi \cos\theta\hat{i} + \sin\varphi\cos\theta\hat{j} - \sin\theta\hat{k}$$

$$\widehat{T}_\varphi = \frac{\partial x}{\partial \varphi}\hat{i} + \frac{\partial y}{\partial \varphi}\hat{j} + \frac{\partial z}{\partial \varphi}\hat{k} = -\sin\varphi\sin\theta\hat{i} + \cos\varphi\sin\theta\hat{J}$$

so

$$\begin{aligned}\widehat{T}_\theta \times \widehat{T}_\varphi &= \begin{vmatrix} \hat{i} & \hat{j} & \hat{k} \\ \cos\varphi\cos\theta & \sin\varphi\cos\theta & -\sin\theta \\ -\sin\varphi\sin\theta & \cos\varphi\sin\theta & 0 \end{vmatrix} \\ &= \cos\varphi\sin^2\theta\hat{i} + \sin\varphi\sin^2\theta\hat{j} + \sin\theta\cos\theta\hat{k}.\end{aligned}$$

Then

$$\begin{aligned}\left\|\widehat{T}_\theta \times \widehat{T}_\varphi\right\|^2 &= \cos^2\varphi\sin^4\theta + \sin^2\varphi\sin^4\theta + \sin^2\theta\cos^2\theta \\ &= \sin^4\theta + \sin^2\theta\cos^2\theta = \sin^2\theta\end{aligned}$$

so

$$\frac{\widehat{T}_\theta \times \widehat{T}_\varphi}{\left\|\widehat{T}_\theta \times \widehat{T}_\varphi\right\|} = \cos\varphi\sin\theta\hat{i} + \sin\varphi\sin\theta\hat{j} + \cos\theta\hat{k}.$$

Now

$$\vec{F}(\Phi(\theta,\varphi)) = \cos\varphi\sin\theta\hat{i} + \sin\varphi\sin\theta\hat{j} + \cos\theta\hat{k}$$

so

$$\vec{F}(\Phi(\theta,\varphi))\cdot\frac{\widehat{T}_\theta \times \widehat{T}_\varphi}{\left\|\widehat{T}_\theta \times \widehat{T}_\varphi\right\|} = \cos^2\varphi\sin^2\theta + \sin^2\varphi\sin^2\theta + \cos^2\theta = 1.$$

Thus the flux over the upper hemisphere is

$$\int_{\theta=0}^{\frac{\pi}{2}}\int_{\varphi=0}^{2\pi}\sin\theta d\varphi d\theta = 2\pi\int_0^{\frac{\pi}{2}}\sin\theta d\theta = 2\pi.$$

EXERCISES

1. If $\nabla f(\widehat{x}) \neq \widehat{0}$, show that $\nabla f(\widehat{x})$ points in the direction in which f is increasing most rapidly.
2. Find the directional derivative of $f(x, y, z) = xyz^3 - 4x$ at the point $(1, -3, 2)$ in the direction $4\widehat{i} - 2\widehat{j} + \widehat{k}$.
3. Find ∇f for the following functions at the given point. Find the direction in which f is increasing most rapidly.
   a. $f(x, y, z) = xe^y + \sin z$ at $(3, -2, 0)$.
   b. $f(r, \theta, z) = zr\cos\theta$ at $(5, \pi/4, 2)$.
   c. $f(r, \theta, \varphi) = r\sin\theta + r\cos\varphi$ at $\left(3, \frac{\pi}{6}, \frac{\pi}{3}\right)$.
4. Let S be the surface consisting of the points (x, y, z) for which $f(x, y, z) = k$ where k is a constant. The equation of the plane tangent to S at the point (x_0, y_0, z_0) is

$$\nabla f(x_0, y_0, z_0) \cdot (x - x_0, y - y_0, z - z_0) = 0.$$

   Find the equation of the plane tangent to the following surfaces at the given point.
   a. $xy^2 + 3z = 7$ at $(1, 2, 1)$.
   b. $e^{3x} + 4yz = 8$ at $(0, 1, 2)$.
5. Find the length of the curve $\sigma(t) = (\cos t, \sin t, t^{3/2})$ $0 \leq t \leq 1$.

Path Integrals

6. Evaluate the following path integrals $\int_\sigma f(x, y, z)ds$.
   a. $f(x, y, z) = x^2 + y^2 + z^2$, $\sigma(t) = (\cos t, \sin t, t)$, $0 \leq t \leq 4$.
   b. $f(x, y, z) = x^2 + y + 3z$, $\sigma(t) = (2t, 5t, 9)$, $-1 \leq t \leq 5$.
   c. $f(x, y, z) = xy$, $\sigma(t) = \left(2t, 3t, \sqrt{3}\,t\right)$, $0 \leq t \leq 4$.

Line Integrals

7. Compute the following line integrals.
   a. $\int_\sigma x^3 y\,dx + xz\,dy + y^2 dz$ where $\sigma(t) = (t^2, t^3, 4)$, $0 \leq t \leq 2$.
   b. $\int_\sigma \sin z dx + \cos z\,dy + 4dz$ where $\sigma(t) = \left(\cos^3 t, \sin^3 t, t\right)$, $0 \leq t \leq \frac{3\pi}{2}$.
   c. $\int_\sigma yz\,dx - zx\,dy + xy\,dz$ where $\sigma(t)$ consists of the line segments from $(2, 0, 0)$ to $(0, 2, 0)$ to $(0, 0, 2)$.
   d. $\int_\sigma (x^2 + y^2 + z^2)d\sigma$ $\sigma(t) = \cos t\widehat{i} + \sin t\widehat{j} + 2k$ from $(1, 0, 0)$ to $(1, 0, 8\pi)$.
   e. $\int x^2 y\,dx + x\,dy$ from $(0, 0)$ to $(1, 1)$ along the path $y = x$.
   f. $\int x^2 y\,dx + x\,dy$ from $(0, 0)$ to $(1, 1)$ along the path $y = x^2$.
   g. $\int_\sigma xydx + (x + y)dy$ $\sigma(t) = \left(1 + \frac{2}{t}, t^3\right)$, $2 \leq t \leq 4$.
8. Find $\int_\sigma \widehat{F} \cdot ds$ where
   a. $\widehat{F} = xy^2\widehat{i} - xy\widehat{j}$; $x = t^3$, $y = t$; $0 \leq t \leq 1$.

b. $\widehat{F} = x^2\widehat{i} + y^2\widehat{j}$; σ is the unit circle, from 0 to 2π.

c. $\widehat{F} = x^2y^2z\widehat{i} + 2yz\widehat{j} - xy\widehat{k}$; $\sigma(t) = t^2\widehat{i} + t\widehat{j} - t^3\widehat{k}$, $0 \le t \le 2$.

9. Find the work done going on upper half of the unit circle from (1, 0) to (−1, 0) when the force is

$$\vec{F} = \frac{-y}{x^2 + y^2}\widehat{i} + \frac{x}{x^2 + y^2}\widehat{j}.$$

10. Calculate the work done by going from (0, 0) to (1, 1) along two different paths where $\vec{F} = (x + y)\widehat{i} + (x - y)\widehat{j}$. What does your answer tell you about $\vec{F}$?

11. Gravitational force is given by

$$\widehat{F} = \frac{1}{r^3}\left(x\widehat{i} + y\widehat{j} + z\widehat{k}\right) \quad \text{where} \quad r = \sqrt{x^2 + y^2 + z^2}.$$

Compute the work done $\int_\sigma \widehat{F} \cdot ds$ where σ is any path for which $\|\sigma(\textit{initial point})\| = R_1$ and $\|\sigma(\text{final point})\| = R_2$.

Integrals of Scalar Functions Over Surfaces

12. Evaluate $\iint_S f(x, y, z)dS$ where $f(x, y, z) = 3x^2$, S is the surface determined by $x^2 + y^2 = 4$, $-2 \le z \le 2$.

13. Evaluate $\iint_S f(x, y, z)dS$ where $f(x, y, z) = 3x^2\cos y$, S is the surface determined by $z = x^2$, $0 \le x \le 2$, $0 \le y \le \pi/2$.

14. Evaluate $\iint_S f(x, y, z)dS$ where $f(x, y, z) = \sqrt{x^2 + y^2}$, S is the surface determined by $z = 9 - x^2 - y^2$, $z \ge 0$.

15. Find the surface area of the paraboloid $z = 4x^2 + 4y^2$, $0 \le z \le 16$.

16. Evaluate $\iint_S (x^2y + 3xyz)dS$ where, S is the plane in the first octet $2x + 4y + z = 8$.

17. Let S be a sphere of radius R. Evaluate

$$\iint_S \frac{1}{\left[(x - x_0)^2 + (y - y_0)^2 + (z - z_0)^2\right]} dS$$

a. in the case $\sqrt{x_0{}^2 + y_0{}^2 + z_0{}^2} < R$

b. in the case $\sqrt{x_0{}^2 + y_0{}^2 + z_0{}^2} > R$.

18. Parameterize the surface of a right circular cone of radius R and height h and find the area of the surface of the cone.

Integrals of Vector Functions Over Surfaces

19. Find the flux of the vector field $\vec{F}(x, y, z) = x\,\widehat{i} + y\,\widehat{j} + z\widehat{k}$ through the upper half of the hemisphere $x^2 + y^2 + z^2 = 4$.

20. Find the flux of the vector field $\vec{F}(x, y, z) = 5\hat{k}$ through the surface $z = 9 - x^2 - y^2$, $z \geq 0$.
21. Find the flux of the vector field $\vec{F}(x, y, z) = y\hat{j} - z\hat{k}$ where S is the surface consisting of two pieces $y = x^2 + z^2$ and $x^2 + z^2 \leq 1$.
22. Find the flux of the vector field $\vec{F}(x, y, z) = 2y\hat{j} + 5z\hat{k}$ and S is the plane $2x + y + 3z = 6$ in the first octet.
23. Find the flux of the vector field $\vec{F}(x, y, z) = 3z^2\hat{i} + 2\hat{j} + xz\hat{k}$ and S is the surface $y = x^2$ $0 \leq x \leq 1$, $0 \leq z \leq 2$.
24. Find the flux of the vector field $\vec{F}(x, y, z) = 2y\hat{j} - z\hat{k}$ and S is the surface bounded by the paraboloid $y = 4x^2 + 4z^2$ and the plane $y = 1$.
25. Find the flux of the vector field $\vec{F}(x, y, z) = x\hat{i} + y\hat{j} + z\hat{k}$ and S is the plane $z = 4 - 2x - y$ in the first octet.
26. Find the flux of the vector field $\vec{F}(x, y, z) = y\hat{i} + x\hat{j} + z\hat{k}$ and S is the surface $z = 4 - x^2 - y^2$, $z \geq 0$.

2.2 THE DIVERGENCE AND CURL

The divergence and curl are two of the most important operators in vector calculus. One way of presenting them is to define them in terms of mathematical formulas. We choose instead to analyze the phenomena from which they arise, and then derive the associated formulas. Both the phenomena describe the action of a vector field at a point. The derivation involves analyzing what happens within a small volume, dividing by the volume to determine the effect per unit volume, and then taking the limit as the volume goes to 0.

DIVERGENCE

Let $\vec{F}$ be a vector field and S be an enclosed surface. Then the flux of $\vec{F}$ through S is

$$\iint_S \vec{F} \cdot \hat{n} dS$$

where $\hat{n}$ is the outward pointing unit vector to the surface of S. Let V be the volume of the region enclosed by S. Let $p = (x_0, y_0, z_0)$ be a point in $\mathbb{R}^3$. The *divergence of $\vec{F}$ at p*, denoted $\operatorname{div} \vec{F}(p)$, is

$$\operatorname{div} \vec{F}(p) = \lim_{\|V\| \to 0,\, p \in V} \frac{1}{V} \iint_S \vec{F} \cdot \hat{n} dS$$

where the limit exists.

Thus, $\operatorname{div} \vec{F}(p)$ is the flux of $\vec{F}$ at p. The formula used to calculate the divergence depends on the coordinate system used. We develop the formulas for Cartesian, cylindrical, and spherical coordinates.

CARTESIAN COORDINATE CASE

Choose the point $p = (x_0, y_0, z_0)$ and enclose p in the center of a rectangular parallelepiped. See Fig. 2.2.1.

Let Δx, Δy and Δz denote the lengths of the sides of the parallelepiped, and let

$$\vec{F}(x, y, z) = F_x(x, y, z)\hat{i} + F_y(x, y, z)\hat{j} + F_z(x, y, z)\hat{k}.$$

This notation will help keep the component functions straight, but we must remember that for the rest of the section, the subscript does not mean a partial derivative.

We assume for the remainder of this chapter that each component function is differentiable. Our parallelepiped has six faces, and we compute

$$\iint_S \vec{F} \cdot \hat{n} dS$$

by computing the integral over each of the faces and then adding the results.

Let S_1 and S_2 denote the faces that are parallel to the y, z plane. See Fig. 2.2.1.

On S_1, $\vec{F} \cdot \hat{n} = \vec{F} \cdot \hat{i} = F_x(x, y, z)$ and on S_2, $\vec{F} \cdot \hat{n} = \vec{F} \cdot (-\hat{i}) = -F_x(x, y, z)$. (Recall that $\hat{n}$ is the unit normal vector that points outward from the surface.) The x coordinate of any point on S_1 is $x_0 + \frac{\Delta x}{2}$. If Δy and Δz are small, then

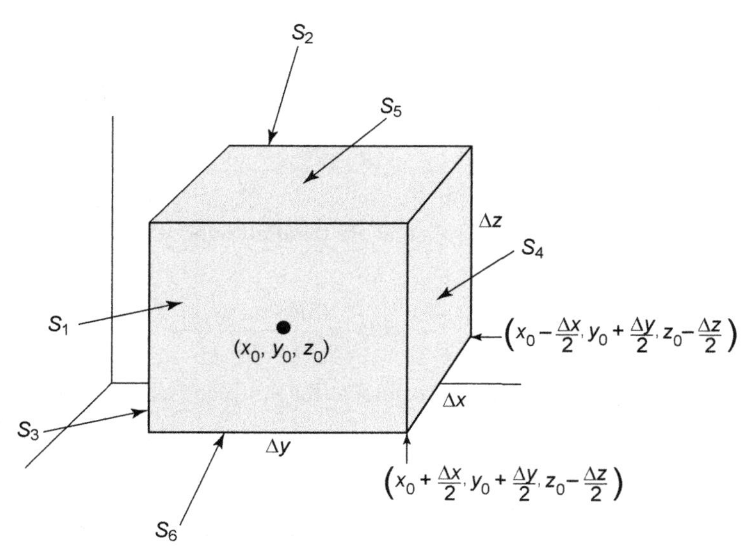

FIGURE 2.2.1

$$\iint_{S_1} \vec{F}\cdot\hat{n}dS = \iint_{S_1} F_x\left(x_0+\frac{\Delta x}{2},y,z\right)dS \approx F_x\left(x_0+\frac{\Delta x}{2},y_0,z_0\right)\iint_{S_1} dydz$$
$$= F_x\left(x_0+\frac{\Delta x}{2},y_0,z_0\right)\Delta y\Delta z.$$

The x coordinate of any point on S_2 is $x_0 - \frac{\Delta x}{2}$. If Δy and Δz are small, then

$$\iint_{S_2} \vec{F}\cdot\hat{n}dS = \iint_{S_2} -F_x\left(x_0-\frac{\Delta x}{2},y,z\right)dS$$
$$\approx -F_x\left(x_0-\frac{\Delta x}{2},y_0,z_0\right)\iint_{S_2} dydz = -F_x\left(x_0-\frac{\Delta x}{2},y_0,z_0\right)\Delta y\Delta z.$$

Thus

$$\iint_{S_1+S_2} \vec{F}\cdot\hat{n}dS \approx \left[F_x\left(x_0+\frac{\Delta x}{2},y_0,z_0\right)-F_x\left(x_0-\frac{\Delta x}{2},y_0,z_0\right)\right]\Delta y\Delta z$$
$$= \frac{\left[F_x\left(x_0+\frac{\Delta x}{2},y_0,z_0\right)-F_x\left(x_0-\frac{\Delta x}{2},y_0,z_0\right)\right]}{\Delta x}\Delta x\Delta y\Delta z.$$

Now $\Delta x\Delta y\Delta z = V$, so

$$\frac{1}{V}\iint_{S_1+S_2} \vec{F}\cdot\hat{n}dS \approx \frac{\left[F_x\left(x_0+\frac{\Delta x}{2},y_0,z_0\right)-F_x\left(x_0-\frac{\Delta x}{2},y_0,z_0\right)\right]}{\Delta x}.$$

In the limit as $\Delta x \to 0$ we have

$$\lim_{\Delta x\to 0}\frac{1}{V}\iint_{S_1+S_2} \vec{F}\cdot\hat{n}dS = \frac{\partial F_x(x_0,y_0,z_0)}{\partial x}.$$

Similarly, if S_3 and S_4 are the faces that are parallel to the x, z plane as shown in Fig. 2.2.1 then

$$\lim_{\Delta y\to 0}\frac{1}{V}\iint_{S_3+S_4} \vec{F}\cdot\hat{n}dS = \frac{\partial F_y(x_0,y_0,z_0)}{\partial y}$$

and if S_5 and S_6 are the faces that are parallel to the x, y plane as shown in Fig. 2.2.1 then

$$\lim_{\Delta z\to 0}\frac{1}{V}\iint_{S_5+S_6} \vec{F}\cdot\hat{n}dS = \frac{\partial F_z(x_0,y_0,z_0)}{\partial z}.$$

Thus,

$$\lim_{\|V\|\to 0}\frac{1}{V}\iint_S \vec{F}\cdot\hat{n}dS = \lim_{\|V\|\to 0}\frac{1}{V}\iint_{S_1+\cdots+S_6}\vec{F}\cdot\hat{n}dS$$

$$= \lim_{\|V\|\to 0}\frac{1}{V}\iint_{S_1+S_2}\vec{F}\cdot\hat{n}dS + \lim_{\|V\|\to 0}\frac{1}{V}\iint_{S_3+S_4}\vec{F}\cdot\hat{n}dS + \lim_{\|V\|\to 0}\frac{1}{V}\iint_{S_5+S_6}\vec{F}\cdot\hat{n}dS$$

$$= \lim_{\Delta x\to 0}\frac{1}{V}\iint_{S_1+S_2}\vec{F}\cdot\hat{n}dS + \lim_{\Delta y\to 0}\frac{1}{V}\iint_{S_3+S_4}\vec{F}\cdot\hat{n}dS + \lim_{\Delta z\to 0}\frac{1}{V}\iint_{S_5+S_6}\vec{F}\cdot\hat{n}dS$$

$$= \frac{\partial F_x(x_0,y_0,z_0)}{\partial x} + \frac{\partial F_y(x_0,y_0,z_0)}{\partial x} + \frac{\partial F_z(x_0,y_0,z_0)}{\partial z} \equiv \text{div}\,\vec{F}$$

and

$$\left(\hat{i}\frac{\partial}{\partial x} + \hat{j}\frac{\partial}{\partial x} + \hat{k}\frac{\partial}{\partial x}\right)\cdot\vec{F}$$

$$= \left(\hat{i}\frac{\partial}{\partial x} + \hat{j}\frac{\partial}{\partial x} + \hat{k}\frac{\partial}{\partial x}\right)\cdot\left(F_x(x,y,z)\hat{i} + F_y(x,y,z)\hat{j} + F_z(x,y,z)\right)\hat{k} \quad (1)$$

$$= \frac{\partial F_x(x,y,z)}{\partial x} + \frac{\partial F_y(x,y,z)}{\partial y} + \frac{\partial F_z(x,y,z)}{\partial z}.$$

Many texts will give the right-hand side of Eq. (1) as the definition of the divergence of a vector field in Cartesian coordinates. It is common to refer to $\hat{i}\frac{\partial}{\partial x} + \hat{j}\frac{\partial}{\partial x} + \hat{k}\frac{\partial}{\partial x}$ as the "del operator."

Example:

The divergence of

$$\vec{F}(x,y,z) = x^2e^y \sin z\hat{i} + 3yz\hat{j} + xy\hat{k}$$

is

$$\frac{\partial}{\partial x}(x^2e^y \sin z) + \frac{\partial}{\partial y}(3yz) + \frac{\partial}{\partial z}(xy) = 2xe^y \sin z + 3z.$$

CYLINDRICAL COORDINATE CASE

In integrating in cylindrical coordinates, the incremental element of volume is shown in Fig. 2.2.2. The volume of the incremental piece is $\Delta V = rd\theta drdz$.

We have the faces of the incremental element and will proceed by grouping the faces in pairs, as we did in the case of Cartesian coordinates. We describe the vector field as

$$\vec{F}(r,\theta,z) = F_r(r,\theta,z)\hat{e}_r + F_\theta(r,\theta,z)\hat{e}_\theta + F_z(r,\theta,z)\hat{e}_z.$$

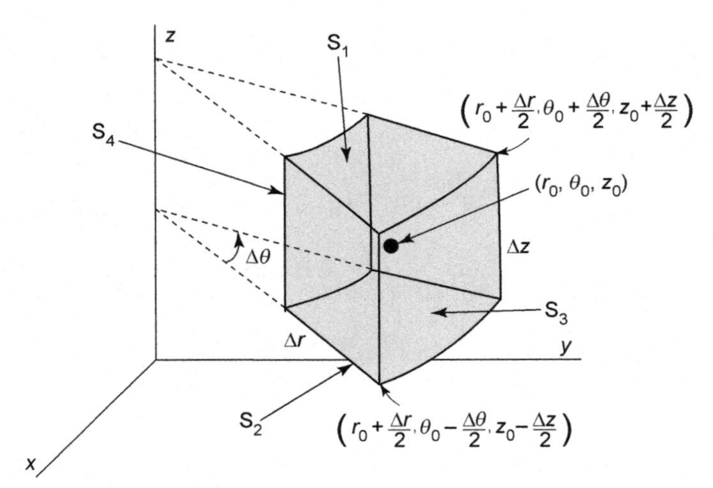

FIGURE 2.2.2

Consider an incremental volume element centered at $p = (r_0, \ \theta_0, \ z_0)$. We compute

$$\iint_{S_1} \vec{F}\cdot\hat{n}dS = \iint_{S_1} \vec{F}\cdot\hat{e}_z dS \approx F_z\left(r_0, \theta_0, z_0 + \frac{\Delta z}{2}\right) r(\Delta\theta)(\Delta r)$$

$$\iint_{S_2} \vec{F}\cdot\hat{n}dS = -\iint_{S_2} \vec{F}\cdot\hat{e}_z dS \approx \ -F_z\left(r_0, \theta_0, z_0 - \frac{\Delta z}{2}\right) r(\Delta\theta)(\Delta r)$$

so

$$\iint_{S_{1+}S_2} \vec{F}\cdot\hat{n}dS \approx \left[F_z\left(r_0, \theta_0, z_0 + \frac{\Delta z}{2}\right) - F_z\left(r_0, \theta_0, z_0 - \frac{\Delta z}{2}\right)\right] \ r(\Delta\theta)(\Delta r)$$

$$= \frac{\left[F_z\left(r_0, \theta_0, z_0 + \frac{\Delta z}{2}\right) - F_z\left(r_0, \theta_0, z_0 - \frac{\Delta z}{2}\right)\right]}{\Delta z} \ r(\Delta\theta)(\Delta r)(\Delta z)$$

$$\approx \frac{\partial F_z(r_0, \theta_0, z_0)}{\partial z}\Delta V.$$

Also,

$$\iint_{S_3} \vec{F}\cdot\hat{n}dS = \iint_{S_3} \vec{F}\cdot\hat{e}_r dS \approx F_r\left(r_0 + \frac{\Delta r}{2}, \theta_0, z_0\right)\left(r_0 + \frac{\Delta r}{2}\right)(\Delta\theta)(\Delta z)$$

$$\iint_{S_4} \vec{F}\cdot\hat{n}dS = -\iint_{S_4} \vec{F}\cdot\hat{e}_r dS \approx \ -F_r\left(r_0 - \frac{\Delta r}{2}, \theta_0, z_0\right)\left(r_0 - \frac{\Delta r}{2}\right)(\Delta\theta)(\Delta z).$$

Thus

$$\iint_{S_3+S_4} \vec{F}\cdot\hat{n}dS \approx \left[F_r\left(r_0+\frac{\Delta r}{2},\theta_0,z_0\right)-F_r\left(r_0-\frac{\Delta r}{2},\theta_0,z_0\right)\right]r_0(\Delta\theta)(\Delta z)$$

$$+F_r(r_0,\theta_0,z_0)(\Delta r)(\Delta\theta)(\Delta z)$$

$$=\frac{\left[F_r\left(r_0+\frac{\Delta r}{2},\theta_0,z_0\right)-F_r\left(r_0-\frac{\Delta r}{2},\theta_0,z_0\right)\right]}{\Delta r}r_0(\Delta\theta)(\Delta r)(\Delta z)$$

$$+\frac{1}{r}F_r(r_0,\theta_0,z_0)(\Delta r)r(\Delta\theta)(\Delta z) \approx \frac{\partial F(r_0,\theta_0,z_0)_r}{\partial r}\Delta V+\frac{1}{r}F_r(r_0,\theta_0,z_0)\Delta V.$$

For the remaining two faces, we have

$$\iint_{S_5}\vec{F}\cdot\hat{n}dS=\iint_{S_5}\vec{F}\cdot\hat{e}_\theta dS \approx F_\theta\left(r_0,\theta_0+\frac{\Delta\theta}{2},z_0\right)(\Delta r)(\Delta z)$$

$$\iint_{S_6}\vec{F}\cdot\hat{n}dS=-\iint_{S_6}\vec{F}\cdot\hat{e}_\theta dS \approx -F_\theta\left(r_0,\theta_0-\frac{\Delta\theta}{2},z_0\right)(\Delta r)(\Delta z)$$

so

$$\iint_{S_5+S_6}\vec{F}\cdot\hat{n}dS \approx \left[F_\theta\left(r_0,\theta_0+\frac{\Delta\theta}{2},z_0\right)-F_\theta\left(r_0,\theta_0-\frac{\Delta\theta}{2},z_0\right)\right](\Delta r)(\Delta z)$$

$$=\frac{1}{r}\frac{\left[F_\theta\left(r_0,\theta_0+\frac{\Delta\theta}{2},z_0\right)-F_\theta\left(r_0,\theta_0-\frac{\Delta\theta}{2},z_0\right)\right]}{\Delta\theta}$$

$$\times(\Delta r)r(\Delta\theta)(\Delta z)\approx\frac{1}{r}\frac{\partial F_\theta(r_0,\theta_0,z_0)}{\partial\theta}\Delta V.$$

Putting the faces together, we have

$$\iint_S\vec{F}\cdot\hat{n}dS=\iint_{S_1+\cdots+S_6}\vec{F}\cdot\hat{n}dS\approx\frac{\partial F_z(r_0,\theta_0,z_0)}{\partial z}\Delta V+\frac{\partial F(r_0,\theta_0,z_0)_r}{\partial r}\Delta V$$

$$+\frac{1}{r}F_r(r_0,\theta_0,z_0)\Delta V+\frac{1}{r}\frac{\partial F_\theta(r_0,\theta_0,z_0)}{\partial\theta}\Delta V$$

so that

$$\frac{1}{\Delta V}\iint_S\vec{F}\cdot\hat{n}dS\approx\frac{1}{\Delta V}\iint_{S_1+\cdots+S_6}\vec{F}\cdot\hat{n}dS\approx\frac{\partial F_z(r_0,\theta_0,z_0)}{\partial z}+\frac{\partial F_r(r_0,\theta_0,z_0)_r}{\partial r}$$

$$+\frac{1}{r}F_r(r_0,\theta_0,z_0)+\frac{1}{r}\frac{\partial F_\theta(r_0,\theta_0,z_0)}{\partial\theta}.$$

As long as the component functions are continuously differentiable, we can take the limit as $\|\Delta V\| \to 0$, and we have in cylindrical coordinates

$$\operatorname{div}\vec{F}(r_0,\theta_0,z_0) = \lim_{\|\Delta V\|\to 0}\frac{1}{\Delta V}\iint_S \vec{F}\cdot\hat{n}dS = \frac{\partial F_z(r_0,\theta_0,z_0)}{\partial z} + \frac{\partial F_r(r_0,\theta_0,z_0)}{\partial r}$$
$$+\frac{1}{r}F_r(r_0,\theta_0,z_0) + \frac{1}{r}\frac{\partial F_\theta(r_0,\theta_0,z_0)}{\partial\theta}$$
$$=\frac{1}{r}\frac{\partial}{\partial r}(rF_r) + \frac{1}{r}\frac{\partial F_\theta}{\partial\theta} + \frac{\partial F_z}{\partial z}.$$

Thus, we say

$$\operatorname{div}\,\vec{F} = \frac{1}{r}\frac{\partial}{\partial r}(rF_r) + \frac{1}{r}\frac{\partial F_\theta}{\partial\theta} + \frac{\partial F_z}{\partial z}.$$

Example:
We compute the divergence of

$$\vec{F}(r,\theta,z) = 5r^2e^z\sin\theta\hat{e}_r + r\tan\theta z^3\hat{e}_\theta + z\hat{e}_z.$$

We have

$$\operatorname{div}\,\vec{F} = \frac{1}{r}\frac{\partial}{\partial r}\left[r\left(5r^2e^z\sin\theta\right)\right] + \frac{1}{r}\frac{\partial}{\partial\theta}\left(r\tan\theta z^3\right) + \frac{\partial}{\partial z}(z)$$
$$=\frac{1}{r}\left(15r^2e^z\sin\theta\right) + \sec^2\theta z^3 + 1 = 15r\,e^z\sin\theta + \sec^2\theta z^3 + 1.$$

SPHERICAL COORDINATE CASE

In integrating in spherical coordinates, the incremental element of volume is shown in Fig. 2.2.3. The volume of the incremental piece is $\Delta V = r^2\sin\theta d\varphi d\theta dr$.

We proceed as in the two previous cases. Consider an incremental volume element centered at $(r_0, \theta_0, \varphi_0)$. Let $\vec{F}(r,\theta,\varphi,) = F_r(r,\theta,\varphi,)\hat{e}_r + F_\theta(r,\theta,\varphi,)\hat{e}_\theta + F_\varphi(r,\theta,\varphi,)\hat{e}_\varphi$. We first compute

$$\iint_{S_1}\vec{F}\cdot\hat{n}dS + \iint_{S_2}\vec{F}\cdot\hat{n}dS = \iint_{S_1}\vec{F}\cdot\hat{e}_\theta dS - \iint_{S_2}\vec{F}\cdot\hat{e}_\theta dS$$
$$\approx F_\theta\left(r_0,\theta_0+\frac{\Delta\theta}{2},\varphi_0\right)\iint_{S_1}dS - F_\theta\left(r_0,\theta_0-\frac{\Delta\theta}{2},\varphi_0\right)\iint_{S_2}dS$$
$$\approx F_\theta\left(r_0,\theta_0+\frac{\Delta\theta}{2},\varphi_0\right)r\sin\left(\theta+\frac{\Delta\theta}{2}\right)\Delta r\Delta\varphi$$
$$-F_\theta\left(r_0,\theta_0-\frac{\Delta\theta}{2},\varphi_0\right)r\sin\left(\theta-\frac{\Delta\theta}{2}\right)\Delta r\Delta\varphi.$$

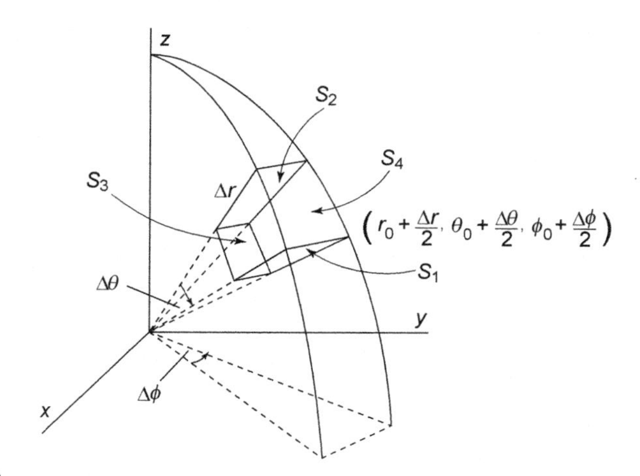

FIGURE 2.2.3

We expand $\sin\left(\theta+\frac{\Delta\theta}{2}\right)$ using the Taylor approximation $f(x+\Delta x) \approx f(x)+f'(x)\Delta x$, so that

$$\sin\left(\theta+\frac{\Delta\theta}{2}\right) \approx \sin\theta+\cos\theta\frac{\Delta\theta}{2}$$

and likewise,

$$\sin\left(\theta-\frac{\Delta\theta}{2}\right) \approx \sin\theta-\cos\theta\frac{\Delta\theta}{2}.$$

Thus

$$F_\theta\left(r_0, \theta_0+\frac{\Delta\theta}{2}, \varphi_0\right) r \sin\left(\theta+\frac{\Delta\theta}{2}\right)\Delta r\Delta\varphi$$

$$\approx F_\theta\left(r_0, \theta_0+\frac{\Delta\theta}{2}, \varphi_0\right) r\left[\sin\theta+\cos\theta\frac{\Delta\theta}{2}\right]\Delta r\Delta\varphi$$

$$F_\theta\left(r_0, \theta_0-\frac{\Delta\theta}{2}, \varphi_0\right) r \sin\left(\theta-\frac{\Delta\theta}{2}\right)\Delta r\Delta\varphi$$

$$\approx F_\theta\left(r_0, \theta_0-\frac{\Delta\theta}{2}, \varphi_0\right) r\left[\sin\theta-\cos\theta\frac{\Delta\theta}{2}\right]\Delta r\Delta\varphi$$

and so

$$\begin{aligned}
&F_\theta\left(r_0, \theta_0 + \frac{\Delta\theta}{2}, \varphi_0\right) r \sin\left(\theta + \frac{\Delta\theta}{2}\right) \Delta r \Delta\varphi \\
&\quad - F_\theta\left(r_0, \theta_0 - \frac{\Delta\theta}{2}, \varphi_0\right) r \sin\left(\theta - \frac{\Delta\theta}{2}\right) \Delta r \Delta\varphi \\
&\approx F_\theta\left(r_0, \theta_0 + \frac{\Delta\theta}{2}, \varphi_0\right) r \sin\theta \Delta r \Delta\varphi - F_\theta\left(r_0, \theta_0 - \frac{\Delta\theta}{2}, \varphi_0\right) r \sin\theta \Delta r \Delta\varphi \\
&\quad + F_\theta\left(r_0, \theta_0 + \frac{\Delta\theta}{2}, \varphi_0\right) r \cos\theta \frac{\Delta\theta}{2} \Delta r \Delta\varphi \\
&\quad + F_\theta\left(r_0, \theta_0 - \frac{\Delta\theta}{2}, \varphi_0\right) r \cos\theta \frac{\Delta\theta}{2} \Delta r \Delta\varphi \\
&\approx \frac{F_\theta\left(r_0, \theta_0 + \frac{\Delta\theta}{2}, \varphi_0\right) - F_\theta\left(r_0, \theta_0 - \frac{\Delta\theta}{2}, \varphi_0\right)}{\Delta\theta} r \sin\theta \Delta r \Delta\varphi \Delta\theta \\
&\quad + F_\theta(r_0, \theta_0, \varphi_0) r \cos\theta \Delta\theta \Delta r \Delta\varphi \\
&= \frac{1}{r} \frac{F_\theta\left(r_0, \theta_0 + \frac{\Delta\theta}{2}, \varphi_0\right) - F_\theta\left(r_0, \theta_0 - \frac{\Delta\theta}{2}, \varphi_0\right)}{\Delta\theta} (r^2 \sin\theta \Delta r \Delta\varphi \Delta\theta) \\
&\quad + \frac{1}{r\sin\theta} F_\theta(r_0, \theta_0, \varphi_0) \cos\theta (r^2 \sin\theta \Delta r \Delta\varphi \Delta\theta) \\
&= \left[\frac{1}{r} \frac{F_\theta\left(r_0, \theta_0 + \frac{\Delta\theta}{2}, \varphi_0\right) - F_\theta\left(r_0, \theta_0 - \frac{\Delta\theta}{2}, \varphi_0\right)}{\Delta\theta} + \frac{1}{r\sin\theta} F_\theta(r_0, \theta_0, \varphi_0) \cos\theta\right] \Delta V \\
&\approx \left[\frac{1}{r} \frac{\partial F_\theta(r_0, \theta_0, \varphi_0)}{\partial\theta} + \frac{1}{r\sin\theta} F_\theta(r_0, \theta_0, \varphi_0) \cos\theta\right] \Delta V.
\end{aligned}$$

In our approximations we have assumed that each component of $\vec{F}$ is continuously differentiable.

The term

$$\frac{1}{r}\frac{\partial F_\theta}{\partial\theta} + \frac{1}{r\sin\theta} F_\theta \cos\theta$$

is often written in the more compact form

$$\frac{1}{r\sin\theta}\left[\frac{\partial}{\partial\theta}(\sin\theta F_\theta)\right].$$

We next compute

$$\iint_{S_3} \vec{F} \cdot \hat{n} dS + \iint_{S_4} \vec{F} \cdot \hat{n} dS = \iint_{S_3} \vec{F} \cdot \hat{e}_\varphi dS - \iint_{S_4} \vec{F} \cdot \hat{e}_\varphi dS$$

$$\approx F_\varphi\left(r_0, \theta_0, \varphi_0 + \frac{\varphi}{2}\right) r \Delta r \Delta\theta - F_\varphi\left(r_0, \theta_0, \varphi_0 - \frac{\varphi}{2}\right) r \Delta r \Delta\theta$$

$$= \frac{F_\varphi\left(r_0, \theta_0, \varphi_0 + \frac{\varphi}{2}\right) - F_\varphi\left(r_0, \theta_0, \varphi_0 - \frac{\varphi}{2}\right)}{\Delta\varphi} \frac{1}{r \sin\theta} r^2 \sin\theta \Delta r \Delta\theta \Delta\varphi$$

$$\approx \frac{1}{r \sin\theta} \frac{\partial F_\varphi(r_0, \theta_0, \varphi_0)}{\partial\varphi} \Delta V.$$

Finally,

$$\iint_{S_5} \vec{F} \cdot \hat{n} dS + \iint_{S_6} \vec{F} \cdot \hat{n} dS = \iint_{S_5} \vec{F} \cdot \hat{e}_r dS - \iint_{S_6} \vec{F} \cdot \hat{e}_r dS$$

$$\approx F_r\left(r_0 + \frac{\Delta r}{2}, \theta_0, \varphi_0\right)\left(r_0 + \frac{\Delta r}{2}\right) \sin\theta \Delta\varphi \left(r_0 + \frac{\Delta r}{2}\right) \Delta\theta$$

$$- F_r\left(r_0 - \frac{\Delta r}{2}, \theta_0, \varphi_0\right)\left(r_0 - \frac{\Delta r}{2}\right) \sin\theta \Delta\varphi \left(r_0 - \frac{\Delta r}{2}\right) \Delta\theta$$

$$= F_r\left(r_0 + \frac{\Delta r}{2}, \theta_0, \varphi_0\right) \sin\theta \Delta\varphi \Delta\theta \left(r_0^2 + r_0 \Delta r + \frac{\Delta r^2}{4}\right)$$

$$- F_r\left(r_0 - \frac{\Delta r}{2}, \theta_0, \varphi_0\right) \sin\theta \Delta\varphi \Delta\theta \left(r_0^2 - r_0 \Delta r + \frac{\Delta r^2}{4}\right)$$

$$\approx \left[F_r\left(r_0 + \frac{\Delta r}{2}, \theta_0, \varphi_0\right) - F_r\left(r_0 - \frac{\Delta r}{2}, \theta_0, \varphi_0\right)\right] r_0^2 \sin\theta \Delta\varphi \Delta\theta$$

$$+ 2F_r(r_0, \theta_0, \varphi_0) r_0 \Delta r \sin\theta \Delta\varphi \Delta\theta$$

$$= \frac{\left[F_r\left(r_0 + \frac{\Delta r}{2}, \theta_0, \varphi_0\right) - F_r\left(r_0 - \frac{\Delta r}{2}, \theta_0, \varphi_0\right)\right]}{\Delta r} \Delta r r_0^2 \sin\theta \Delta\varphi \Delta\theta$$

$$+ \frac{2}{r_0} F_r(r_0, \theta_0, \varphi_0) r_0^2 \Delta r \sin\theta \Delta\varphi \Delta\theta \approx \frac{\partial F_r(r_0, \theta_0, \varphi_0)}{\partial r} \Delta V$$

$$+ \frac{2}{r_0} F_r(r_0, \theta_0, \varphi_0) \Delta V = \left[\frac{\partial F_r(r_0, \theta_0, \varphi_0)}{\partial r} + \frac{2}{r_0} F_r(r_0, \theta_0, \varphi_0)\right] \Delta V.$$

The expression

$$\frac{\partial F_r}{\partial r} + \frac{2}{r} F_r$$

is often written in the final formula for the divergence as

$$\frac{1}{r^2}\left[2rF_r + r^2\frac{\partial F_r}{\partial r}\right].$$

Thus,

$$\iint_S \vec{F} \cdot \widehat{n} dS = \iint_{S_1+\cdots+S_6} \vec{F} \cdot \widehat{n} dS \approx \frac{1}{r^2}\left[2rF_r + r^2\frac{\partial F_r}{\partial r}\right]\Delta V + \frac{1}{r \sin\theta}\frac{\partial F_\varphi}{\partial \varphi}\Delta V + \frac{1}{r\sin\theta}\left[\frac{\partial}{\partial\theta}(\sin\theta F_\theta)\right]\Delta V$$

and so

$$\lim_{\|\Delta V\| \to 0} \frac{1}{\Delta V}\iint_S \vec{F} \cdot \widehat{n} dS = \frac{1}{r^2}\left[2rF_r + r^2\frac{\partial F_r}{\partial r}\right] + \frac{1}{r \sin\theta}\frac{\partial F_\varphi}{\partial \varphi} + \frac{1}{r\sin\theta}\left[\frac{\partial}{\partial\theta}(\sin\theta F_\theta)\right].$$

Finally, we remark that it is important to remember that the divergence is a property of a vector field and not a formula.

Example:

We compute the divergence of

$$\vec{F}(r,\theta,\varphi) = r^2 \sin\theta \cos\varphi \widehat{e}_r + e^r \cos\theta \sin\varphi \widehat{e}_\theta + \cos\theta \cos\varphi \widehat{e}_\varphi.$$

We have

$$F_r = r^2 \sin\theta \cos\varphi, \quad F_\theta = e^r \cos\theta \sin\varphi, \quad F_\varphi = \cos\theta \cos\varphi$$

so that

$$\frac{\partial F_r}{\partial r} = 2r \sin\theta \cos\varphi,$$

$$\frac{\partial}{\partial\theta}(\sin\theta F_\theta) = \frac{\partial}{\partial\theta}\sin\theta e^r \cos\theta \sin\varphi = e^r \sin\varphi(-\sin^2\theta + \cos^2\theta),$$

$$\frac{\partial F_\varphi}{\partial\varphi} = -\cos\theta \sin\varphi.$$

Then

$$\operatorname{div}\left(\vec{F}\right) = \frac{1}{r^2}\left[2rF_r + r^2\frac{\partial F_r}{\partial r}\right] + \frac{1}{r\sin\theta}\frac{\partial F_\varphi}{\partial\varphi} + \frac{1}{r\sin\theta}\left[\frac{\partial}{\partial\theta}(\sin\theta F_\theta)\right]$$

$$= \frac{1}{r^2}\left[2r\left(r^2\sin\theta\cos\varphi\right) + r^2(2r\sin\theta\cos\varphi)\right] + \frac{1}{r\sin\theta}(-\cos\theta\sin\varphi)$$

$$+ \frac{1}{r\sin\theta}e^r\sin\varphi\left(-\sin^2\theta + \cos^2\theta\right)$$

$$= 4r\sin\theta\cos\varphi - \frac{1}{r\sin\theta}\cos\theta\sin\varphi + \frac{1}{r\sin\theta}e^r\sin\varphi\left(-\sin^2\theta + \cos^2\theta\right).$$

THE CURL

The line integral of a vector field over a path gives the tendency of the vector field to follow that path. This is often called the *circulation* of the vector field along the path. A positive (negative) circulation indicates that we move with (against) the direction of the vector field.

To compute the curl of a vector field at a point we compute the circulation of a vector field per unit area in an infinitesimally small circle about a particular point in a particular plane. This is equal to the component of the curl of the vector field in the direction normal to plane. We then say that we have found the tendency of a vector field to rotate or "curl" about the point.

Said another way, the curl of a vector field measures the tendency of a vector field to cause rotation.

THE CURL IN CARTESIAN COORDINATES

Let the vector field be denoted

$$\vec{F}(x,y,z) = F_x(x,y,z)\hat{i} + F_y(x,y,z)\hat{j} + F_z(x,y,z)\hat{k}.$$

Choose (x_0, y_0, z_0). We form three rectangular loops centered around (x_0, y_0, z_0), and compute the circulation around each. The first loop is parallel to the x,y plane (so that it is normal to the vector $\hat{k}$). The coordinates of the corners of the loop are $\left(x_0 - \frac{\Delta x}{2}, y_0 - \frac{\Delta y}{2}, z_0\right)$, $\left(x_0 + \frac{\Delta x}{2}, y_0 - \frac{\Delta y}{2}, z_0\right)$, $\left(x_0 + \frac{\Delta x}{2}, y_0 + \frac{\Delta y}{2}, z_0\right)$, $\left(x_0 - \frac{\Delta x}{2}, y_0 + \frac{\Delta y}{2}, z_0\right)$. See Fig. 2.2.4.

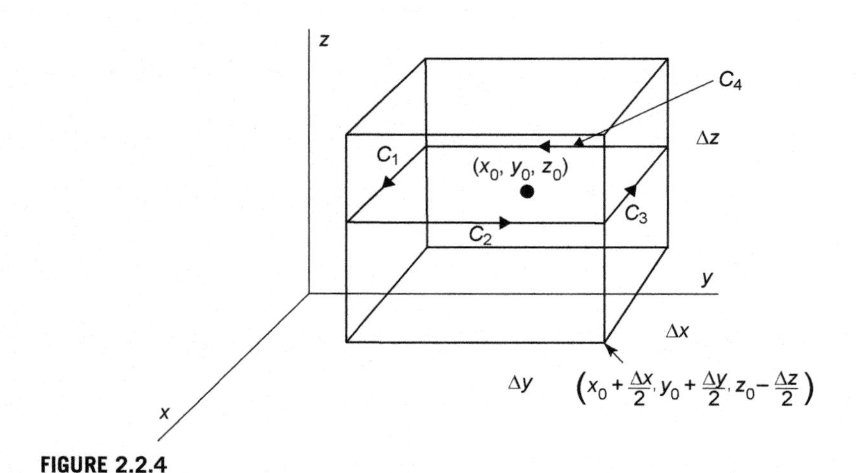

FIGURE 2.2.4

In this discussion $\widehat{t}$ will be either $\widehat{i}$, $\widehat{j}$, or $\widehat{k}$, whichever is parallel to the path. We compute the path integral of $\vec{F}$ around the loop. On C_1 and C_3 the only contribution of $\vec{F}$ is due to $F_x\widehat{i}$. Thus

$$\int_{C_1} \vec{F}\cdot\widehat{t}ds = \int_{x_0-\frac{\Delta x}{2}}^{x_0+\frac{\Delta x}{2}} F_x\left(x,\ y_0-\frac{\Delta y}{2},\ z_0\right)dx$$

we approximate F_x on C_1 by $F_x\left(x_0, y_0-\frac{\Delta y}{2}, z_0\right)$. So

$$\int_{C_1} \vec{F}\cdot\widehat{t}ds \approx F_x\left(x_0, y_0-\frac{\Delta y}{2}, z_0\right)\Delta x.$$

Because the direction of the path C_3, we have

$$\int_{C_3} \vec{F}\cdot\widehat{t}ds \approx \int_{x_0+\frac{\Delta x}{2}}^{x_0-\frac{\Delta x}{2}} F_x\left(x,\ y_0+\frac{\Delta y}{2},\ z_0\right)dx \approx -F_x\left(x_0, y_0+\frac{\Delta y}{2}, z_0\right)\Delta x$$

so

$$\int_{C_1+C_3} \vec{F}\cdot\widehat{t}ds \approx \left[F_x\left(x_0, y_0-\frac{\Delta y}{2}, z_0\right) - F_x\left(x_0, y_0+\frac{\Delta y}{2}, z_0\right)\right]\Delta x$$

$$= \frac{-\left[F_x\left(x_0, y_0+\frac{\Delta y}{2}, z_0\right) - F_x\left(x_0, y_0-\frac{\Delta y}{2}, z_0\right)\right]}{\Delta y}\Delta x\Delta y$$

$$\approx -\frac{\partial F_x(x_0, y_0, z_0)}{\partial y}\Delta x\Delta y = -\frac{\partial F_x(x_0, y_0, z_0)}{\partial y}\Delta A$$

where $\Delta A = \Delta x\Delta y$ is the area of the loop.

On C_2 we have

$$\int_{C_2} \vec{F} \cdot \hat{t} ds = \int_{y_0-\frac{\Delta y}{2}}^{y_0+\frac{\Delta y}{2}} F_y\left(x_0 + \frac{\Delta x}{2}, y, z_0\right) dy \approx F_y\left(x_0 + \frac{\Delta x}{2}, y_0, z_0\right)\Delta y$$

and on C_4

$$\int_{C_4} \vec{F} \cdot \hat{t} ds = \int_{y_0+\frac{\Delta y}{2}}^{y_0-\frac{\Delta y}{2}} F_y\left(x_0 - \frac{\Delta x}{2}, y, z_0\right) dy \approx -F_y\left(x_0 - \frac{\Delta x}{2}, y_0, z_0\right)\Delta y$$

so

$$\begin{aligned}\int_{C_2+C_4} \vec{F} \cdot \hat{t} ds &\approx F_y\left(x_0 + \frac{\Delta x}{2}, y_0, z_0\right)\Delta y - F_y\left(x_0 - \frac{\Delta x}{2}, y_0, z_0\right)\Delta y \\ &= \frac{\left[F_y\left(x_0 + \frac{\Delta x}{2}, y_0, z_0\right) - F_y\left(x_0 - \frac{\Delta x}{2}, y_0, z_0\right)\right]}{\Delta x}\Delta x \Delta y \\ &\approx \frac{\partial F_y(x_0, y_0, z_0)}{\partial x}\Delta A.\end{aligned}$$

Thus the circulation around this loop is

$$\int_{C_1+\cdots+C_4} \vec{F} \cdot \hat{t} ds \approx \left(\frac{\partial F_y(x_0, y_0, z_0)}{\partial x} - \frac{\partial F_x(x_0, y_0, z_0)}{\partial y}\right)\Delta A.$$

This is the circulation normal to the $\hat{k}$ vector. We divide by ΔA and take the limit as $\|\Delta A\| \to 0$ to get

$$\lim_{\|\Delta A\| \to 0} \frac{1}{\|\Delta A\|} \int_{C_1+\cdots+C_4} \vec{F} \cdot \hat{t} ds = \frac{\partial F_y(x_0, y_0, z_0)}{\partial x} - \frac{\partial F_x(x_0, y_0, z_0)}{\partial y}$$

is the circulation about (x_0, y_0, z_0) in the x,y plane.

We repeat this procedure for the loop parallel to the x,z plane, so it will be perpendicular to the $\hat{j}$ vector. Our explanation will be more brief. The loop perpendicular to the $\hat{j}$ vector is shown in Fig. 2.2.5.

In this case we have

$$\int_{C_1} \vec{F} \cdot \hat{t} ds = \int_{x_0-\frac{\Delta x}{2}}^{x_0+\frac{\Delta x}{2}} F_x\left(x, y_0, z_0 - \frac{\Delta z}{2}\right) dx \approx F_x\left(x_0, y_0, z_0 - \frac{\Delta z}{2}\right)\Delta x$$

$$\int_{C_3} \vec{F} \cdot \hat{t} ds = \int_{x_0+\frac{\Delta x}{2}}^{x_0-\frac{\Delta x}{2}} F_x\left(x, y_0, z_0 + \frac{\Delta z}{2}\right) dx \approx -F_x\left(x_0, y_0, z_0 + \frac{\Delta z}{2}\right)\Delta x$$

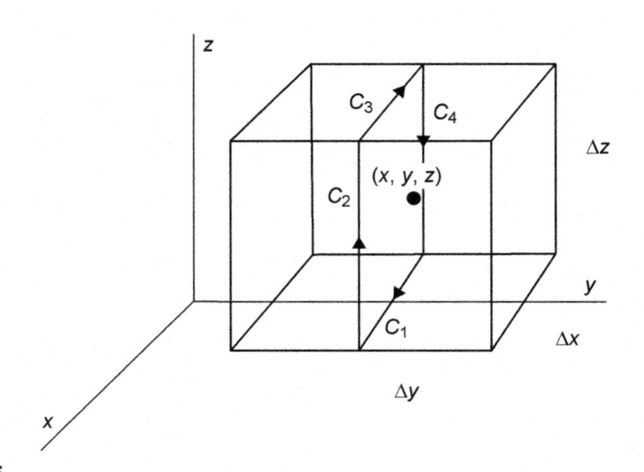

FIGURE 2.2.5

so

$$\int_{C_1+C_3} \vec{F}\cdot\hat{t}ds \approx \left[F_x\left(x_0,\ y_0,\ z_0-\frac{\Delta z}{2}\right)-F_x\left(x_0,\ y_0,\ z_0+\frac{\Delta z}{2}\right)\right]\Delta x$$

$$= \frac{-\left[F_x\left(x_0,\ y_0,\ z_0+\frac{\Delta z}{2}\right)-F_x\left(x_0,\ y_0,\ z_0-\frac{\Delta z}{2}\right)\right]}{\Delta z}\Delta x\Delta z$$

$$\approx -\frac{\partial F_x(x_0,\ y_0,\ z_0)}{\partial z}\Delta x\Delta z = -\frac{\partial F_x(x_0,\ y_0,\ z_0)}{\partial z}\Delta A$$

where $\Delta A = \Delta x\Delta z$ is the area of the loop.

We also have

$$\int_{C_2} \vec{F}\cdot\hat{t}ds = \int_{z_0-\frac{\Delta z}{2}}^{z_0+\frac{\Delta z}{2}} F_z\left(x_0-\frac{\Delta x}{2},\ y_0,\ z\right)dz \approx F_z\left(x_0-\frac{\Delta x}{2},\ y_0,\ z_0\right)\Delta z$$

$$\int_{C_4} \vec{F}\cdot\hat{t}ds = \int_{z_0+\frac{\Delta z}{2}}^{z_0-\frac{\Delta z}{2}} F_z\left(x_0-\frac{\Delta x}{2},\ y_0,\ z_0\right)dz \approx -F_z\left(x_0+\frac{\Delta x}{2},\ y_0,\ z_0\right)\Delta x$$

so

$$\int_{C_2+C_4} \vec{F}\cdot\hat{t}ds \approx F_z\left(x_0-\frac{\Delta x}{2},\ y_0,\ z_0\right)\Delta z - F_z\left(x_0+\frac{\Delta x}{2},\ y_0,\ z_0\right)\Delta z$$

$$= -\frac{\left[F_z\left(x_0+\frac{\Delta x}{2},\ y_0,\ z_0\right)-F_z\left(x_0-\frac{\Delta x}{2},\ y_0,\ z_0\right)\right]}{\Delta x}(\Delta x)(\Delta z)$$

$$\approx -\frac{\partial F_z(x_0, y_0, z_0)}{\partial x}\Delta A.$$

Thus the circulation around this loop is

$$\int_{C_1+\cdots+C_4} \vec{F}\cdot\hat{t}ds \approx \left(\frac{\partial F_x(x_0,\ y_0,\ z_0)}{\partial z} - \frac{\partial F_z(x_0, y_0, z_0)}{\partial x}\right)\Delta A.$$

We divide by ΔA and take the limit as $\|\Delta A\| \to 0$ to get

$$\lim_{\|\Delta A\|\to 0} \frac{1}{\Delta A}\int_{C_1+\cdots+C_4} \vec{F}\cdot\hat{t}ds = \frac{\partial F_x(x_0,\ y_0,\ z_0)}{\partial z} - \frac{\partial F_z(x_0, y_0, z_0)}{\partial x}$$

is the circulation about (x_0, y_0, z_0) in the x,z plane, which is the circulation normal to the $\hat{j}$ vector.

The loop perpendicular to the $\hat{i}$ vector is shown in Fig. 2.2.6.

We leave it as an exercise to show

$$\lim_{\|\Delta A\|\to 0} \frac{1}{\Delta A}\int_{C_1+\cdots+C_4} \vec{F}\cdot\hat{t}ds = \frac{\partial F_z(x_0,\ y_0,\ z_0)}{\partial y} - \frac{\partial F_y(x_0, y_0, z_0)}{\partial z}$$

is the circulation about (x_0, y_0, z_0) in the y,z plane, which is the circulation normal to the $\hat{i}$ vector.

The curl of $\vec{F} = F_x\hat{i} + F_y\hat{j} + F_z\hat{k}$, denoted $\nabla\times\vec{F}$, in Cartesian coordinates is defined to be

$$\nabla\times\vec{F} = \left(\frac{\partial F_z}{\partial y} - \frac{\partial F_y}{\partial z}\right)\hat{i} + \left(\frac{\partial F_x}{\partial z} - \frac{\partial F_z}{\partial x}\right)\hat{j} + \left(\frac{\partial F_y}{\partial x} - \frac{\partial F_x}{\partial y}\right)\hat{k}.$$

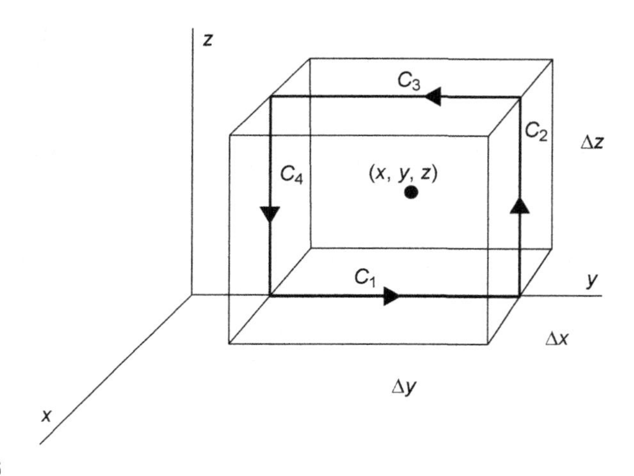

FIGURE 2.2.6

The formula defies memorization, but there is an easy way to do the computation, namely

$$\nabla \times \vec{F} = \begin{vmatrix} \hat{i} & \hat{j} & \hat{k} \\ \dfrac{\partial}{\partial x} & \dfrac{\partial}{\partial y} & \dfrac{\partial}{\partial z} \\ F_x & F_y & F_z \end{vmatrix}.$$

Example:
The curl of

$$\vec{F}(x, y, z) = x^2 e^y \sin z\hat{i} + 3yz\hat{j} + xy\hat{k}$$

is

$$\nabla \times \vec{F} = \begin{vmatrix} \hat{i} & \hat{j} & \hat{k} \\ \dfrac{\partial}{\partial x} & \dfrac{\partial}{\partial y} & \dfrac{\partial}{\partial z} \\ x^2 e^y \sin z & 3yz & xy \end{vmatrix} = (x - 3y)\hat{i} + \left(x^2 e^y \cos z - y\right)\hat{j} - x^2 e^y \sin z\hat{k}.$$

THE CURL IN CYLINDRICAL COORDINATES

In cylindrical coordinates we denote the vector field

$$\vec{F}(r, \theta, z) = F_r(r, \theta, z)\hat{e}_r + F_\theta(r, \theta, z)\hat{e}_\theta + F_z(r, \theta, z)\hat{e}_z.$$

Choose a point (r_0, θ_0, z_0) an construct a basic increment of volume in cylindrical coordinates centered at (r_0, θ_0, z_0). As in the Cartesian coordinate case, we construct three loops around the incremental piece, each centered at (r_0, θ_0, z_0) and each perpendicular to one of the vectors $\hat{e}_r, \hat{e}_\theta, \hat{e}_z$. We compute the circulation around each loop.

We first consider the loop perpendicular to $\hat{e}_z$. The coordinates of the corners of the loop are $\left(r_0 - \frac{\Delta r}{2}, \theta_0 - \frac{\Delta\theta}{2}, z_0\right)$, $\left(r_0 + \frac{\Delta r}{2}, \theta_0 - \frac{\Delta\theta}{2}, z_0\right)$, $\left(r_0 + \frac{\Delta r}{2}, \theta_0 + \frac{\Delta\theta}{2}, z_0\right)$ and $\left(r_0 - \frac{\Delta r}{2}, \theta_0 + \frac{\Delta\theta}{2}, z_0\right)$. See Fig. 2.2.7.

The path integral on C_1

$$\int_{r_0 - \frac{\Delta r}{2}}^{r_0 + \frac{\Delta r}{2}} F_r\left(r, \theta_0 - \frac{\Delta\theta}{2}, z_0\right) dr \approx F_r\left(r_0, \theta_0 - \frac{\Delta\theta}{2}, z_0\right) \Delta r$$

where we have approximated $F_r\left(r, \theta_0 + \frac{\Delta\theta}{2}, z_0\right)$ on C_1 by $F_r\left(r_0, \theta_0 + \frac{\Delta\theta}{2}, z_0\right)$.

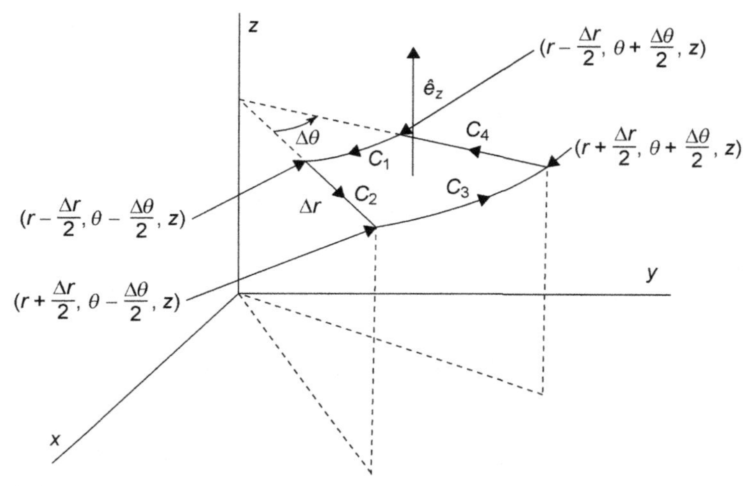

FIGURE 2.2.7

The path integral on C_3 is

$$\int_{r_0+\frac{\Delta r}{2}}^{r_0-\frac{\Delta r}{2}} F_r\left(r\ , \theta_0 + \frac{\Delta\theta}{2}, z_0\right) dr \approx\ - F_r\left(r_0, \theta_0 + \frac{\Delta\theta}{2}, z_0\right)\Delta r$$

so the sum of the two integrals is approximately

$$\left[F_r\left(r_0, \theta_0 - \frac{\Delta\theta}{2}, z_0\right) - F_r\left(r_0, \theta_0 + \frac{\Delta\theta}{2}, z_0\right)\right]\Delta r$$
$$= -\frac{\left[F_r\left(r_0, \theta_0 + \frac{\Delta\theta}{2}, z_0\right) - F_r\left(r_0, \theta_0 - \frac{\Delta\theta}{2}, z_0\right)\right]}{\Delta\theta}\Delta\theta\Delta r.$$

The area enclosed by the loop, denoted ΔA, is approximately

$$\Delta A \approx r_0(\Delta\theta)(\Delta r)$$

so

$$-\frac{\left[F_r\left(r_0, \theta_0 + \frac{\Delta\theta}{2}, z_0\right) - F_r\left(r_0, \theta_0 - \frac{\Delta\theta}{2}, z_0\right)\right]}{\Delta\theta}\frac{1}{r_0}r_0\Delta\theta\Delta r \approx\ -\frac{1}{r_0}\frac{\partial F_r(r_0, \theta_0, z_0)}{\partial\theta}\Delta A.$$

The path integral on the path C_3 is

$$\int_{\theta_0-\frac{\Delta\theta}{2}}^{\theta_0+\frac{\Delta\theta}{2}} F_\theta\left(r_0+\frac{\Delta r}{2},\theta,z_0\right)\left(r_0+\frac{\Delta r}{2}\right)d\theta$$

$$\approx F_\theta\left(r_0+\frac{\Delta r}{2},\theta_0,z_0\right)\left(r_0+\frac{\Delta r}{2}\right)\Delta\theta$$

and the path integral on the path C_4 is

$$\int_{\theta_0+\frac{\Delta\theta}{2}}^{\theta_0-\frac{\Delta\theta}{2}} F_\theta\left(r_0-\frac{\Delta r}{2},\theta,z_0\right)\left(r_0-\frac{\Delta r}{2}\right)d\theta$$

$$\approx -F_\theta\left(r_0-\frac{\Delta r}{2},\theta_0,z_0\right)\left(r_0-\frac{\Delta r}{2}\right)\Delta\theta.$$

Summing the integrals on C_3 and C_4 gives

$$F_\theta\left(r_0+\frac{\Delta r}{2},\theta_0,z_0\right)\left(r_0+\frac{\Delta r}{2}\right)\Delta\theta - F_\theta\left(r_0-\frac{\Delta r}{2},\theta_0,z_0\right)\left(r_0-\frac{\Delta r}{2}\right)\Delta\theta$$

$$\approx \left[F_\theta\left(r_0+\frac{\Delta r}{2},\theta_0,z_0\right)-F_\theta\left(r_0-\frac{\Delta r}{2},\theta_0,z_0\right)\right]r_0\Delta\theta + F_\theta(r_0,\theta_0,z_0)\Delta r\Delta\theta$$

$$=\frac{\left[F_\theta\left(r_0+\frac{\Delta r}{2},\theta_0,z_0\right)-F_\theta\left(r_0-\frac{\Delta r}{2},\theta_0,z_0\right)\right]}{\Delta r}r_0\Delta\theta\Delta r+\frac{1}{r_0}F_\theta(r_0,\theta_0,z_0)r_0\Delta r\Delta\theta$$

$$\approx \frac{\partial F_\theta(r_0,\theta_0,z_0)}{\partial r}\Delta A+\frac{1}{r_0}F_\theta(r_0,\theta_0,z_0)\Delta A.$$

Combining the integrals over the four paths gives

$$-\frac{1}{r_0}\frac{\partial F_r(r_0,\theta_0,z_0)}{\partial\theta}\Delta A+\frac{\partial F_\theta(r_0,\theta_0,z_0)}{\partial r}\Delta A+F_\theta(r_0,\theta_0,z_0)\Delta A.$$

We next consider the loop perpendicular to $\widehat{e}_\theta$. The coordinates of the corners of the loop are $\left(r_0-\frac{\Delta r}{2},\theta_0,z_0-\frac{\Delta z}{2}\right)$, $\left(r_0+\frac{\Delta r}{2},\theta_0,z_0-\frac{\Delta z}{2}\right)$, $\left(r_0+\frac{\Delta r}{2},\theta_0,z_0+\frac{\Delta z}{2}\right)$ and $\left(r_0-\frac{\Delta r}{2},\theta_0,z_0+\frac{\Delta z}{2}\right)$. The area of the loop is $\Delta A=(\Delta r)\cdot(\Delta z)$. See Fig. 2.2.8.

The path integral on C_1 is approximately $-F_r\left(r_0,\theta_0,z_0-\frac{\Delta z}{2}\right)\Delta r$, the path integral on C_3 is approximately $F_r\left(r_0,\theta_0,z_0+\frac{\Delta z}{2}\right)\Delta r$ so the combined value is

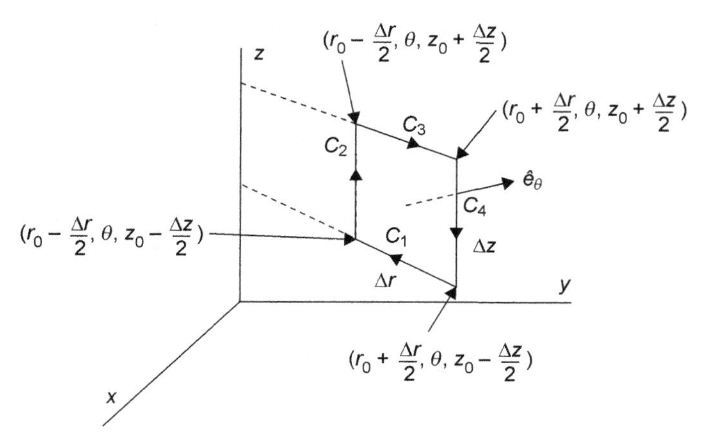

FIGURE 2.2.8

$$
\begin{aligned}
&F_r\left(r_0, \theta_0, z_0 + \frac{\Delta z}{2}\right)\Delta r - F_r\left(r_0, \theta_0, z_0 - \frac{\Delta z}{2}\right)\Delta r \\
&= \frac{\left[F_r\left(r_0, \theta_0, z_0 + \frac{\Delta z}{2}\right) - F_r\left(r_0, \theta_0, z_0 - \frac{\Delta z}{2}\right)\right]}{\Delta z}\Delta r \Delta z \\
&\approx \frac{\partial F_r(r_0, \theta_0, z_0)}{\partial z}\Delta r \Delta z = \frac{\partial F_r(r_0, \theta_0, z_0)}{\partial z}\Delta A.
\end{aligned}
$$

The path integral on C_2 is approximately $-F_z\left(r_0 + \frac{\Delta r}{2}, \theta_0, z_0\right)\Delta z$, the path integral on C_4 is approximately $F_z\left(r_0 - \frac{\Delta r}{2}, \theta_0, z_0\right)\Delta z$ so the combined value is

$$
\begin{aligned}
&-F_z\left(r_0 + \frac{\Delta r}{2}, \theta_0, z_0\right)\Delta z + F_z\left(r_0 - \frac{\Delta r}{2}, \theta_0, z_0\right)\Delta z \\
&= \frac{-\left[F_z\left(r_0 + \frac{\Delta r}{2}, \theta_0, z_0\right) - F_z\left(r_0 - \frac{\Delta r}{2}, \theta_0, z_0\right)\right]}{\Delta r}\Delta r \Delta z \\
&\approx \frac{-\partial F_z(r_0, \theta_0, z_0)}{\partial r}\Delta A.
\end{aligned}
$$

Combining the integrals on the four paths gives the integral around the loop. This is

$$
\frac{\partial F_r(r_0, \theta_0, z_0)}{\partial z}\Delta A - \frac{\partial F_z(r_0, \theta_0, z_0)}{\partial r}\Delta A.
$$

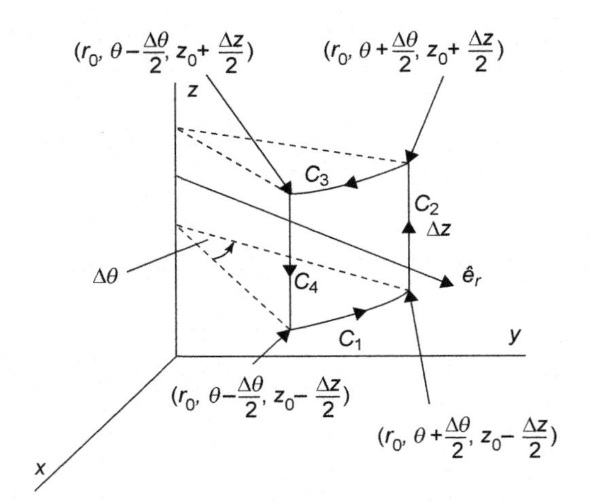

FIGURE 2.2.9

We next consider the loop perpendicular to $\hat{e}_r$. The coordinates of the corners of the loop are $\left(r_0, \theta_0 - \frac{\Delta\theta}{2}, z_0 - \frac{\Delta z}{2}\right)$, $\left(r_0, \theta_0 - \frac{\Delta\theta}{2}, z_0 + \frac{\Delta z}{2}\right)$, $\left(r_0, \theta_0 + \frac{\Delta\theta}{2}, z_0 + \frac{\Delta z}{2}\right)$ and $\left(r_0, \theta_0 + \frac{\Delta\theta}{2}, z_0 - \frac{\Delta z}{2}\right)$. See Fig. 2.2.9. The length of one side of the loop is $r_0\Delta\theta$, and the length of the other side is Δz so the area of the loop is $\Delta A = r_0\Delta\theta\Delta z$.

Note that the length of C_1 is $r_0\Delta\theta$ so the path integral on C_1 is approximately $F_\theta\left(r_0, \theta_0, z_0 - \frac{\Delta z}{2}\right)r_0\Delta\theta$, the path integral on C_3 is approximately $-F_\theta\left(r_0, \theta_0, z_0 + \frac{\Delta z}{2}\right)r_0\Delta\theta$, so the combined value is

$$\begin{aligned}
&F_\theta\left(r_0, \theta_0, z_0 - \frac{\Delta z}{2}\right)r_0\Delta\theta - F_\theta\left(r_0, \theta_0, z_0 + \frac{\Delta z}{2}\right)r_0\Delta\theta \\
&= \frac{\left[F_\theta\left(r_0, \theta_0, z_0 - \frac{\Delta z}{2}\right) - F_\theta\left(r_0, \theta_0, z_0 + \frac{\Delta z}{2}\right)\right]}{\Delta z} r_0\Delta z\Delta\theta \\
&= -\frac{\left[F_\theta\left(r_0, \theta_0, z_0 + \frac{\Delta z}{2}\right) - F_\theta\left(r_0, \theta_0, z_0 - \frac{\Delta z}{2}\right)\right]}{\Delta z} r_0\Delta z\Delta\theta \\
&\approx -\frac{\partial F_\theta(r_0, \theta_0, z_0)}{\partial z} r_0\Delta z\Delta\theta \\
&= -\frac{\partial F_\theta(r_0, \theta_0, z_0)}{\partial z}\Delta A.
\end{aligned}$$

The path integral on C_2 is approximately $F_z\left(r_0, \theta_0 + \frac{\Delta\theta}{2}, z_0\right)\Delta z$, the path integral on C_4 is approximately $-F_z\left(r_0, \theta_0 - \frac{\Delta\theta}{2}, z_0\right)\Delta z$ so the combined value is

$$\begin{aligned}
&F_z\left(r_0, \theta_0 + \frac{\Delta\theta}{2}, z_0\right)\Delta z - F_z\left(r_0, \theta_0 - \frac{\Delta\theta}{2}, z_0\right)\Delta z \\
&= \frac{1}{r_0}\frac{\left[F_z\left(r_0, \theta_0 + \frac{\Delta\theta}{2}, z_0\right) - F_z\left(r_0, \theta_0 - \frac{\Delta\theta}{2}, z_0\right)\right]}{\Delta\theta} r_0\Delta\theta\Delta z \\
&\approx \frac{1}{r_0}\frac{\partial F_z(r_0, \theta_0, z_0)}{\partial\theta}\Delta A.
\end{aligned}$$

Thus, the path integral around the loop is

$$\frac{1}{r_0}\frac{\partial F_z(r_0, \theta_0, z_0)}{\partial\theta}\Delta A - \frac{\partial F_\theta(r_0, \theta_0, z_0)}{\partial z}\Delta A.$$

Finally, we have that the curl of $\vec{F}(r, \theta, z) = F_r(r, \theta, z)\widehat{e}_r + F_\theta(r, \theta, z)\widehat{e}_\theta + F_z(r, \theta, z)\widehat{e}_z$ in cylindrical coordinates is

$$\nabla \times \vec{F} = \left(\frac{1}{r}\frac{\partial F_z}{\partial\theta} - \frac{\partial F_\theta}{\partial z}\right)\widehat{e}_r + \left(\frac{\partial F_r}{\partial z} - \frac{\partial F_z}{\partial r}\right)\widehat{e}_\theta + \left[\frac{1}{r}\frac{\partial}{\partial r}(rF_\theta) - \frac{1}{r}\frac{\partial F_r}{\partial\theta}\right]\widehat{e}_z.$$

Example:
We compute the curl of

$$\vec{F}(r, \theta, z) = 5r^2e^z \sin\theta\widehat{e}_r + r\tan\theta z^3\widehat{e}_\theta + z\widehat{e}_z.$$

We have

$$F_z = z, \quad \text{so} \quad \frac{\partial F_z}{\partial\theta} = 0 \quad \text{and} \quad \frac{\partial F_z}{\partial r} = 0$$

$$F_\theta = r(\tan\theta)z^3 \quad \text{so} \quad \frac{\partial F_\theta}{\partial z} = 3z^2 r\tan\theta \quad \text{and} \quad \frac{\partial}{\partial r}(rF_\theta) = 2rz^3\tan\theta$$

$$F_r = 5r^2e^z\sin\theta \quad \text{so} \quad \frac{\partial F_r}{\partial z} = 5r^2e^z\sin\theta \quad \text{and} \quad \frac{\partial F_r}{\partial\theta} = 5r^2e^z\cos\theta.$$

Thus,

$$\left(\frac{1}{r}\frac{\partial F_z}{\partial\theta} - \frac{\partial F_\theta}{\partial z}\right) = -3z^2r\tan\theta, \quad \left(\frac{\partial F_r}{\partial z} - \frac{\partial F_z}{\partial r}\right) = 5r^2e^z\sin\theta,$$

$$\frac{1}{r}\frac{\partial}{\partial r}(rF_\theta) - \frac{1}{r}\frac{\partial F_r}{\partial\theta} = 2\,z^3\tan\theta - 5r\,e^z\cos\theta$$

so

$$\nabla \times \vec{F} = \left(\frac{1}{r}\frac{\partial F_z}{\partial \theta} - \frac{\partial F_\theta}{\partial z}\right)\widehat{e}_r + \left(\frac{\partial F_r}{\partial z} - \frac{\partial F_z}{\partial r}\right)\widehat{e}_\theta + \left[\frac{1}{r}\frac{\partial}{\partial r}(rF_\theta) - \frac{1}{r}\frac{\partial F_r}{\partial \theta}\right]\widehat{e}_z$$
$$= -3z^2 r \tan\theta \widehat{e}_r + 5r^2 e^z \sin\theta \widehat{e}_\theta + \left(2z^3 \tan\theta - 5r e^z \cos\theta\right)\widehat{e}_z.$$

THE CURL IN SPHERICAL COORDINATES

In spherical coordinates, we define the vector field to be $\vec{F} = F_r\widehat{e}_r + F_\theta\widehat{e}_\theta + F_\varphi\widehat{e}_\varphi$. We construct an incremental element of volume in spherical coordinates about the point $(r_0, \theta_0, \varphi_0)$ as shown in Fig. 2.2.10. We construct paths about $(r_0, \theta_0, \varphi_0)$ perpendicular to the vectors $\widehat{e}_r$, $\widehat{e}_\theta$, and $\widehat{e}_\varphi$.

We compute the path integral of $\vec{F}$ around the path that is perpendicular to $\widehat{e}_r$. A diagram of the path is shown in Fig. 2.2.11.

The area enclosed by the path is $\Delta A \approx r_0 \sin\theta_0(\Delta\varphi) r_0(\Delta\theta) = r_0^2 \sin\theta_0(\Delta\varphi)(\Delta\theta)$.

We have the length of $C_1 = r_0 \sin\left(\theta_0 + \frac{\Delta\theta}{2}\right)\Delta\varphi$; the length of $C_2 = r_0\Delta\theta$; the length of $C_3 = r_0 \sin\left(\theta_0 - \frac{\Delta\theta}{2}\right)\Delta\varphi$; the length of $C_4 = r_0\Delta\theta$.

From Taylor's theorem, we have $f(x + \Delta x) \approx f(x) + f'(x)\Delta x$, so $\sin\left(\theta_0 + \frac{\Delta\theta}{2}\right) \approx \sin\theta_0 + \cos\theta_0\frac{\Delta\theta}{2}$ and $\sin\left(\theta_0 - \frac{\Delta\theta}{2}\right) \approx \sin\theta_0 - \cos\theta_0\frac{\Delta\theta}{2}$.

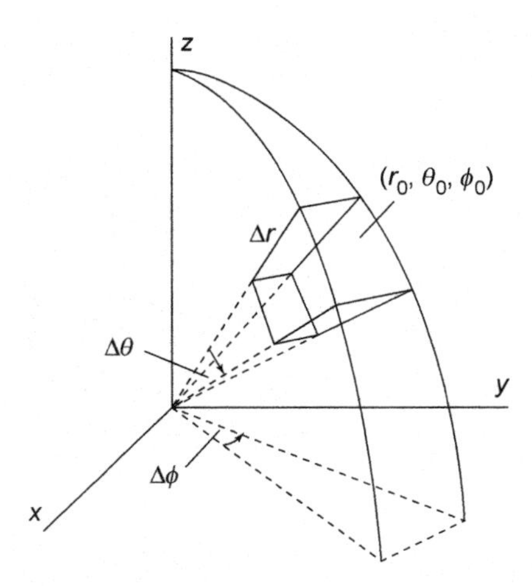

FIGURE 2.2.10

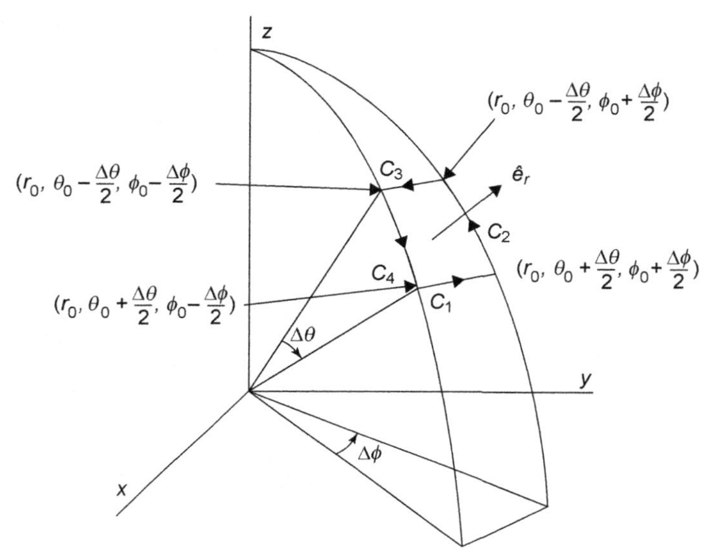

FIGURE 2.2.11

Thus,

$$\int_{C_1} \vec{F} \cdot \hat{t} ds \approx F_\varphi \left(r_0, \theta_0 + \frac{\Delta\theta}{2}, \varphi_0 \right) r_0 \sin\left(\theta_0 + \frac{\Delta\theta}{2} \right) \Delta\varphi$$

$$\approx F_\varphi \left(r_0, \theta_0 + \frac{\Delta\theta}{2}, \varphi_0 \right) r_0 \left[\sin \theta_0 + \frac{\Delta\theta}{2} \cos \theta_0 \right] \Delta\varphi$$

$$= F_\varphi \left(r_0, \theta_0 + \frac{\Delta\theta}{2}, \varphi_0 \right) r_0 \sin \theta_0 \Delta\varphi$$

$$+ F_\varphi \left(r_0, \theta_0 + \frac{\Delta\theta}{2}, \varphi_0 \right) r_0 \cos \theta_0 \frac{\Delta\theta}{2} \Delta\varphi$$

$$\int_{C_3} \vec{F} \cdot \hat{t} ds \approx -\left\{ F_\varphi \left(r_0, \theta_0 - \frac{\Delta\theta}{2}, \varphi_0 \right) r_0 \sin\left(\theta_0 - \frac{\Delta\theta}{2} \right) \Delta\varphi \right\}$$

$$\approx -\left\{ F_\varphi \left(r_0, \theta_0 - \frac{\Delta\theta}{2}, \varphi_0 \right) r_0 \sin \theta_0 \Delta\varphi \right.$$

$$\left. - F_\varphi \left(r_0, \theta_0 - \frac{\Delta\theta}{2}, \varphi_0 \right) r_0 \cos \theta_0 \frac{\Delta\theta}{2} \Delta\varphi \right\}.$$

So

$$\int_{C_1+C_3} \vec{F}\cdot\hat{t}ds \approx F_\varphi\left(r_0, \theta_0+\frac{\Delta\theta}{2}, \varphi_0\right) r_0 \sin\theta_0 \Delta\varphi$$

$$+F_\varphi\left(r_0, \theta_0+\frac{\Delta\theta}{2}, \varphi_0\right) r_0 \cos\theta_0 \frac{\Delta\theta}{2}\Delta\varphi$$

$$-\left\{F_\varphi\left(r_0, \theta_0-\frac{\Delta\theta}{2}, \varphi_0\right) r_0 \sin\theta_0 \Delta\varphi\right.$$

$$\left.-F_\varphi\left(r_0, \theta_0-\frac{\Delta\theta}{2}, \varphi_0\right) r_0 \cos\theta_0 \frac{\Delta\theta}{2}\Delta\varphi\right\}$$

$$= \left[F_\varphi\left(r_0, \theta_0+\frac{\Delta\theta}{2}, \varphi_0\right) r_0 \sin\theta_0 \Delta\varphi\right.$$

$$\left.-F_\varphi\left(r_0, \theta_0-\frac{\Delta\theta}{2}, \varphi_0\right) r_0 \sin\theta_0 \Delta\varphi\right]$$

$$+F_\varphi\left(r_0, \theta_0+\frac{\Delta\theta}{2}, \varphi_0\right) r_0 \cos\theta_0 \Delta\theta\Delta\varphi$$

$$= \frac{F_\varphi\left(r_0, \theta_0+\frac{\Delta\theta}{2}, \varphi_0\right) - F_\varphi\left(r_0, \theta_0-\frac{\Delta\theta}{2}, \varphi_0\right)}{\Delta\theta} r_0 \sin\theta_0 \Delta\theta\Delta\varphi$$

$$+F_\varphi\left(r_0, \theta_0+\frac{\Delta\theta}{2}, \varphi_0\right) r_0 \cos\theta_0 \Delta\theta\Delta\varphi$$

$$\approx \frac{\partial F_\varphi(r_0, \theta_0, \varphi_0)}{\partial\theta} r_0 \sin\theta_0 \Delta\varphi\Delta\theta + F_\varphi(r_0, \theta_0, \varphi_0) r_0 \cos\theta_0 \Delta\theta\Delta\varphi$$

$$= \frac{1}{r_0}\frac{\partial F_\varphi(r_0, \theta_0, \varphi_0)}{\partial\theta} r_0^2 \sin\theta_0 \Delta\varphi\Delta\theta + \frac{1}{r_0 \sin\theta_0} F_\varphi(r_0, \theta_0, \varphi_0) r_0^2 \sin\theta_0 \cos\theta_0 \Delta\varphi\Delta\theta$$

$$= \left[\frac{1}{r_0}\frac{\partial F_\varphi(r_0, \theta_0, \varphi_0)}{\partial\theta} + \frac{1}{r_0 \sin\theta_0} F_\varphi(r_0, \theta_0, \varphi_0)\cos\theta_0\right]\Delta A.$$

Note that

$$\frac{1}{r}\frac{\partial F_\varphi}{\partial\theta} + \frac{1}{r\sin\theta}\cos\theta\, F_\varphi = \frac{1}{r\sin\theta}\frac{\partial}{\partial\theta}(\sin\theta F_\varphi),$$

and we shall use the more compact form of this expression in the final formula.

Next,

$$\int_{C_2} \vec{F}\cdot\hat{t}ds \approx -F_\theta\left(r_0, \theta_0+\frac{\Delta\theta}{2}, \varphi_0\right) r_0 \Delta\theta$$

and

$$\int_{C_4} \vec{F} \cdot \hat{t} ds \approx F_\theta\left(r_0, \theta_0 - \frac{\Delta\theta}{2}, \varphi_0\right) r_0 \Delta\theta$$

so

$$\begin{aligned}
\int_{C_2+C_4} \vec{F} \cdot \hat{t} ds &\approx -\left[F_\theta\left(r_0, \theta_0 + \frac{\Delta\theta}{2}, \varphi_0\right) - F_\theta\left(r_0, \theta_0 - \frac{\Delta\theta}{2}, \varphi_0\right)\right] r_0 \Delta\theta \\
&= -\frac{1}{r_0 \sin \theta_0}\left[\frac{F_\theta\left(r_0, \theta_0 + \frac{\Delta\theta}{2}, \varphi_0\right) - F_\theta\left(r_0, \theta_0 - \frac{\Delta\theta}{2}, \varphi_0\right)}{\Delta\varphi}\right] r_0^2 \sin \theta_0 \Delta\varphi \Delta\theta \\
&\approx -\frac{1}{r_0 \sin \theta_0} \frac{\partial F_\theta(r_0, \theta_0, \varphi_0)}{\partial \varphi} \Delta A.
\end{aligned}$$

Thus, we have

$$\begin{aligned}
\int_{C_1+C_2+C_3+C_4} \vec{F} \cdot \hat{t} ds \approx \Bigg\{&\left[\frac{1}{r_0} \frac{\partial F_\varphi(r_0, \theta_0, \varphi_0)}{\partial \theta} + \frac{1}{r_0 \sin \theta_0} F_\varphi(r_0, \theta_0, \varphi_0)\right] \\
&- \frac{1}{r_0 \sin \theta_0} \frac{\partial F_\theta(r_0, \theta_0, \varphi_0)}{\partial \varphi}\Bigg\} \Delta A.
\end{aligned}$$

Dividing by ΔA and taking the limit as $\|\Delta A\| \to 0$, we get the rotational component of $\vec{F}$ in the direction of $\hat{e}_r$ is

$$\begin{aligned}
\lim_{\|\Delta A\| \to 0} \frac{1}{\|\Delta A\|} \int_{C_1+C_2+C_3+C_4} \vec{F} \cdot \hat{t} ds &= \frac{1}{r} \frac{\partial F_\varphi}{\partial \theta} + \frac{1}{r \sin \theta} F_\varphi - \frac{1}{r \sin \theta} \frac{\partial F_\theta}{\partial \varphi} \\
&= \frac{1}{r \sin \theta} \frac{\partial}{\partial \theta}(\sin \theta F_\varphi) - \frac{1}{r \sin \theta} \frac{\partial F_\theta}{\partial \varphi}.
\end{aligned}$$

We next compute the path integral of $\vec{F}$ around the path that is perpendicular to $\hat{e}_\varphi$. A diagram of the path is shown in Fig. 2.2.12.

The area the path is $\Delta A \approx r_0(\Delta\theta) \cdot (\Delta r)$. The length of C_1 is $\left(r_0 - \frac{\Delta r}{2}\right)\Delta\theta$, and the length of C_3 is $\left(r_0 + \frac{\Delta r}{2}\right)\Delta\theta$. The length of C_2 is Δr, as is length of C_4. Now

$$\begin{aligned}
\int_{C_1} \vec{F} \cdot \hat{t} ds &\approx -F_\theta\left(r_0 - \frac{\Delta r}{2}, \theta_0, \varphi_0\right)\left(r_0 - \frac{\Delta r}{2}\right)\Delta\theta \\
&= -F_\theta\left(r_0 - \frac{\Delta r}{2}, \theta_0, \varphi_0\right) r_0 \Delta\theta + F_\theta\left(r_0 - \frac{\Delta r}{2}, \theta_0, \varphi_0\right) \frac{\Delta r}{2} \Delta\theta
\end{aligned}$$

$$\begin{aligned}
\int_{C_3} \vec{F} \cdot \hat{t} ds &\approx F_\theta\left(r_0 + \frac{\Delta r}{2}, \theta_0, \varphi_0\right)\left(r_0 + \frac{\Delta r}{2}\right)\Delta\theta \\
&= F_\theta\left(r_0 + \frac{\Delta r}{2}, \theta_0, \varphi_0\right) r_0 \Delta\theta + F_\theta\left(r_0, \theta_0 + \frac{\Delta\theta}{2}, \varphi_0\right) \frac{\Delta r}{2} \Delta\theta
\end{aligned}$$

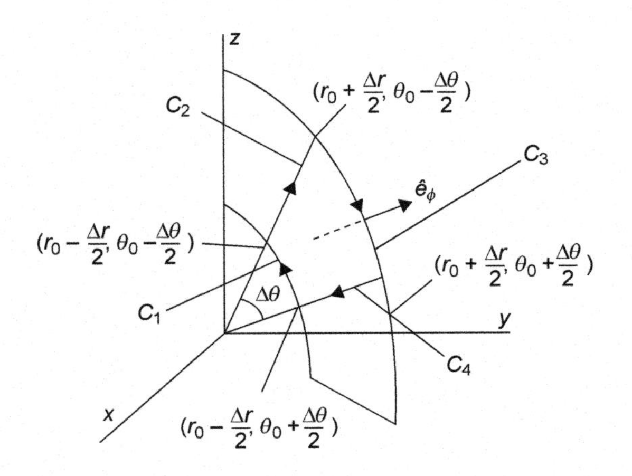

FIGURE 2.2.12

so

$$\begin{aligned}\int_{C_1+C_3} \vec{F}\cdot\hat{t}ds &\approx \left[F_\theta\left(r_0+\frac{\Delta r}{2},\theta_0,\varphi_0\right)-F_\theta\left(r_0-\frac{\Delta r}{2},\theta_0,\varphi_0\right)\right]r_0\Delta\theta \\ &\quad + F_\theta(r_0,\theta_0,\varphi_0)\Delta r\Delta\theta \\ &= \frac{\left[F_\theta\left(r_0+\frac{\Delta r}{2},\theta_0,\varphi_0\right)-F_\theta\left(r_0-\frac{\Delta r}{2},\theta_0,\varphi_0\right)\right]}{\Delta r}(\Delta r)r_0\Delta\theta \\ &\quad + \frac{1}{r_0}F_\theta(r_0,\theta_0,\varphi_0)(\Delta r)r_0(\Delta\theta) \\ &\approx \frac{\partial F_\theta(r_0,\theta_0,\varphi_0)}{\partial r}\Delta A + \frac{1}{r_0}F_\theta(r_0,\theta_0,\varphi_0)\Delta A.\end{aligned}$$

Also,

$$\int_{C_2} \vec{F}\cdot\hat{t}ds \approx F_r\left(r_0,\theta_0-\frac{\Delta\theta}{2},\varphi_0\right)\Delta r \qquad \int_{C_4} \vec{F}\cdot\hat{t}ds \approx -F_r\left(r_0,\theta_0+\frac{\Delta\theta}{2},\varphi_0\right)\Delta r$$

so

$$\begin{aligned}\int_{C_2+C_4} \vec{F}\cdot\hat{t}ds &\approx F_r\left(r_0,\theta_0-\frac{\Delta\theta}{2},\varphi_0\right)\Delta r - F_r\left(r_0,\theta_0+\frac{\Delta\theta}{2},\varphi_0\right)\Delta r \\ &= -\frac{\left[F_r\left(r_0,\theta_0+\frac{\Delta\theta}{2},\varphi_0\right)-F_r\left(r_0,\theta_0-\frac{\Delta\theta}{2},\varphi_0\right)\right]}{\Delta\theta}(\Delta\theta)(\Delta r) \\ &\approx -\frac{1}{r_0}\frac{\partial F_r(r_0,\theta_0,\varphi_0)}{\partial\theta}r_0(\Delta\theta)(\Delta r) = -\frac{1}{r_0}\frac{\partial F_r(r_0,\theta_0,\varphi_0)}{\partial\theta}\Delta A.\end{aligned}$$

Dividing by ΔA and taking the limit as $\|\Delta A\| \to 0$, we get the rotational component of $\vec{F}$ in the direction of $\widehat{e}_\varphi$ is

$$\lim_{\|\Delta A\|\to 0} \frac{1}{\|\Delta A\|} \int_{C_1+C_2+C_3+C_4} \vec{F}\cdot\widehat{t}ds = \frac{\partial F_\theta}{\partial r} + \frac{1}{r}\frac{\partial F_\theta}{\partial r} - \frac{1}{r}\frac{\partial F_r}{\partial \theta} = \frac{1}{r}\frac{\partial}{\partial r}(rF_\theta) - \frac{1}{r}\frac{\partial F_r}{\partial \theta}.$$

We next compute the path integral of $\vec{F}$ around the path that is perpendicular to $\widehat{e}_\theta$. A diagram of the path is shown in Fig. 2.2.13.

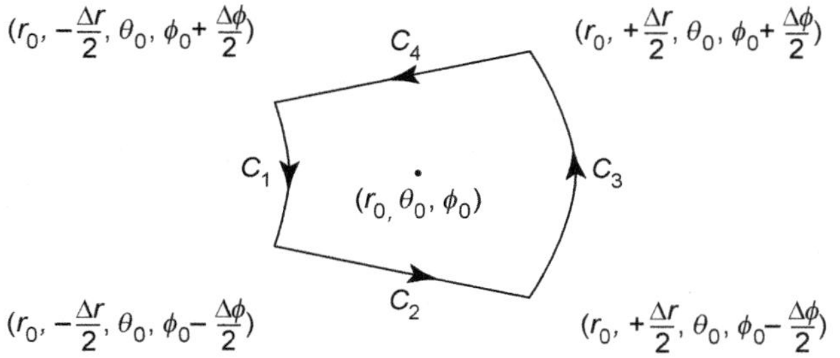

FIGURE 2.2.13

The area the path is $\Delta A \approx r_0 \sin\theta_0(\Delta\varphi)(\Delta r)$. The length of C_1 is $\left(r_0 - \frac{\Delta r}{2}\right)\sin\theta_0(\Delta\varphi)$ and the length of C_3 is $\left(r_0 + \frac{\Delta r}{2}\right)\sin\theta_0(\Delta\varphi)$. The length of C_2 is Δr, as is length of C_4. Now

$$\begin{aligned}\int_{C_1} \vec{F}\cdot\widehat{t}ds &\approx F_\varphi\left(r_0 - \frac{\Delta r}{2}, \theta_0, \varphi_0\right)\left(r_0 - \frac{\Delta r}{2}\right)\sin\theta_0(\Delta\varphi)\\ &= F_\varphi\left(r_0 - \frac{\Delta r}{2}, \theta_0, \varphi_0\right) r_0 \sin\theta_0(\Delta\varphi)\\ &\quad - F_\varphi\left(r_0 - \frac{\Delta r}{2}, \theta_0, \varphi_0\right)\sin\theta_0\frac{\Delta r}{2}(\Delta\varphi)\end{aligned}$$

and

$$\begin{aligned}\int_{C_3} \vec{F}\cdot\widehat{t}ds &\approx -F_\varphi\left(r_0 + \frac{\Delta r}{2}, \theta_0, \varphi_0\right)\left(r_0 + \frac{\Delta r}{2}\right)\sin\theta_0(\Delta\varphi)\\ &= -F_\varphi\left(r_0 + \frac{\Delta r}{2}, \theta_0, \varphi_0\right) r_0 \sin\theta_0(\Delta\varphi)\\ &\quad - F_\varphi\left(r_0, \theta_0 + \frac{\Delta\theta}{2}, \varphi_0\right)\frac{\Delta r}{2}\sin\theta_0(\Delta\varphi)\end{aligned}$$

so

$$\int_{C_1+C_3} \vec{F}\cdot\hat{t}ds \approx -\left[F_\varphi\left(r_0+\frac{\Delta r}{2},\theta_0,\varphi_0\right)-F_\varphi\left(r_0-\frac{\Delta r}{2},\theta_0,\varphi_0\right)\right]r_0\sin\theta_0(\Delta\varphi)$$

$$-F_\varphi(r_0,\theta_0,\varphi_0)\Delta r\sin\theta_0(\Delta\varphi)$$

$$=-\frac{\left[F_\varphi\left(r_0+\frac{\Delta r}{2},\theta_0,\varphi_0\right)-F_\varphi\left(r_0-\frac{\Delta r}{2},\theta_0,\varphi_0\right)\right]}{\Delta r}(\Delta r)r_0\sin\theta_0(\Delta\varphi)$$

$$-\frac{1}{r_0}F_\varphi(r_0,\theta_0,\varphi_0)r_0(\Delta r)\sin\theta_0(\Delta\varphi)$$

$$\approx-\frac{\partial F_\varphi(r_0,\theta_0,\varphi_0)}{\partial r}\Delta A-\frac{1}{r_0}F_\varphi(r_0,\theta_0,\varphi_0)\Delta A.$$

Also,

$$\int_{C_2}\vec{F}\cdot\hat{t}ds\approx F_r\left(r_0,\theta_0,\varphi_0+\frac{\Delta\varphi}{2}\right)\Delta r$$

$$\int_{C_4}\vec{F}\cdot\hat{t}ds\approx -F_r\left(r_0,\theta_0,\varphi_0-\frac{\Delta\varphi}{2}\right)\Delta r$$

so

$$\int_{C_2+C_4}\vec{F}\cdot\hat{t}ds\approx F_r\left(r_0,\theta_0,\varphi_0+\frac{\Delta\varphi}{2}\right)\Delta r-F_r\left(r_0,\theta_0,\varphi_0-\frac{\Delta\varphi}{2}\right)\Delta r$$

$$=\frac{\left[F_r\left(r_0,\theta_0,\varphi_0+\frac{\Delta\varphi}{2}\right)-F_r\left(r_0,\theta_0,\varphi_0-\frac{\Delta\varphi}{2}\right)\right]}{\Delta\varphi}(\Delta\varphi)(\Delta r)$$

$$\approx\frac{1}{r_0\sin\theta_0}\frac{\partial F_r(r_0,\theta_0,\varphi_0)}{\partial\varphi}r_0\sin\theta_0(\Delta\varphi)(\Delta r)$$

$$\approx\frac{1}{r_0\sin\theta_0}\frac{\partial F_r(r_0,\theta_0,\varphi_0)}{\partial\varphi}\Delta A.$$

Dividing by ΔA and taking the limit as $\|\Delta A\|\to 0$, we get the rotational component of $\vec{F}$ in the direction of $\hat{e}_\theta$ is

$$\lim_{\|\Delta A\|\to 0}\frac{1}{\|\Delta A\|}\int_{C_1+C_2+C_3+C_4}\vec{F}\cdot\hat{t}ds=-\frac{\partial F_\varphi}{\partial r}A-\frac{1}{r}F_\varphi+\frac{1}{r\sin\theta}\frac{\partial F_r}{\partial\varphi}$$

$$=\frac{1}{r\sin\theta}\frac{\partial F_r}{\partial\varphi}-\frac{1}{r}\frac{\partial}{\partial r}(rF_\varphi).$$

Thus, in spherical coordinates, if $\vec{F} = F_r\hat{e}_r + F_\theta\hat{e}_\theta + F_\varphi\hat{e}_\varphi$, then

$$\nabla \times \vec{F} = \left(\frac{1}{r\sin\theta}\frac{\partial}{\partial\theta}(\sin\theta F_\varphi) - \frac{1}{r\sin\theta}\frac{\partial F_\theta}{\partial\varphi}\right)\hat{e}_r + \left(\frac{1}{r\sin\theta}\frac{\partial F_r}{\partial\varphi} - \frac{1}{r}\frac{\partial}{\partial r}(rF_\varphi)\right)\hat{e}_\theta + \left(\frac{1}{r}\frac{\partial}{\partial r}(rF_\theta) - \frac{1}{r}\frac{\partial F_r}{\partial\theta}\right)\hat{e}_\varphi.$$

Example:
We compute the curl of

$$\vec{F}(r,\theta,\varphi)) = r^2\sin\theta\cos\varphi\hat{e}_r + e^r\cos\theta\sin\varphi\hat{e}_\theta + \cos\theta\cos\varphi\hat{e}_\varphi.$$

We have

$$F_r = r^2\sin\theta\cos\varphi,\quad F_\theta = e^r\cos\theta\sin\varphi,\quad F_\varphi = \cos\theta\cos\varphi$$

so

$$\frac{\partial}{\partial\theta}(\sin\theta F_\varphi) = \frac{\partial}{\partial\theta}\sin\theta\cos\theta\cos\varphi = (-\sin^2\theta + \cos^2\theta)\cos\varphi$$

$$\frac{\partial F_\theta}{\partial\varphi} = e^r\cos\theta\cos\varphi,\quad \frac{\partial}{\partial r}(rF_\theta) = (re^r + e^r)\cos\theta\sin\varphi$$

$$\frac{\partial F_r}{\partial\theta} = r^2\cos\theta\cos\varphi,\quad \frac{\partial F_r}{\partial\varphi} = -r^2\sin\theta\sin\varphi,\quad \frac{\partial}{\partial r}(rF_\varphi) = \cos\theta\cos\varphi.$$

Thus, the curl of $\vec{F}$ is

$$\left(\frac{1}{r\sin\theta}\frac{\partial}{\partial\theta}(\sin\theta F_\varphi) - \frac{1}{r\sin\theta}\frac{\partial F_\theta}{\partial\varphi}\right)\hat{e}_r + \left(\frac{1}{r\sin\theta}\frac{\partial F_r}{\partial\varphi} - \frac{1}{r}\frac{\partial}{\partial r}(rF_\varphi)\right)\hat{e}_\theta$$

$$+\left(\frac{1}{r}\frac{\partial}{\partial r}(rF_\theta) - \frac{1}{r}\frac{\partial F_r}{\partial\theta}\right)\hat{e}_\varphi$$

$$= \left[\frac{1}{r\sin\theta}(-\sin^2\theta + \cos^2\theta)\cos\varphi - \frac{1}{r\sin\theta}e^r\cos\theta\cos\varphi\right]\hat{e}_r$$

$$+\left[\left(\frac{-1}{r\sin\theta}\right)r^2\sin\theta\sin\varphi - \frac{1}{r}\cos\theta\cos\varphi\right]\hat{e}_\theta$$

$$+\left[\frac{1}{r}(re^r + e^r)\cos\theta\sin\varphi - \frac{1}{r}r^2\cos\theta\cos\varphi\right]\hat{e}_\varphi.$$

EXERCISES

1. Compute the gradient of
   a. $f(r, \theta, z) = z^2e^r\sin\theta + 2rz$.
   b. $f(r, \theta, \varphi) = r\sin\varphi\cos\theta - \tan\varphi$.
   c. $f(r, \theta, z) = r^2 + rz\sec\theta$.

d. $f(r, \theta, \varphi) = r\,\theta\,\varphi$.
e. $f(r, \theta, \varphi) = \ln(r\varphi) + \sin\theta$.
f. $f(r, \theta, z) = z/r\cos\theta$.

2. Compute the curl and divergence of
a. $\underset{\rightharpoonup}{F}(r,\theta,\varphi) = -\frac{1}{r\tan\theta}\widehat{e}_\varphi$.
b. $\underset{\rightharpoonup}{F}(r,\theta,\varphi) = \cos\theta\widehat{e}_r + \sin\theta\widehat{e}_\theta + r^2\tan\varphi\widehat{e}_\varphi$.
c. $\underset{\rightharpoonup}{F}(r,\theta,z) = r\widehat{e}_\theta$.
d. $\underset{\rightharpoonup}{F}(r,\theta,\varphi) = \sin\theta\widehat{e}_\theta + r\varphi\widehat{e}_\varphi$.
e. $\underset{\rightharpoonup}{F}(r,\theta,z) = r^2\tan\theta\widehat{e}_r + z\widehat{e}_\theta + e^r z\sin\theta\widehat{e}_z$.
f. $F(r,\theta,z) = \ln r\widehat{e}_r - \cos\theta\widehat{e}_\theta + 4\widehat{e}_z$.

3.

a. In cylindrical coordinates, let $\vec{F} = F_r\widehat{e}_r + F_\theta\widehat{e}_\theta + F_z\widehat{e}_z$. Show that the curl of can be computed using the formula

$$\nabla\times\vec{F} = \begin{vmatrix} \frac{1}{r}\widehat{e}_r & \widehat{e}_\theta & \frac{1}{r}\widehat{e}_z \\ \frac{\partial}{\partial r} & \frac{\partial}{\partial\theta} & \frac{\partial}{\partial z} \\ F_r & rF_\theta & F_z \end{vmatrix}.$$

b. In spherical coordinates, let $\vec{F} = F_r\widehat{e}_r + F_\theta\widehat{e}_\theta + F_\varphi\widehat{e}_\varphi$. Show that the curl of can be computed using the formula

$$\nabla\times\vec{F} = \begin{vmatrix} \frac{1}{r^2\sin\theta}\widehat{e}_r & \frac{1}{r\sin\theta}\widehat{e}_\theta & \frac{1}{r}\widehat{e}_\varphi \\ \frac{\partial}{\partial r} & \frac{\partial}{\partial\theta} & \frac{\partial}{\partial\varphi} \\ F_r & rF_\theta & r\sin\theta F_\varphi \end{vmatrix}.$$

2.3 GREEN'S THEOREM, THE DIVERGENCE THEOREM, AND STOKES' THEOREM

In this section, our main focus is the three major theorems of integral vector calculus—Green's theorem, the divergence theorem (Gauss' theorem), and Stokes' theorem. In a sense, each of these is an extension of the fundamental theorem of calculus. We begin with Green's theorem, which relates the line integral over the boundary of a region to the area enclosed by the region. (This relates to the fundamental theorem of calculus,

$$\int_a^b f'(x)dx = f(b) - f(a)$$

in that the left-hand side is an expression over a region $[a, b]$ and the right-hand side is an expression over the boundary of the region.)

We assume that the boundary of the region is a simple closed curve with positive orientation. A simple closed curve is the one that does not intersect itself. A positively oriented curve is a curve that is traversed in the counterclockwise direction, so that the region lies on the left of the direction of transversal. See Fig. 2.3.1.

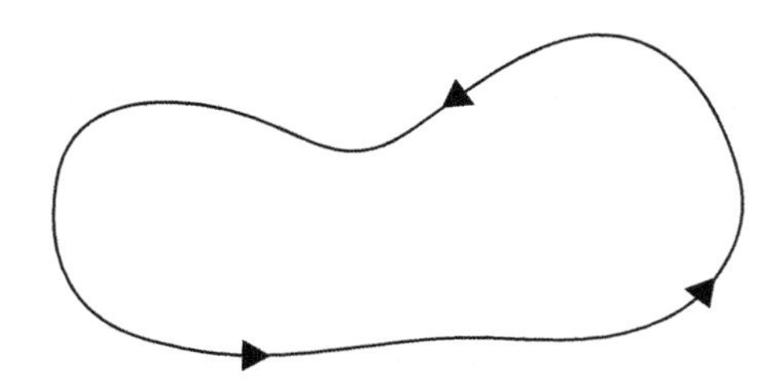

FIGURE 2.3.1

Recall that to compute

$$\int_C Pdx + Qdy$$

we parameterize C by a function $\sigma(t) = (x(t), y(t))$, $a \leq t \leq b$ and compute

$$\int_{t=a}^{b} \left[P(x(t), y(t)) \frac{dx}{dt} + Q(x(t), y(t)) \frac{dy}{dt} \right] dt.$$

We assume that $\sigma'(t)$ is piecewise continuous.

Theorem (Green's theorem):

Let C be a simple closed curve in $\mathbb{R}^2$ that is positively oriented and piecewise smooth. Let D be the region that is enclosed by C. Suppose that $P: \mathbb{R}^2 \to \mathbb{R}$ and $Q: \mathbb{R}^2 \to \mathbb{R}$ have continuous partial derivatives on an open set containing D. Then

$$\int_C P(x, y)dx + Q(x, y)dy = \iint_D \left(\frac{\partial Q(x, y)}{\partial x} - \frac{\partial P(x, y)}{\partial y} \right) dxdy.$$

We present a major idea in the proof for a simplified region. Consider the region shown in Fig. 2.3.2.

We show that in this case

$$\iint_D \left(\frac{\partial P(x, y)}{\partial y} \right) dydx = -\int_C P(x, y)dx.$$

Referring to Fig. 2.3.2, we have

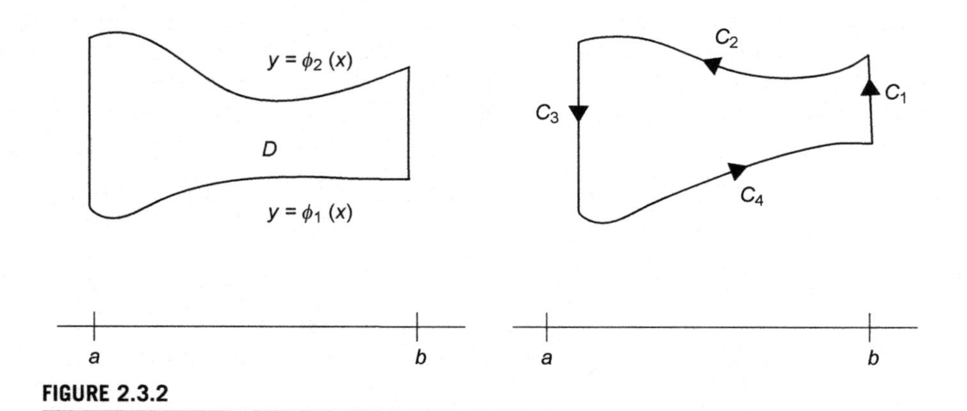

FIGURE 2.3.2

$$\iint_D \left(\frac{\partial P(x,y)}{\partial y}\right) dydx$$

$$= \int_{x=a}^{b} \int_{y=\varphi_1(x)}^{\varphi_2(x)} \left(\frac{\partial P(x,y)}{\partial y}\right) dydx$$

$$= \int_a^b [P(x,\varphi_2(x)) - P(x,\varphi_1(x))]dx.$$

We also have

$$\int_C P(x,y)dx = \int_{C_1} P(x,y)dx + \int_{C_2} P(x,y)dx + \int_{C_3} P(x,y)dx + \int_{C_4} P(x,y)dx.$$

Now $dx = 0$ on C_1 and C_3, so

$$\int_C P(x,y)dx = \int_{C_2} P(x,y)dx + \int_{C_4} P(x,y)dx.$$

Also

$$\int_{C_2} P(x,y)dx = \int_{x=b}^{a} P(x,\varphi_2(x))dx = -\int_{x=a}^{b} P(x,\varphi_2(x))dx$$

and

$$\int_{C_4} P(x,y)dx = \int_{x=a}^{b} P(x,\varphi_1(x))dx$$

so

$$\int_C P(x,y)dx = -\int_{x=a}^{b} P(x,\varphi_2(x))dx + \int_{x=a}^{b} P(x,\varphi_1(x))dx$$
$$= -\left[\int_{x=a}^{b} P(x,\varphi_2(x)) - P(x,\varphi_1(x))dx\right].$$

Thus

$$\iint_D \left(\frac{\partial P(x,y)}{\partial y}\right) dydx = \int_a^b [P(x,\varphi_2(x)) - P(x,\varphi_1(x))]dx = -\int_C P(x,y)dx.$$

In Exercise 2 we show

$$\int_C Q(x,y)dy = \iint_D \frac{\partial Q(x,y)}{\partial x} dxdy$$

for a region of a particular type. Putting these results together gives Green's theorem.

Vector form of Green's theorem:

If $\vec{F}$ is a vector field and D is a region in the $x-y$ plane with boundary C as mentioned above, then

$$\int_C \vec{F}\cdot d\hat{r} = \iint_D \left(\nabla \times \vec{F}\right)\cdot \hat{k}dxdy.$$

This relates to the first statement of Green's theorem by taking $\vec{F} = P(x,y)\hat{i} + Q(x,y)\hat{j}$.

The next example illustrates that it is often easier to compute the line integral around a simple closed curve using $\iint_D \left(\frac{\partial Q(x,y)}{\partial x} - \frac{\partial P(x,y)}{\partial y}\right) dxdy$ rather than $\int_C P(x,y)dx + Q(x,y)dy$. On the other hand, areas can sometimes be more easily computed using $\int_C P(x,y)dx + Q(x,y)dy$.

Example:

We verify Green's theorem in the case $P(x, y) = xy$, $Q(x, y) = y$, and D is the region $x^2 + y^2 \le 1$.

To compute

$$\int_C Pdx + Qdy$$

we parameterize C by $\sigma(\theta) = (\cos\theta, \sin\theta)\ 0 \le \theta \le 2\pi$, so that

$$x = \cos\theta, y = \sin\theta, \frac{dx}{d\theta} = -\sin\theta, \frac{dy}{d\theta} = \cos\theta,\ \ P(x,y) = \cos\theta\sin\theta, Q(x,y) = \sin\theta.$$

Then,

$$\int_C Pdx + Qdy = \int_0^{2\pi} (-\sin^2\theta \cos\theta + \sin\theta \cos\theta)d\theta = \frac{-\sin^3\theta}{3} + \frac{\sin^2\theta}{2}\bigg|_0^{2\pi} = 0.$$

To compute

$$\iint_D \left(\frac{\partial Q(x,y)}{\partial x} - \frac{\partial P(x,y)}{\partial y}\right) dxdy$$

note that

$$\frac{\partial Q(x,y)}{\partial x} = 0, \qquad \frac{\partial P(x,y)}{\partial y} = x$$

so that in polar coordinates

$$\iint_D \left(\frac{\partial Q(x,y)}{\partial x} - \frac{\partial P(x,y)}{\partial y}\right) dxdy = -\int_{\theta=0}^{2\pi} \left(\int_{r=0}^{1} \cos\theta r^2 dr\right) d\theta = -\frac{1}{3}\int_0^{2\pi} \cos\theta d\theta = 0.$$

An Application:

If we take $P = -y$ and $Q = x$, then

$$\iint_D \left(\frac{\partial Q(x,y)}{\partial x} - \frac{\partial P(x,y)}{\partial y}\right) dxdy = 2\iint_D dx\, dy = 2(\text{Area of } D)$$

so the area of D can be computed according to its line integral using

$$\int_C -ydx + xdy = 2\iint_D dx\, dy = 2(\text{Area of } D)$$

or

$$\text{Area of } D = \frac{1}{2}\int_C -ydx + xdy.$$

We use this idea to compute the area of the ellipse

$$\frac{x^2}{a^2} + \frac{y^2}{b^2} = 1.$$

We parameterize the boundary using

$$x = a\cos\theta, \quad y = b\sin\theta, \quad 0 \le \theta \le 2\pi$$

so that

$$\frac{dx}{d\theta} = -a\sin\theta, \qquad \frac{dy}{d\theta} = b\cos\theta$$

and

$$\frac{1}{2}\int_C -ydx + xdy = \frac{1}{2}\int_0^{2\pi}[(-b\sin\theta)(-a\sin\theta) + (a\cos\theta)(b\cos\theta)]d\theta$$

$$= \frac{1}{2}ab\int_0^{2\pi}\left[\sin^2\theta + \cos^2\theta\right]d\theta = \frac{1}{2}ab(2\pi) = \pi ab.$$

Green's theorem can be extended to regions with "holes" such as the one shown in Fig. 2.3.3A. Note that each part of the boundary has been assigned an orientation. The idea is to divide the region into parts as shown in Fig. 2.3.3B, and compute the integral of each piece. What makes it work is that parts of the integrals cancel each other as Fig. 2.3.3B shows.

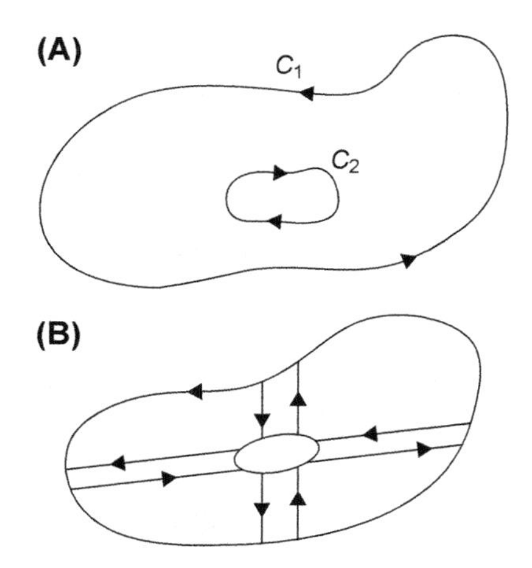

FIGURE 2.3.3A AND B

The divergence (Gauss') theorem:

We give the statement of the divergence theorem in two and three dimensions but give the proof only in the two dimensional case. The intuition of the theorem is better described in the three-dimensional case, and this is where we begin.

Recall that in Cartesian coordinates, the divergence of a vector field in three dimensions $\vec{F}(x,y,z) = F_1(x,y,z)\hat{i} + F_2(x,y,z)\hat{j} + F_3(x,y,z)\hat{k}$ is given by

$$\text{div}\left(\vec{F}\right) = \nabla\cdot\vec{F} = \left(\frac{\partial}{\partial x}\hat{i} + \frac{\partial}{\partial y}\hat{j} + \frac{\partial}{\partial z}\hat{k}\right)\cdot\left(F_1(x,y,z)\hat{i} + F_2(x,y,z)\hat{j} + F_3(x,y,z)\hat{k}\right)$$

$$= \frac{\partial F_1}{\partial x} + \frac{\partial F_2}{\partial y} + \frac{\partial F_3}{\partial z}.$$

Suppose the flow of a fluid in a region is described by the vector field $\vec{F}$. The amount of fluid that flows through a membrane in time t depends on the area of the surface, the orientation of the surface with the direction of the flow, and the velocity of the flow. See Fig. 2.3.4.

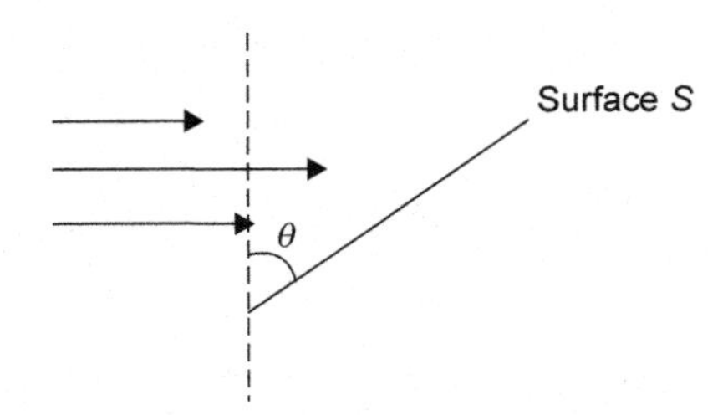

FIGURE 2.3.4

If the surface is perpendicular to the direction of flow, then the amount of flow through the surface in time t is

$$\text{velocity} \times \text{area} \times \text{time}$$

and if the normal to the surface makes an angle θ with the direction of flow, then the amount of flow through the surface in time t is

$$\text{velocity} \times \text{area} \times \text{time} \times \cos\theta.$$

If $\widehat{n}$ is the unit vector normal to the surface, then the amount of fluid that crosses one unit of area of the surface in one unit of time is

$$\text{velocity} \times \cos\theta = \vec{F} \cdot \widehat{n}. \tag{1}$$

Now consider an infinitesimal element of volume ΔV. For specificity, suppose this is a parallelepiped $\Delta x \Delta y \Delta z$ and the fluid flows through six faces. See Fig. 2.3.5.

Consider the two surfaces parallel to the $y-z$ plane that we have labeled S1 and S2, and suppose the direction of flow is as indicated by the arrows. The net outflow per unit time through surfaces S1 and S2 is the outflow per unit area per unit time through

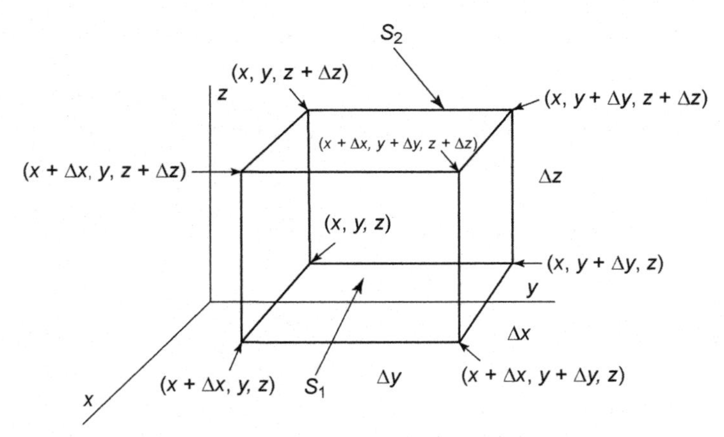

FIGURE 2.3.5

surface S2 minus the inflow per unit area per unit time through surface S1 multiplied by the area of each surface, which is $\Delta y \Delta z$. In terms of a formula, this is

$$[F_1(x+\Delta x, y, z) - F_1(x, y, z)]\Delta y \Delta z = \frac{[F_1(x+\Delta x, y, z) - F_1(x, y, z)]}{\Delta x}\Delta x \Delta y \Delta z$$
$$\approx \frac{\partial F_1}{\partial x}\Delta V.$$

Similar relations hold for the other pairs of faces. Summing, we get that the net outflow through the volume ΔV is

$$\left(\frac{\partial F_1}{\partial x} + \frac{\partial F_2}{\partial y} + \frac{\partial F_3}{\partial z}\right)\Delta V = \left(\nabla \cdot \vec{F}\right) dV.$$

The outflow through the total volume V is

$$\iiint_V \left(\nabla \cdot \vec{F}\right) dV.$$

But from Eq. (1), we get that this outflow is also

$$\iint_S \vec{F} \cdot \widehat{n}\, dS.$$

The divergence theorem is the statement that these two quantities are the same.

Theorem (divergence theorem in two dimensions):

Suppose that D is a region in $\mathbb{R}^2$, and C is a piecewise smooth simple closed curve that encloses the region D. Let $\widehat{n}$ be the unit normal vector that points outward from C (so that $\widehat{n}$ changes from point to point along C). Suppose that $\widehat{F}$ is a continuously differentiable vector field on an open set that contains $C \cup D$. Then

$$\int_C \widehat{F} \cdot \widehat{n}\, ds = \iint_D (\nabla \cdot \widehat{F})\, dA.$$

Remark:

Notice that the expression on the right is in a sense the integral of the derivative over a region and the expression on the left is the integral of the function over the boundary.

Proof:

Let $\sigma(t) = (x(t), y(t))$ be a parameterization of C that is positively oriented. Then

$$\widehat{n} = \frac{(y'(t),\ -x'(t))}{\sqrt{[x'(t)]^2 + [y'(t)]^2}}$$

is a unit normal vector that points outward from C, as we show in Exercise 3. Let

$$\widehat{F}(x, y) = \widehat{F}(x(t), y(t)) = Q(x(t), y(t))\widehat{\imath} - P(x(t), y(t))\widehat{\jmath}.$$

(We make this unusual looking choice of functions so that at the end of the computations it will be apparent how we are using Green's theorem in the proof.)

Now

$$ds = \sqrt{[x'(t)]^2 + [y'(t)]^2}dt$$

so

$$\hat{F}\cdot\hat{n}\,ds = \left\{\frac{Q(x(t),y(t))y'(t) + P(x(t),y(t))x'(t)}{\sqrt{[x'(t)]^2 + [y'(t)]^2}}\right\}\sqrt{[x'(t)]^2 + [y'(t)]^2}\,dt$$
$$= [Q(x(t),y(t))y'(t) + P(x(t),y(t))x'(t)]dt.$$

Thus

$$\int_C \hat{F}\cdot\hat{n}\,ds = \int_C Qdy + Pdx.$$

By Green's theorem

$$\int_C Pdx + Qdy = \iint_D \left(\frac{\partial Q}{\partial x} - \frac{\partial P}{\partial y}\right)dx\,dy.$$

For our choice of $\hat{F}$,

$$\nabla\cdot\hat{F} = \frac{\partial Q}{\partial x} - \frac{\partial P}{\partial y}$$

so

$$\iint_D (\nabla\cdot\hat{F})dxdy = \iint_D \left(\frac{\partial Q}{\partial x} - \frac{\partial P}{\partial y}\right)dx\,dy = \int_C Qdy + Pdx = \int_C \hat{F}\cdot\hat{n}\,ds.$$

Example:
We evaluate $\iint_D (\nabla\cdot\hat{F})dxdy$ and $\int_C \hat{F}\cdot\hat{n}\,ds$ in the case

$$\hat{F}(x,y) = x\hat{i} + y\hat{j} \text{ and } C:\ x^2 + y^2 = 16.$$

We have $\nabla\cdot\hat{F} = 2$, so $\iint_D (\nabla\cdot\hat{F})dxdy = 2\iint_D 1dxdy = 2(16\pi) = 32\pi$, since the area of the circle enclosed by C is 16π.

To find $\int_C \hat{F}\cdot\hat{n}\,ds$ we parameterize C using $x = 4\cos\theta$, $y = 4\sin\theta$, $0 \le \theta \le 2\pi$. Then

$$\hat{n} = \frac{4\cos\theta\hat{i} + 4\sin\theta\hat{j}}{\sqrt{(-4\sin\theta\ )^2 + (4\cos\theta)^2}} = \cos\theta\hat{i} + \sin\theta\hat{j} \text{ and } ds = 4d\theta.$$

So

$$\hat{F}\cdot\hat{n}\,ds = \left(4\cos\theta\hat{i} + 4\sin\theta\hat{j}\right)\cdot\left(\cos\theta\hat{i} + \sin\theta\hat{j}\right)4d\theta = 16d\theta$$

and thus

$$\int_C \hat{F}\cdot\hat{n}\,ds = \int_0^{2\pi} 16d\theta = 32\pi.$$

Theorem (divergence theorem in three dimensions):

Suppose that V is a closed bounded region in $\mathbb{R}^3$ with piecewise smooth boundary S, and $\widehat{F}(x,y,z)$ is a continuously differentiable vector field on an open set containing V. Then

$$\iiint_V (\nabla\cdot\widehat{F})dV = \iint_S \widehat{F}\cdot\widehat{n}\,dS$$

where $\widehat{n}$ is the unit vector that points outward from the surface S.

The next example highlights the idea that it is often easier to compute the flux through a closed surface using $\iiint_\Omega(\nabla\cdot F)dV$ rather than $\iint_{\partial\Omega}F\cdot dS$, which is an important application of the divergence theorem.

Example:

Let Ω be the cylinder whose base radius is a, which is centered at the origin of the x,y plane, and whose height is h. Let F be the vector field $F = xi + y\widehat{j} + z\widehat{k}$. We first compute $\iiint_\Omega(\nabla\cdot F)dV$. We have

$$\nabla\cdot F = \frac{\partial x}{\partial x} + \frac{\partial y}{\partial y} + \frac{\partial z}{\partial z} = 3$$

so that

$$\iiint_\Omega (\nabla\cdot F)dV = \iiint_\Omega (3)dV = 3\cdot\text{volume of the cylinder} = 3\pi a^2 h.$$

To compute

$$\iint_{\partial\Omega} F\cdot dS = \iint_{\partial\Omega} (F\cdot n)dS,$$

we must consider three surfaces. Let

$$\begin{aligned}\partial\Omega_1 &= \text{top of the cylinder},\\ \partial\Omega_2 &= \text{bottom of the cylinder},\\ \partial\Omega_3 &= \text{side of the cylinder}.\end{aligned}$$

On $\partial\Omega_1$, $z = h$, so $F = x\widehat{i} + y\widehat{j} + h\widehat{k}$ and the outward pointing unit normal is $n = \widehat{k}$. Thus

$$\iint_{\partial\Omega_1} (F\cdot n)dS = \iint_{\partial\Omega_1} h dS = h\cdot\text{Area of the top} = h\cdot\pi a^2.$$

On $\partial\Omega_2$, $z = 0$, so $F = xi + y\widehat{j} + 0\widehat{k}$ and the outward pointing unit normal is $n = -\widehat{k}$. Thus

$$\iint_{\partial\Omega_2} (F\cdot n)dS = \iint_{\partial\Omega_2} 0 dS = 0.$$

On $\partial\Omega_3$, the curved surface, $x^2 + y^2 = a^2$, the vector

$$\nabla(x^2 + y^2) = 2x\widehat{i} + 2y\widehat{j}$$

is normal to the surface, so

$$\frac{2x\hat{i} + 2y\hat{j}}{\sqrt{4x^2 + 4y^2}} = \frac{2x\hat{i} + 2y\hat{j}}{2a} = \hat{n}$$

is a unit vector normal to the surface. Then

$$\iint_{\partial\Omega_3} (F \cdot n) dS = \iint_{\partial\Omega_3} \left(\left(xi + y\hat{j} + z\hat{k} \right) \cdot \left(\frac{2x\hat{i} + 2y\hat{j}}{2a} \right) \right) dS$$

$$= \iint_{\partial\Omega_3} \frac{2x^2 + 2y^2}{2a} dS = \iint_{\partial\Omega_3} \frac{2a^2}{2a} dS = a \cdot \text{area of } \partial\Omega_3 = a \cdot 2\pi a h$$

$$= 2\pi a^2 h.$$

Thus

$$\iint_{\partial\Omega} (F \cdot n) dS$$

$$= \iint_{\partial\Omega_1} (F \cdot n) dS$$

$$+ \iint_{\partial\Omega_2} (F \cdot n) dS + \iint_{\partial\Omega_3} (F \cdot n) dS = h \cdot \pi a^2 + 0 + 2\pi a^2 h$$

$$= 3\pi a^2 h.$$

Example:
Evaluate

$$\iint_S x^3 dydz + x^2 y dxdz + x^2 z dxdy$$

over the surface of the cylinder bounded by $x^2 + y^2 = 16$, $z = 0$, $z = 3$.
Solution:
We use the divergence theorem with

$$F = x^3 \hat{i} + x^2 y \hat{j} + x^2 z \hat{k}.$$

Then

$$\nabla \cdot F = 3x^2 + x^2 + x^2 = 5x^2$$

so that

$$\iint_S x^3 dydz + x^2 y dxdz + x^2 z dxdy = \iiint_V 5x^2 dxdydz.$$

Converting to cylindrical coordinates,

$$\iiint_V 5x^2 dxdydz = \int_{\theta=0}^{2\pi}\int_{r=0}^{4}\int_{z=0}^{3} 5r^2\cos^2\theta r\, dzdrd\theta$$
$$= 5\cdot 3\cdot\frac{4^4}{4}\int_{\theta=0}^{2\pi}\cos^2\theta d\theta = 960\pi.$$

Example:
Gauss' law for inverse–square fields
An inverse–square field $\vec{F}$ is a vector field for which

$$\left\|\vec{F}\right\| = \frac{c}{\|\hat{r}\|^2} \quad \text{and} \quad \vec{F}(\hat{r}) = \frac{c}{\|\hat{r}\|^3}\hat{r}.$$

In $\mathbb{R}^3$

$$\vec{F}(x,y,z) = \frac{c}{(x^2+y^2+z^2)^{\frac{3}{2}}}\left(x\hat{i}+y\hat{j}+z\hat{k}\right).$$

We show in Exercise 30 that for such a vector field div $\vec{F}(x,y,z) = 0$. Gauss' law for inverse–square fields says that if σ is a closed orientable surface that surrounds the origin, then the outward flux of $\vec{F}$ across σ is

$$\iint_\sigma \vec{F}\cdot\hat{n}dS = 4\pi c.$$

Let G be the region enclosed by σ. To prove the result, we cannot use the divergence theorem directly because $\vec{F}$ is not continuous at the origin. We circumvent this problem by constructing a sphere of radius ε about the origin where ε is small enough so that the sphere is contained in G. See Fig. 2.3.6. Let σ_ε denote the surface of this sphere.

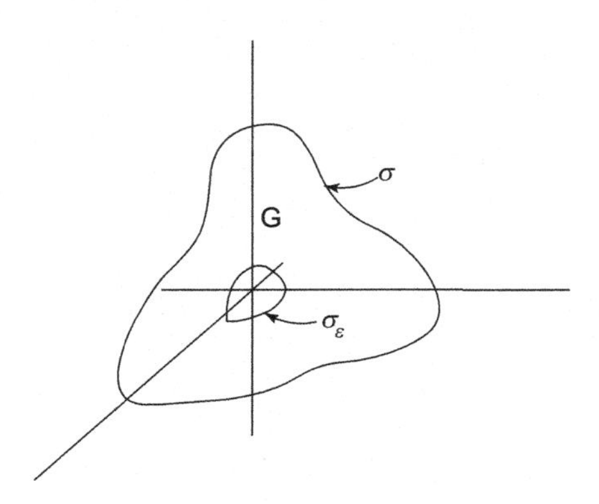

FIGURE 2.3.6

Let H_ε denote the solid that lies between σ and σ_ε. This is a solid with a hole, but the divergence theorem applies to such regions by a proof identical in spirit to the proof that Green's theorem applies to a region with holes. Thus, we have

$$\iiint_{H_\varepsilon} \operatorname{div} \vec{F}\, dV = \iint_\sigma \vec{F}\cdot\hat{n}dS + \iint_{\sigma_\varepsilon} \vec{F}\cdot\hat{n}dS$$

but since div $\vec{F} = 0$, we have

$$\iint_\sigma \vec{F}\cdot\hat{n}dS = -\iint_{\sigma_\varepsilon} \vec{F}\cdot\hat{n}dS.$$

Now

$$\iint_{\sigma_\varepsilon} \vec{F}\cdot\hat{n}dS = \iint_{\sigma_\varepsilon} \frac{c}{\|\hat{r}\|^3}\hat{r}\cdot\left(-\frac{\hat{r}}{\|\hat{r}\|}\right)dS = -c\iint_{\sigma_\varepsilon} \frac{\|\hat{r}\|^2}{\|\hat{r}\|^4}dS.$$

On σ_ε, $\|\hat{r}\| = \varepsilon$, so

$$-c\iint_{\sigma_\varepsilon} \frac{\|\hat{r}\|^2}{\|\hat{r}\|^4}dS = -\frac{c}{\varepsilon^2}\iint_{\sigma_\varepsilon} 1dS = -\frac{c}{\varepsilon^2}\cdot 4\pi\varepsilon^2 = -4\pi c$$

since the surface area of a sphere of radius ε is $4\pi\varepsilon^2$. Thus

$$\iint_\sigma \vec{F}\cdot\hat{n}dS = 4\pi c.$$

We return to the problem introduced in Section 1.2 of finding the Laplacian in a general orthogonal coordinate system. We denote the coordinates by (u_1, u_2, u_3) and the scaling factors by h_1, h_2, h_3. Consider an infinitesimal cube with sides parallel to the curvilinear coordinate axes as shown in Fig. 2.3.7. The lengths of the edges of the cube are h_1du_1, h_2du_2 and h_3du_3.

The gradient of the function Ψ in the u_1 direction is $\frac{1}{h_1}\frac{\partial \Psi}{\partial u_1}$. The divergence of a vector field $\vec{F} = F_1\hat{e}_1 + F_2\hat{e}_2 + F_3\hat{e}_3$ at a point p is the flux through the six faces of the infinitesimal cube centered at p divided by the volume of the cube. We consider the flux through the pair of parallel faces labeled Face 1 and Face 2. As in the proof of the divergence theorem, the net flow across the two faces is

$$\frac{\partial}{\partial u_1}(h_2h_3F_1)du_1du_2du_3.$$

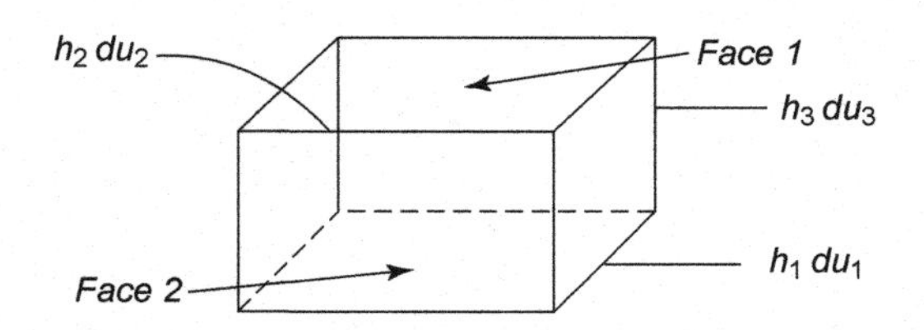

FIGURE 2.3.7

The flow across the other pairs of faces has analogous expressions. According to the divergence theorem,

$$\left(\nabla\cdot\vec{F}\right)(\text{volume}) = \frac{\partial}{\partial u_1}(h_2h_3F_1)du_1du_2du_3 + \frac{\partial}{\partial u_2}(h_1h_3F_2)du_1du_2du_3 + \frac{\partial}{\partial u_3}(h_1h_2F_3)du_1du_2du_3$$

so

$$\left(\nabla\cdot\vec{F}\right)h_1du_1h_2du_2h_3du_3$$

$$= \frac{\partial}{\partial u_1}(h_2h_3F_1)du_1du_2du_3 + \frac{\partial}{\partial u_2}(h_1h_3F_2)du_1du_2du_3 + \frac{\partial}{\partial u_3}(h_1h_2F_3)du_1du_2du_3$$

and so

$$\left(\nabla\cdot\vec{F}\right) = \frac{1}{h_1h_2h_3}\left\{\frac{\partial}{\partial u_1}(h_2h_3F_1) + \frac{\partial}{\partial u_2}(h_1h_3F_2) + \frac{\partial}{\partial u_3}(h_1h_2F_3)\right\}. \quad (2)$$

To determine an expression for the Laplacian, we let $\vec{F} = \nabla f$ in Eq. (2) to get

$$\nabla\cdot\nabla f = \Delta f\left(\text{or } \nabla^2 f\right)$$

$$= \frac{1}{h_1h_2h_3}\left\{\frac{\partial}{\partial u_1}\left(h_2h_3\frac{\partial f}{\partial u_1}\right) + \frac{\partial}{\partial u_2}\left(h_1h_3\frac{\partial f}{\partial u_2}\right) + \frac{\partial}{\partial u_3}\left(h_1h_2\frac{\partial f}{\partial u_3}\right)\right\}.$$

Stokes' theorem:

Stokes' theorem is a generalization of Green's theorem to $\mathbb{R}^3$. In Stokes' theorem we relate an integral over a surface to a line integral over the boundary of the surface. We assume that the surface is two sided (a Mobius strip is an example of a one-sided surface) that consists of a finite number of pieces, each of which has a normal vector at each point. This allows us to consider surfaces such as cubes that do not have normal vectors at the edges. We assume an orientation of the surface, which means we agree on the direction of the unit normal vector. This is necessary because there will be two unit normal vectors at each point p on the surface, $\widehat{n}(p)$ and $-\widehat{n}(p)$. This induces an orientation for the boundary of the surface.

Theorem (Stokes' theorem):

Let S be an oriented surface, and let ∂S denote the oriented boundary of S. If $\vec{F}$ is a continuously differentiable vector field on S, then

$$\iint_S \left(\nabla\times\vec{F}\right)dS = \int_{\partial S}\vec{F}\cdot ds.$$

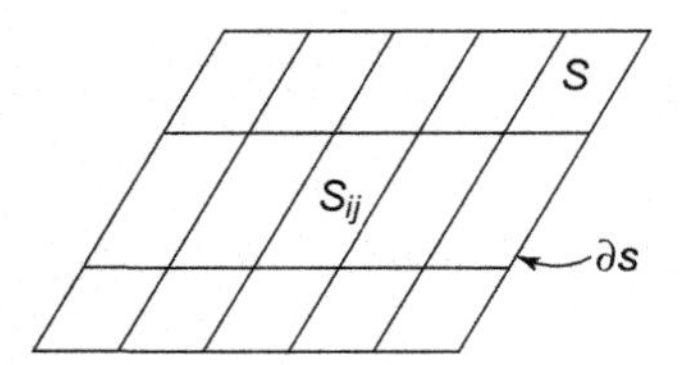

FIGURE 2.3.8

We give a sketch of the central idea in the proof of Stokes' theorem, which is simply Green's theorem. We begin by dividing the surface S into small squares, and focus on one square, that we call S_{ij}. See Fig. 2.3.8.

Let $\vec{F} = F_x\widehat{i} + F_y\widehat{j} + F_z\widehat{k}$ be a vector field and $\widehat{s} = x\widehat{i} + y\widehat{j} + z\widehat{k}$, so that $d\vec{s} = dx\widehat{i} + dy\widehat{j} + dz\widehat{k}$. We compute

$$\oint_{\partial S_{ij}} \vec{F} \cdot d\vec{s}$$

by which we mean the line integral around the line integral around the boundary of S_{ij}. We choose the axes so that S_{ij} is perpendicular to the z-axis. Then

$$\vec{F} \cdot d\vec{s} = F_x dx + F_y dy + F_z dz.$$

Consider S_{ij} as shown in Fig. 2.3.9.

We have

$$\int_{C_1} \vec{F} \cdot d\vec{s} = \int_{x=x_1}^{x_1+\Delta x} F_x(x, y_1)dx$$

and

$$\int_{C_3} \vec{F} \cdot d\vec{s} = \int_{x=x_1+\Delta x}^{x_1} F_x(x, y_1 + \Delta y)dx = -\int_{x=x_1}^{x_1+\Delta x} F_x(x, y_1 + \Delta y)dx$$

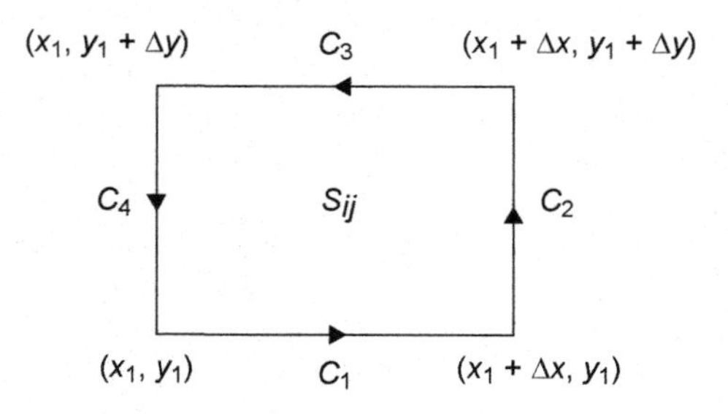

FIGURE 2.3.9

so

$$\int_{C_1} \vec{F} \cdot d\vec{s} + \int_{C_3} \vec{F} \cdot d\vec{s} = \int_{x=x_1}^{x_1+\Delta x} F_x(x, y_1)dx - \int_{x=x_1}^{x_1+\Delta x} F_x(x, y_1 + \Delta y)dx$$
$$= \int_{x=x_1}^{x_1+\Delta x} [F_x(x, y_1) - F_x(x, y_1 + \Delta y)]dx$$
$$= -\int_{x=x_1}^{x_1+\Delta x} [F_x(x, y_1 + \Delta y) - F_x(x, y_1)]dx.$$

Also

$$\int_{C_2} \vec{F} \cdot d\vec{s} = \int_{y=y_1}^{y_1+\Delta y} F_y(x_1 + \Delta x, y)dy$$

and

$$\int_{C_4} \vec{F} \cdot d\vec{s} = \int_{y=y_1+\Delta y}^{y_1} F_y(x_1, y)dy = -\int_{y=y_1}^{y_1+\Delta y} F_y(x_1, y)dy$$

so

$$\int_{C_2} \vec{F} \cdot d\vec{s} + \int_{C_4} \vec{F} \cdot d\vec{s} = \int_{y=y_1}^{y_1+\Delta y} [F_y(x_1 + \Delta x, y) - F_y(x_1, y)]dy.$$

By Green's theorem we have

$$\oint_{\partial S_{ij}} \vec{F} \cdot d\vec{s} = \int_{C_1} \vec{F} \cdot d\vec{s} + \int_{C_3} \vec{F} \cdot d\vec{s} + \int_{C_2} \vec{F} \cdot d\vec{s} + \int_{C_4} \vec{F} \cdot d\vec{s}$$
$$= \oint_{\partial S_{ij}} F_x dx + F_y dy = \iint_{S_{ij}} \left(\frac{\partial F_y}{\partial x} - \frac{\partial F_x}{\partial y} \right) dxdy.$$

Recall that

$$(\nabla \times \vec{F}) \cdot \widehat{k} = \frac{\partial F_y}{\partial x} - \frac{\partial F_x}{\partial y},$$

so that in this special case

$$\oint_{\partial S_{ij}} \vec{F} \cdot d\vec{s} = \iint_S (\nabla \times \vec{F}) \cdot \widehat{k} \, dS.$$

In the general case, the square will not be in the x–y plane but if we replace $\widehat{k}$ by the unit normal vector $\widehat{n}$, we get

$$\oint_{\partial S_{ij}} \vec{F} \cdot d\vec{s} = \iint_S (\nabla \times \vec{F}) \cdot \widehat{n} \, dS.$$

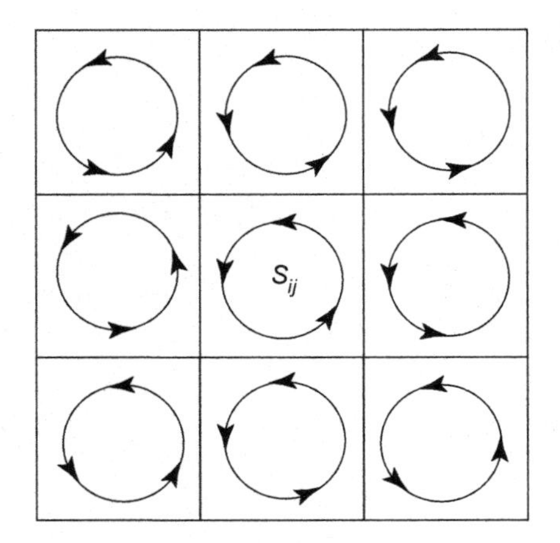

FIGURE 2.3.10

For a second way to get the same result, note that

$$\int_{C_2} \vec{F} \cdot d\vec{s} + \int_{C_4} \vec{F} \cdot d\vec{s} = \int_{y=y_1}^{y_1+\Delta y} [F_y(x_1+\Delta x, y) - F_y(x_1, y)]dy \approx \frac{\partial F_y}{\partial x}\Delta x \Delta y$$

and

$$\int_{C_1} \vec{F} \cdot d\vec{s} + \int_{C_3} \vec{F} \cdot d\vec{s} = -\int_{x=x_1}^{x_1+\Delta x} [F_x(x, y_1+\Delta y) - F_x(x, y_1)]dx \approx -\frac{\partial F_x}{\partial y}\Delta x \Delta y.$$

Now we put the squares together. Fig. 2.3.10 describes the idea.

For adjoining squares, the contributions at the common boundary cancel one another because they are in opposite directions. Thus, when we sum up all squares the only pieces that do not cancel are those on the boundary of S. Thus we get

$$\iint_S \left(\nabla \times \vec{F}\right) dS = \int_{\partial S} \vec{F} \cdot ds.$$

Example:

We verify Stokes' theorem in the case $\vec{F} = 3y\hat{i} + 4x\hat{j} + (2x+2z)\hat{k}$ where S is the upper half of the hemisphere $x^2 + y^2 + z^2 = 1$.

Now ∂S is $x^2 + y^2 = 1$, so we let

$$\sigma(t) = (\cos t, \sin t, 0)$$

and then on ∂S

$$\sigma'(t) = (-\sin t, \cos t, 0)$$
$$\vec{F}(\sigma(t)) = (3 \sin t, 4 \cos t, 2 \cos t)\ .$$
$$\vec{F}(\sigma(t)) \cdot \sigma'(t) = -3 \sin^2 t + 4 \cos^2 t$$

We then have

$$\int_{\partial S} \vec{F} \cdot ds = \int_0^{2\pi} (-3\sin^2 t + 4\cos^2 t)dt = \pi.$$

We next compute

$$\iint_S \left(\nabla \times \vec{F}\right) dS.$$

We have

$$\nabla \times \vec{F} = \begin{vmatrix} \hat{i} & \hat{j} & \hat{k} \\ \dfrac{\partial}{\partial x} & \dfrac{\partial}{\partial y} & \dfrac{\partial}{\partial z} \\ 3y & 4x & 2x + 2z \end{vmatrix} = -2\hat{j} + \hat{k}.$$

We parameterize S with spherical coordinates

$$x = \sin\theta\cos\varphi, \quad y = \sin\theta\sin\varphi, \quad z = \cos\theta$$

so that

$$T_\theta = \cos\theta\cos\varphi\hat{i} + \cos\theta\sin\varphi\hat{j} - \sin\theta\hat{k}$$
$$T_\varphi = -\sin\theta\sin\varphi\hat{i} + \sin\theta\cos\varphi\hat{j}$$

and

$$T_\theta \times T_\varphi = \sin^2\theta\cos\varphi\hat{i} + \sin^2\theta\sin\varphi\hat{j} + \sin\theta\cos\theta\hat{k}$$

is normal to the surface. Then

$$\left(\nabla \times \vec{F}\right) \cdot \hat{n} = \left(\nabla \times \vec{F}\right) \cdot (T_\theta \times T_\varphi) = -2\sin^2\theta\sin\varphi + \sin\theta\cos\theta$$

and

$$\iint_S \left(\nabla \times \vec{F}\right) \cdot \hat{n}\, dS = \int_{\theta=0}^{\pi/2} \left(\int_{\varphi=0}^{2\pi} (-2\sin^2\theta\sin\varphi + \sin\theta\cos\theta) d\varphi \right) d\theta = \pi.$$

Note that we could have computed the normal vector using $T_\varphi \times T_\theta$ instead of $T_\theta \times T_\varphi$, in which case we would have

$$\iint_S \left(\nabla \times \vec{F}\right) \cdot \hat{n}\, dS = -\pi.$$

To make the signs of the integrals agree in Stokes' theorem, we must follow the "right-hand rule" that determines the outward pointing normal for a surface that is not closed according to the direction of the line integral of the boundary. If the signs are opposite, you should have taken the cross product in the reverse order.

The following theorem gives results that will be important in the solution of partial differential equations.

Theorem (Green's identities):

1. (Green's first identity) If f and g are scalar functions with continuous partial derivatives in a region R, and if V is a region within R with surface S, then

$$\iiint_V (f\nabla^2 g + \nabla f \cdot \nabla g)\,dV = \iint_S f\frac{\partial g}{\partial n}\,dS$$

where $\frac{\partial g}{\partial n} = \nabla g \cdot \widehat{n}$ the directional derivative of g in the direction that is outward normal to the surface S.

2. (Green's second identity) If f and g are scalar functions with continuous partial derivatives in a region R, and if V is a region within R with surface S, then

$$\iiint_V (f\nabla^2 g - g\nabla^2 f)\,dV = \iint_S \left(f\frac{\partial g}{\partial n} - g\frac{\partial f}{\partial n}\right)dS.$$

We leave the proof of the theorem to Exercises 12 and 13.

An application of Stokes' theorem:

We can use Stokes' theorem to show that Ampere's law is equivalent to one of Maxwell's equations. Consider a group of wires that are carrying current $\widehat{I}$, and the wires are enclosed by a closed loop C, as shown in Fig. 2.3.11.

Suppose the current induces a magnetic field $\widehat{B}$. Ampere's law says

$$\oint_C \widehat{B} \cdot d\widehat{l} = \widehat{I}.$$

Let $\widehat{J}$ denote the current density, which is the current crossing a unit area that is perpendicular to $\widehat{J}$. Then $\widehat{J} \cdot \widehat{n}\, dS$ is the current across the surface element dS, where $\widehat{n}$ is a unit vector parallel to $\widehat{I}$. Then

$$\iint_S \widehat{J} \cdot \widehat{n}\, dS$$

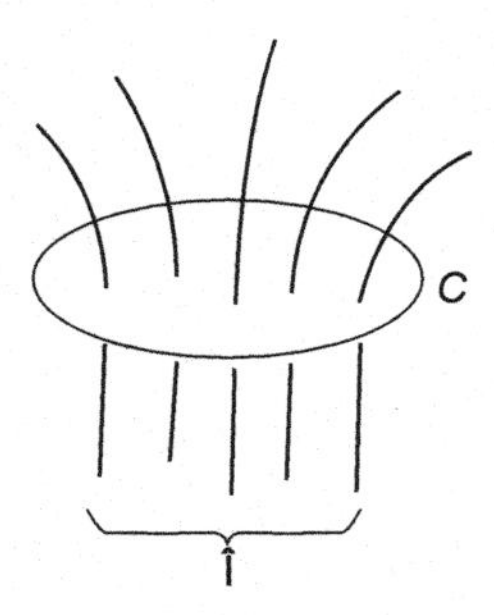

FIGURE 2.3.11

is the current across any surface S that is bounded by C. So we have

$$\oint_C \widehat{B} \cdot d\widehat{l} = \iint_S \widehat{J} \cdot \widehat{n}\, dS.$$

By Stokes' theorem

$$\oint_C \widehat{B} \cdot d\widehat{l} = \iint_S (\nabla \times \widehat{B}) \cdot \widehat{n}\, dS$$

so

$$\iint_S \widehat{J} \cdot \widehat{n}\, dS = \iint_S (\nabla \times \widehat{B}) \cdot \widehat{n}\, dS.$$

Since this holds for any surface S, we have $\nabla \times \widehat{B} = \widehat{J}$, which is one of Maxwell's equations.

Note that we could have begun with $\nabla \times \widehat{B} = \widehat{J}$ and arrived at Ampere's law by reversing the steps.

An application of the divergence theorem:

Coulomb's law says that the electric field $\vec{E}$ at the point r due to a point charge q located at the origin is

$$\vec{E} = \frac{q}{4\pi\varepsilon r^2} \widehat{e}_r,$$

where $\widehat{e}_r$ is the outward pointing unit normal vector in the direction from the origin to r.

The electric displacement $\vec{D}$ is defined by

$$\vec{D} = \varepsilon \vec{E} = \frac{q}{4\pi r^2} \widehat{e}_r.$$

If $\widehat{n}$ is the outward pointing normal to the sphere centered at the origin at the point r, then

$$\vec{D} \cdot \widehat{n} = \frac{q}{4\pi r^2}.$$

If S is a sphere of radius $|r|$ centered at the origin, and ∂S is the boundary of S, then

$$\oiint_{\partial S} \vec{D} \cdot \widehat{n} dS = \oiint_{\partial S} \frac{q}{4\pi r^2} dS = q. \tag{3}$$

Gauss' law says that Eq. (3) holds for any surface that has only the charge q in its interior. If there is more than one charge in S, then

$$\oiint_{\partial S} \vec{D} \cdot \widehat{n} dS = \sum q_i.$$

Now suppose that at each point x interior to S we have a charge density $\rho(x)$. Then in the infinitesimal volume dV centered at x the quantity of charge in dV is approximately $\rho(x)dV$

Then

$$\oiint_{\partial S} \vec{D} \cdot \hat{n}\, dS = \iiint_S \rho(x) dV.$$

But by the divergence theorem

$$\iiint_S \rho(x) dV = \oiint_{\partial S} \vec{D} \cdot \hat{n}\, dS = \iiint_S \nabla \cdot \vec{D} dV.$$

Since this is true for any surface, we have $\nabla \cdot \vec{D} = \rho$, which is another of Maxwell's equations.

Note that we could have begun with this Maxwell's equation and derived Gauss' law.

CONSERVATIVE FIELDS

In this section we suppose that $\vec{F}$ is an $\mathbb{R}^3$ vector field. One way to create a vector field is by taking the gradient of a function. Such vector fields are called gradient fields or conservative fields. These vector fields are prominent in physics because if f represents a potential, such as gravitational or electrical potential, then ∇f represents a force. The next theorem characterizes conservative fields.

Definition:

A vector field satisfying the conditions in the theorem below is said to be *conservative*.

Theorem:

Let $\vec{F}$ be a differentiable $\mathbb{R}^3$ vector field. The following conditions are equivalent:

1. There is a function f for which $\vec{F} = \nabla f$.
2. $\nabla \times \vec{F} = 0$.
3. For any oriented simple closed curve C, $\int_C \vec{F} \cdot d\vec{s} = 0$.
4. If C_1 and C_2 are two simple oriented curves that have the same end points, then

$$\int_{C_1} \vec{F} \cdot d\vec{s} = \int_{C_2} \vec{F} \cdot d\vec{s}.$$

Proof:

(1) ⇒ (2). In Exercise 15 we show that for any function f with continuous second partial derivatives, $\nabla \times \nabla f = 0$.

(2) ⇒ (3). Suppose that C is an oriented simple closed curve, and S is a one-sided surface whose boundary is C. Let σ be a path that represents C. Then by Stokes' theorem

$$\int_C \vec{F} \cdot d\vec{s} = \int_\sigma \vec{F} \cdot d\vec{s} = \int_S \left(\nabla \times \vec{F} \right) \cdot dS.$$

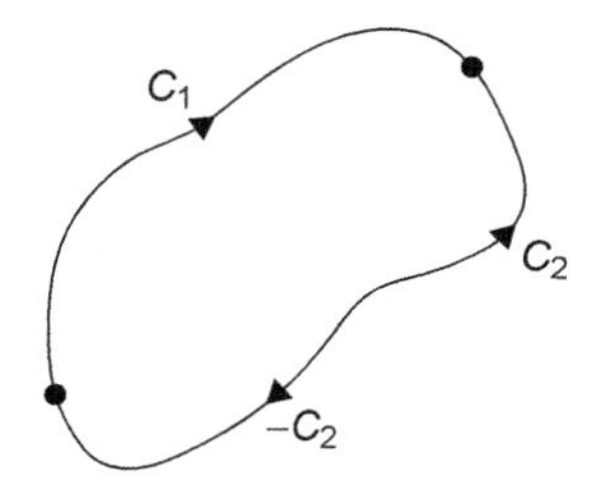

FIGURE 2.3.12

But $\nabla \times \vec{F} = 0$, so $\int_S \left(\nabla \times \vec{F}\right) \cdot dS = \int_C \vec{F} \cdot d\vec{s} = 0$.

(3)⇒(4). Suppose C_1 and C_2 are two simple oriented curves that have the same end points. Let $-C_2$ denote the curve with the same image, but opposite orientation as C_2. See Fig. 2.3.12.

Then $C = C_1 \cup (-C_2)$ is a closed curve. The curve C may not be simple, but if it is, then

$$\begin{aligned} 0 = \int_C \vec{F} \cdot d\vec{s} = \int_{C_1 \cup (-C_2)} \vec{F} \cdot d\vec{s} &= \int_{C_1} \vec{F} \cdot d\vec{s} + \int_{-C_2} \vec{F} \cdot d\vec{s} \\ &= \int_{C_1} \vec{F} \cdot d\vec{s} - \int_{C_2} \vec{F} \cdot d\vec{s} = 0 \end{aligned}$$

so

$$\int_{C_1} \vec{F} \cdot d\vec{s} = \int_{C_2} \vec{F} \cdot d\vec{s}.$$

If $C = C_1 \cup (-C_2)$ is not a simple closed curve, a more elaborate proof, that is beyond the scope of this text, is required.

(4)⇒(1). Let $\vec{F}(x, y, z) = F_1(x, y, z)\hat{i} + F_2(x, y, z)\hat{j} + F_3(x, y, z)\hat{k}$. For $(x_0, y_0, z_0) \in \mathbb{R}^3$, let σ be a path from (0,0,0) to (x_0, y_0, z_0) and define

$$f(x_0, y_0, z_0) = \int_\sigma \vec{F} \cdot d\sigma.$$

Since we assume (4) holds, the definition of $f(x_0, y_0, z_0)$ is independent of the path. We shall show that

$$\frac{\partial f}{\partial x} = F_1, \quad \frac{\partial f}{\partial y} = F_2, \quad \frac{\partial f}{\partial z} = F_3.$$

We first show $\frac{\partial f}{\partial z} = F_3$. Let σ be the path from (0, 0, 0) to (x_0, y_0, z_0) given by $\sigma = \sigma_1 + \sigma_2 + \sigma_3$ where

σ_1 is the path along the x-axis from (0, 0, 0) to $(x_0, 0, 0)$
σ_2 is the path parallel to the y-axis from $(x_0, 0, 0)$ to $(x_0, y_0, 0)$
σ_3 is the path parallel to the z-axis from $(x_0, y_0, 0)$ to (x_0, y_0, z_0). See Fig. 2.3.13.

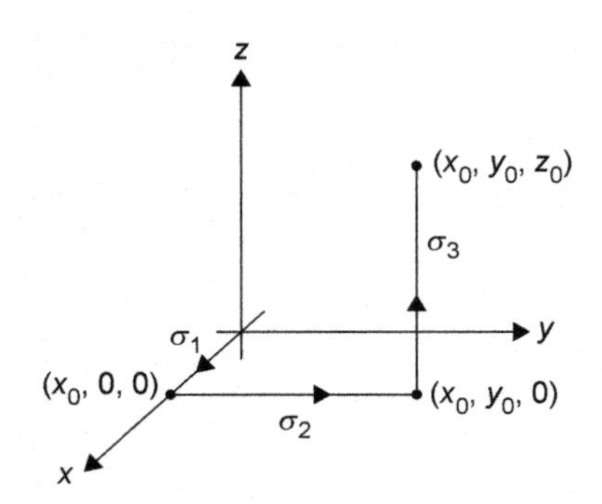

FIGURE 2.3.13

Then

$$\int_{\sigma} \vec{F} \cdot d\sigma = \int_{\sigma_1} \vec{F} \cdot d\sigma_1 + \int_{\sigma_2} \vec{F} \cdot d\sigma_2 + \int_{\sigma_3} \vec{F} \cdot d\sigma_3.$$

Now

$$\begin{aligned} \vec{F} \cdot d\sigma_1 &= \left(F_1(x,y,z)\hat{i} + F_2(x,y,z)\hat{j} + F_3(x,y,z)\hat{k}\right) \cdot \left(dx\,\hat{i} + 0\hat{j} + 0\hat{k}\right) \\ &= F_1(x,y,z)dx \end{aligned}$$

and similarly,
$\vec{F} \cdot d\sigma_2 = F_2(x,y,z)dy$ and $\vec{F} \cdot d\sigma_3 = F_3(x,y,z)dz$.
Note that on σ_1, $y = 0$ and $z = 0$ so

$$\int_{\sigma_1} \vec{F} \cdot d\sigma_1 = \int_{t=0}^{x} F_1(t,0,0)dt;$$

on σ_2, $x = x_0$ and $z = 0$, so

$$\int_{\sigma_2} \vec{F} \cdot d\sigma_2 = \int_{t=0}^{y} F_2(x_0,t,0)dt;$$

and on σ_3, $x = x_0$ and $y = y_0$, so

$$\int_{\sigma_3} \vec{F} \cdot d\sigma_3 = \int_{t=0}^{z} F_3(x_0,y_0,t)dt.$$

Thus,

$$\begin{aligned} f(x_0,y_0,z_0) &= \int_{\sigma} \vec{F} \cdot d\sigma \\ &= \int_{\sigma_1} \vec{F} \cdot d\sigma_1 + \int_{\sigma_2} \vec{F} \cdot d\sigma_2 + \int_{\sigma_3} \vec{F} \cdot d\sigma_3 \\ &= \int_{t=0}^{x_0} F_1(t,0,0)dt + \int_{t=0}^{y_0} F_2(x_0,t,0)dt + \int_{t=0}^{z_0} F_3(x_0,y_0,t)dt \end{aligned}$$

so

$$f(x_0, y_0, z_0 + \Delta z) = \int_{t=0}^{x_0} F_1(t, 0, 0)dt + \int_{t=0}^{y_0} F_2(x_0, t, 0)dt + \int_{t=0}^{z_0+\Delta z} F_3(x_0, y_0, t)dt.$$

Then

$$\begin{aligned} f(x_0, y_0, z_0 + \Delta z) - f(x_0, y_0, z_0) &= \int_{t=0}^{z_0+\Delta z} F_3(x_0, y_0, t)dt - \int_{t=0}^{z_0} F_3(x_0, y_0, t)dt \\ &= \int_{t=z_0}^{z_0+\Delta z} F_3(x_0, y_0, t)dt = F_3(x_0, y_0, \bar{z})\Delta z \end{aligned}$$

for some $\bar{z} \in [z_0, z_0 + \Delta z]$. So

$$\frac{f(x_0, y_0, z_0 + \Delta z) - f(x_0, y_0, z_0)}{\Delta z} = F_3(x_0, y_0, \bar{z})$$

and

$$\frac{\partial f}{\partial z} = \lim_{\Delta z \to 0} \frac{f(x, y, z + \Delta z) - f(x, y, z)}{\Delta z} = F_3(x, y, z).$$

To show that $\dfrac{\partial f}{\partial y} = F_2(x, y, z)$, let τ be the path from (0, 0, 0) to (x_0, y_0, z_0) given by $\tau = \tau_1 + \tau_2 + \tau_3$ where

τ_1 is the path along the z-axis from (0, 0, 0) to (0, 0, z_0).
τ_2 is the path parallel to the x-axis from (0, 0, z_0) to (x_0, 0, z_0).
τ_3 is the path parallel to the z-axis from (x_0, 0, z_0) to (x_0, y_0, z_0). See Fig. 2.3.14.

Similar to the analysis above, we get

$$f(x_0, y_0, z_0) = \int_{t=0}^{z_0} F_3(0, 0, t)dt + \int_{t=0}^{x_0} F_1(t, 0, z_0)dt + \int_{t=0}^{y_0} F_3(x_0, t, z_0)dt$$

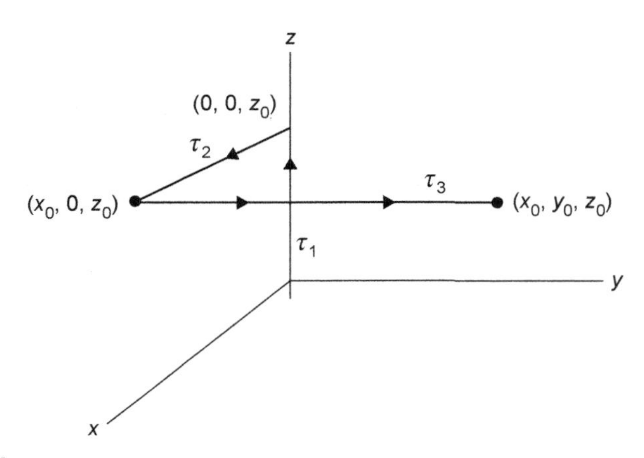

FIGURE 2.3.14

and

$$f(x_0, y_0 + \Delta y, z) = \int_{t=0}^{z_0} F_3(0,0,t)dt + \int_{t=0}^{x_0} F_1(t,0,z_0)dt + \int_{t=0}^{y_0+\Delta y} F_3(x_0,t,z_0)dt$$

so

$$f(x_0, y_0 + \Delta y, z_0) - f(x_0, y_0, z_0) = \int_{t=y_0}^{y_0+\Delta y} F_3(x_0,t,z_0)dt.$$

From this, we can arrive at $\frac{\partial f}{\partial y} = F_2(x,y,z)$.

In Exercise 29, we show $\frac{\partial f}{\partial x} = F_1(x,y,z)$.

Example:

Show that the vector field

$$\vec{F}(x,y,z) = (2xyz + \sin x)\hat{i} + x^2z\hat{j} + x^2y\hat{k}$$

is conservative, and find a function $f(x, y, z)$ for which $\vec{F} = \nabla f$.

We show that $\vec{F}$ is conservative by showing that $\nabla \times \vec{F} = 0$. We have

$$\nabla \times \vec{F} = \begin{vmatrix} \hat{i} & \hat{j} & \hat{k} \\ \frac{\partial}{\partial x} & \frac{\partial}{\partial y} & \frac{\partial}{\partial z} \\ 2xyz + \sin x & x^2z & x^2y \end{vmatrix}$$

$$= \left[\frac{\partial}{\partial y}(x^2y) - \frac{\partial}{\partial z}(x^2z)\right]\hat{i} - \left[\frac{\partial}{\partial x}(x^2y) - \frac{\partial}{\partial z}(2xyz + \sin x)\right]\hat{j}$$

$$+ \left[\frac{\partial}{\partial x}(x^2z) - \frac{\partial}{\partial y}(2xyz + \sin x)\right]\hat{k}$$

$$= (x^2 - x^2)\hat{i} - (2xy - 2xy)\hat{j} + (2xz - 2xz)\hat{k} = \hat{0}.$$

We seek a function f for which $\nabla f = \frac{\partial f}{\partial x}\hat{i} + \frac{\partial f}{\partial y}\hat{j} + \frac{\partial f}{\partial z}\hat{k} = \vec{F}$. If we have

$$\frac{\partial f}{\partial x}\hat{i} + \frac{\partial f}{\partial y}\hat{j} + \frac{\partial f}{\partial z}\hat{k} = \vec{F} = F_1\hat{i} + F_2\hat{j} + F_3\hat{k}.$$

Then we must have

$$\frac{\partial f}{\partial x} = F_1, \quad \frac{\partial f}{\partial y} = F_2, \quad \frac{\partial f}{\partial z} = F_3.$$

In our example, this is

$$\frac{\partial f}{\partial x} = 2xyz + \sin x, \quad \frac{\partial f}{\partial y} = x^2 z, \quad \frac{\partial f}{\partial z} = x^2 y.$$

We begin our analysis with $\frac{\partial f}{\partial y} = x^2 z$. From this, we get

$$f(x, y, z) = \int \frac{\partial f}{\partial y} dy = \int (x^2 z) dy.$$

Here, we must be careful in integrating with respect to y because the constant of integration could be a function of x or z because $\frac{\partial}{\partial y} g(x, z) = 0$. Thus,

$$f(x, y, z) = \int (x^2 z) dy = x^2 yz + g(x, z).$$

Next we use the condition $\frac{\partial f}{\partial x} = 2xyz + \sin x$. We have

$$\frac{\partial f}{\partial x} = 2yz + \frac{\partial}{\partial x} g(x, z) \text{ and } \frac{\partial f}{\partial x} = F_1 = 2xyz + \sin x.$$

Thus $\frac{\partial}{\partial x} g(x, z) = \sin x$, and so $g(x, z) = -\cos x + h(z)$. Thus, we write

$$f(x, y, z) = x^2 yz - \cos x + h(z).$$

Finally, we use the condition $\frac{\partial f}{\partial z} = x^2 y$. We have

$$\frac{\partial f}{\partial z} = x^2 y + \frac{d}{dz} h(z) \text{ and } \frac{\partial f}{\partial z} = F_3 = x^2 y.$$

Thus $\frac{d}{dz} h(z) = 0$, so h is a constant. We conclude

$$f(x, y, z) = x^2 yz - \cos x + C.$$

Theorem:

Suppose that $\vec{F}$ is a conservative vector field for which $\vec{F} = \nabla f$. Then

$$\int_{(x_0, y_0, z_0)}^{(x_1, y_1, z_1)} \vec{F} \cdot d\hat{r} = \int_{(x_0, y_0, z_0)}^{(x_1, y_1, z_1)} \nabla f \cdot d\hat{r} = f(x_1, y_1, z_1) - f(x_0, y_0, z_0).$$

Example:

The work done by the force field $\vec{F}(x, y, z) = (2xyz + \sin x)\hat{i} + x^2 z\hat{j} + x^2 y\hat{k}$ in moving a particle from (10, −1, 4) to (10, 3, 4) is

$$\int_{(0, -1, 4)}^{(0, 3, 4)} \left[(2xyz + \sin x)\hat{i} + x^2 z\hat{j} + x^2 y\hat{k}\right] \cdot d\hat{r} = x^2 yz - \cos x\Big|_{(10, -1, 4)}^{(10, 3, 4)}$$
$$= 1200 + 400$$

since $(2xyz + \sin x)\hat{i} + x^2 z\hat{j} + x^2 y\hat{k} = \nabla(x^2 yz - \cos x)$.

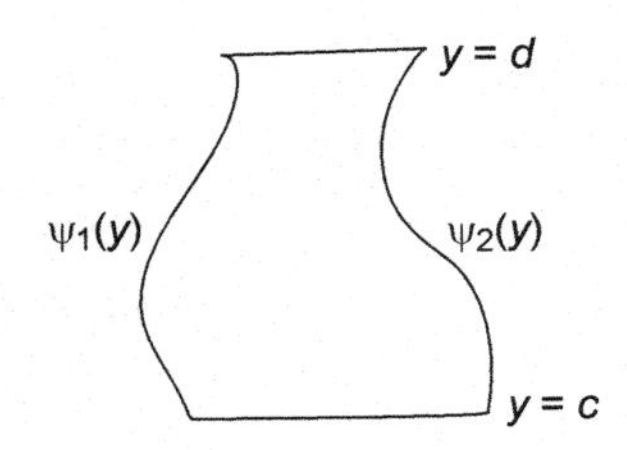

FIGURE 2.3.15

EXERCISES

1. Find a unit normal vector to the following surfaces at the point (x_0, y_0):
   a. $z = 9 - x^2 - y^2$.
   b. $x^2 - y^2 + z^2 = 1$.
   c. $z = \sqrt{1 - x^2}$.
2. Derive an expression for a unit normal vector to the surface $y = g(x, z)$.
3. Show that for the region shown in Fig. 2.3.15

$$\oint_C Q(x,y)dy = \iint_S \frac{\partial Q(x,y)}{\partial x} dxdy.$$

4. Show that in $\mathbb{R}^3$ if $S_\varepsilon = \{x | \|x\| < \varepsilon\}$ and $f : \mathbb{R}^3 \to \mathbb{R}$ is bounded, then

$$\lim_{\varepsilon \downarrow 0} \iiint_{S_\varepsilon} \|x\|^{-2} f(x) dS = 0.$$

5. Show that

$$\widehat{n} = \frac{(y'(t), -x'(t))}{\sqrt{(x'(t))^2 + (y'(t))^2}}$$

is a unit vector normal to $\sigma'(t)$ where $\sigma(t) = (x(t), y(t))$.

6. If

$$\widehat{u} = u_1 \widehat{i} + u_2 \widehat{j} + u_3 \widehat{k} \quad \text{and} \quad \widehat{n} = \cos \alpha \widehat{i} + \cos \beta \widehat{j} + \cos \gamma \widehat{k}$$

where α, β, and γ are the angles that $\widehat{n}$ makes with the positive x, y, and z axes respectively, then

$$\iint_S (u_1 \cos \alpha + u_2 \cos \beta + u_3 \cos \gamma) dS = \iint_S (u_1 dydz + u_2 dxdz + u_3 dxdy).$$

Problems on Green's theorem

7. Verify Green's theorem for the following examples. Each boundary C is assumed to be positively oriented.

a. $\oint_C x^2y\,dx + xy^3dy$ where C is the rectangle whose vertices are (0, 0), (2, 0), (2, 2), and (0, 2).

b.

i. $\oint_C y^3\,dx - x^3dy$ where C is the circle $x^2+y^2=1$.

ii. $\oint_C y^3\,dx - x^3dy$ where the region is between $x^2+y^2=1$ and $x^2+y^2=4$.

c. $\oint_C y^2\,dx + x^2dy$ where the region is between $y=x$ and $y=x^2$.

8. Evaluate using Green's theorem:

a. $\oint_C 2xy\,dx + (x+y)dy$ where the region is between $y=0$ and $y=1-x^2$.

b. $\oint_C e^x\cos y\,dx - e^x\sin y dy$ where C is the circle $x^2+y^2=1$.

c. $\oint_C (2y+3x)dx + (y-4x)dy$ where C is the circle $x^2+y^2=9$.

9. Suppose that $f(x,y)$ has continuous second partial derivatives and

$$\frac{\partial^2 f}{\partial x^2} + \frac{\partial^2 f}{\partial y^2} = 0.$$

Show that

$$\int_{\partial D} \frac{\partial f}{\partial y} dx - \frac{\partial f}{\partial x} dy = 0$$

where D is any circle and ∂D is the boundary of D.

Problems on Stokes' Theorem

10. Verify Stokes' theorem in the following cases:

a. $\hat{F} = y\hat{i} + 2x\hat{j} + \hat{k},\ S: z = \sqrt{9-x^2-y^2},\quad z \geq 0.$

b. $\hat{F} = 2y\hat{i} + 3x\hat{j} - z\hat{k},\ S:$ upper half of the sphere $x^2+y^2+z^2=16$.

c. $\hat{F} = z\hat{i} + x\hat{j},\ S$ is the square $0 \leq x \leq 4,\quad 0 \leq y \leq 4,\quad z = 2.$

d. $\hat{F} = -y^2\hat{i} + x^2\hat{j},\quad S: x^2+y^2 \leq 25,\quad z = 0.$

e. $\hat{F} = -y\hat{i} + 2x\hat{j} + (x+z)\hat{k},\ S:$ upper half of the sphere $x^2+y^2+z^2=1$.

11. (Green's first identity) Show that if f and g are scalar functions with continuous partial derivatives in a region R, and if V is a region within R with surface S, then

$$\iiint_V (f\nabla^2 g + \nabla f \cdot \nabla g)dV = \iint_S f\frac{\partial g}{\partial n} dS$$

where $\frac{\partial g}{\partial n} = \nabla g \cdot n$, the directional derivative of g in the direction that is outward normal to the surface S.

Use this result to show

$$\iiint_V |\nabla f|^2 dV = \iint_S \frac{\partial f}{\partial n} dS.$$

12. (Green's second identity) Show that if f and g are scalar functions with continuous partial derivatives in a region R, and if V is a region within R with surface S, then

$$\iiint_V (f\nabla^2 g - g\nabla^2 f) dV = \iint_S \left(f\frac{\partial g}{\partial n} - g\frac{\partial f}{\partial n}\right) dS.$$

13. The flux of a vector field $\widehat{F}$ is the flow rate of a quantity through a unit area. If S is a surface and $\widehat{n}$ is the outward directed unit normal to the surface, then the rate of flow through the surface S is given by

$$\iint_S \widehat{F}\cdot\widehat{n} dS \equiv \iint_S \widehat{F}\cdot d\widehat{S}.$$

Suppose we have a material whose density is ρ that flows with velocity $v\widehat{k}$ through a hemisphere whose equation is

$$z = \sqrt{R^2 - x^2 - y^2}.$$

a. Show that the unit normal to the surface is

$$\widehat{n} = \frac{1}{R}\left(x\widehat{i} + y\widehat{j} + z\widehat{k}\right).$$

b. Show that $\widehat{F}\cdot\widehat{n} = \rho\, vz/R$.

c. Show that the flow rate through S is $\rho v\pi R^2$.

14. Show that if S is any closed surface and $\widehat{F}$ is a vector field, then

$$\iint_S \left(\text{curl } \widehat{F}\right)\cdot\widehat{n} dS = 0.$$

15. Show that for a differentiable function $f(x, y, z)$, $\nabla \times \nabla f = 0$.

16. Verify that $\widehat{F}$ is a conservative force, and find a function f for which $\widehat{F} = \nabla f$ in the following cases;

a. $\widehat{F} = \left(2xz^3 + 6y\right)\widehat{i} + (6x - 2yz)\widehat{j} + \left(3x^2z^2 - y^2\right)\widehat{k}$.

b. $\widehat{F} = (2xy + 3)\widehat{i} + \left(x^2 - 4z\right)\widehat{j} + 4y\widehat{k}$.

17. Use Stokes' theorem to evaluate $\oint_C \widehat{F}\cdot d\widehat{r}$ in the following cases:

a. $\widehat{F} = y\widehat{i} + 2x\widehat{j} + 2z\widehat{k}$, $C: x^2 + y^2 = 4$ in the $z = 1$ plane.

b. $\widehat{F} = x^3y^4\widehat{i} + 3\widehat{j} + 2z\widehat{k}$, $C: x^2 + y^2 = 9$ in the $z = 0$ plane.

c. $\widehat{F} = (2x - y)\widehat{i} + (4x + 3z)\widehat{j} + (2y - z)\widehat{k}$, C is the boundary of the triangle whose vertices are (0, 0, 2), (0, 2, 0), (2, 0, 0).

18. Faraday's law states that if a current I passes through a loop C encloses that encloses a region S, then the electromotive force around the loop is the negative of the rate of change of the magnetic flux through the surface. As an equation, this is stated

$$\oint_C \widehat{E}\cdot d\widehat{l} = -\frac{\partial}{\partial t}\iint_S \widehat{B}\cdot\widehat{n}\,dS$$

where $\widehat{n}$ is the normal vector to S.

Use Stokes' theorem to show that Faraday's law implies

$$\nabla\times\widehat{E} = -\frac{\partial\widehat{B}}{\partial t}$$

which is one of the Maxwell's equations.

19. The magnetic field $\widehat{B}$ due to a current loop is given by $\widehat{B} = \text{curl}\left(\widehat{A}\right)$ where

$$\widehat{A} = -\frac{y}{(x^2+y^2+z^2)^{\frac{3}{2}}}\widehat{i} + \frac{x}{(x^2+y^2+z^2)^{\frac{3}{2}}}\widehat{j}.$$

For C a circle of radius a centered at $(0, 0, \alpha)$ that is parallel to the xy-plane, use Stokes' theorem to calculate the flux of $\widehat{B}$ through C for α large.

20. A solenoid is a wound spiral of wire (see Fig. 2.3.16). A magnetic field $\widehat{B}$ is created when current flows through the wire. We orient the solenoid so that the axis lies along the z-axis. If the solenoid is infinitely long, then $\widehat{B}$ is zero outside the solenoid, and $\widehat{B} = \beta\widehat{k}$ inside the solenoid, where β is a constant that depends on the current strength and the spacing of the turns of the coil. Suppose that the radius of the coil is a.

Show that $\widehat{B} = \text{curl}\left(\widehat{A}\right)$ where

$$\widehat{A} = \begin{cases} \frac{1}{2}a^2\beta\left(-\frac{y}{r^2}\widehat{i} + \frac{x}{r^2}\widehat{j}\right) & \text{if } r > a \\ \frac{1}{2}\beta\left(-y\widehat{i} + x\widehat{j}\right) & \text{if } r < a \end{cases}$$

where $r = \sqrt{x^2+y^2}$.

21. Show that the flux of $\widehat{B}$ through a circle in the xy-plane of radius $r < a$ is $\pi r^2\beta$.

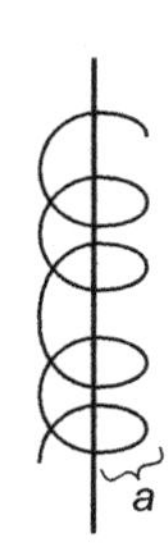

FIGURE 2.3.16

Problems on Divergence Theorem

22. Verify the divergence theorem in the following cases:

a. $\widehat{F} = y^2\widehat{i} + 2xy\widehat{j} + z^2\widehat{k}$ and S is the surface of the cylinder bounded by $x^2 + y^2 = 9$ and the planes $z = -2$ and $z = 3$.

b. $\widehat{F} = x\widehat{i} + y\widehat{j} + z\,\widehat{k}$ and S is the surface of the cube bounded by the planes $x = 0$ and $x = 1$, $y = 0$ and $y = 1$, and $z = 0$, and $z = 1$.

23. Evaluate the following integrals using the divergence theorem:

a. $\int (x+z)dydz + (y+z)dxdz + (x+y)dxdy$ where S is the surface of problem 10(a).

b. $\int x^2dydz + y^2dxdz + z^2dxdy$ where S is the surface of problem 10(b).

24. A function $f(x, y)$ is harmonic on S if

$$\frac{\partial^2 f}{\partial x^2} + \frac{\partial^2 f}{\partial y^2} = 0$$

on S. Show that if $f(x, y)$ is harmonic on S, then

$$\iint_S \frac{\partial f}{\partial n} dS = 0.$$

25. By letting $\widehat{F} = \nabla f$ in

$$\text{div } \vec{F}(p) = \lim_{V \to 0,\, p \in V} \frac{1}{V} \iint_S \vec{F} \cdot \widehat{n} dS$$

show that

$$\nabla^2 f = \lim_{V \to 0,\, p \in V} \frac{1}{V} \iint_S \frac{\partial f}{\partial n} dS.$$

Problems on Conservative Field

26. Show that the gravitational field

$$\vec{F}(x, y, z) = -\frac{GMm}{r^2}\widehat{e}_r$$

is conservative. To convert to Cartesian coordinates

$$\widehat{e}_r = \left(\frac{x}{\sqrt{x^2+y^2+z^2}}, \frac{y}{\sqrt{x^2+y^2+z^2}}, \frac{z}{\sqrt{x^2+y^2+z^2}} \right).$$

27. Show that the vector field $\vec{F}(x, y, z) = \left(xy^2 + 3x^2y\right)\widehat{i} + \left(x^3 + x^2y\right)\widehat{j} + 2\widehat{k}$ is conservative, and find a function $f(x, y, z)$ for which $\vec{F} = \nabla f$.

28. Show that the vector field $\vec{F}(x, y, z) = (6xy \pm 2z)\hat{i} + 3x^2\hat{j} + 2x\hat{k}$ is conservative, and find a function $f(x, y, z)$ for which $\vec{F} = \nabla f$.

29. Show that in the proof of the theorem of the equivalence of conditions for a vector field, $\dfrac{\partial f}{\partial x} = F_1(x, y, z)$.

30. Show that the divergence of the vector field

$$\vec{F}(x, y, z) = \frac{c}{(x^2 + y^2 + z^2)^{\frac{3}{2}}}\left(x\hat{i} + y\hat{j} + z\hat{k}\right)$$

is 0.

CHAPTER

Green's Functions

3

3.1 INTRODUCTION

In this chapter we discuss how to solve a nonhomogeneous linear second-order ordinary differential equation with given boundary conditions by presenting the solution in terms of an integral. More precisely, we find a function $G(x,t)$, for which the solution to

$$L[y] = a_0(x)y''(x) + a_1(x)y'(x) + a_0(x)y(x) = -f(x) \quad 0 < x < 1$$
$$y(0) = 0, \quad y(1) = 0 \tag{1}$$

is expressed as

$$y(x) = \int_0^1 G(x,t)f(t)dt.$$

CAUTION:

Some authors construct the solution for $L[y] = f(x)$. This means the Green's function we derive will be the negative of theirs. Our approach follows that given in Weinberger, *First Course in Partial Differential Equations* and Pinsky, *Partial Differential Equations and Boundary-Value Problems with Applications*: Third Edition among others.

The function $G(x,t)$ is called Green's function after the English mathematician George Green, who pioneered work in this area in the 1830s.

We shall demonstrate three methods of constructing Green's functions: using (1) the Dirac-delta function, (2) variation of parameters, and (3) eigenfunction expansions.

What we shall do is analogous to what one does in solving matrix equations in linear algebra. Namely, suppose A is an $n \times n$ matrix and we want to solve

$$A\widehat{u} = \widehat{f}, \tag{2}$$

where $\widehat{f}$ is a known $n \times 1$ vector and $\widehat{u}$ is an unknown vector to be determined. If A is invertible, then regardless of what $\widehat{f}$ is, Eq. (2) will have a unique solution given by

$$\widehat{u} = A^{-1}\widehat{f}.$$

Mathematical Physics with Partial Differential Equations. https://doi.org/10.1016/B978-0-12-814759-7.00003-X

In this setting A^{-1} exists if and only if 0 is not an eigenvalue of A.

The same situation will be true for Green's functions; that is, if 0 is not an eigenvalue of L, then we shall be able to construct the Green's function that provides the unique solution to $L[y] = f(x)$ with appropriate boundary conditions.

Unless stated otherwise, we assume that 0 *is not an eigenvalue of* L.

3.2 CONSTRUCTION OF GREEN'S FUNCTION USING THE DIRAC-DELTA FUNCTION

We first provide some intuitive motivation for this approach.

Let

$$L[y(x)] = a_0(x)y''(x) + a_1(x)y'(x) + a_2(x)y(x), \quad 0 \le x \le 1.$$

We want to solve

$$L[y(x)] = -f(x).$$

Fix t, $0 < t < 1$. Suppose there is a function $G(x,t)$ for which

$$L[G(x,t)] = -\delta(x-t). \tag{1}$$

Then, multiplying both sides of Eq. (1) by $f(t)$ and integrating with respect to t from $t = 0$ to $t = 1$, we get

$$\int_0^1 \{L[G(x,t)]\}f(t)dt = -\int_0^1 \delta(x-t)f(t)dt = -f(x).$$

Suppose further that

$$\int_0^1 \{L[G(x,t)]\}f(t)dt = L\left[\int_0^1 G(x,t)f(t)dt\right].$$

Then

$$L\left[\int_0^1 G(x,t)f(t)dt\right] = \int_0^1 \{L[G(x,t)]\}f(t)dt = -f(x).$$

So, if

$$y(x) = \int_0^1 G(x,t)f(t)dt$$

then

$$L[y(x)] = L\left[\int_0^1 G(x,t)f(t)dt\right] = -f(x).$$

The steps in the heuristic argument we have just given may seem plausible, but we are doing operations on $G(x,t)$ without knowing what $G(x,t)$ is. Next we hypothesize some conditions that $G(x,t)$ should satisfy, and show that these conditions are sufficient to determine $G(x,t)$.

After solving a problem using this method, we should check that the solution is valid.

To construct the Green's function $G(x,t)$ using the Dirac-delta function, we first fix t with $0 < t < 1$.

We then find $G(x,t)$ so that $G(x,t)$ satisfies the homogeneous equation $L(G(x,t)) = 0$ for $0 \le x < t$ and $t < x \le 1$. This means that $G(x,t)$ will have two parts: one for $x < t$ and one for $x > t$. To do this, one finds two linearly independent solutions to the homogeneous equation $L[y] = 0$. Suppose these two solutions are $y_1(x)$ and $y_2(x)$. Then there are constants, $c_i(t)$, $i = 1,\ldots,4$ that depend on t, for which

$$G(x,t) = \begin{cases} c_1(t)y_1(x) + c_2(t)y_2(x) & 0 \le x < t \\ c_3(t)y_1(x) + c_4(t)y_2(x) & t < x \le 1 \end{cases}.$$

Once we have found each $c_i(t)$, we know $G(x,t)$.

The procedure to find the $c_i(t)$ can be organized as follows:

1. The boundary conditions, $G(0,t) = y(0) = 0$ and $G(1,t) = y(1) = 0$ enable us to establish some relations among the $c_i(t)$.
2. $G(x,t)$ is continuous in x; in particular at $x = t$. This means

$$\lim_{x \uparrow t} G(x,t) = \lim_{x \downarrow t} G(x,t).$$

This establishes another relation among the $c_i(t)$.

3. $\dfrac{dG(x,t)}{dx}$ has a "jump" at $x = t$.

We now derive the magnitude of the jump.

Consider

$$a_0(x)y''(x) + a_1(x)y'(x) + a_2(x)y(x) = -\delta(x-t)$$

where the functions $a_i(x)$ are continuous and $a_0(x) \neq 0$. Let $\varepsilon > 0$ be given. Consider

$$\int_{t-\varepsilon}^{t+\varepsilon} [a_0(x)y''(x) + a_1(x)y'(x) + a_2(x)y(x)]dx = \int_{t-\varepsilon}^{t+\varepsilon} -\delta(x-t)dx = -1.$$

Since the functions $a_i(x)$ are continuous, $a_i(x) \approx a_i(t)$ if $x \ \varepsilon \ [t-\varepsilon, t+\varepsilon]$ and ε is small. In that case,

$$\int_{t-\varepsilon}^{t+\varepsilon} [a_0(x)y''(x) + a_1(x)y'(x) + a_2(x)y(x)]dx$$

$$\approx a_0(t)\int_{t-\varepsilon}^{t+\varepsilon} y''(x)dx + a_1(t)\int_{t-\varepsilon}^{t+\varepsilon} y'(x)dx + a_2(t)\int_{t-\varepsilon}^{t+\varepsilon} y(x)dx = -1.$$

Now

$$\int_{t-\varepsilon}^{t+\varepsilon} y'(x)dx = y(t+\varepsilon) - y(t-\varepsilon) \approx 0$$

since $y'(x)$ and $y(x)$ are continuous and ε is small. Likewise,

$$\int_{t-\varepsilon}^{t+\varepsilon} y(x)dx \approx 0.$$

Also

$$a_0(t)\int_{t-\varepsilon}^{t+\varepsilon} y''(x)dx = a_0(t)[y'(t+\varepsilon) - y'(t-\varepsilon)].$$

Taking the limit as $\varepsilon \to 0$ in

$$\int_{t-\varepsilon}^{t+\varepsilon} [a_0(x)y''(x) + a_1(x)y'(x) + a_2(x)y(x)]dx = \int_{t-\varepsilon}^{t+\varepsilon} -\delta(x-t)dx = -1$$

we get

$$\lim_{\varepsilon\to 0} a_0(t)[y'(t+\varepsilon) - y'(t-\varepsilon)] = a_0(t)\left[\lim_{t\downarrow x}\frac{dG(x,t)}{dx} - \lim_{t\uparrow x}\frac{dG(x,t)}{dx}\right] = -1.$$

This gives the jump condition on $G(x,t)$; namely

$$\lim_{x\downarrow t}\frac{dG(x,t)}{dx} - \lim_{x\uparrow t}\frac{dG(x,t)}{dx} = -\frac{1}{a_0(t)}.$$

A final condition is that $G(x,t)$ be symmetric; i.e., $G(x,t) = G(t,x)$.

Recapping, there are five conditions that determine $G(x,t)$ for the differential equation given by Eq. (1):

1. $G(0,t) = 0.$
2. $G(1,t) = 0.$
3. $\lim_{x\uparrow t} G(x,t) = \lim_{x\downarrow t} G(x,t).$
4. $\lim_{x\downarrow t}\frac{dG(x,t)}{dx} - \lim_{x\uparrow t}\frac{dG(x,t)}{dx} = -\frac{1}{a_0(t)}.$
5. $G(x,t) = G(t,x).$

There is an intuitive way to see how the solution to $L[y] = -\delta(x-t)$ yields the solution

$$y(t) = -\int_0^1 f(x)G(x,t)dx$$

using the example

$$y''(x) = -f(x) \quad 0 < x < 1, \quad y(0) = y(1) = 0. \tag{2}$$

The explanation we give follows that in Stakgold, *Green's Functions and Boundary-Value Problems*.

A physical interpretation of Eq. (2) is we have a wire stretched between the points $x = 0$ and $x = 1$. At a point x between $x = 0$ and $x = 1$ a vertical pressure $f(x)$ is applied. (We assume that $f(x)$ is a continuous function.) The function $y(x)$ is the deflection of the string at the point x.

Choose a positive integer N and divide the interval [0,1] into N equal parts. The length of each part is then $1/N$. Let x_k denote the center of the kth subinterval and let Δx denote the length of each interval. We assume the deflection of the string due to the continuous pressure $f(x)$ is approximately equal to the deflection of the string due to the sum of forces of magnitude $f(x_k)\Delta x$ each concentrated at the point x_k. The defection at $x = t$ due to the single force $f(x_k)\Delta x$ concentrated at x_k is

$$G(t, x_k)f(x_k)\Delta x$$

and, by the principle of superposition, the deflection due to the sum of the concentrated forces is

$$\sum_{k=1}^{N} G(t, x_k)f(x_k)\Delta x. \tag{3}$$

The limit of expression (3) as $N \to \infty$ (equivalently, as $\Delta x \to 0$) is

$$\int_0^1 G(t, x)f(x)dx.$$

Note: For our examples we use the boundary conditions $y(0) = 0$, $y(1) = 0$. But analogous ideas work if we use boundary conditions such as $y'(0) = 0$, $y'(1) = 0$ as we do in some of the exercises.

Next, we give two examples that demonstrate the computations involved in finding the Green's function.

Example:

We find the Green's function for

$$y''(x) = -f(x), \quad y(0) = 0, \quad y(1) = 0$$

using the Dirac-delta function.

Two linearly independent solutions to the associated homogeneous equation

$$y''(x) = 0$$

are

$$y_1(x) = 1 \text{ and } y_2(x) = x.$$

Thus

$$G(x, t) = \begin{cases} c_1(t)1 + c_2(t)x & 0 \le x < t \\ c_3(t)1 + c_4(t)x & t < x \le 1 \end{cases}.$$

Since $G(0,t) = y(0) = 0$, we have $c_1(t) = 0$. Since $G(1,t) = y(1) = 0$ we have $c_3(t)1 + c_4(t)1 = 0$, so $c_4(t) = -\,c_3(t)$.

To this point we have

$$G(x,t) = \begin{cases} c_2(t)x & 0 \le x < t \\ c_3(t) - c_3(t)x & t < x \le 1 \end{cases}.$$

We now apply the continuity condition $\lim_{x \uparrow t} G(x,t) = \lim_{x \downarrow t} G(x,t)$ to get

$$c_2(t)t = c_3(t) - c_3(t)t$$

so

$$c_2(t) = \frac{c_3(t) - c_3(t)t}{t}.$$

Then

$$G(x,t) = \begin{cases} \left(\dfrac{c_3(t) - c_3(t)t}{t}\right)x & 0 \le x < t \\ c_3(t) - c_3(t)x & t < x \le 1 \end{cases}.$$

Finally, we apply the jump condition on the derivative of $G(x,t)$. In this example, $a_0(x) = 1$, so

$$\lim_{x \downarrow t} \frac{dG(x,t)}{dx} - \lim_{x \uparrow t} \frac{dG(x,t)}{dx} = -\frac{1}{a_0(t)} = -1.$$

Thus

$$-c_3(t) - \left(\frac{c_3(t) - c_3(t)t}{t}\right) = -1$$

so

$$c_3(t)t + (c_3(t) - c_3(t)t) = t \text{ and thus } c_3(t) = t.$$

Finally, we have

$$G(x,t) = \begin{cases} \left(\dfrac{t - t^2}{t}\right)x & 0 \le x < t \\ t - tx & t < x \le 1 \end{cases} = \begin{cases} (1-t)x & 0 \le x < t \\ t(1-x) & t < x \le 1 \end{cases}.$$

We demonstrate that the answer is correct in the case $f(x) = x^2$. We assert that

$$y(x) = \int_{t=0}^{x} t(1-x)t^2 dt + \int_{t=x}^{1} (1-t)xt^2 dt$$

solves the initial value problem

$$y''(x) = -x^2, \quad y(0) = 0, \quad y(1) = 0.$$

Now

$$\int_{t=0}^{x} t(1-x)t^2 dt = \frac{x^4(1-x)}{4}$$

and

$$\int_{t=x}^{1} (1-t)xt^2 dt = \frac{x(3x^4 - 4x^3 + 1)}{12}.$$

Then

$$y(x) = \frac{x^4(1-x)}{4} + \frac{x(3x^4 - 4x^3 + 1)}{12} = \frac{x - x^4}{12}$$

and

$$y''(x) = -x^2, \quad y(0) = 0, \quad y(1) = 0.$$

Example:
We find the Green's function for Bessel's equation.
The homogeneous form of Bessel's equation is

$$[xy'(x)]' - \frac{m^2}{x}y(x) = xy''(x) + y'(x) - \frac{m^2}{x}y(x) = 0, \quad y(0) = y(1) = 0. \quad (4)$$

Note: Even though $a_0(0) = 0$, the technique is valid for this example because $a_0(t) \neq 0$ if $t \neq 0$.

We find the Green's function for Bessel's equation by following the steps given above. We first find two linearly independent solutions of Eq. (4). We set $y(x) = x^k$ in Eq. (4) to get

$$\begin{aligned} &xk(k-1)x^{k-2} + kx^{k-1} - m^2x^{k-1} \\ &\quad = x^{k-1}\left[k(k-1) + k - m^2\right] = x^{k-1}\left(k^2 - m^2\right) = 0. \end{aligned}$$

The solutions to this equation are $k = \pm m$.
Thus two linearly independent solutions to Eq. (4) are

$$y_1(x) = x^m \text{ and } y_2(x) = x^{-m}.$$

The Green's function is then

$$G(x,t) = \begin{cases} c_1(t)x^m + c_2(t)x^{-m} & 0 \leq x < t \\ c_3(t)x^m + c_4(t)x^{-m} & t < x \leq 1 \end{cases}.$$

The boundary condition $G(0, t) = y(0) = 0$ forces $c_2(t) = 0$. Since $G(1, t) = y(1) = 0$, we have

$$c_3(t) + c_4(t) = 0, \text{ or } c_4(t) = -c_3(t).$$

To this point, we have

$$G(x,t)=\begin{cases} c_1(t)x^m & 0\le x<t \\ c_3(t)x^m-c_3(t)x^{-m} & t<x\le 1 \end{cases}.$$

We now use the fact that $G(x,t)$ is continuous at $x=t$; that is,

$$c_1(t)t^{\,m}=\lim_{x\uparrow t}G(x,t)=\lim_{x\downarrow t}G(x,t)=c_3(t)t^m-c_3(t)t^{-m}$$

to get

$$c_1(t)=c_3(t)-c_3(t)t^{-2m}.$$

Now we have

$$G(x,t)=\begin{cases} \left[c_3(t)-c_3(t)t^{-2m}\right]x^m & 0\le x<t \\ c_3(t)x^m-c_3(t)x^{-m} & t<x\le 1 \end{cases}.$$

We next use the jump condition on the derivative of $G(x,t)$.
We have

$$\frac{dG(x,t)}{dx}=\begin{cases} mx^{m-1}\left[c_3(t)-c_3(t)t^{-2m}\right] & 0\le x<t \\ mx^{m-1}c_3(t)+mx^{-m-1}c_3(t) & t<x\le 1 \end{cases}$$

so that

$$\lim_{x\downarrow t}\frac{dG(x,t)}{dx}=mt^{m-1}c_3(t)+mt^{-m-1}c_3(t)\ \text{ and}$$

$$\lim_{x\uparrow t}\frac{dG(x,t)}{dx}=mt^{m-1}c_3(t)-\ c_3(t)mt^{-m-1}.$$

Then

$$\begin{aligned}-\frac{1}{a_0(t)}=-\frac{1}{t}&=\lim_{x\downarrow t}\frac{dG(x,t)}{dx}-\lim_{x\uparrow t}\frac{dG(x,t)}{dx}\\ &=\left[mt^{m-1}c_3(t)+mt^{-m-1}c_3(t)\right]\\ &-\left[mt^{m-1}c_3(t)-c_3(t)mt^{-m-1}\right]=2c_3(t)mt^{-m-1}.\end{aligned}$$

So

$$c_3(t)=-\frac{t^m}{2m}.$$

Finally, we have

$$G(x,t)=\begin{cases} \left[c_3(t)-c_3(t)t^{-2m}\right]x^m & 0\le x<t \\ c_3(t)x^m-c_3(t)x^{-m} & t<x\le 1 \end{cases}=\begin{cases} x^m\left(\dfrac{t^{-m}-t^m}{2m}\right) & 0\le x<t \\ t^m\left(\dfrac{x^{-m}-x^m}{2m}\right) & t<x\le 1 \end{cases}.$$

We demonstrate that the Green's function gives the solution in the case $m = 3$ and $f(x) = x^2$. We have

$$\int_{t=0}^{1} G(x,t)f(t)dt = \int_{t=0}^{x} t^3\left(\frac{x^{-3} - x^3}{6}\right)t^2dt + \int_{t=x}^{1} x^3\left(\frac{t^{-3} - t^3}{6}\right)t^2dt.$$

Now

$$\int_{t=0}^{x} t^3\left(\frac{x^{-3} - x^3}{6}\right)t^2dt = \frac{x^3(1-x^6)}{36} \quad \text{and}$$

$$\int_{t=x}^{1} x^3\left(\frac{t^{-3} - t^3}{6}\right)t^2dt = \frac{x^3(x^6-1)}{36} - \frac{x^3 \ln x}{6}$$

so

$$y(x) = \frac{x^3(1-x^6)}{36} + \frac{x^3(x^6-1)}{36} - \frac{x^3 \ln x}{6}.$$

One can check that

$$x\frac{d^2}{dx^2}\left[\frac{x^3(1-x^6)}{36} + \frac{x^3(x^6-1)}{36} - \frac{x^3 \ln x}{6}\right]$$
$$+\frac{d}{dx}\left[\frac{x^3(1-x^6)}{36} + \frac{x^3(x^6-1)}{36} - \frac{x^3 \ln x}{6}\right]$$
$$-\frac{9}{x}\left[\frac{x^3(1-x^6)}{36} + \frac{x^3(x^6-1)}{36} - \frac{x^3 \ln x}{6}\right] = -x^2$$

and

$$\lim_{x\to 0} y(x) = 0 \quad \text{and} \quad y(1) = 0.$$

In Exercise 7 we show that the limit of the above Green's function as m approaches zero from above gives the Green's function for

$$[xy'(x)]' = 0.$$

EXERCISES

1. Find Green's function for $y''(x) + y(x) = -f(x)$ with the following initial conditions:
   a. $y(0) = y(1) = 0$
   b. $y(0) = y'(1) = 0$
   c. $y'(0) = y'(1) = 0.$
2. Solve $y''(x) + y(x) = x^2$, $y'(0) = y'(1) = 0$. Verify that your solution is valid.

3. Find Green's function for $y''(x) = -f(x)$ with the following initial conditions:
   a. $y(0) = y'(1) = 0$
   b. $y'(0) = y(1) = 0$.
4. Find the Green's function for $y''(x) + \frac{1}{4}y(x) = x,\ y'(0) = y(1) = 0$. Write the expression for the solution. If you have a computer algebra system (CAS), simplify the expression and verify the solution.
5. Find the Green's function for $y''(x) - y(x) = -f(x), \quad -\infty < x < \infty \ \lim_{x\to\pm\infty} y(x) = 0$.
6. Find the Green's function for $y''(x) - y(x) = -f(x),\ y(0) = y'(1) = 0$.
7. Does the limit of the Green's function for the Bessel equation as m approaches zero from above give the Green's function for $[xy'(x)]' = 0,\ y(0) = y(1) = 0$?

3.3 GREEN'S FUNCTION USING VARIATION OF PARAMETERS

A second method for constructing Green's function is based on the following theorem that uses the technique called variation of parameters to find a particular solution to certain second-order differential equations. A proof to the theorem can be found in many ordinary differential equation textbooks including Boyce and DiPrima (2008).

Theorem:

Suppose that $y_1(x)$ and $y_2(x)$ are linearly independent solutions to

$$[p(x)y'(x)]' + q(x)y(x) = p(x)y''(x) + p'(x)y'(x) + q(x)y(x) = 0.$$

A particular solution to

$$[p(x)y'(x)]' + q(x)y(x) = -f(x)$$

is given by

$$y_p(x) = \int_0^x \left(\frac{y_1(x)y_2(t) - y_1(t)y_2(x)}{(y_1(t)y_2'(t) - y_2(t)y_1'(t))p(t)} \right) f(t)dt.$$

One can use this to construct the Green's function for

$$[p(x)y'(x)]' + q(x)y(x) = -f(x)$$

using the following steps:

1. Find two linearly independent solutions, $y_1(x)$ and $y_2(x)$ to

$$[p(x)y'(x)]' + q(x)y(x) = 0.$$

2. The general solution, $y(x)$, to

$$[p(x)y'(x)]' + q(x)y(x) = -f(x)$$

is

$$\begin{aligned} y(x) &= c_1 y_1(x) + c_2 y_2(x) + y_p(x) \\ &= c_1 y_1(x) + c_2 y_2(x) + \int_0^x \left(\frac{y_1(x)y_2(t) - y_1(t)y_2(x)}{(y_1(t)y_2'(t) - y_2(t)y_1'(t))p(t)} \right) f(t)dt. \end{aligned}$$

where the constants c_1 and c_2 are determined by the boundary conditions.

3. The expression for $y(x)$ is combined into a single integral.

Example:
We find the Green's function for

$$y''(x) = -f(x), \quad y(0) = 0, \quad y(1) = 0$$

using variation of parameters.

We have seen that two linearly independent solutions to the homogeneous equation are

$$y_1(x) = 1 \quad \text{and} \quad y_2(x) = x$$

and since $p(x) = 1$, we have

$$\frac{y_1(x)y_2(t) - y_1(t)y_2(x)}{(y_1(t)y_2'(t) - y_2(t)y_1'(t))p(t)} = \frac{1 \cdot t - 1 \cdot x}{1 - 0} = t - x.$$

Then

$$\begin{aligned} y(x) &= c_1 y_1(x) + c_2 y_2(x) + \int_0^x \left(\frac{y_1(x)y_2(t) - y_1(t)y_2(x)}{(y_1(t)y_2'(t) - y_2(t)y_1'(t))p(t)} \right) f(t)dt \\ &= c_1 + c_2 x + \int_0^x (t - x) f(t)dt. \end{aligned}$$

Since $y(0) = 0$, then $c_1 = 0$, so

$$y(x) = c_2 x + \int_0^x (t - x) f(t)dt.$$

Now

$$\begin{aligned} y(1) = 0 = c_2 \cdot 1 + \int_0^1 (t - 1) f(t)dt \quad \text{so} \\ c_2 = -\int_0^1 (t - 1) f(t)dt = \int_0^1 (1 - t) f(t)dt. \end{aligned}$$

Thus

$$y(x) = x\int_0^1 (1-t)f(t)dt + \int_0^x (t-x)f(t)dt$$
$$= x\int_0^x (1-t)f(t)dt + x\int_x^1 (1-t)f(t)dt + \int_0^x (t-x)f(t)dt$$
$$= \int_0^x [x(1-t) + (t-x)]f(t)dt + \int_x^1 x(1-t)f(t)dt$$
$$= \int_0^x [x - xt + t - x]f(t)dt + \int_x^1 x(1-t)f(t)dt$$
$$= \int_0^x [t(1-x)]f(t)dt + \int_x^1 x(1-t)f(t)dt.$$

Finally, we have

$$G(x,t) = \begin{cases} t(1-x) & t < x \\ x(1-t) & t > x \end{cases}.$$

Example:
We construct Green's function for Bessel's equation

$$[xy'(x)]' - \frac{m^2}{x}y(x) = xy''(x) + y'(x) - \frac{m^2}{x}y(x) = -f(x), y(0) = y(1) = 0$$

using variation of parameters.

We have seen that two linearly independent solutions to the associated homogeneous equation are

$$y_1(x) = x^m \quad \text{and} \quad y_2(x) = x^{-m}$$

so

$$y_1'(x) = mx^{m-1} \quad \text{and} \quad y_2'(x) = -mx^{-m-1}.$$

Then

$$y_1(t)y_2'(t) - y_2(t)y_1'(t) = (t^m)\left(-mt^{-m-1}\right) - \left(t^{-m}\right)\left(mt^{m-1}\right) = -2mt^{-1}.$$

So, since $p(x) = x$,

$$\left(y_1(t)y_2'(t) - y_2(t)y_1'(t)\right)p(t) = \left(y_1(t)y_2'(t) - y_2(t)y_1'(t)\right)t = -2m.$$

Also

$$y_1(x)y_2(t) - y_1(t)y_2(x) = x^m t^{-m} - t^m x^{-m}.$$

So the particular solution, $y_p(x)$, is

$$y_p(x) = \int_0^x \frac{\left(t^m x^{-m} - x^m t^{-m}\right)}{2m} f(t)dt$$

and the general solution is

$$\begin{aligned} y(x) &= c_1 y_1(x) + c_2 y_2(x) + y_p(x) \\ &= c_1 x^m + c_2 x^{-m} + \int_0^x \frac{(t^m x^{-m} - x^m t^{-m})}{2m} f(t)dt. \end{aligned}$$

We use the boundary conditions to determine c_1 and c_2.
Since $y(0) = 0$, then $c_2 = 0$, so

$$y(x) = c_1 x^m + \int_0^x \frac{(t^m x^{-m} - x^m t^{-m})}{2m} f(t)dt.$$

Since $y(1) = 0$,

$$0 = c_1 + \int_0^1 \frac{1}{2m}(t^m - t^{-m}) f(t)dt$$

and

$$c_1 = \int_0^1 \frac{1}{2m}(t^{-m} - t^m) f(t)dt.$$

Thus

$$y(x) = x^m \int_0^1 \frac{1}{2m}(t^{-m} - t^m) f(t)dt + \int_0^x \frac{(t^m x^{-m} - x^m t^{-m})}{2m} f(t)dt.$$

We have

$$x^m \int_0^1 \frac{1}{2m}(t^{-m} - t^m) f(t)dt = \int_0^x \frac{x^m}{2m}(t^{-m} - t^m) f(t)dt + \int_x^1 \frac{x^m}{2m}(t^{-m} - t^m) f(t)dt$$

so

$$\begin{aligned} y(x) &= \int_0^x \frac{x^m}{2m}(t^{-m} - t^m) f(t)dt + \int_x^1 \frac{x^m}{2m}(t^{-m} - t^m) f(t)dt \\ &\quad + \int_0^x \frac{(t^m x^{-m} - x^m t^{-m})}{2m} f(t)dt \\ &= \int_0^x \frac{1}{2m}(x^m t^{-m} - x^m t^m + t^m x^{-m} - x^m t^{-m}) f(t)dt \\ &\quad + \int_x^1 \frac{1}{2m} x^m (t^{-m} - t^m) f(t)dt \\ &= \int_0^x \frac{1}{2m} t^m (x^m - x^{-m}) f(t)dt + \int_x^1 \frac{1}{2m} x^m (t^{-m} - t^m) f(t)dt \\ &= \int_0^1 G(x,t) f(t)dt \end{aligned}$$

where

$$G(x,t) = \begin{cases} \frac{1}{2m} t^m \left(x^m - x^{-m}\right) & t < x \\ \frac{1}{2m} x^m \left(t^{-m} - t^m\right) & x < t \end{cases}.$$

This appears different from the solution we derived using the Dirac-delta function. But, because $G(x,t) = G(t,x)$, they are equal.

EXERCISES

Use variation of parameters to solve the following problems.

1. Find Green's function for $y''(x) + y(x) = -f(x)$ with the following initial conditions:
   a. $y(0) = y(1) = 0$
   b. $y(0) = y'(1) = 0$
   c. $y'(0) = y'(1) = 0$.
2. Solve $y''(x) + y(x) = x^2$, $y'(0) = y'(1) = 0$.
3. Find Green's function for $y''(x) = -f(x)$ with the following initial conditions:
   a. $y(0) = y'(1) = 0$
   b. $y'(0) = y'(1) = 0$.
4. Solve $y''(x) + \frac{1}{4}y(x) = \sin x, \quad y(0) = y(1) = 0$.
5. Find the Green's function for $y''(x) - y(x) = -f(x)$, $y(0) = y(1) = 0$.

3.4 CONSTRUCTION OF GREEN'S FUNCTION FROM EIGENFUNCTIONS

We continue to consider the operator

$$L[y(x)] = [p(x)y'(x)]' + q(x)y(x).$$

We want to solve

$$L[y(x)] = -f(x)$$
$$y(0) = 0, \quad y(1) = 0.$$

Suppose that $\{\phi_n\}$ is a complete orthonormal basis for the vector space consisting of eigenvectors of L that satisfy the boundary conditions, and that λ_n is the eigenvalue of ϕ_n. We assume the inner product

$$\langle f(x), g(x) \rangle = \int_0^1 f(x)g(x)dx.$$

Then we have

$$f(x) = \sum f_n\phi_n(x), \quad y(x) = \sum y_n\phi_n(x)$$

where

$$f_n = \langle f(x), \phi_n(x)\rangle = \int_0^1 f(x)\phi_n(x)dx.$$

We must find the y_n's.

Since L is a linear operator and $\phi_n(x)$ is an eigenvector of L with eigenvalue λ_n for each n, we have

$$L[y(x)] = L\Big(\sum y_n\phi_n(x)\Big) = \sum y_n\lambda_n\phi_n(x).$$

So, from $L[y(x)] = -f(x)$, we get

$$\sum y_n\lambda_n\phi_n(x) = -\sum f_n\phi_n(x),$$

and, since $\{\phi_n(x)\}$ is a basis,

$$y_n\lambda_n = -f_n.$$

Now, 0 is not an eigenvalue, so

$$y_n = -\frac{f_n}{\lambda_n}$$

and thus

$$y(x) = \sum y_n\phi_n(x) = -\sum \frac{f_n}{\lambda_n}\phi_n(x).$$

We now want to find $G(x,t)$ so that

$$y(x) = \int_0^1 G(x,t)f(t)dt;$$

that is, we want to find the Green's function.

Recall

$$f_n = \int_0^1 f(t)\phi_n(t)dt$$

so

$$y(x) = -\sum \frac{\phi_n(x)}{\lambda_n}f_n = -\sum \frac{\phi_n(x)}{\lambda_n}\int_0^1 f(t)\phi_n(t)dt = -\sum \int_0^1 \frac{\phi_n(x)\phi_n(t)}{\lambda_n}f(t)dt$$

$$= -\int_0^1 \Big(\sum \frac{\phi_n(x)\phi_n(t)}{\lambda_n}\Big)f(t)dt,$$

where we have assumed moving the summation inside the integral is legitimate.

Thus,

$$G(x,t) = -\sum \frac{\phi_n(x)\phi_n(t)}{\lambda_n}. \tag{1}$$

Clearly, to find the Green's function using this method, the crucial step is to find the eigenvalues and eigenfunctions for L that satisfy the initial conditions.

We note that some authors consider the problem

$$\overline{L}[y(x)] + \mu y(x) = [p(x)y'(x)]' + \overline{q}(x)y(x) + \mu y(x) = -f(x)$$

where μ is not an eigenvalue of $\overline{L}$ (but now 0 may be an eigenvalue of $\overline{L}$) and obtain the Green's function

$$\overline{G}(x,t) = -\sum \frac{\phi_n(x)\phi_n(t)}{\lambda_n - \mu}. \tag{2}$$

By adjusting either $q(x)$ or $\overline{q}(x)$, either problem can be changed into the other. The preference would be for the version for which the eigenvalues/eigenvectors are easier to find.

Example:

Use the eigenfunction expansion to find the Green's function for

$$y''(x) + y(x) = -f(x), \quad y(0) = 0, \quad y(1) = 0.$$

We consider the second form of the problem with $L[y(x)] = y''(x)$ and $\mu = 1$. The eigenvalues and eigenfunctions for

$$L[y(x)] = y''(x)$$

are

$$y(x) = \sin(\alpha x) \quad y(x) = \cos(\alpha x)$$

each with eigenvalue $-\alpha^2$. We now find the eigenvalues for which the eigenfunctions satisfy the initial conditions. Suppose

$$y(x) = A \ \sin(\alpha x) + B \ \cos(\alpha x)$$

and $y(0) = 0$. Then $B = 0$. If $y(1) = 0$, then

$$y(1) = 0 = A \sin \alpha$$

so $\alpha = n\pi$ where n is an integer. (Otherwise, we have only the trivial solution.) The eigenvalue of L for the eigenfunction $\psi_n(x) = \sin(n\pi x)$ is $-(n\pi)^2$. Since

$$\int_0^1 \sin^2(n\pi x)dx = \frac{1}{2}, \quad \text{then } \|\psi_n(x)\| = \frac{1}{\sqrt{2}}.$$

Thus $\{\sqrt{2}\sin(n\pi x)\} = \{\phi_n(x)\}$ is an orthonormal set of eigenfunctions for L that satisfy the initial conditions. So, according to Eq. (2), the Green's function for this example is

$$G(x,t) = -\sum \frac{\phi_n(x)\phi_n(t)}{\lambda_n - \mu} = -\sum \frac{[\sqrt{2}\sin(n\pi x)][\sqrt{2}\sin(n\pi t)]}{-(n\pi)^2 - 1}.$$

EXERCISES

Then use eigenfunction expansion to find the Green's function for the following problems.

1. Find Green's function for $y''(x) + y(x) = -f(x)$ with the following initial conditions:
   a. $y(0) = y'(1) = 0$
   b. $y'(0) = y'(1) = 0$.
2. Solve $y''(x) + y(x) = x^2$, $y'(0) = y'(1) = 0$.

3.5 MORE GENERAL BOUNDARY CONDITIONS

In our construction of the Green's functions we have restricted the initial conditions to the case $y(0) = 0$, $y(1) = 0$ to arrive at the solution to

$$[p(x)y'(x)]' + q(x)y(x) = -f(x), \quad y(0) = 0, \quad y(1) = 0 \tag{1}$$

is given by

$$y(x) = \int_0^1 G(x,t)f(t)dt.$$

Green's functions are also used to solve the more general problem

$$[p(x)y'(x)]' + q(x)y(x) = -f(x),$$

with boundary conditions

$$y(0) = a$$
$$y(1) = b.$$

To solve the more general problem, suppose that $y_1(x)$ and $y_2(x)$ are the linearly independent solutions we used to construct the Green's function. We want to find numbers α and β so that

$$\alpha y_1(0) + \beta y_2(0) = a$$
$$\alpha y_1(1) + \beta y_2(1) = b.$$

Then

$$y(x) = \int_0^1 G(x,t)f(t)dt + \alpha y_1(x) + \beta y_2(x)$$

is a solution to

$$[p(x)y'(x)]' + q(x)y(x) = -f(x); \quad y(0) = a, \ y(1) = b.$$

Example:
We have seen in problem 1(a) of Section 3.2 that the Green's function for

$$y''(x) + y(x) = -f(x), y(0) = 0, y(1) = 0$$

is

$$G(x,t) = \begin{cases} -\dfrac{\sin(t-1)}{\sin 1}\sin x, & 0 \le x < t \\ -\dfrac{\sin t}{\sin 1}\sin(x-1), & t < x \le 1 \end{cases}$$

where $y_1(x) = \cos x$ and $y_2(x) = \sin x$. Suppose we want to solve

$$y''(x) + y(x) = -f(x), \;\; y(0) = 3, \;\; y(1) = 2.$$

Then we need to find numbers α and β so that

$$\alpha \cos 0 + \beta \sin 0 = 3$$

$$\alpha \cos 1 + \beta \sin 1 = 2.$$

We find

$$\alpha = 3, \;\; \beta = \frac{2 - 3\cos 1}{\sin 1}$$

so that the solution is given by

$$\begin{aligned} y(x) &= \int_0^1 G(x,t)f(t)dt + \alpha y_1(x) + \beta y_2(x) \\ &= \int_0^x -\frac{\sin t}{\sin 1}\sin(x-1)f(t)dt + \int_x^1 -\frac{\sin(t-1)}{\sin 1}\sin x f(t)dt \\ &\quad + 3\cos x + \left(\frac{2-3\cos 1}{\sin 1}\right)\sin x. \end{aligned}$$

If you have a CAS, it is interesting to verify this for some choice of $f(t)$.

EXERCISES

Find the solution for the following initial value problems using Green's function. Note that the Green's function was determined in exercises from the previous section.

1. Find solution for $y''(x) + y(x) = -f(x)$ with the following initial conditions:
   a. $y(0) = 4$, $y(1) = -2$
   b. $y(0) = 6$, $y(1) = 4$.
2. Solve $y''(x) + y(x) = x^2$, $y(0) = 1$, $2y(1) = 5$.

3. Find solution for $y''(x) = -f(x)$ with the following initial conditions:
   a. $y(0) = 4,\ y(1) = 2$
   b. $y(0) = -3,\ y(1) = 7$.
4. Solve $y''(x) + \frac{1}{4}y(x) = \sin x, \quad y(0) = 3, y(1) = -1$.
5. Find the solution for $y''(x) - y(x) = -f(x),\ y(0) = 1,\ y(1) = -2$.
6. Describe how you would solve $y''(x) + y(x) = -f(x),\ \ y'(0) = 3,\ \ y'(1) = 5$.

3.6 THE FREDHOLM ALTERNATIVE (OR, WHAT IF 0 IS AN EIGENVALUE?)

Throughout this chapter, we have assumed that 0 is not an eigenvalue of the operator L. As was mentioned at the beginning of the chapter, if 0 is an eigenvalue of L then

$$L[y] = a_0(x)y''(x) + a_1(x)y'(x) + a_0(x)y(x) = -f(x)$$

may not have a solution. This case is known as Fredholm's alternative. To understand the problem at a more elementary level, we consider an example from linear algebra.

Suppose that $T : \mathbb{R}^3 \to \mathbb{R}^3$ is a linear operator with eigenvalue 0. Suppose that T is self-adjoint so we may choose a basis for which the matrix representation of T is diagonal. (This is not necessary, but it will make the ideas easier to follow.) Suppose that in this basis the matrix representation of T is

$$A = \begin{pmatrix} 1 & 0 & 0 \\ 0 & 2 & 0 \\ 0 & 0 & 0 \end{pmatrix}.$$

If $\widehat{b}$ is a nonzero vector in the eigenspace of 0, say

$$\widehat{b} = \begin{pmatrix} 0 \\ 0 \\ 5 \end{pmatrix},$$

then $A\widehat{x} = \widehat{b}$, that is

$$\begin{pmatrix} 1 & 0 & 0 \\ 0 & 2 & 0 \\ 0 & 0 & 0 \end{pmatrix}\begin{pmatrix} x \\ y \\ z \end{pmatrix} = \begin{pmatrix} 0 \\ 0 \\ 5 \end{pmatrix} \quad \text{or} \quad \begin{pmatrix} x \\ 2y \\ 0z \end{pmatrix} = \begin{pmatrix} 0 \\ 0 \\ 5 \end{pmatrix}$$

has no solution. If $\widehat{b}$ is in the orthogonal complement of the eigenspace of 0, say

$$\widehat{b} = \begin{pmatrix} 3 \\ 7 \\ 0 \end{pmatrix},$$

then $A\hat{x} = \hat{b}$, that is

$$\begin{pmatrix} 1 & 0 & 0 \\ 0 & 2 & 0 \\ 0 & 0 & 0 \end{pmatrix}\begin{pmatrix} x \\ y \\ z \end{pmatrix} = \begin{pmatrix} 3 \\ 7 \\ 0 \end{pmatrix} \quad \text{or} \quad \begin{pmatrix} x \\ 2y \\ 0z \end{pmatrix} = \begin{pmatrix} 3 \\ 7 \\ 0 \end{pmatrix}$$

has infinitely many solutions. There will be exactly one solution $\hat{x}$ if we add the requirement that the inner product of $\hat{x}$ and the eigenvector of 0 is 0.

The essential points of this example that will carry over to our setting are:

1. When 0 is an eigenvalue of the operator L, we may not be able to solve the equation

$$Ly = f$$

for y. We can, however, solve the equation

$$Ly = -(f - g) \tag{1}$$

where g is the projection of f onto the eigenspace of 0.

2. In the case that the dimension of the eigenspace of 0 is one, g is determined by

$$g = \langle f, \psi \rangle \psi$$

where ψ is a normalized eigenvector of 0.

3. The solution to Eq. (1) is not unique, but if we also require that $\langle y, \psi \rangle = 0$, then the solution is unique.

The following example demonstrates these ideas for the case

$$L[y(x)] = [p(x)y'(x)]' + q(x)y(x).$$

Suppose

$$y''(x) = f(x) \quad 0 < x < 1.$$

Solutions to the homogeneous equation are

$$y_1(x) = x \quad y_2(x) = 1$$

so that any function $\psi(x) = ax + b$ is an eigenfunction of $L[y(x)] = y''(x)$ with eigenvalue 0. We want to specify initial conditions that cause $\psi(x)$ to have only one parameter. We then choose a value of that parameter so that $\psi(x)$ will have norm 1. Some of the possible choices include

$$y(0) = 0, \ \ y'(1) = y(1) \ \text{ gives } \ \psi(x) = ax;$$

$$y'(0) = 0, \ \ y(0) = y(1) \ \text{ gives } \ \psi(x) = b;$$

$$y(0) = y'(0) \ \text{ gives } \ \psi(x) = ax + a.$$

We analyze the case $y(0) = y'(0)$. Then $\psi(x) = ax + a = a(x+1)$ is an eigenvector corresponding to the initial conditions. We determine a so that $\psi(x)$ will be a unit vector. We have

$$\langle \psi(x), \psi(x) \rangle = a^2 \int_0^1 (x+1)^2 dx = a^2 \frac{(x+1)^3}{3}\bigg|_0^1 = \frac{7a^2}{3}$$

so

$$\langle \psi(x), \psi(x) \rangle = 1 \quad \text{if } a = \pm\sqrt{\frac{3}{7}}.$$

We take $\psi(x) = \sqrt{\frac{3}{7}}(x+1)$. Now

$$\begin{aligned} f(x), \langle \psi(x)\psi(x) \rangle &= \psi(x) \int_0^1 f(t)\psi(t)dt = a(x+1) \int_0^1 f(t)a(t+1)dt \\ &= a^2(x+1) \int_0^1 (t+1)f(t)dt. \end{aligned}$$

Thus, we have

$$y''(x) = -\left[f(x) - a^2(x+1) \int_0^1 (t+1)f(t)dt \right].$$

Let

$$b = \int_0^1 (t+1)f(t)dt.$$

Then

$$y''(x) = -f(x) + a^2 b(x+1).$$

Note that if $y''(x) = F(x)$, then

$$y'(x) = \int_0^x F(t)dt + C$$

so

$$y'(x) = -\int_0^x f(z)dz + a^2 b \int_0^x (z+1)dz + C.$$

Then $y'(0) = C$. Now

$$\begin{aligned} y'(x) &= -\int_0^x f(z)dz + a^2 b \int_0^x (z+1)dz + C \\ &= -\int_0^x f(z)dz + a^2 b \left(\frac{x^2}{2} + x \right) + C. \end{aligned}$$

Also, if

$$y'(x) = G(x) = \int_0^x h(z)dz$$

then

$$y(x) = \int_0^x G(w)dw = \int_0^x \left(\int_0^w h(z)dz \right) dw + D.$$

Thus,

$$\begin{aligned} y(x) &= -\int_0^x \left(\int_0^w f(z)dz \right) dw + a^2 b \int_0^x \left(\frac{z^2}{2} + z \right) dz + Cx + D \\ &= -\int_0^x \left(\int_0^w f(z)dz \right) dw + a^2 b \left(\frac{x^3}{6} + \frac{x^2}{2} \right) + Cx + D. \end{aligned} \tag{2}$$

Note that $y(0) = D$. Since $y'(0) = C$ and $y(0) = y'(0)$, we have $C = D$. Next, we reverse the limits of integration in

$$\int_0^x \left(\int_0^w f(z)dz \right) dw.$$

The region of integration for the integral is shown in Fig. 3.6.1.

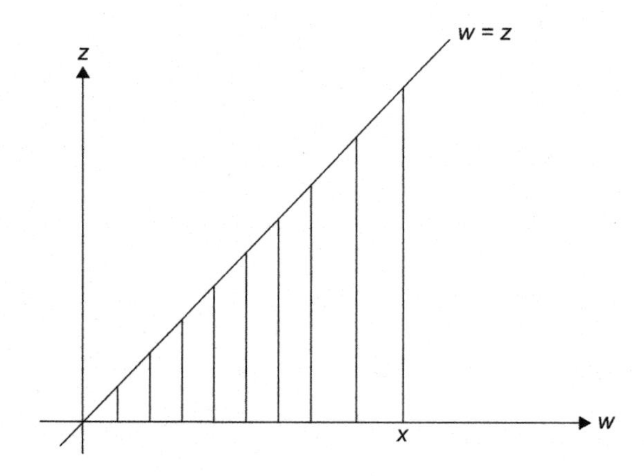

FIGURE 3.6.1

Region of integration

We have

$$\int_0^x \left(\int_0^w f(z)dz \right) dw = \int_0^x \left(\int_z^x f(z)dw \right) dz = \int_0^x (x-z)f(z)dz.$$

So, from Eq. (2) we have

$$\begin{aligned} y(x) &= -\int_0^x (x-z)f(z)dz + a^2 b\left(\frac{x^3}{6} + \frac{x^2}{2}\right) + Cx + C \\ &= \int_0^z (z-x)f(z)dz + a^2 b\left(\frac{x^3}{6} + \frac{x^2}{2}\right) + Cx + C. \end{aligned} \tag{3}$$

To determine C we use the orthogonality condition

$$\int_0^1 y(x)\psi(x)dx = 0.$$

This gives

$$\int_0^1 \left[\int_0^x (z-x)f(z)dz + a^2 b\left(\frac{x^3}{6} + \frac{x^2}{2}\right) + C(x+1) \right] a(x+1)dx = 0$$

or

$$\int_0^1 \left[\int_0^x (z-x)f(z)dz + a^2 b\left(\frac{x^3}{6} + \frac{x^2}{2}\right) + C(x+1) \right] (x+1)dx = 0.$$

We evaluate

$$\int_0^1 \left[\int_{z=0}^x (z-x)f(z)dz \right] (x+1)dx$$

by reversing the limits of integration. The region of integration is shown in Fig. 3.6.2.

We have

$$\int_0^1 \left[\int_{z=0}^x (z-x)f(z)dz \right] (x+1)dx = \int_{z=0}^1 \left[\int_{x=z}^1 (z-x)(x+1)dx \right] f(z)dz.$$

Now

$$\int_{x=z}^1 (z-x)(x+1)dx = -\frac{z^3}{6} - \frac{z^2}{2} + \frac{3}{2}z - \frac{5}{6}.$$

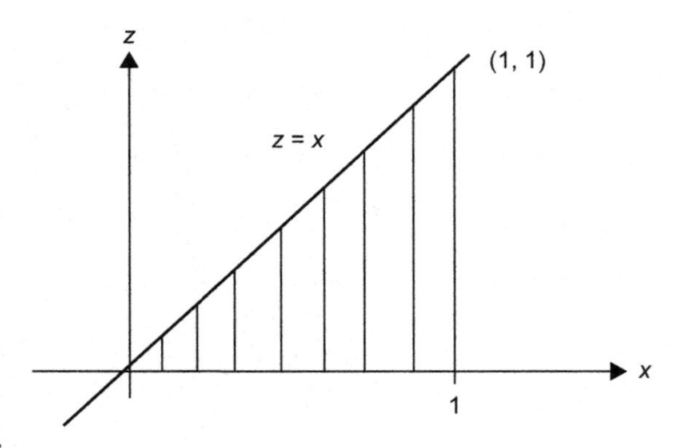

FIGURE 3.6.2
Region of integration

So,

$$\int_0^1 \left[\int_0^x (z-x)f(z)dz + a^2 b\left(\frac{x^3}{6}+\frac{x^2}{2}\right) + C(x+1) \right](x+1)dx$$

$$= \int_{z=0}^1 \left(-\frac{z^3}{6}-\frac{z^2}{2}+\frac{3}{2}z-\frac{5}{6}\right)f(z)dz$$

$$+ a^2 b\int_{x=0}^1 \left(\frac{x^3}{6}+\frac{x^2}{2}\right)(x+1)dx + C\int_{x=0}^1 (x+1)^2 dx = 0.$$

Now

$$\int_0^1 \left(\frac{x^3}{6}+\frac{x^2}{2}\right)(x+1)dx = \frac{11}{30} \quad \text{and} \quad \int_0^1 (x+1)^2 dx = \frac{7}{3}$$

so we have

$$\int_{z=0}^1 \left(-\frac{z^3}{6}-\frac{z^2}{2}+\frac{3}{2}z-\frac{5}{6}\right)f(z)dz$$

$$+a^2 b\int_{x=0}^1 \left(\frac{x^3}{6}+\frac{x^2}{2}\right)(x+1)dx + C\int_{x=0}^1 (x+1)^2 dx$$

$$= \int_{z=0}^1 \left(-\frac{z^3}{6}-\frac{z^2}{2}+\frac{3}{2}z-\frac{5}{6}\right)f(z)dz + \frac{11}{30}a^2 b + \frac{7}{3}C = 0.$$

Also,

$$\frac{11}{30}a^2b = \frac{11}{30}\cdot\frac{3}{7}\int_0^1 (t+1)f(t)dt = \frac{11}{70}\int_0^1 (t+1)f(t)dt.$$

We then have

$$\frac{7}{3}C = \int_{z=0}^1 \left(\frac{z^3}{6} + \frac{z^2}{2} - \frac{3}{2}z + \frac{5}{6}\right)f(z)dz - \frac{11}{70}\int_0^1 (t+1)f(t)dt$$

so

$$C = \frac{3}{7}\int_{z=0}^1 \left(\frac{z^3}{6} + \frac{z^2}{2} - \frac{174}{105}z + \frac{71}{105}\right)f(z)dz.$$

Thus,

$$y(x) = \int_0^x (z-x)f(z)dz + a^2b\left(\frac{x^3}{6} + \frac{x^2}{2}\right) + C(x+1)$$

$$= \int_0^x (z-x)f(z)dz + \frac{3}{7}\left(\frac{x^3}{6} + \frac{x^2}{2}\right)\left[\int_0^x (z+1)f(z)dz + \int_x^1 (z+1)f(z)dz\right] + (x+1)$$

$$+ \left[\frac{3}{7}\int_{z=0}^x \left(\frac{z^3}{6} + \frac{z^2}{2} - \frac{174}{105}z + \frac{71}{105}\right)f(z)dz + \frac{3}{7}\int_{z=x}^1 \left(\frac{z^3}{6} + \frac{z^2}{2} - \frac{174}{105}z + \frac{71}{105}\right)f(z)dz\right].$$

Thus, the Green's function is

$$G(x,z) = \begin{cases} (z-x) + \frac{3}{7}\left(\frac{x^3}{6} + \frac{x^2}{2}\right)(z+1) + \frac{3}{7}(x+1)\left(\frac{z^3}{6} + \frac{z^2}{2} - \frac{174}{105}z + \frac{71}{105}\right) & z < x \\ \frac{3}{7}\left(\frac{x^3}{6} + \frac{x^2}{2}\right)(z+1) + \frac{3}{7}(x+1)\left(\frac{z^3}{6} + \frac{z^2}{2} - \frac{174}{105}z + \frac{71}{105}\right) & z > x \end{cases}.$$

In the case where 0 is not an eigenvalue for the operator L we found that the Green's function for L is

$$G(x,t) = \sum_n \frac{\psi_n(x)\psi_n(t)}{\lambda_n}$$

where $\{\psi_n\}$ are orthonormal eigenvectors of L with the eigenvalue of ψ_n equal to λ_n.

We give an example to demonstrate that if 0 is an eigenvalue for L, then the modified Green's function is

$$\overline{G}(x,t) = \sum_{\lambda_n \neq 0} \frac{\psi_n(x)\psi_n(t)}{\lambda_n}.$$

Suppose L is an operator with an orthonormal basis of eigenvectors $\{\psi_1, \psi_2, \psi_3\}$ and

$$L\psi_1 = 2\psi_1, \quad L\psi_2 = 3\psi_2, \quad L\psi_3 = 0.$$

Now, $L[y] = f$ has a solution y if and only if $f = \alpha\psi_1 + \beta\psi_2$. We solve

$$L[y] = 6\psi_1 - 5\psi_2.$$

Let $y = a\psi_1 + \mathrm{b}\psi + c\psi_3$. We shall find a, b, c.
Now

$$L[y] = L(a\psi_1 + b\psi_2 + c\psi_3) = 2a\psi_1 + 3b\psi_2 = 6\psi_1 - 5\psi_2$$

so

$$a = \frac{6}{2} = \frac{\langle\psi_1, f\rangle}{\lambda_1} \qquad b = \frac{-5}{3} = \frac{\langle\psi_2, f\rangle}{\lambda_2}$$

and

$$y = \sum_{n=1}^{2} \frac{\langle\psi_n, f\rangle}{\lambda_n}\psi_n$$

which corresponds to

$$y(x) = \sum_{n=1}^{2} \frac{1}{\lambda_n}\psi_n(x) \int_0^1 \psi_n(t) f(t)dt = \int_0^1 \left(\sum_{n=1}^{2} \frac{\psi_n(x)\psi_n(t)}{\lambda_n} \right) f(t)dt.$$

The point of this example is if f is a function for which $L[y] = f$ has a solution, then the eigenvector whose eigenvalue is 0 does not appear in the Green's function.

EXERCISES

1. Repeat the example we computed for the initial conditions $y'(0) = 0$, $y(0) = y(1)$.
2. Repeat the example we computed for the initial conditions $y(0) = 0$, $y'(1) = y(1)$.

3.7 GREEN'S FUNCTION FOR THE LAPLACIAN IN HIGHER DIMENSIONS

One of the most frequently encountered operators is the negative Laplacian, $-\Delta$. In this section we derive the fundamental solution or the free-space Green's function (the situation where we have no boundary conditions) for the negative Laplacian in two and three dimensions.

We first note that

$$\Delta G(\hat{x}, \hat{x}_0) = \delta(\hat{x} - \hat{x}_0)$$

has no angular dependence, so Laplace's equation is invariant under rotation. Thus we have

$$G(\widehat{x}, \widehat{x}_0) = G(r)$$

where $r = |\widehat{x} - \widehat{x}_0|$.

We use the divergence theorem to derive the free-space Green's function for the negative Laplacian in two and three dimensions.

Consider

$$-\Delta E = -\left(\frac{\partial^2}{\partial x_1^2} + \frac{\partial^2}{\partial x_2^2} + \frac{\partial^2}{\partial x_3^2}\right)E = \delta(x_1, x_2, x_3).$$

A common interpretation for E is the electrostatic potential at a point (x_1, x_2, x_3) due to a unit charge at (0, 0, 0). This interpretation leads us to pose the problem in spherical coordinates and to assume that E depends only on r. We derived the formula for the Laplacian in spherical coordinates earlier, and if E depends only on r, then

$$\Delta E = \frac{1}{r^2}\frac{\partial}{\partial r}\left(r^2\frac{\partial E}{\partial r}\right).$$

Setting $\Delta E = 0$ yields

$$\frac{\partial}{\partial r}\left(r^2\frac{\partial E}{\partial r}\right) = 0 \quad \text{or} \quad r^2\frac{\partial E}{\partial r} = C.$$

So $-\Delta E = 0$ says

$$\frac{\partial E}{\partial r} = -\frac{C}{r^2},$$

and thus

$$E(r) = \frac{C}{r} + D.$$

If we assume (as with electrostatic potential) that $\lim_{r\to\infty} E(r) = 0$, then $E(r) = \frac{C}{r}$. To determine C, we shall appeal to the divergence theorem.

Recall that in spherical coordinates, if f is a function, then

$$\nabla f = \frac{\partial f}{\partial r}\widehat{e}_r + \frac{1}{r}\frac{\partial f}{\partial \theta}\widehat{e}_\theta + \frac{1}{r\sin\theta}\frac{\partial f}{\partial \phi}\widehat{e}_\varphi.$$

Since E depends only on r,

$$\nabla E = \frac{\partial E}{\partial r}\widehat{e}_r.$$

We relate this particular situation to the divergence theorem. The region Ω will be a sphere of radius ε, centered at the origin, which we denote R_ε. The boundary of

Ω, $\partial\Omega$, is the surface of this sphere and will be denoted ∂R_ε. The vector field F will be $\frac{\partial E}{\partial r}\widehat{e}_r$. Since our surface is that of sphere centered at the origin, the normal unit vector to the surface is $\widehat{e}_r$. Thus

$$\vec{F}\cdot\vec{n} = \left(\frac{\partial E}{\partial r}\right)\widehat{e}_r\cdot\widehat{e}_r = \frac{\partial E}{\partial r},$$

and since Ω is R_ε, and we shall be integrating where $r=\varepsilon$, so we shall be integrating $\left(\frac{\partial E}{\partial r}\right)_{r=\varepsilon}$. Thus, relating our problem to one of the integrals in the divergence theorem, we have

$$\iint_{\partial\Omega}(F\cdot n)dS = \int_{\partial R_\varepsilon}\left(\frac{\partial E}{\partial r}\right)_{r=\varepsilon}dS.$$

We want to find fundamental solution for

$$-\Delta E = \delta(x).$$

In our setting,

$$\text{div } F = \nabla\cdot F = \nabla\cdot\nabla E \equiv \Delta E,$$

so

$$\iiint_\Omega(\nabla\cdot F)dV = \iint_{\partial\Omega}F\cdot dS$$

is expressed in our setting as

$$-\iiint_{R_\varepsilon}\Delta E dV = -\iint_{\partial R_\varepsilon}\left(\frac{\partial E}{\partial r}\right)_{r=\varepsilon}dS.$$

But we have

$$-\Delta E = \delta(x),$$

so

$$-\iint_{\partial R_\varepsilon}\left(\frac{\partial E}{\partial r}\right)_{r=\varepsilon}dS = -\iiint_{R_\varepsilon}\Delta E dV = \iiint_{R_\varepsilon}\delta(x)dV = 1.$$

Note that the integral on the left is a surface integral and the singularity at $r=0$ does not come into play. The idea is the same as the example of Gauss's law for inverse square fields.

Back to the three-dimensional case. Since

$$E = \frac{C}{r},$$

we have

$$\left(\frac{\partial E}{\partial r}\right)_{r=\varepsilon} = \left(-\frac{C}{r^2}\right)_{r=\varepsilon} = -\frac{C}{\varepsilon^2},$$

so

$$-\frac{C}{\varepsilon^2}\iint_{\partial R_\varepsilon} 1\, dS = -\frac{C}{\varepsilon^2}\cdot 4\pi\varepsilon^2 = -4\pi C.$$

Thus,

$$-\iint_{\partial R_\varepsilon}\left(\frac{\partial E}{\partial r}\right)_{r=\varepsilon} dS = 4\pi C = 1,$$

so $C = \frac{1}{4\pi}$, and we have in three dimensions

$$-\Delta E = \delta(x_1, x_2, x_3) = \delta(\widehat{r}_0) \quad \text{if } E(\widehat{r}) = \frac{1}{4\pi|\widehat{r} - \widehat{r}_0|}.$$

We note an important connection. What we have just done is the equivalent of the jump condition in the one-dimensional case. Here is a sketch of the intuition: In both cases, we let the region about 0 $(\text{or } \widehat{0})$ collapse toward the origin. The property of the $\delta(x)$ function forces the integral of this region to always be 1. Speaking in an intuitive manner, in the one-dimensional case

$$\int_{-\varepsilon}^{\varepsilon} \frac{d^2 g}{dx^2} = g'(\varepsilon) - g'(-\varepsilon) = \int_{-\varepsilon}^{\varepsilon} \delta(x) = 1,$$

and in higher dimensions,

$$\int_{R_\varepsilon} \Delta g dV = \int_{\partial R_\varepsilon} (\nabla g)_{r=\varepsilon} dS = \int \delta(x) dV = 1.$$

In two dimensions, using polar coordinates and again assuming E depends only on r, we have

$$\Delta E = \frac{1}{r}\frac{\partial}{\partial r}\left(r\frac{\partial E}{\partial r}\right) = 0,$$

so

$$\frac{\partial}{\partial r}\left(r\frac{\partial E}{\partial r}\right) = 0 \quad \text{and} \quad r\frac{\partial E}{\partial r} = C.$$

Thus

$$\frac{\partial E}{\partial r} = \frac{C}{r} \quad \text{so} \quad E(r) = \int C\frac{dr}{r} = C\ln r + D.$$

If $D = 0$, the divergence theorem gives

$$-\iint_{R_\varepsilon} \Delta E dV = -\int_{\partial R_\varepsilon} \left(\frac{\partial E}{\partial r}\right)_{r=\varepsilon} dS = -\frac{C}{\varepsilon}\int_{\partial R_\varepsilon} dS = -\frac{C}{\varepsilon}2\pi\varepsilon = -C2\pi.$$

But

$$-\iint_{R_\varepsilon} \Delta E dV = \iint_{R_\varepsilon} \delta(x) dV = 1,$$

so

$$-C2\pi = 1 \quad \text{and} \quad C = -\frac{1}{2\pi}.$$

Finally, we have

$$E(r) = C \ln r = -\frac{1}{2\pi}\ln r = \frac{1}{2\pi}\ln\frac{1}{r}.$$

Summarizing, the equation $-\Delta E = \delta(x)$, assuming that E depends only on r, gives

$$E(\widehat{r}) = \begin{cases} \dfrac{1}{4\pi|\widehat{r} - \widehat{r}_0|} & \text{in three dimensions} \\ \dfrac{1}{2\pi}\ln\dfrac{1}{|\widehat{r} - \widehat{r}_0|} & \text{in two dimensions} \end{cases}.$$

This gives that the fundamental solution in three dimensions is

$$G(\widehat{r}, \widehat{r}_0) = \frac{1}{4\pi|\widehat{r} - \widehat{r}_0|}$$

and in two dimensions is

$$G(\widehat{r}, \widehat{r}_0) = \frac{1}{2\pi}\ln\frac{1}{|\widehat{r} - \widehat{r}_0|}.$$

One can extend this idea to higher dimensions. Suppose that

$$\Delta u(\widehat{x}) = 0, \quad \widehat{x} \in \mathbb{R}^n.$$

We again use the fact that the Laplacian is invariant under rotations to let

$$v(r) = \Delta u(\widehat{x}) \quad \text{where } r = \|\widehat{x}\|.$$

Now

$$\Delta u(x_1, \cdots, x_n) = \frac{\partial^2 u}{\partial x_1^2} + \cdots + \frac{\partial^2 u}{\partial x_n^2}.$$

We have

$$\frac{\partial u}{\partial x_i} = \frac{dv}{dr}\frac{\partial r}{\partial x_i}$$

and

$$\frac{\partial r}{\partial x_i} = \frac{\partial\sqrt{x_1^2 + \cdots x_n^2}}{\partial x_i} = \frac{x_i}{\sqrt{x_1^2 + \cdots x_n^2}} = \frac{x_i}{r}$$

so

$$\frac{\partial u}{\partial x_i} = v'(r)\frac{x_i}{r}.$$

Now

$$\frac{\partial^2 u}{\partial x_i^2} = \frac{\partial}{\partial x_i}\left(\frac{\partial u}{\partial x_i}\right) = \frac{\partial}{\partial x_i}\left(v'(r)\frac{x_i}{r}\right) = \frac{\partial}{\partial x_i}\left[\left(\frac{v'(r)}{r}\right)x_i\right]$$
$$= \left(\frac{v'(r)}{r}\right) + x_i\frac{\partial}{\partial x_i}\left(\frac{v'(r)}{r}\right).$$

Also,

$$\frac{\partial}{\partial x_i}\left(\frac{v'(r)}{r}\right) = \frac{r\frac{\partial}{\partial x_i}(v'(r)) - v'(r)\frac{\partial}{\partial x_i}(r)}{r^2}$$

and

$$\frac{\partial}{\partial x_i}(v'(r)) = v''(r)\frac{x_i}{r}, \quad \frac{\partial}{\partial x_i}(r) = \frac{x_i}{r}.$$

Thus

$$x_i\frac{\partial}{\partial x_i}\left(\frac{v'(r)}{r}\right) = x_i\left[\frac{rv''(r)\frac{x_i}{r} - v'(r)\frac{x_i}{r}}{r^2}\right] = x_i^2\left[\frac{1}{r^2}v''(r) - \frac{v'(r)}{r^3}\right]$$

so

$$\frac{\partial^2 u}{\partial x_i^2} = \left(\frac{v'(r)}{r}\right) + x_i^2\left[\frac{1}{r^2}v''(r) - \frac{v'(r)}{r^3}\right]$$

and we have

$$\Delta u(x_1,\cdots,x_n)=\frac{\partial^2 u}{\partial x_1^2}+\cdots+\frac{\partial^2 u}{\partial x_n^2}=\sum_{i=1}^{n}\left\{\left(\frac{v'(r)}{r}\right)+x_i^2\left[\frac{1}{r^2}v''(r)-\frac{v'(r)}{r^3}\right]\right\}$$

$$=n\left(\frac{v'(r)}{r}\right)+\sum_{i=1}^{n}x_i^2\frac{1}{r^2}v''(r)-\sum_{i=1}^{n}x_i^2\frac{v'(r)}{r^3}=(n-1)\left(\frac{v'(r)}{r}\right)+v''(r).$$

So to solve $\Delta u=0$, we must solve

$$(n-1)\left(\frac{v'(r)}{r}\right)+v''(r)=0.$$

Then

$$v''(r)=\frac{1-n}{r}v'(r)\quad\text{so}\quad\frac{v''(r)}{v'(r)}=(1-n)\frac{1}{r}$$

and

$$\ln v'(r)=(1-n)\ln r+C.$$

Thus

$$v'(r)=\frac{D}{r^{n-1}}$$

and we have

$$v(r)=\frac{D}{(2-n)r^{n-2}}+E.$$

If we assume that $v(r)\to 0$ as $r\to\infty$, then $E=0$. If we then apply the divergence theorem as we did in the two- and three-dimensional cases we get

$$u(\widehat{x})=v(r)=\frac{1}{(n-2)S(n)}\frac{1}{r^{n-2}},\quad n\geq 3$$

where $S(n)$ is the surface area of the sphere in n dimensions.

EXERCISES

In the exercises, you may use the fact that if E depends only on R, then in N dimensions

$$\Delta E=\frac{1}{r^{N-1}}\frac{\partial}{\partial r}\left(r^{N-1}\frac{\partial E}{\partial r}\right).$$

1. In four dimensions, the volume and surface area of a sphere of radius R are computed by

$$V = \frac{1}{2}\pi^2 R^4 \quad S = 2\pi^2 R^3.$$

Solve $-\Delta E = \delta(x)$, assuming that E depends only on r in four dimensions.

2. In five dimensions, the volume and surface area of a sphere of radius R are computed by

$$V = \frac{8}{15}\pi^2 R^5 \quad S = \frac{8}{3}\pi^2 R^4.$$

Solve $-\Delta E = \delta(x)$, assuming that E depends only on r in five dimensions.

CHAPTER

Fourier Series

4

4.1 INTRODUCTION

One technique that is often used in solving partial differential equations is separation of variables. This method yields a set of ordinary differential equations whose solutions are "pasted together" to provide a solution to the partial differential equation.

In the problems that we shall consider, each ordinary differential equation can be considered as an eigenvalue/eigenfunction problem where the differential operator is self-adjoint. An important question is, given the set of eigenfunctions $\{\phi_k\}$ and a function $f(x)$, do there exist constants c_k so that the sequence of functions $\{S_n(x)\}$ defined by

$$\{S_n(x)\} = \sum_{k=1}^{n} c_k \phi_k$$

converges to $f(x)$.

There are three senses of convergence that we shall consider:

1. Pointwise convergence
2. Uniform convergence
3. L^2convergence (also called convergence in the mean).

   L^2 convergence means

$$\lim_{N\to\infty} \langle S_N(x) - f(x), S_N(x) - f(x) \rangle = 0.$$

An advantage of L^2 convergence that we know from Section 1.1 is that the constants c_n should be the Fourier coefficients.

In the text, we shall consider the set of trigonometric functions $\{\sin nx, \cos nx\}$, Bessel functions, and Legendre polynomials. Each of these sets arises as a set of eigenfunctions for a particular self-adjoint differential operator. In this section we analyze trigonometric Fourier series, which are series of trigonometric functions. We choose this set because the mathematics is the best developed and is the simplest to demonstrate. It will serve to illustrate the basic questions that need to be addressed for each system.

Mathematical Physics with Partial Differential Equations. https://doi.org/10.1016/B978-0-12-814759-7.00004-1

4.2 BASIC DEFINITIONS

We have seen that the operator $L = \frac{d^2}{dx^2}$ is self-adjoint on the vector space

$$V = \{f | f \text{ is periodic of period } 2\pi \text{ and } f'' \text{ is integrable}\}$$

with inner product

$$\langle f, g \rangle = \int_{-\pi}^{\pi} f(x)g(x)dx.$$

The eigenfunctions of L are

$$\{1, \sin(nx), \cos(nx) | n \in \mathbb{Z}^+\}.$$

The eigenvalue of $\sin(nx)$ and $\cos(nx)$ is $-n^2$ and the eigenvalue of 1 is 0. Since

$$\langle \sin(nx), \cos(nx) \rangle = \int_{-\pi}^{\pi} \sin(nx)\cos(nx)dx = \frac{1}{2}\int_{-\pi}^{\pi} \sin(2nx)dx = 0$$

and since eigenfunctions of a self-adjoint operator with distinct eigenvalues are orthogonal, the set

$$\{1, \sin(nx), \cos(nx) | n \in \mathbb{Z}^+\}$$

is an orthogonal set of eigenvectors for L. Next we compute the norm of each of these eigenvectors.

We have

$$\begin{aligned} 2\pi &= \int_{-\pi}^{\pi} 1dx = \int_{-\pi}^{\pi} \sin^2(nx) + \cos^2(nx)dx \\ &= \int_{-\pi}^{\pi} \sin^2(nx)dx + \int_{-\pi}^{\pi} \cos^2(nx)dx \end{aligned}$$

and

$$\int_{-\pi}^{\pi} \sin^2(nx)dx = \int_{-\pi}^{\pi} \cos^2(nx)dx$$

so

$$\langle \sin(nx), \sin(nx) \rangle = \int_{-\pi}^{\pi} \sin^2(nx)dx = \pi$$

and

$$\langle \cos(nx), \cos(nx) \rangle = \int_{-\pi}^{\pi} \cos^2(nx)dx = \pi.$$

Also

$$\langle 1, 1 \rangle = \int_{-\pi}^{\pi} 1^2 \, dx = 2\pi.$$

Thus

$$\|\sin(nx)\| = \|\cos(nx)\| = \sqrt{\pi} \text{ and } \|1\| = \sqrt{2\pi}.$$

In Exercise 1, we show that the Fourier expansion for $f(x)$ is

$$\frac{a_0}{2} + \sum_{n=1}^{\infty} (a_n \cos nx + b_n \sin nx)$$

where

$$a_n = \frac{1}{\pi} \int_{-\pi}^{\pi} f(x) \cos(nx) dx \quad a_0 = \frac{1}{\pi} \int_{-\pi}^{\pi} f(x) dx,$$

so

$$\pi a_n = \int_{-\pi}^{\pi} f(x) \cos(nx) \, dx \quad \pi a_0 = \int_{-\pi}^{\pi} f(x) dx$$

and

$$b_n = \frac{1}{\pi} \int_{-\pi}^{\pi} f(x) \sin(nx) dx, \text{ so } \pi b_n = \int_{-\pi}^{\pi} f(x) \cos(nx) dx.$$

For the trigonometric polynomials, the function $S_N(x)$ given earlier is

$$S_N(x) = \frac{a_0}{2} + \sum_{n=1}^{N} (a_n \cos nx + b_n \sin nx).$$

We begin with preliminary work on L^2 convergence. We show that

$$\int_{-\pi}^{\pi} [f(x) - S_N(x)]^2 dx = \int_{-\pi}^{\pi} [f(x)]^2 dx - \pi \left(\frac{a_0^2}{2} + \sum_{n=1}^{N} \left(a_n^2 + b_n^2 \right) \right).$$

Now

$$\int_{-\pi}^{\pi} [f(x) - S_N(x)]^2 dx = \int_{-\pi}^{\pi} [f(x)]^2 dx - 2 \int_{-\pi}^{\pi} f(x) S_N(x) dx + \int_{-\pi}^{\pi} [S_N(x)]^2 dx.$$

We have

$$\int_{-\pi}^{\pi} f(x)S_N(x)dx = \int_{-\pi}^{\pi} f(x)\frac{a_0}{2}dx + \int_{-\pi}^{\pi} f(x)\left(\sum_{n=1}^{N}(a_n \cos nx)\right) + \int_{-\pi}^{\pi} f(x)\left(\sum_{n=1}^{N}(b_n \sin nx)\right)$$

$$= \frac{a_0}{2}\int_{-\pi}^{\pi} f(x)dx + \sum_{n=1}^{N}\left(a_n \int_{-\pi}^{\pi} f(x)\cos nx\, dx\right) + \sum_{n=1}^{N}\left(b_n \int_{-\pi}^{\pi} f(x)\sin nxdx\right)$$

$$= \frac{a_0}{2}(\pi a_0) + \sum_{n=1}^{N} a_n(\pi a_n) + \sum_{n=1}^{N} b_n(\pi b_n)$$

$$= \pi\left(\frac{a_0^2}{2} + \sum_{n=1}^{N}\left(a_n^2 + b_n^2\right)\right).$$

Also

$$\int_{-\pi}^{\pi} [S_N(x)]^2 dx = \int_{-\pi}^{\pi}\left[\frac{a_0}{2} + \sum_{n=1}^{N}(a_n \cos nx + b_n \sin nx)\right]^2 dx$$

$$= \int_{-\pi}^{\pi}\frac{a_0^2}{4}dx + \sum_{n=1}^{N} a_n^2 \int_{-\pi}^{\pi} \cos^2(nx)dx + \sum_{n=1}^{N} b_n^2 \int_{-\pi}^{\pi} \sin^2(nx)dx$$

(by the orthogonality of $\{1, \sin(nx), \cos(nx)\}$)

$$= 2\pi\frac{a_0^2}{4} + \pi\sum_{n=1}^{N} a_n^2 + \pi\sum_{n=1}^{N} b_n^2 = \pi\left(\frac{a_0^2}{2} + \sum_{n=1}^{N}\left(a_n^2 + b_n^2\right)\right).$$

Thus

$$0 \le \int_{-\pi}^{\pi} [f(x) - S_N(x)]^2 dx$$

$$= \int_{-\pi}^{\pi} [f(x)]^2 dx - 2\int_{-\pi}^{\pi} f(x)S_N(x)dx + \int_{-\pi}^{\pi} [S_N(x)]^2 dx$$

$$= \int_{-\pi}^{\pi} [f(x)]^2 dx - 2\left[\pi\left(\frac{{a_0}^2}{2} + \sum_{n=1}^{N}({a_n}^2 + {b_n}^2)\right)\right] + \pi\left(\frac{{a_0}^2}{2} + \sum_{n=1}^{N}({a_n}^2 + {b_n}^2)\right)$$

$$= \int_{-\pi}^{\pi} [f(x)]^2 dx - \pi\left(\frac{{a_0}^2}{2} + \sum_{n=1}^{N}({a_n}^2 + {b_n}^2)\right). \tag{1}$$

From this computation, we get some important results:
Theorem (Bessel's Inequality):
For every positive integer N, we have

$$\pi\left(\frac{a_0^2}{2}+\sum_{n=1}^{N}\left(a_n^2+b_n^2\right)\right)\leq\int_{-\pi}^{\pi}[f(x)]^2dx.$$

Proof:
From (1), we have

$$0\leq\int_{-\pi}^{\pi}[f(x)-S_N(x)]^2dx=\int_{-\pi}^{\pi}[f(x)]^2dx-\pi\left(\frac{a_0{}^2}{2}+\sum_{n=1}^{N}(a_n{}^2+b_n{}^2)\right).$$

Corollary (Riemann–Lebesgue Lemma):
If $\int_{-\pi}^{\pi}[f(x)]^2dx$ is finite and $f(x)$ is integrable, then

$$\lim_{n\to\infty}\int_{-\pi}^{\pi}f(x)\cos nx=0\ \text{ and }\ \lim_{n\to\infty}\int_{-\pi}^{\pi}f(x)\sin nx\,dx=0.$$

Proof:
Since

$$\sum_{n=1}^{N}a_n{}^2\leq\int_{-\pi}^{\pi}[f(x)]^2dx<\infty\quad\text{for every } N$$

the sequence of partial sums for the series $\sum_{n=1}^{\infty}a_n{}^2$ is monotone increasing and bounded, so it converges.
Since the series converges, the terms must go to 0; i.e.,

$$\lim_{n\to\infty}a_n{}^2=0.$$

But then

$$\lim_{n\to\infty}a_n=\lim_{n\to\infty}\frac{1}{\pi}\int_{-\pi}^{\pi}f(x)\cos nx\,dx=0.$$

Similarly,

$$\lim_{n\to\infty}b_n=\lim_{n\to\infty}\frac{1}{\pi}\int_{-\pi}^{\pi}f(x)\sin nx\,dx=0.$$

EXERCISES

1. Show that the Fourier expansion for $f(x)$ is $\frac{a_0}{2} + \sum_{n=1}^{\infty} (a_n \cos nx + b_n \sin nx)$ where

$$a_n = \frac{1}{\pi} \int_{-\pi}^{\pi} f(x)\cos(nx)dx \quad a_0 = \frac{1}{\pi} \int_{-\pi}^{\pi} f(x)dx$$

so

$$\pi a_n = \int_{-\pi}^{\pi} f(x)\cos(nx)dx \quad \pi a_0 = \int_{-\pi}^{\pi} f(x)dx$$

and

$$b_n = \frac{1}{\pi} \int_{-\pi}^{\pi} f(x)\sin(nx)dx, \text{ so } \pi b_n = \int_{-\pi}^{\pi} f(x)\sin(nx)dx.$$

2. The Fourier series for $f(x)$ is given by $\frac{a_0}{2} + \sum_{n=1}^{\infty} (a_n \cos nx + b_n \sin nx)$ where formulas for a_n and b_n have been derived.
   a. Show that if $f(x)$ is even; i.e., if $f(x) = f(-x)$, then $b_n = 0$.
   b. Show that if $f(x)$ is odd; i.e., if $f(x) = -f(-x)$, then $a_n = 0$.
3. The function $f(x) = |\sin x|$ is an even function, so the Fourier coefficients b_n will be 0. However, $\sin x = |\sin x|$ for $0 \le x \le \pi$. Show that

$$\sin x = \frac{2}{\pi} - \frac{4}{\pi} \sum_{n=1}^{\infty} \frac{\cos(2nx)}{4n^2 - 1} \qquad 0 \le x \le \pi.$$

4.
   a. Using the same reasoning as in Problem 3, show that

$$x = \frac{\pi}{2} - \frac{4}{\pi} \sum_{n=1}^{\infty} \frac{\cos(2n-1)x}{(2n-1)^2} \qquad 0 \le x \le \pi.$$

   b. Use the result of part (a) to show

$$\sum_{n=1}^{\infty} \frac{1}{(2n-1)^2} = \frac{\pi^2}{8}.$$

5.

a. Show that the Fourier series for $f(x) = x + \frac{1}{4}x^2$ on $(-\pi, \pi)$ is

$$\frac{\pi^2}{12} + \sum_{n=1}^{\infty} (-1)^n \left(\frac{\cos nx}{n^2} - 2\frac{\sin nx}{n} \right).$$

b. Assuming that the Fourier series converges to the function, use the answer to part (a) to find

$$\sum_{n=1}^{\infty} \frac{(-1)^n}{n^2}.$$

6.

a. Show that the Fourier series for $f(x) = x$ on $(-\pi,\pi)$ is

$$2\sum_{n=1}^{\infty} \frac{(-1)^{n+1}}{n} \sin nx.$$

b. Assuming that the Fourier series converges to the function, use the answer to part (a) to find

$$\sum_{n=1}^{\infty} \frac{(-1)^{n+1}}{2n-1}.$$

7. Show that a linear combination of even functions is even and a linear combination of odd functions is odd.

8. For $S_N(x) = \frac{a_0}{2} + \sum_{n=1}^{N} (a_n \cos nx + b_n \sin nx)$ show why

$$\int_{-\pi}^{\pi} [S_N(x)]^2 dx = \pi \left(\frac{a_0^2}{2} + \sum_{n=1}^{N} \left(a_n^2 + b_n^2 \right) \right).$$

9. Multiply each function by the appropriate constant to change $\{1, \cos x, \sin x, \cos 2x, \sin 2x, \ldots\}$ from an orthogonal basis to an orthonormal basis.

4.3 METHODS OF CONVERGENCE OF FOURIER SERIES

We next address the question stated earlier: given a function $f(x)$, when does

$$S_N(x) = \frac{a_0}{2} + \sum_{n=1}^{N} (a_n \cos nx + b_n \sin nx)$$

converge to $f(x)$ as $N \to \infty$.

As we noted, there are three senses of convergence that we shall consider:

1. Pointwise convergence
2. Uniform convergence
3. L^2 convergence.

In Fourier analysis, it is common to work with the set of functions

$$\mathcal{F} = \{\text{functions that are piecewise continuous of period } 2\pi\}.$$

Piecewise continuous functions are those that are continuous except at a finite number of points, and where a discontinuity exists, the left and right hand limits exist and are finite.

We let

$$R_N(x) = f(x) - S_N(x)$$

so that $R_N(x)$ is a "remainder term" that measures how much $S_N(x)$ differs from $f(x)$.

Note that $S_N(x)$ converges to $f(x)$ pointwise (uniformly, in the L^2 sense) exactly when $R_N(x)$ converges to 0 in the same sense.

In this section we shall show:

1. If $f(x)$ is differentiable at x_0, then the Fourier series for $f(x)$ converges to $f(x)$ at x_0. Thus if $f(x)$ is differentiable, then the Fourier series for $f(x)$ converges pointwise to $f(x)$.
2. If $f(x)$ is continuous and $f'(x)$ is piecewise continuous, then the Fourier series for $f(x)$ converges uniformly to $f(x)$.
3. If $f(x)$ is continuous, then the Fourier series of $f(x)$ converges to $f(x)$ in the L^2 sense.

We begin by determining an explicit formula for $R_N(x)$.

Definition:

The *Dirichlet kernel* of index N is the function $D_N(x)$ defined by

$$D_N(x) = \frac{1}{2} + \cos x + \cos(2x) + \cdots + \cos(Nx).$$

The graphs of $D_5(x)$and $D_{10}(x)$ are shown in Fig. 4.3.1. As n becomes larger, the peak of $D_n(x)$ at $x = 0$ becomes more pronounced.

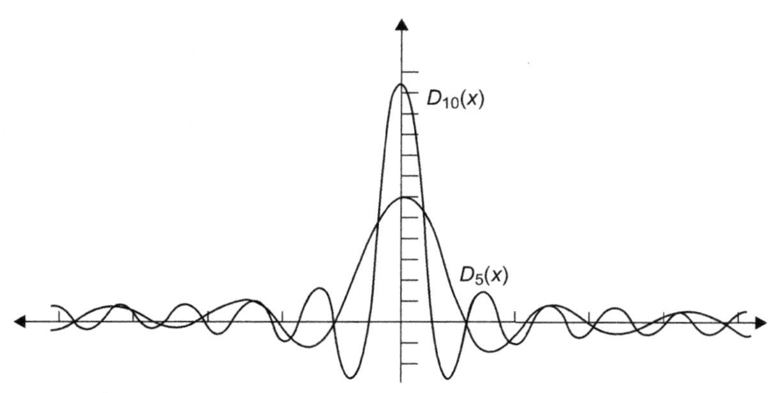

FIGURE 4.3.1

In Exercise 1, we show that

$$D_N(x) = \begin{cases} \dfrac{\sin\left(N+\dfrac{1}{2}\right)x}{2\sin\left(\dfrac{x}{2}\right)}, & \text{if } x \neq 2\pi m, \ m \text{ an integer} \\ N+\dfrac{1}{2} & \text{if } x = 2\pi m \end{cases}.$$

Note that

$$\frac{1}{\pi}\int_{-\pi}^{\pi} D_N(t)\,dt = 1$$

for every N.

Theorem:

Let $f \in \mathcal{F}$ and $S_N(x)$, $R_N(x)$, and $D_N(x)$ be as previously defined. Then

$$\begin{aligned} S_N(x) &= \frac{1}{\pi}\int_{-\pi}^{\pi} f(u)D_N(x-u)du \\ &= \frac{1}{\pi}\int_{-\pi}^{\pi} f(x-t)D_N(t)\,dt \\ &= \frac{1}{\pi}\int_{-\pi}^{\pi} f(x+t)D_N(t)\,dt \end{aligned}$$

and

$$R_N(x) = \frac{1}{\pi}\int_{-\pi}^{\pi}[f(x)-f(x-t)]D_N(t)\,dt.$$

Proof:
We have

$$\frac{1}{\pi}\int_{-\pi}^{\pi} f(u)D_N(x-u)du = \frac{1}{\pi}\int_{-\pi}^{\pi} f(u)\left[\frac{1}{2}+\sum_{k=1}^{N}(\cos k(x-u)\right]du$$

$$=\frac{1}{\pi}\int_{-\pi}^{\pi} f(u)\left[\frac{1}{2}+\sum_{k=1}^{N}(\cos kx\cos ku+\sin kx\sin ku)\right]du$$

$$=\frac{1}{2\pi}\int_{-\pi}^{\pi} f(u)du + \sum_{k=1}^{N}\cos kx\left(\frac{1}{\pi}\int_{-\pi}^{\pi} f(u)\cos ku\,du\right)$$

$$+\sum_{k=1}^{N}\sin kx\left(\frac{1}{\pi}\int_{-\pi}^{\pi} f(u)\sin ku\,du\right)$$

$$=\frac{a_0}{2}+\sum_{k=1}^{N}(a_k\cos kx+b_k\sin kx)=S_N(x).$$

To show

$$\int_{-\pi}^{\pi} f(u)D_N(x-u)du = \int_{-\pi}^{\pi} f(x-t)D_N(t)dt$$

we make the change of variables $u = x - t$. Then $du = -dt$. (Remember that x is fixed.)

If $u = \pi$, then $t = x - \pi$ and if $u = -\pi$, then $t = x + \pi$. Thus

$$\int_{-\pi}^{\pi} f(u)D_N(x-u)du = -\int_{x+\pi}^{x-\pi} f(x-t)D_N(t)dt = \int_{x-\pi}^{x+\pi} f(x-t)D_N(t)dt.$$

Now $f(x)$ and $D_N(x)$ are periodic of period 2π. Thus the integrals of the product of these functions over any interval of length 2π are equal. So we have

$$\int_{x-\pi}^{x+\pi} f(x-t)D_N(t)dt = \int_{-\pi}^{\pi} f(x-t)D_N(t)dt.$$

Showing that

$$\int_{-\pi}^{\pi} f(u)D_N(x-u)du = \int_{-\pi}^{\pi} f(x-t)D_N(t)dt$$

is done in a similar manner and is left for Exercise 3.

Now

$$R_N(x) = f(x) - S_N(x) \text{ and } \frac{1}{\pi}\int_{-\pi}^{\pi} D_N(t)dt = 1$$

so

$$f(x) = \frac{1}{\pi}\int_{-\pi}^{\pi} f(x)D_N(t)dt$$

and thus

$$R_N(x) = \frac{1}{\pi}\int_{-\pi}^{\pi} [f(x) - f(x-t)]D_N(t)\, dt.$$

We now give a sufficient condition for convergence of the Fourier series of a function to the value of the function at a particular point.

Theorem:

Let $f \in \mathcal{F}$, and suppose that f is differentiable at x_0. Then

$$f(x_0) = \lim_{N\to\infty}\left[\frac{a_0}{2} + \sum_{n=1}^{N}(a_n \cos nx_0 + b_n \sin nx_0)\right]$$

where a_n and b_n are the Fourier coefficients of f.

Proof:

We have

$$R_N(x) = \frac{1}{\pi}\int_{-\pi}^{\pi} [f(x) - f(x-t)]D_N(t)dt$$

$$= \frac{1}{\pi}\int_{-\pi}^{\pi}\left[[f(x) - f(x-t)]\frac{\sin\left(N+\frac{1}{2}\right)t}{2\sin\left(\frac{t}{2}\right)}\right]dt$$

$$= \frac{1}{\pi}\int_{-\pi}^{\pi}\frac{f(x) - f(x-t)}{2\sin\left(\frac{t}{2}\right)}\left(\sin Nt \cos\frac{t}{2} + \cos Nt \sin\frac{t}{2}\right)dt$$

$$= \frac{1}{\pi}\int_{-\pi}^{\pi}\frac{f(x) - f(x-t)}{2\sin(t/2)}\left(\sin Nt \cos\frac{t}{2}\right)dt + \frac{1}{\pi}\int_{-\pi}^{\pi}\frac{f(x) - f(x-t)}{2}(\cos Nt)dt.$$

By the Riemann–Lebesgue lemma

$$\lim_{N\to\infty}\frac{1}{\pi}\int_{-\pi}^{\pi}\frac{f(x) - f(x-t)}{2}(\cos Nt)dt = 0 \quad \text{if } f \in \mathcal{F}.$$

Next we consider

$$\int_{-\pi}^{\pi} \frac{f(x)-f(x-t)}{2\sin(t/2)}\left(\sin Nt \cos\frac{t}{2}\right)dt.$$

Since $f \in \mathcal{F}$ and $\sin(t/2)$ is continuous, the function

$$\frac{f(x)-f(x-t)}{2\sin(t/2)}$$

is piecewise continuous (in t) except possibly at $t=0$. When $t=0$, the function is of the form 0/0, but if f is differentiable at x, as a function of t, we have

$$\lim_{t\to 0}\frac{f(x)-f(x-t)}{2\sin(t/2)} = \lim_{t\to 0}\frac{f(x)-f(x-t)}{t}\frac{t/2}{\sin(t/2)} = f'(x)$$

since

$$\lim_{\theta\to 0}\frac{\sin\theta}{\theta} = 1.$$

So if f is differentiable at x, as a function of t, the function

$$\frac{f(x)-f(x-t)}{2\sin(t/2)}\cos\frac{t}{2}$$

is in $\mathcal{F}$, and by the Riemann–Lebesgue lemma

$$\lim_{N\to\infty}\int_{-\pi}^{\pi} \frac{f(x)-f(x-t)}{2\sin(t/2)}\left(\sin Nt \cos\frac{t}{2}\right)dt = 0.$$

Thus we have that the Fourier series of a function converges to the value of the function at points where the function is differentiable, and so if $f(x)$ is differentiable, the Fourier series of $f(x)$ converges pointwise to $f(x)$.

To begin to address the question of uniform convergence, we recall the Schwarz inequality for real numbers.

SCHWARZ INEQUALITY

If $a_1, a_2, \ldots, a_n$ and $b_1, b_2, \ldots, b_n$ are real numbers, then

$$\sum_{i=1}^{n} a_i b_i \le \left(\sum_{i=1}^{n} a_i^2\right)^{1/2}\left(\sum_{i=1}^{n} b_i^2\right)^{1/2}.$$

Theorem:

Suppose that $f(x)$ is a continuous function of period 2π, and suppose that $f' \in \mathcal{F}$. Then the Fourier series of $f(x)$ converges uniformly to f.

Proof:

Suppose that $\sum(a_n \cos nx + b_n \sin nx)$ is the Fourier series of $f(x)$ and $\sum(c_n \cos nx + d_n \sin nx)$ is the Fourier series of $f'(x)$. Then $\sum a_n^2$, $\sum b_n^2$, $\sum c_n^2$, and $\sum d_n^2$ are finite.

Now

$$c_n = \frac{1}{\pi}\int_{-\pi}^{\pi} f'(x)\cos nx\,dx \text{ and } d_n = \frac{1}{\pi}\int_{-\pi}^{\pi} f'(x)\sin nx\,dx.$$

Integrating the expression for c_n by parts gives

$$c_n = \frac{1}{\pi}\left(f(x)\cos nx|_{-\pi}^{\pi} + n\int_{-\pi}^{\pi} f(x)\sin nx\,dx \right) = \frac{n}{\pi}\int_{-\pi}^{\pi} f(x)\sin nx\,dx = na_n$$

since $f(-\pi)\cos(-n\pi) = f(\pi)\cos(n\pi)$.

Thus $\sum_{n=1}^{\infty} n^2 a_n^2 < \infty$ and similarly $\sum_{n=1}^{\infty} n^2 b_n^2 < \infty$. By the Schwarz inequality

$$\sum_{n=1}^{N} |x_n y_n| \le \left(\sum_{n=1}^{N} x_n^2\right)^{1/2}\left(\sum_{n=1}^{N} y_n^2\right)^{1/2} \quad \text{for every } N$$

so that convergence of the series $\sum x_n^2$ and $\sum y_n^2$ implies convergence of the series $\sum |x_n y_n|$. Thus

$$\sum_{n=1}^{\infty} |a_n| = \sum_{n=1}^{\infty}\left|\left(\frac{1}{n}\right)(na_n)\right| \le \left[\sum_{n=1}^{\infty}\left(\frac{1}{n^2}\right)\right]^{1/2}\left[\sum_{n=1}^{\infty}(n^2 a_n^2)\right]^{1/2}$$

and both $\sum_{n=1}^{\infty}\left(\frac{1}{n^2}\right)$ and $\sum_{n=1}^{\infty}(n^2 a_n^2)$ are convergent. Thus $\sum_{n=1}^{\infty} |a_n|$ converges. For every x, we have $|a_n \cos nx| \le |a_n|$, so the series $\sum_{n=1}^{\infty} |a_n \cos nx|$ converges uniformly by the Weierstrass M-test. Similarly, $\sum_{n=1}^{\infty} |b_n \sin nx|$ converges uniformly. Thus the Fourier series converges absolutely and uniformly. Since $f(x)$ is differentiable at all but a finite number of points, the Fourier series of $f(x)$ converges to $f(x)$ except possibly at those points. However, the uniform limit of a series of continuous functions is continuous, so it follows that the Fourier series of $f(x)$ converges uniformly to $f(x)$.

Next we determine a condition on $f(x)$ that will ensure L^2 convergence. This occurs if and only if $\lim_{N\to\infty} \langle R_N(x), R_N(x)\rangle = 0$.

In the proof of Bessel's inequality, we found that if $\{c_n\}$ is the set of Fourier coefficients of $f(x)$ then

$$0 \le \langle f - S_N, f - S_N\rangle = \int_{-\pi}^{\pi} [f(x)]^2 dx - \sum_{n=1}^{N} c_n^2$$

and so

$$\sum_{n=1}^{N} c_n^2 \le \int_{-\pi}^{\pi} [f(x)]^2 dx.$$

Thus a necessary and sufficient condition for $\lim_{N\to\infty} \langle R_N(x), R_N(x)\rangle = 0$ is that

$$\sum_{n=1}^{N} c_n^2 = \int_{-\pi}^{\pi} [f(x)]^2 dx.$$

We determine conditions on $f(x)$ that will ensure this.
Here is an outline of how we shall proceed:

1. Let $f \in \mathcal{F}$ have the Fourier series

$$\frac{a_0}{2} + \sum_{n=1}^{\infty} (a_n \cos nx + b_n \sin nx)$$

and let

$$S_m(x) = \frac{a_0}{2} + \sum_{n=1}^{m} (a_n \cos nx + b_n \sin nx).$$

We form the Cesaro sum

$$\sigma_m(x) = \frac{S_0(x) + S_1(x) + \cdots + S_m(x)}{m+1}.$$

(Note: Some authors, Apostol "Mathematical Analysis," for example, define

$$\sigma_m(x) = \frac{S_0(x) + S_1(x) + \cdots + S_{m-1}(x)}{m}.$$

We follow the definition used by Rudin in "Principles of Mathematical Analysis").

2. We form an approximate identity called the Fejer kernel, $\{K_n(t)\}$. To define the Fejer kernel, recall that the Dirichlet kernel of index N, $D_N(t)$, is given by

$$D_N(t) = \frac{1}{2} + \cos t + \cos(2t) + \cdots + \cos(Nt).$$

The Fejer kernel of index N is defined by

$$K_N(t) = \frac{D_0(t) + D_1(t) + \cdots + D_N(t)}{N+1} \tag{2}$$

which is the Cesaro sum for the Dirichlet kernel.

3. Derive properties of the Fejer kernel that allow us to conclude that $\{\sigma_m(x)\}$ converges to $f(x)$ uniformly.
Graphs of the Fejer kernel for $N = 5$ and $N = 10$ are shown in Figure 4.3.2.

Theorem:
Let

$$K_N(t) = \frac{D_0(t) + D_1(t) + \cdots + D_N(t)}{N+1}.$$

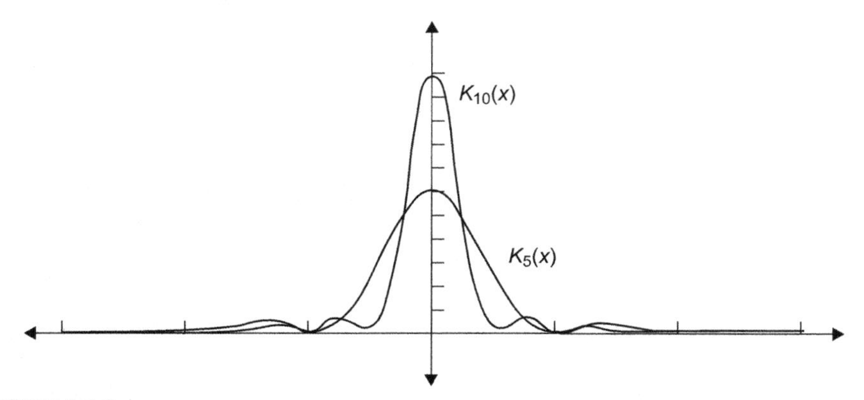

FIGURE 4.3.2

Then

1. $\frac{1}{\pi}\int_{-\pi}^{\pi} K_N(t)dt = 1.$
2. $K_N(x) = \frac{1}{2(N+1)} \frac{\sin^2[(N+1)x/2]}{\sin^2\left(\frac{x}{2}\right)} \geq 0.$
3. $K_N(x) \leq \frac{1}{(N+1)(1-\cos\delta)}$ for $0 < \delta \leq |x| \leq \pi.$
4. $\lim_{x\to 0} K_N(x) = \frac{N+1}{2}.$

The proof is left as Exercise 2. Note that (1) and (3) imply that $\{K_n(x)\}$ when restricted to the interval $[-\pi,\pi]$ is an approximate identity. (More precisely, the sequence of functions

$$J_n(x) = \begin{cases} K_n(x) & -\pi \leq x \leq \pi \\ 0 & \text{otherwise} \end{cases}$$

is an approximate identity.)

This means that $\{\sigma_m(x)\}$ converges pointwise to $f(x)$. However, pointwise convergence is not sufficient to ensure L^2 convergence.

Theorem (Fejer's Theorem):

Let $f(x)$ be a continuous function of period 2π, and let

$$S_m(x) = \frac{a_0}{2} + \sum_{n=1}^{m}(a_n \cos nx + b_n \sin nx)$$

be the mth partial sum of the Fourier series generated by f. Let

$$\sigma_m(x) = \frac{S_0(x) + S_1(x) + \cdots + S_m(x)}{m+1}.$$

Then $\{\sigma_m(x)\}$ converges to $f(x)$ uniformly.

Proof:

Let $\varepsilon > 0$ be given. We must show there is a number $N(\varepsilon)$ such that if $n > N(\varepsilon)$, then

$$|\sigma_n(x) - f(x)| < \varepsilon \text{ for every } x \in [-\pi, \pi].$$

Since $f(x)$ is continuous on $[-\pi,\pi]$, it is bounded and uniformly continuous on $[-\pi,\pi]$. Thus there is a number M such that $|f(x)| \le M$ if $x \in [-\pi, \pi]$ and a $\delta > 0$ such that if $|x' - x''| < \delta$, then $|f(x') - f(x'')| < \varepsilon$ for $x', x'' \in [-\pi, \pi]$.

Now

$$S_k(x) = \frac{1}{\pi} \int_{-\pi}^{\pi} f(x-t) D_k(t) dt$$

so

$$\begin{aligned}
\sigma_n(x) &= \frac{S_0(x) + S_1(x) + \cdots + S_n(x)}{n+1} \\
&= \frac{1}{\pi} \int_{-\pi}^{\pi} f(x-t) \left[\frac{D_0(t) + D_1(t) + \cdots + D_n(t)}{n+1} \right] dt \\
&= \frac{1}{\pi} \int_{-\pi}^{\pi} f(x-t) K_n(t) dt.
\end{aligned}$$

Since

$$\frac{1}{\pi} \int_{-\pi}^{\pi} K_n(t) dt = 1, \quad \text{then } f(x) = \frac{1}{\pi} \int_{-\pi}^{\pi} f(x) K_n(t) dt.$$

Thus

$$|\sigma_n(x) - f(x)| = \left| \frac{1}{\pi} \int_{-\pi}^{\pi} [f(x-t) - f(x)] K_n(t) dt \right| \le \frac{1}{\pi} \int_{-\pi}^{\pi} |f(x-t) - f(x)| K_n(t) dt$$

$$= \frac{1}{\pi} \left[\int_{-\pi}^{-\delta} |f(x-t) - f(x)| K_n(t) dt + \int_{-\delta}^{\delta} |f(x-t) - f(x)| K_n(t) dt \right.$$

$$\left. + \int_{\delta}^{\pi} |f(x-t) - f(x)| K_n(t) dt \right] \le \frac{1}{\pi} \left[2M \left(\int_{-\pi}^{-\delta} K_n(t) dt + \int_{\delta}^{\pi} K_n(t) dt \right) \right.$$

$$\left. + \int_{-\delta}^{\delta} |f(x-t) - f(x)| K_n(t)\, dt \right] \quad \text{for } 0 < \delta < \pi.$$

Now choose δ so that if $|t| < \delta$, then $|f(x-t) - f(x)| < \varepsilon/2$.

This will make

$$\int_{-\delta}^{\delta} |f(x-t) - f(x)| K_n(t)dt < \frac{\varepsilon}{2} \int_{-\delta}^{\delta} K_n(t)dt < \frac{\varepsilon}{2} \quad \text{for every } n.$$

Since

$$K_N(x) \leq \frac{1}{(N+1)(1-\cos\delta)} \quad \text{for } 0 < \delta \leq |x| \leq \pi,$$

after δ has been chosen, there is a number $N(\varepsilon)$ such that if $n > N(\varepsilon)$, then $K_n(x) < \frac{\varepsilon\pi}{8M}$ if $\delta \leq |x| \leq \pi$. Thus if $n > N(\varepsilon)$, then

$$\begin{aligned} &|\sigma_n(x) - f(x)| \\ &\leq \frac{1}{\pi}\left[2M\left(\int_{-\pi}^{-\delta} K_n(t)dt + \int_{\delta}^{\pi} K_n(t)dt\right) + \int_{-\delta}^{\delta} |f(x-t) - f(x)| K_n(t)\, dt\right] \\ &< \frac{1}{\pi}\left[2M\left(\frac{\varepsilon\pi}{8M} + \frac{\varepsilon\pi}{8M}\right) + \frac{\varepsilon}{2}\right] = \frac{1}{\pi}\left(\frac{\varepsilon\pi}{4} + \frac{\varepsilon\pi}{4} + \frac{\varepsilon}{2}\right) = \varepsilon \quad \text{if } x \in [-\pi, \pi]. \end{aligned}$$

Corollary:

If $f(x)$ is piecewise continuous, then

$$\lim_{n\to\infty} \langle f(x) - S_n(x), f(x) - S_n(x) \rangle = 0.$$

Proof:

We have

$$\langle f(x) - S_n(x), f(x) - S_n(x) \rangle \leq \langle f(x) - \sigma_n(x), f(x) - \sigma_n(x) \rangle.$$

Since $\{\sigma_n(x)\}$ converges to $f(x)$ uniformly, then

$$\lim_{n\to\infty} \langle f(x) - \sigma_n(x), f(x) - \sigma_n(x) \rangle = \lim_{n\to\infty} \int_{-\pi}^{\pi} |f(x) - \sigma_n(x)|^2 dx = 0.$$

so

$$\lim_{n\to\infty} \langle f(x) - S_n(x), f(x) - S_n(x) \rangle = 0.$$

Corollary (Parseval's Formula):

If $f(x)$ is piecewise continuous, then

$$\frac{1}{\pi} \int_{-\pi}^{\pi} |f(x)|^2 dx = \frac{{a_0}^2}{2} + \sum_{n=1}^{\infty} ({a_n}^2 + {b_n}^2).$$

We return to a question we posed early in the section: when can a function be represented as a Fourier series.

Definition:
A set of eigenfunctions $\{\phi_n(x)\}$ is *complete* if

$$\lim_{N\to\infty}\langle f(x)-S_N(x),f(x)-S_N(x)\rangle=0$$

for every function for which $\langle f,f\rangle$ is finite. As always, if $\{\phi_n(x)\}$ is an orthonormal basis, then

$$S_N(x)=\sum_{n=1}^{N}\langle\phi_n(x),f(x)\rangle\phi_n(x).$$

We now show that the trigonometric polynomials are complete. To do so, we use the fact that if $f(x)$ is a function for which

$$\int_{-\pi}^{\pi}|f(x)|^2dx<\infty,$$

then given $\varepsilon>0$, there is a continuously differentiable function $f_\varepsilon(x)$ for which

$$\langle f(x)-f_\varepsilon(x),f(x)-f_\varepsilon(x)\rangle=\int_{-\pi}^{\pi}|f(x)-f_\varepsilon(x)|^2dx<\frac{\varepsilon}{4}.$$

(See Rudin's "Real and Complex Analysis," for instance.)

Since $f_\varepsilon(x)$ is continuously differentiable, there is a number $N(\varepsilon)$ so that if $n>N(\varepsilon)$ then

$$\langle S_n^\varepsilon(x)-f_\varepsilon(x),S_n^\varepsilon(x)-f_\varepsilon(x)\rangle=\int_{-\pi}^{\pi}\left|S_n^\varepsilon(x)-f_\varepsilon(x)\right|^2dx<\frac{\varepsilon}{4}$$

where $S_n^\varepsilon(x)$ is the nth partial sum of the Fourier expansion of $f_\varepsilon(x)$. Now,

$$\langle f(x)-S_N(x),f(x)-S_N(x)\rangle\le\langle f(x)-S_n^\varepsilon(x),f(x)-S_n^\varepsilon(x)\rangle$$
$$=\int_{-\pi}^{\pi}\left|f(x)-S_n^\varepsilon(x)\right|^2dx=\int_{-\pi}^{\pi}\left|f(x)-f_\varepsilon(x)+f_\varepsilon(x)-S_n^\varepsilon(x)\right|^2dx.$$

We now bound

$$\left|f(x)-f_\varepsilon(x)+f_\varepsilon(x)-S_n^\varepsilon(x)\right|^2=\left[(f(x)-f_\varepsilon(x))+(f_\varepsilon(x)-S_n^\varepsilon(x))\right]^2.$$

Note that

$$(a+b)^2=a^2+2ab+b^2 \text{ and } 2ab\le a^2+b^2 \left(\text{since } (a-b)^2\ge 0\right).$$

Thus

$$(a+b)^2\le 2a^2+2b^2$$

and so

$$\int_{-\pi}^{\pi} \left|f(x) - f_\varepsilon(x) + f_\varepsilon(x) - S_n^\varepsilon(x)\right|^2 dx$$

$$\leq 2\left(\int_{-\pi}^{\pi} (f(x) - f_\varepsilon(x))^2 dx + \int_{-\pi}^{\pi} \left(f_\varepsilon(x) - S_n^\varepsilon(x)\right)^2 dx\right)$$

$$< 2\left(\frac{\varepsilon}{4} + \frac{\varepsilon}{4}\right) = \varepsilon \quad \text{if } n > N(\varepsilon).$$

Fourier Series on Arbitrary Intervals:

If instead of having our functions being periodic of period 2π they are periodic of period $2L$, the formulas for Fourier coefficients for $f(x)$ would change to

$$a_n = \frac{1}{L}\int_{-L}^{L} f(x)\cos\left(\frac{n\pi x}{L}\right)dx \quad n = 0, 1, 2, \ldots$$

$$b_n = \frac{1}{L}\int_{-L}^{L} f(x)\sin\left(\frac{n\pi x}{L}\right)dx \quad n = 1, 2, \ldots$$

and the formula the Fourier series for $f(x)$ would change to

$$f(x) \sim \frac{a_0}{2} + \sum_{n=1}^{\infty} a_n \cos\left(\frac{n\pi x}{L}\right) + b_n \sin\left(\frac{n\pi x}{L}\right)$$

as we verify in Exercise 4.

EXERCISES

1. Show that

$$D_N(x) = \frac{1}{2} + \cos x + \cos(2x) + \cdots + \cos(Nx)$$

$$= \begin{cases} \dfrac{\sin\left(N + \frac{1}{2}\right)x}{2\sin\left(\frac{x}{2}\right)}, & \text{if } x \neq 2\pi m, \ m \text{ an integer} \\[2ex] N + \dfrac{1}{2} & \text{if } x = 2\pi m \end{cases}.$$

2. Show that

$$A\cos t + B\sin t = \sqrt{A^2 + B^2}\sin(t + \varphi)$$

where $\varphi = \tan^{-1}(A/B)$. This yields another way to express a Fourier series.

3. Let

$$K_N(t) = \frac{D_0(t) + D_1(t) + \cdots + D_N(t)}{N+1}.$$

Show that

a. $\frac{1}{\pi}\int_{-\pi}^{\pi} K_N(t)dt = 1.$

b. $K_N(x) = \frac{1}{2(N+1)} \frac{\sin^2[(N+1)x/2]}{\sin^2\left(\frac{x}{2}\right)} \geq 0.$

c. $K_N(x) \leq \frac{1}{(N+1)(1-\cos\delta)}$ for $0 < \delta \leq |x| \leq \pi.$

d. $\lim_{x\to 0} K_N(x) = \frac{N+1}{2}.$

4. Show that

$$\int_{-\pi}^{\pi} f(u)D_N(x-u)du = \int_{-\pi}^{\pi} f(x-t)D_N(t)dt.$$

5. Suppose that $f(x)$ is piecewise continuous and periodic of period $2L$. Mimic the case for a function that is periodic of period 2π to show

$$a_n = \frac{1}{L}\int_{-L}^{L} f(x)\cos\left(\frac{n\pi x}{L}\right)dx \quad n = 0, 1, 2, \ldots$$

$$b_n = \frac{1}{L}\int_{-L}^{L} f(x)\sin\left(\frac{n\pi x}{L}\right)dx \quad n = 1, 2, \ldots$$

and the formula the Fourier series for $f(x)$ would change to

$$f(x) \sim \frac{a_0}{2} + \sum_{n=1}^{\infty} a_n \cos\left(\frac{n\pi x}{L}\right) + b_n \sin\left(\frac{n\pi x}{L}\right).$$

6.

a. Find the Fourier coefficients for

$$f(x) = x.$$

b. Use the answer to part (a) and Parseval's theorem to show $\sum_{n=1}^{\infty} \frac{1}{n^2} = \frac{\pi^2}{6}.$

7. Find the Fourier series for

$$f(x) = \begin{cases} x & 0 \leq x \leq \pi \\ 0 & -\pi \leq x < 0 \end{cases}.$$

8. Find the Fourier series for

$$f(x) = \begin{cases} x & 0 \le x \le \pi \\ -x & -\pi \le x < 0 \end{cases}.$$

Use this to show

$$\sum_{n=1}^{\infty} \frac{1}{(2n-1)^2} = \frac{\pi^2}{8}.$$

Note the result is same as problem 4, Section 3.1, but the technique is different.

9. Find the Fourier series for

$$f(x) = \begin{cases} \cos x & 0 \le x \le \pi \\ 0 & -\pi \le x < 0 \end{cases}.$$

10. Use a CAS to plot the functions in problems 6–9 and the first n terms of the Fourier series for $n = 3$, 10, and 25.

11. Find the Fourier series for $f(x) = x^2$, $-\pi < x < \pi$ and use it along with Parseval's theorem to show

$$\sum_{n=1}^{\infty} \frac{1}{n^4} = \frac{\pi^4}{90}.$$

4.4 THE EXPONENTIAL FORM OF FOURIER SERIES

In later sections, we shall sometimes find it convenient to express the Fourier series of a function in exponential form; that is,

$$f(x) = \sum_{n=-\infty}^{\infty} c_n e^{inx}.$$

This representation is possible because of Euler's formula

$$e^{inx} = \cos(nx) + i\sin(nx).$$

We will determine a formula for the coefficients c_n and a relationship between c_n and the a_n and b_n we defined previously. We have

$$\int_{-\pi}^{\pi} f(x)e^{ikx}\,dx = \int_{-\pi}^{\pi}\left(\sum_{n=-\infty}^{\infty} c_n e^{inx}\right)e^{ikx}dx = \sum_{n=-\infty}^{\infty} c_n\left(\int_{-\pi}^{\pi} e^{i(n+k)x}dx\right).$$

Now

$$\int_{-\pi}^{\pi} e^{i(n+k)x}dx = \begin{cases} 0 & \text{if } n+k \neq 0;\text{ i.e., unless } n=-k \\ 2\pi & \text{if } n+k=0;\text{ i.e., if } n=-k \end{cases}.$$

Thus

$$\int_{-\pi}^{\pi} f(x)e^{ikx}dx = \int_{-\pi}^{\pi}\left(\sum_{n=-\infty}^{\infty} c_n e^{inx}\right)e^{ikx}dx = \int_{-\pi}^{\pi} c_{-k}dx = 2\pi c_{-k}$$

and so

$$c_{-k} = \frac{1}{2\pi}\int_{-\pi}^{\pi} f(x)e^{ikx}dx \text{ or } c_k = \frac{1}{2\pi}\int_{-\pi}^{\pi} f(x)e^{-ikx}dx.$$

Example:
Find the exponential form of the Fourier series for

$$f(x) = \begin{cases} 0 & 0 \le x \le \pi \\ x & -\pi \le x < 0 \end{cases}.$$

We have for $k > 0$

$$c_k = \frac{1}{2\pi}\int_{-\pi}^{\pi} f(x)e^{-ikx}dx = \frac{1}{2\pi}\int_{-\pi}^{\pi} xe^{-ikx}dx.$$

Using integration by parts or a CAS

$$\int_{-\pi}^{0} xe^{-ikx}dx = -\frac{\cos(\pi k)}{k^2} - \pi\frac{\sin(\pi k)}{k} + \frac{1}{k^2} + i\left(\frac{\pi\cos(\pi k)}{k} - \frac{\sin(\pi k)}{k^2}\right).$$

For k an integer, this is

$$-\frac{\cos(\pi k)}{k^2} + \frac{1}{k^2} + i\left(\frac{\pi\cos(\pi k)}{k}\right) = \begin{cases} \pi\left(\dfrac{i}{k}\right) & \text{if } k \text{ is even} \\ \left(\dfrac{-i\pi}{k} + \dfrac{2}{k^2}\right) & \text{if } k \text{ is odd}. \end{cases}$$

Thus,

$$c_k = \begin{cases} \dfrac{1}{2\pi}\left(\dfrac{i\pi}{k}\right) & \text{if } k>0 \text{ is even} \\ \dfrac{1}{2\pi}\left(-i\dfrac{\pi}{k} + \dfrac{2}{k^2}\right) & \text{if } k>0 \text{ is odd} \end{cases}$$

$$c_{-k} = \frac{1}{2\pi}\int_{-\pi}^{\pi} f(x)e^{ikx}dx = \frac{1}{2\pi}\int_{-\pi}^{0} xe^{ikx}dx.$$

Now

$$\int_{-\pi}^{0} xe^{ikx}dx = -\frac{\cos(\pi k)}{k^2} - \pi\frac{\sin(\pi k)}{k} + \frac{1}{k^2} + i\left(-\frac{\pi\cos(\pi k)}{k} + \frac{\sin(\pi k)}{k^2}\right).$$

For k an integer, this is

$$-\frac{\cos(\pi k)}{k^2} + \frac{1}{k^2} + i\left(-\frac{\pi\cos(\pi k)}{k}\right)$$

$$-\frac{\cos(\pi k)}{k^2} + \frac{1}{k^2} + i\left(\frac{\pi\cos(\pi k)}{k}\right) = \begin{cases} \pi\left(\dfrac{i}{k}\right) & \text{if } k \text{ is even} \\ \left(\dfrac{-i\pi}{k} + \dfrac{2}{k^2}\right) & \text{if } k \text{ is odd}. \end{cases}$$

Thus,

$$c_{-k} = \begin{cases} \dfrac{1}{2\pi}\left(\dfrac{i\pi}{k}\right) & \text{if } k \text{ is even} \\ \dfrac{1}{2\pi}\left(-i\dfrac{\pi}{k} + \dfrac{2}{k^2}\right) & \text{if } k \text{ is odd.} \end{cases}$$

We now relate c_n and the a_n and b_n we defined previously. We have

$$a_n\cos(nx) = \frac{a_n}{2}\left(e^{inx} + e^{-inx}\right), \quad b_n\sin(nx) = \frac{b_n}{2i}\left(e^{inx} - e^{-inx}\right).$$

So

$$\begin{aligned}\sum_{n=0}^{\infty} a_n\cos(nx) + b_n\sin(nx) &= a_0 + (a_1\cos(x) + b_1\sin(x)) \\ &\quad (a_2\cos(2x) + b_2\sin(2x)) + \cdots \\ &= a_0 + \frac{a_1}{2}\left(e^{ix} + e^{-ix}\right) + \frac{a_2}{2}\left(e^{i2x} + e^{-i2x}\right) + \cdots \\ &\quad + \frac{b_1}{2i}\left(e^{ix} - e^{-ix}\right) + \frac{b_2}{2i}\left(e^{i2x} - e^{-i2x}\right) \\ &= a_0 + e^{ix}\left(\frac{a_1}{2} + \frac{b_1}{2i}\right) + e^{2ix}\left(\frac{a_2}{2} + \frac{b_2}{2i}\right) + \cdots \\ &\quad + e^{-ix}\left(\frac{a_1}{2} - \frac{b_1}{2i}\right) + e^{-2ix}\left(\frac{a_2}{2} - \frac{b_2}{2i}\right) + \cdots\end{aligned}$$

$$\sum_{n=-\infty}^{\infty} c_n e^{inx} = c_0 + c_1e^{i1x} + c_2e^{i2x} + \cdots + c_{-1}e^{-ix} + c_{-2}e^{-2ix} + \cdots.$$

If

$$\sum_{n=0}^{\infty} a_n \cos(nx) + b_n \sin(nx) = \sum_{n=-\infty}^{\infty} c_n e^{inx}$$

then we must have

$$c_0 = a_0, c_1 = \left(\frac{a_1}{2} + \frac{b_1}{2i}\right), c_2 = \left(\frac{a_1}{2} + \frac{b_2}{2i}\right), c_{-1} = \left(\frac{a_1}{2} - \frac{b_1}{2i}\right), c_{-2} = \left(\frac{a_1}{2} - \frac{b_2}{2i}\right).$$

So

$$c_j = \left(\frac{a_j}{2} + \frac{b_j}{2i}\right), c_{-j} = \left(\frac{a_j}{2} - \frac{b_j}{2i}\right).$$

From these equations, we also have

$$c_j + c_{-j} = \left(\frac{a_j}{2} + \frac{b_j}{2i}\right) + \left(\frac{a_j}{2} - \frac{b_j}{2i}\right) = a_j$$

and

$$c_j - c_{-j} = \left(\frac{a_j}{2} + \frac{b_j}{2i}\right) - \left(\frac{a_j}{2} - \frac{b_j}{2i}\right) = \frac{1}{i} b_j \quad \text{so } b_j = i(c_j - c_{-j}).$$

Note: if trigonometric series is as above, then $c_0 = a_0$, but we often give the first term as $a_0/2$ in which case $c_0 = a_0/2$.

EXERCISES

1. Find the exponential form of the Fourier series for the Dirac delta function, given that the Fourier series for $\delta(x)$ is

$$\frac{1}{2\pi} + \frac{1}{\pi} \sum_{n=1}^{\infty} \cos nx.$$

2. Find the exponential form of the Fourier series for

$$f(x) = \begin{cases} 1 & 0 \le x \le \pi \\ -1 & -\pi \le x < 0 \end{cases}.$$

3. Find the exponential form of the Fourier series for

$$f(x) = \begin{cases} x & 0 \le x \le \pi \\ 0 & -\pi \le x < 0 \end{cases}.$$

4. Find the exponential form of the Fourier series for

$$f(x) = \begin{cases} x & 0 \le x \le \pi \\ -x & -\pi \le x < 0 \end{cases}.$$

5. Find the exponential form of the Fourier series for

$$f(x) = \begin{cases} \cos x & 0 \le x \le \pi \\ 0 & -\pi \le x < 0 \end{cases}.$$

6. Find the exponential form of the Fourier series for $f(x) = |\sin x|$.

7. Find a_k and b_k in terms of c_k and c_{-k}.

8. This exercise develops the Poisson integral formula in polar coordinates. We suppose that $u(r,\theta)$ is a solution to $\Delta u = 0$; $u(1,\theta) = f(\theta)$, which is Laplace's equation on a circle of radius 1.
Suppose that

$$u(r,\theta) = \frac{1}{2\pi}\int_{-\pi}^{\pi} f(\varphi)d\varphi + \sum_{n=1}^{\infty} r^n \left[\int_{-\pi}^{\pi} f(\varphi)\cos n\theta \cos n\varphi \, d\varphi + \int_{-\pi}^{\pi} f(\varphi)\sin n\theta \sin n\varphi \, d\varphi\right].$$

a. Show that $u(r,\theta) = \frac{1}{\pi}\int_{-\pi}^{\pi} f(\varphi)\left[\frac{1}{2} + \sum_{n=1}^{\infty} r^n \cos[n(\theta-\varphi)]\right]d\varphi$.

b. Show that $\left(r^2 + 1 - 2r\cos\theta\right)\left[\sum_{n=1}^{\infty} r^n \cos n\theta\right] = r\cos\theta - r^2$.

c. Show that $\left(r^2 + 1 - 2r\cos\theta\right)\left[\frac{1}{2} + \sum_{n=1}^{\infty} r^n \cos n\theta\right] = \frac{1}{2}\left(1 - r^2\right)$.

d. Replace θ by $(\theta - \varphi)$ and conclude

$$u(r,\theta) = \frac{1-r^2}{2\pi}\int_{-\pi}^{\pi} \frac{f(\varphi)}{1 + r^2 - 2r\cos(\theta - \varphi)} d\varphi.$$

In the case the radius of the circle is R, the formula changes to

$$u(r,\theta) = \frac{R^2 - r^2}{2\pi} \int_{-\pi}^{\pi} \frac{f(\varphi)}{R^2 + r^2 - 2rR\cos(\theta - \varphi)} d\varphi.$$

Either of these formulas is the Poisson integral formula.

4.5 FOURIER SINE AND COSINE SERIES

Suppose that $f(x)$ is a function defined on $(0,\pi]$. We can extend $f(x)$ to be an even function on $[-\pi,\pi]$, which we denote by $f_1(x)$, by defining

$$f_1(x) = \begin{cases} f(x) & \text{if } 0 < x \le \pi \\ a_0 & \text{if } x = 0 \\ f(-x) & \text{if } -\pi \le x < 0 \end{cases}.$$

We can then extend $f_1(x)$ to be periodic of period 2π on $(-\infty,\infty)$ by defining $f_1(x + 2\pi) = f_1(x)$. Likewise, we can extend $f(x)$ to be an odd function on $[-\pi,\pi]$, which we denote by $f_2(x)$, by defining

$$f_2(x) = \begin{cases} f(x) & \text{if } 0 < x \le \pi \\ 0 & \text{if } x = 0 \\ -f(-x) & \text{if } -\pi \le x < 0 \end{cases}.$$

We can then extend $f_2(x)$ to be periodic of period 2π on $(-\infty,\infty)$ by defining $f_2(x + 2\pi) = f_2(x)$.

If either $f_1(x)$ or $f_2(x)$ can be expressed as a Fourier series, then both can be expressed as a Fourier series. Suppose that this is the case. Then $f_1(x)$ being an even function can be written in its Fourier series

$$f_1(x) = \frac{a_0}{2} + \sum_{n=1}^{\infty} a_n \cos nx$$

and $f_2(x)$ being an odd function can be written in its Fourier series

$$f_2(x) = \sum_{n=1}^{\infty} b_n \sin nx.$$

The point of this argument is that if $f(x)$ is a suitably well-behaved function on $(0,\pi]$, then it can be expressed as either a Fourier cosine series or a Fourier sine series. We now determine a formula for the Fourier coefficients. Consider

$$f_2(x) = \sum_{n=1}^{\infty} b_n \sin nx.$$

Then

$$b_n = \frac{1}{\pi} \int_{-\pi}^{\pi} f_2(x) \sin nx \, dx.$$

Since both $f_2(x)$ and $\sin nx$ are odd functions, their product is an even function, so

$$b_n = \frac{1}{\pi} \int_{-\pi}^{\pi} f_2(x) \sin nx \, dx = \frac{2}{\pi} \int_{0}^{\pi} f_2(x) \sin nx \, dx.$$

Thus we can expand $f(x)$ on $(0,\pi]$ into a sine series $\sum_{n=1}^{\infty} b_n \sin nx$ where

$$b_n = \frac{2}{\pi} \int_{0}^{\pi} f(x) \sin nx \, dx.$$

If instead of the interval being $(0,\pi]$ it is $(0,L]$, the formula for b_n will change to

$$b_n = \frac{2}{L} \int_{0}^{L} f(x) \sin\left(\frac{n\pi x}{L}\right) dx.$$

Similarly, we can expand $f(x)$ on $(0,\pi]$ into a cosine series

$$\frac{a_0}{2} + \sum_{n=1}^{\infty} a_n \cos nx$$

where

$$a_n = \frac{2}{\pi} \int_{0}^{\pi} f(x) \cos nx \, dx$$

if the interval is $(0,\pi]$ and

$$a_n = \frac{2}{L} \int_{0}^{L} f(x) \cos\left(\frac{n\pi x}{L}\right) dx$$

if the interval is $(0,L]$.

Example:

Let $f(x) = \sin x$ on $(0,\pi]$. We find the cosine series for $f(x)$. We have

$$a_n = \frac{2}{\pi} \int_{0}^{\pi} f(x) \cos nx \, dx = \frac{2}{\pi} \int_{0}^{\pi} \sin x \cos nx \, dx.$$

Now

$$\sin(x + nx) = \sin x \cos nx - \sin nx \cos x$$

$$\sin(x - nx) = \sin x \cos nx + \sin nx \cos x$$

so

$$\sin x \cos nx = \frac{1}{2}[\sin(x+nx) + \sin(x-nx)]$$

and

$$a_n = \frac{2}{\pi}\int_0^{\pi} \sin x \cos nx dx = \frac{1}{\pi}\int_0^{\pi} [\sin(x+nx) + \sin(x-nx)]dx.$$

If $n = 1$ then

$$a_1 = \frac{1}{\pi}\int_0^{\pi} [\sin(x+x) + \sin(x-x)]dx = \frac{1}{\pi}\int_0^{\pi} [\sin(2x)]dx = 0.$$

Otherwise,

$$a_n = \frac{1}{\pi}\left[\int_0^{\pi} \sin(1+n)xdx + \int_0^{\pi} \sin(1-n)xdx\right].$$

Now, if $n \neq 1$

$$\int_0^{\pi} \sin(1+n)xdx = -\frac{1}{n+1}\cos(1+n)x|_{x=0}^{\pi} = -\frac{1}{n+1}[\cos(1+n)\pi - 1]$$

and

$$\int_0^{\pi} \sin(1-n)xdx = -\frac{1}{1-n}\cos(1-n)x|_{x=0}^{\pi} = \frac{1}{n-1}[\cos(1-n)\pi - 1].$$

Now

$$\cos(1+n)\pi - 1 = \cos(1-n)\pi - 1 = \begin{cases} -2 & \text{if } n \text{ is even} \\ 0 & \text{if } n \text{ is odd} \end{cases}$$

so

$$\int_0^{\pi} \sin(1+n)xdx + \int_0^{\pi} \sin(1-n)xdx$$
$$= \begin{cases} \dfrac{2}{n+1} - \dfrac{2}{n-1} = \dfrac{-4}{n^2-1} = \dfrac{4}{1-n^2} & \text{if } n \text{ is even} \\ 0 & \text{if } n \text{ is odd} \end{cases}.$$

Thus

$$a_n = \frac{1}{\pi}\left[\int_0^{\pi} \sin(1+n)xdx + \int_0^{\pi} \sin(1-n)xdx\right] = \frac{2}{\pi}\cdot\frac{1+(-1)^n}{1-n^2}.$$

Hence, the cosine series for $f(x) = \sin x$ on $(0,\pi]$ is

$$\frac{a_0}{2} + \sum_{n=1}^{\infty} a_n \cos nx = \frac{2}{\pi} + \frac{2}{\pi}\sum_{n=2}^{\infty} \frac{1+(-1)^n}{1-n^2}\cos nx = \frac{2}{\pi} - \frac{4}{\pi}\sum_{n=1}^{\infty} \frac{\cos(2\,nx)}{4n^2-1}.$$

EXERCISES

1. Let $f(x) = \cos x$ on $(0,\pi]$. Find the sine series for $f(x)$.
2. Find the sine and cosine series for the following functions on $(0,\pi]$.
   a. $f(x) = 1$.
   b. $f(x) = x$.
   c. $f(x) = x^2$.

4.6 DOUBLE FOURIER SERIES

In later applications we shall want to express a function of two variables $f(x,y)$, $0 \le x \le \pi$, $0 \le y \le \pi$ as a double series of sine functions. That is, we want to find constants B_{mn} so that

$$f(x,y) = \sum_{n=1}^{\infty}\left(\sum_{m=1}^{\infty} B_{mn} \sin my\right)\sin nx.$$

We show how the numbers B_{mn} are determined.

Fix $\bar{y} \in [0,\pi]$. Then $f(x,\bar{y})$ is a function of x. If $f(x,y)$ is continuous in both x and y, then

$$f(x,\bar{y}) = \sum_{n=1}^{\infty} b_n(\bar{y}) \sin nx \tag{1}$$

where the notation $b_n(\bar{y})$ is used to emphasize that the constants depend on the fixed value of y that we have chosen. We know that $b_n(\bar{y})$ is computed according to

$$b_n(\bar{y}) = \frac{2}{\pi}\int_0^{\pi} f(x,\bar{y}) \sin nx.$$

Now

$$f(x,\bar{y}) = \sum_{n=1}^{\infty}\left(\sum_{m=1}^{\infty} B_{mn} \sin m\bar{y}\right)\sin nx. \tag{2}$$

From Eqs. (1) and (2) we get

$$b_n(\bar{y}) = \left(\sum_{m=1}^{\infty} B_{mn} \sin m\bar{y}\right)$$

so

$$\frac{2}{\pi}\int_0^{\pi} f(x,\bar{y})\sin nx = \left(\sum_{m=1}^{\infty} B_{mn}\sin m\bar{y}\right)$$

for each $\bar{y} \in [0, \pi]$.

Now fix n and define a function of y, denoted $F_n(y)$, by

$$F_n(y) = \frac{2}{\pi}\int_0^{\pi} f(x,y)\sin nx dx \tag{3}$$

so that

$$F_n(y) = \left(\sum_{m=1}^{\infty} B_{mn}\sin my\right). \tag{4}$$

From Eq. (4) we get

$$\begin{aligned} B_{mn} &= \frac{2}{\pi}\int_0^{\pi} F_n(y)\sin my\, dy \\ &= \frac{2}{\pi}\int_0^{\pi}\left[\frac{2}{\pi}\int_0^{\pi} f(x,y)\sin nx\, dx\right]\sin my\, dy \\ &= \frac{4}{\pi^2}\int_0^{\pi}\int_0^{\pi} f(x,y)\sin nx \sin my\, dxdy. \end{aligned}$$

EXERCISES

1. Find the coefficients for the double Fourier sine series for the following functions for $0 \le x, y \le \pi$

   a. $f(x,y) = x + y$.

   b. $f(x,y) = xy$.

   c. $f(x,y) = x^2y^2$.

CHAPTER

Three Important Equations

5

5.1 INTRODUCTION

Partial differential equations (PDEs) are extremely important in both mathematics and physics. A major purpose of this text is to give an introduction to some of the simplest and most important PDEs in both disciplines, and techniques for their solution. Accordingly, we focus on three equations:

1. the heat equation

$$\frac{\partial u(t, \widehat{x})}{\partial t} = \kappa \Delta u(t, \widehat{x})$$

2. the wave equation

$$\frac{\partial^2 u(t, \widehat{x})}{\partial t^2} = \alpha^2 \Delta u(t, \widehat{x})$$

3. Laplace's equation

$$\Delta u(x, y) = 0.$$

It would appear that we are severely limiting ourselves by examining only three equations. However, these encompass any PDE of the form

$$A \frac{\partial^2 u(t, x)}{\partial t^2} + B \frac{\partial^2 u(t, x)}{\partial x \partial t} + C \frac{\partial^2 u(t, x)}{\partial x^2} = 0 \tag{1}$$

because by a linear transformation of variables, any equation of the form of Eq. (1) can be transformed into one of the three equations. We show at the end of the section that the transformation is exactly the same as transforming an equation of the form

$$Ax^2 + Bxy + Cy^2 = 0 \tag{2}$$

into the standard form of a parabola, hyperbola, or an ellipse by a rotation of axes according to the sign of $B^2 - 4AC$. Following the nomenclature of the geometrical

Mathematical Physics with Partial Differential Equations. https://doi.org/10.1016/B978-0-12-814759-7.00005-3

figures, if $B^2 - 4AC < 0$ the PDE is said to be parabolic, if $B^2 - 4AC = 0$ the equation is elliptic, and if $B^2 - 4AC > 0$ the equation is hyperbolic. Thus the heat equation is the prototypical parabolic equation, the wave equation is the prototypical wave equation, and Laplace's equation is the prototypical elliptical equation. We shall see that for the heat equation, the initial conditions diffuse in time whereas in the wave equation initial conditions are propagated, changing position but not shape.

The Laplacian appears in each of these equations, and we begin by exploring the physical significance of that operator. Consider the heat equation. Here, $u(t,x,y,z)$ is the temperature of a solid homogenous body at the point (x,y,z) at time t. Then $\frac{\partial u(t,x,y,z)}{\partial t}$ is the rate of change of the temperature at the point (x,y,z). The temperature at (x,y,z) will undergo a change if and only if the temperature in the immediate vicinity at (x,y,z) is different than at (x,y,z). We consider the case in one space dimension.

Suppose we want to compare the value of a function $f(x)$ with the average value of the function at $x+h$ and $x-h$; i.e., we compare $f(x)$ with $[f(x+h)+f(x-h)]/2$. How does $f''(x)$ enter in? By the definition of the second derivative

$$f''(x) = \lim_{h\to 0} \frac{f'(x+h) - f'(x)}{h} = \lim_{h\to 0} \frac{f'(x) - f'(x-h)}{h}. \tag{3}$$

We use the second limit. Approximate $f'(x)$ and $f'(x-h)$ using

$$f'(x) = \lim_{h\to 0} \frac{f(x+h) - f(x)}{h}$$

and

$$f'(x-h) = \lim_{h\to 0} \frac{f(x) - f(x-h)}{h}.$$

Substitute the approximations into Eq. (3) to get

$$f''(x) \approx \frac{\dfrac{f(x+h)-f(x)}{h} - \dfrac{f(x)-f(x-h)}{h}}{h} = \frac{f(x+h) - 2f(x) + f(x-h)}{h^2}$$

$$= \frac{2}{h^2}\left\{\frac{1}{2}[f(x+h) + f(x-h)] - f(x)\right\}.$$

Thus $f''(x)$ is a measure of the difference between the value of a function at a point and the average of the values of the function in the immediate vicinity.

5.2 LAPLACE'S EQUATION

The simplest second-order PDE is Laplace's equation, which in two variables is

$$\Delta u(x,y) = u_{xx} + u_{yy} = 0.$$

Solutions to Laplace's equation are called harmonic functions, which play a key role in complex analysis. We review some facts from complex analysis that will be important for us.

An analytic function $f(x,y)$ is of the form

$$f(x,y) = u(x,y) + iv(x,y)$$

where $u(x,y)$ and $v(x,y)$ are harmonic functions. It is not true that any combination of two harmonic functions is an analytic function. For $f(x,y)$ to be analytic, $v(x,y)$ must be a harmonic conjugate of (x,y). This means $u(x,y)$ and $v(x,y)$ must satisfy the Cauchy–Riemann equations, which are

$$\frac{\partial u}{\partial x} = \frac{\partial v}{\partial y} \text{ and } \frac{\partial u}{\partial y} = -\frac{\partial v}{\partial x}.$$

A domain in the complex plane is a connected open set. A simply connected domain is a domain with no holes. If D is a simply connected domain and $u(x,y)$ is a harmonic function, then the function $v(z)$ is defined by

$$v(z) = \int_{z_0}^{z} u_x dy - u_y dx$$

for any path in D that connects z_0 and z is a harmonic conjugate of $u(x,y)$ and

$$f(x,y) = u(x,y) + iv(x,y)$$

is analytic in D. From complex analysis, we have the following result:

MAXIMUM MODULUS PRINCIPLE

If $f(z)$ is analytic and not constant on a domain D, then $|f(z)|$ does not attain its maximum on D. That is, the maximum of $|f(z)|$ on the closure of D is attained on the boundary of D.

Corollary:
If $f(z)$ is analytic and not constant on a domain D and if $f(z) \neq 0$ on D, then $|f(z)|$ does not attain its minimum on D.

Proof:
Let $g(z) = \frac{1}{f(z)}$. Then $g(z)$ is analytic and $|g(z)|$ does not attain its maximum on D.

Corollary:
If

$$f(x,y) = u(x,y) + iv(x,y)$$

is analytic and not constant on a domain D, then $u(x,y)$ does not attain its maximum on D.

Proof:
We have

$$\left|e^{f(z)}\right| = \left|e^{u(x,y)+iv(x,y)}\right| = \left|e^{u(x,y)}\right|\left|e^{iv(x,y)}\right| = e^{u(x,y)}$$

since

$$\left|e^{i\theta}\right| = 1 \text{ and } \left|e^{u(x,y)}\right| = e^{u(x,y)}$$

and if $f(z)$ is analytic, then $e^{f(z)}$ is analytic.

EXERCISES

1. Determine whether the following functions are harmonic:
   a. $f(x,y) = x^2y - y^2x$.
   b. $f(x,y) = x^3y - y^3x$.
   c. $f(x,y) = e^x \sin y$.
   d. $f(x,y) = \sin x - \cos y$.
   e. $f(x,y) = \ln(x^2 + y^2)$
   f. $f(x,y) = \arctan\left(\frac{y}{x}\right)$.
2. Show that $u(r,\theta) = \ln r$, $r > 0$, $0 < \theta < 2\pi$ is harmonic.
3. Determine whether $u(r,\theta) = \sqrt{r}e^{i\theta/2}$, $r > 0$, $0 < \theta < 2\pi$ is harmonic.

5.3 DERIVATION OF THE HEAT EQUATION IN ONE DIMENSION

Before presenting the heat equation, we review the concept of heat. Energy transfer that takes place because of temperature difference is called heat flow. The energy transferred in this way is called heat. Thus heat refers to the transfer of energy, not the amount of energy contained within a system. An example of a unit of heat is the calorie. One calorie is the amount of heat required to raise 1 g of water from 14.5°C to 15.5°C.

The quantity of heat H required to raise a body of mass m from T_1 to T_2 is approximately proportional to $\Delta T = T_1 - T_2$, and is proportional to m. It is also dependent on the material of the body. This quality is called the specific heat of the material—typically denoted c. Thus we have $H = mc\Delta T$. The amount of heat required for an infinitesimal change in temperature dT is denoted dH. To summarize, heat is energy in transit, and dH does not represent the change in the amount of heat in the body inasmuch as "the amount of heat in a body" has no meaning. When a quantity of heat dH is transferred in time dt, then the rate of energy transfer is $\frac{dH}{dt}$.

We now derive the heat equation in one dimension. Suppose that we have a rod of length L. While the derivation will be for the case that the rod is one-dimensional, it is advantageous to visualize the rod as having a cross-sectional area of one square unit. Let

$u(x,t)$ = the temperature of the rod at the point x $(0 \le x \le L)$ at time $t (t \ge 0)$.
$\widehat{q}(x,t)$ = the heat flow at point x at time t (a vector quantity)
ρ = the density of the material (assumed to be constant)
c = the specific heat of the material
ΔQ = change in internal energy
Δu = change in temperature.

For $\widehat{n}$, a unit vector, $\widehat{q}\cdot\widehat{n}$, is the heat flux in the direction of $\widehat{n}$. The two laws that we use in our derivation of the heat equation are conservation of energy and Fourier's law. Fourier's law is

$$\widehat{q}(x,t) = -k\nabla u(x,t)$$

which, in one dimension, is

$$\widehat{q}(x,t) = -k\frac{\partial u(x,t)}{\partial x}\widehat{i}. \tag{1}$$

The number k is a constant of the material called the thermal conductivity. The reason for the negative sign is that heat flows from higher to lower temperatures and $\frac{\partial u(x,t)}{\partial x}$ is positive if $u(x,t)$ is increasing as x increases.

Consider the small segment $[x, x+\Delta x]$. (See Fig. 5.2.1.) The amount of heat energy passing through the segment at point x in time Δt is approximately $\widehat{q}(x,t)\cdot\widehat{i}(\Delta t)$, and the amount of heat energy passing through the segment at point $x+\Delta x$ in time Δt is approximately $\widehat{q}(x+\Delta x,t)\cdot\widehat{i}(\Delta t)$. Depending on the direction of heat flow and in the absence of a heat source or a heat sink, one of these quantities will represent an addition to the internal energy of the segment, and the other will represent a removal of internal energy from the segment. For definiteness, suppose energy is being added at x and removed at $x+\Delta x$. Then change in internal energy is

$$\Delta Q = \left[\widehat{q}(x,t)\cdot\widehat{i} - \widehat{q}(x+\Delta x,t)\cdot\widehat{i}\right]\Delta t. \tag{2}$$

When the amount of energy ΔQ is added to a body of mass m and specific heat c, the temperature of the body rises according to

$$\Delta Q = mc\Delta u. \tag{3}$$

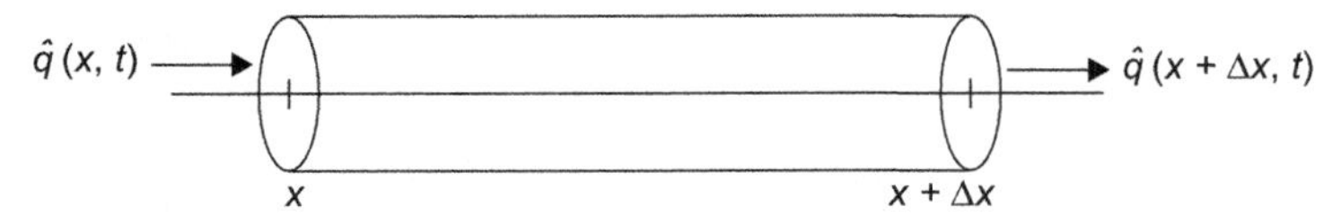

FIGURE 5.2.1

Thus from Eqs. (2) and (3) we get

$$\left[\widehat{q}(x,t)\cdot\widehat{i}-\widehat{q}(x+\Delta x,t)\cdot\widehat{i}\right]\Delta t = mc\Delta u$$

or

$$\widehat{q}(x,t)\cdot\widehat{i}-\widehat{q}(x+\Delta x,t)\cdot\widehat{i} = mc\frac{\Delta u}{\Delta t}.$$

Taking the limit as $\Delta t \to 0$ gives

$$\widehat{q}(x,t)\cdot\widehat{i}-\widehat{q}(x+\Delta x,t)\cdot\widehat{i} = mc\frac{\partial u}{\partial t}.$$

So, from Eqs. (1) and (2) we get

$$\widehat{q}(x,t)\cdot\widehat{i}-\widehat{q}(x+\Delta x,t)\cdot\widehat{i} = -k\left[\frac{\partial u(x,t)}{\partial x}-\frac{\partial u(x+\Delta x,t)}{\partial x}\right] = mc\frac{\partial u}{\partial t} = \rho(\Delta x)c\frac{\partial u}{\partial t}$$

or

$$\frac{-k\left[\frac{\partial u(x,t)}{\partial x}-\frac{\partial u(x+\Delta x,t)}{\partial x}\right]}{\Delta x} = \rho c\frac{\partial u}{\partial t}.$$

Taking the limit as $\Delta x \to 0$ gives

$$k\frac{\partial^2 u(x,t)}{\partial x^2} = \rho c\frac{\partial u}{\partial t}.$$

This is often written

$$\frac{\partial^2 u(x,t)}{\partial x^2} = \frac{1}{\alpha^2}\frac{\partial u}{\partial t},$$

where $\alpha^2 = \frac{k}{\rho c}$.

EXERCISES

1. Show that if there is a heat source $f(x)$, then the heat equation becomes

$$\frac{1}{\alpha^2}\frac{\partial u}{\partial t} = \frac{\partial^2 u(x,t)}{\partial x^2} + f(x).$$

5.4 DERIVATION OF THE WAVE EQUATION IN ONE DIMENSION

Suppose we have a string stretched along the x-axis between $x = 0$ and $x = L$ with tension T. The string is fixed at the endpoints. We distort the string in the vertical direction by plucking it, which will cause the string to vibrate in the vertical direction. We assume there is no damping and no external forces, and we want to find the equation that governs the dynamics of motion.

Consider Fig. 5.3.1 below that depicts the forces on a small element of string in the interval $[x, x + \Delta x]$. We let

$u(x,t)$ = the vertical displacement of the string from the x-axis at point x at time t
$\theta(x,t)$ = the angle the string makes with a horizontal line at point x at time t
$T(x,t)$ = the tension in the string at point x at time t
$\rho(x)$ = the mass density of the string at point x.

We assume that the string is perfectly flexible, so that the forces that parts of the string exert on one another are tangential to the string. This means that $T(x,t)$ is tangent to the string at (x,t) so

$$\tan[\theta(x,t)] = \text{the slope of the tangent line} = \frac{\partial T(x,t)}{\partial x}.$$

We assume that there is no net force in the horizontal direction so that the string vibrates vertically. This means

$$T(x,t)\cos[\theta(x,t)] = T(x + \Delta x, t)\cos[\theta(x + \Delta x, t)];$$

i.e., the horizontal force to the left is equal to the horizontal force to the right.

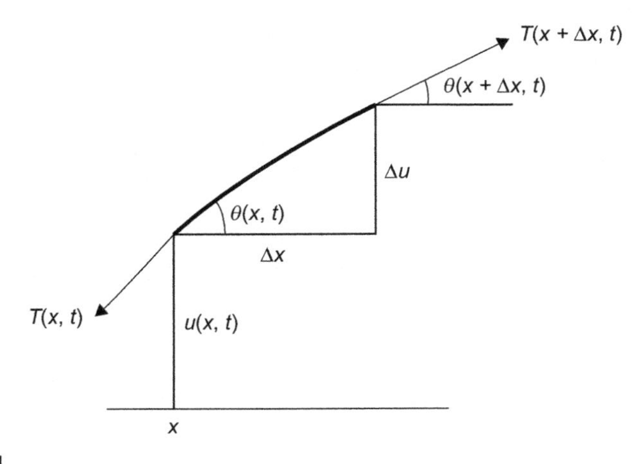

FIGURE 5.3.1

Analysis of forces in the u (vertical) direction:

The net force in the vertical direction is not zero, and we apply Newton's law, Force = mass × acceleration. The mass of the string between x and $x+\Delta x$ is approximately

$$\rho(x)\sqrt{(\Delta x)^2+(\Delta u)^2}.$$

The acceleration is the second derivative with respect to time, which is $\frac{\partial^2 u(x,t)}{\partial t^2}$ at the point (x,t).

The force in the u-direction is the sum of $T(x,t)\sin[\theta(x,t)]$ and $T(x+\Delta x,t)\ \sin[\theta(x+\Delta x,t)]$, which, because of the direction of $T(x,t)$ and $T(x+\Delta x,t)$, is

$$T(x+\Delta x,t)\sin[\theta(x+\Delta x,t)]-T(x,t)\sin[\theta(x,t)].$$

Thus the governing equation in the u-direction is

$$\begin{aligned}\rho(x)\sqrt{(\Delta x)^2+(\Delta u)^2}\ \frac{\partial^2 u(x,t)}{\partial t^2}&\\ = T(x+\Delta x,t)\sin[\theta(x+\Delta x,t)]&-T(x,t)\sin[\theta(x,t)].\end{aligned} \tag{1}$$

This is the equation that connects the tension and the vertical displacement from equilibrium.

Divide Eq. (1) by Δx to get

$$\rho(x)\sqrt{1+\left(\frac{\Delta u}{\Delta x}\right)^2}\ \frac{\partial^2 u(x,t)}{\partial t^2}=\frac{T(x+\Delta x,t)\sin(x+\Delta x,t)-T(x,t)\sin[\theta(x,t)]}{\Delta x}. \tag{2}$$

We now work with the right-hand side of Eq. (2). Recall that

$$T(x,t)\cos[\theta(x,t)]=T(x+\Delta x,t)\cos[\theta(x+\Delta x,t)]$$

so the right-hand side of Eq. (1) is equal to

$$\begin{aligned}&\left(\frac{\dfrac{T(x+\Delta x,t)\sin[\theta(x+\Delta x,t)]}{T(x+\Delta x,t)\cos[\theta(x+\Delta x,t)]}-\dfrac{T(x,t)\sin[\theta(x,t)]}{T(x,t)\cos[\theta(x,t)]}}{\Delta x}\right)T(x,t)\cos[\theta(x,t)]\\ &=\frac{\tan[\theta(x+\Delta x,t)]-\tan[\theta(x,t)]}{\Delta x}T(x,t)\cos[\theta(x,t)].\end{aligned} \tag{3}$$

Now

$$\tan[\theta(x,t)]=\frac{\partial T(x,t)}{\partial x}$$

so the right-hand side of Eq. (3) can be written

$$\left(\frac{\dfrac{\partial T(x+\Delta x,t)}{\partial x}-\dfrac{\partial T(x,t)}{\partial x}}{\Delta x}\right)T(x,t)\cos\left[\theta(x,t)\right].$$

Thus Eq. (2) can be written

$$\rho(x)\sqrt{1+\left(\frac{\Delta u}{\Delta x}\right)^2}\,\frac{\partial^2 u(x,t)}{\partial t^2} = \left(\frac{\dfrac{\partial T(x+\Delta x,t)}{\partial x}-\dfrac{\partial T(x,t)}{\partial x}}{\Delta x}\right)T(x,t)\cos\left[\theta(x,t)\right]. \tag{4}$$

Taking the limit as $\Delta x \to 0$ in Eq. (4) gives

$$\rho(x)\sqrt{1+\left(\frac{\partial u}{\partial x}\right)^2}\,\frac{\partial^2 u(x,t)}{\partial t^2} = \frac{\partial^2 T(x,t)}{\partial x^2}T(x,t)\cos\left[\theta(x,t)\right]. \tag{5}$$

We want to express the right-hand side of Eq. (5) in terms of $u(x,t)$. To do this, we refer back to Fig. 5.3.1 and note that

$$\frac{\partial T(x,t)}{\partial x} = \tan\left[\theta(x,t)\right] = \lim_{\Delta x\to 0}\frac{\Delta u}{\Delta x} = \frac{\partial u}{\partial x} \text{ and } \frac{\partial^2 T(x,t)}{\partial x^2} = \frac{\partial^2 u}{\partial x^2}.$$

So we have

$$\rho(x)\sqrt{1+\left(\frac{\partial u}{\partial x}\right)^2}\,\frac{\partial^2 u(x,t)}{\partial t^2} = \frac{\partial^2 u}{\partial x^2}T(x,t)\cos\left[\theta(x,t)\right]. \tag{6}$$

Eq. (6) is not solvable, so all derivations of the equation known as the wave equation make simplifying assumptions. (Actually, we have ignored some factors already, including elasticity.) For a more complete derivation that includes these factors, see Weinberger, *A First Course in Partial Differential Equations*. We assume that the mass density of the string is constant and replace $\rho(x)$ by ρ and assume that θ is small so that

$$T(x,t)\cos\left[\theta(x,t)\right] \approx T(x,t) \text{ and } \tan\left[\theta(x,t)\right] \approx 0$$

and thus

$$\sqrt{1+\left(\frac{\partial u}{\partial x}\right)^2} \approx 1.$$

If we assume that $T(x,t)$ is constant, $T(x,t) = T$, with these assumptions and approximations, Eq. (6) becomes

$$\rho \frac{\partial^2 u(x,t)}{\partial t^2} = T \frac{\partial^2 u(x,t)}{\partial x^2}$$

or

$$\frac{\partial^2 u(x,t)}{\partial t^2} = c^2 \frac{\partial^2 u(x,t)}{\partial x^2} \tag{7}$$

where $c^2 = \frac{T}{\rho}$.

Eq. (7) is the wave equation with no external forces. If there were external forces present whose sum was $F(x,t)$, then we would modify Eq. (7) to

$$\rho(x)\sqrt{(\Delta x)^2 + (\Delta u)^2}\ \frac{\partial^2 u(x,t)}{\partial t^2}$$
$$= T(x+\Delta x,t)\sin\left[\theta(x+\Delta x,t)\right] - T(x,t)\sin\left[\theta(x,t)\right] + F(x,t)$$

and continue the analysis as we did. The result would be the equation

$$\rho \frac{\partial^2 u(x,t)}{\partial t^2} = T \frac{\partial^2 u(x,t)}{\partial x^2} + F(x,t).$$

EXERCISES

1. Complete the derivation of the wave equation in the case there is an external force $F(x,t)$.
2. Show that a solution to

$$\frac{\partial^2 u(\xi,\eta)}{\partial \xi \partial \eta} = 0$$

is $u(\xi,\eta) = f(\xi) + g(\eta)$. (We use this in the next section.)

3. Show that a solution to

$$\frac{\partial^2 u(x,t)}{\partial x^2} = \frac{1}{v^2}\frac{\partial^2 u(x,t)}{\partial x^2}$$

is $u(x,t) = f(x - vt) + g(x + vt)$ where f and g have continuous second derivatives with respect to x and t.

5.5 AN EXPLICIT SOLUTION OF THE WAVE EQUATION

The wave equation in one dimension

$$\frac{\partial^2 u(x,t)}{\partial t^2} - c^2 \frac{\partial^2 u(x,t)}{\partial x^2} = 0$$

can be solved explicitly by making a change of variables that will convert the equation into the form

$$\frac{\partial^2 u(\xi, \eta)}{\partial \xi \partial \eta} = 0.$$

We demonstrate that the change of variables

$$\xi = x + ct$$

$$\eta = x - ct$$

accomplishes the desired transformation. To do this, we use the chain rule several times. We have

$$\frac{\partial u}{\partial x} = \frac{\partial u}{\partial \xi}\cdot\frac{\partial \xi}{\partial x} + \frac{\partial u}{\partial \eta}\cdot\frac{\partial \eta}{\partial x} = \frac{\partial u}{\partial \xi}\cdot 1 + \frac{\partial u}{\partial \eta}\cdot 1.$$

We also have

$$\frac{\partial^2 u}{\partial x^2} = \frac{\partial}{\partial x}\left(\frac{\partial u}{\partial x}\right) = \frac{\partial}{\partial x}\left(\frac{\partial u}{\partial \xi} + \frac{\partial u}{\partial \eta}\right).$$

Now

$$\frac{\partial}{\partial x}\left(\frac{\partial u}{\partial \xi}\right) = \frac{\partial}{\partial \xi}\left(\frac{\partial u}{\partial \xi}\right)\cdot\frac{\partial \xi}{\partial x} + \frac{\partial}{\partial \eta}\left(\frac{\partial u}{\partial \xi}\right)\cdot\frac{\partial \eta}{\partial x} = \frac{\partial^2 u}{\partial \xi^2}\cdot 1 + \frac{\partial^2 u}{\partial \xi \partial \eta}\cdot 1 = \frac{\partial^2 u}{\partial \xi^2} + \frac{\partial^2 u}{\partial \xi \partial \eta}$$

and

$$\frac{\partial}{\partial x}\left(\frac{\partial u}{\partial \eta}\right) = \frac{\partial}{\partial \xi}\left(\frac{\partial u}{\partial \eta}\right)\cdot\frac{\partial \xi}{\partial x} + \frac{\partial}{\partial \eta}\left(\frac{\partial u}{\partial \eta}\right)\cdot\frac{\partial \eta}{\partial x} = \frac{\partial^2 u}{\partial \xi \partial \eta} + \frac{\partial^2 u}{\partial \eta^2}.$$

Thus

$$\frac{\partial^2 u}{\partial x^2} = \frac{\partial^2 u}{\partial \xi^2} + 2\frac{\partial^2 u}{\partial \xi \partial \eta} + \frac{\partial^2 u}{\partial \eta^2}.$$

We also have

$$\frac{\partial u}{\partial t} = \frac{\partial u}{\partial \xi}\cdot\frac{\partial \xi}{\partial t} + \frac{\partial u}{\partial \eta}\cdot\frac{\partial \eta}{\partial t} = \frac{\partial u}{\partial \xi}\cdot c + \frac{\partial u}{\partial \eta}\cdot(-c) = c\left(\frac{\partial u}{\partial \xi} - \frac{\partial u}{\partial \eta}\right)$$

and

$$\frac{\partial^2 u}{\partial t^2} = c\frac{\partial}{\partial t}\left(\frac{\partial u}{\partial \xi} - \frac{\partial u}{\partial \eta}\right).$$

Now

$$\frac{\partial}{\partial t}\left(\frac{\partial u}{\partial \xi}\right) = \frac{\partial}{\partial \xi}\left(\frac{\partial u}{\partial \xi}\right)\cdot\frac{\partial \xi}{\partial t} + \frac{\partial}{\partial \eta}\left(\frac{\partial u}{\partial \xi}\right)\cdot\frac{\partial \eta}{\partial t} = \frac{\partial^2 u}{\partial \xi^2}\cdot c + \frac{\partial^2 u}{\partial \xi \partial \eta}\cdot(-c)$$

and

$$\frac{\partial}{\partial t}\left(\frac{\partial u}{\partial \eta}\right) = \frac{\partial}{\partial \xi}\left(\frac{\partial u}{\partial \eta}\right)\cdot\frac{\partial \xi}{\partial t} + \frac{\partial}{\partial \eta}\left(\frac{\partial u}{\partial \eta}\right)\cdot\frac{\partial \eta}{\partial t} = \frac{\partial^2 u}{\partial \xi \partial \eta}\cdot c + \frac{\partial^2 u}{\partial \eta^2}\cdot(-c).$$

Thus

$$\begin{aligned}\frac{\partial^2 u}{\partial t^2} &= c\left[\frac{\partial^2 u}{\partial \xi^2}\cdot c + \frac{\partial^2 u}{\partial \xi \partial \eta}\cdot(-c) - \frac{\partial^2 u}{\partial \xi \partial \eta}\cdot c + \frac{\partial^2 u}{\partial \eta^2}\cdot(-c)\right]\\ &= c^2\left(\frac{\partial^2 u}{\partial \xi^2} - 2\frac{\partial^2 u}{\partial \xi \partial \eta} + \frac{\partial^2 u}{\partial \eta^2}\right)\end{aligned}$$

and we have

$$\begin{aligned}0 &= \frac{\partial^2 u}{\partial t^2} - c^2\frac{\partial^2 u}{\partial x^2}\\ &= c^2\left(\frac{\partial^2 u}{\partial \xi^2} - 2\frac{\partial^2 u}{\partial \xi \partial \eta} + \frac{\partial^2 u}{\partial \eta^2} - \frac{\partial^2 u}{\partial \xi^2} - 2\frac{\partial^2 u}{\partial \xi \partial \eta} - \frac{\partial^2 u}{\partial \eta^2}\right)\\ &= -4c^2\frac{\partial^2 u}{\partial \xi \partial \eta}.\end{aligned}$$

We now solve

$$\frac{\partial^2 u}{\partial \xi \partial \eta} = 0.$$

Integrating $\frac{\partial^2 u}{\partial \xi \partial \eta}$ with respect to ξ gives

$$\frac{\partial u}{\partial \eta} = \int \frac{\partial^2 u}{\partial \xi \partial \eta} d\xi = \int 0\, d\xi = \Psi(\eta)$$

because

$$\frac{d}{d\xi}(\Psi(\eta)) = 0.$$

Integrating $\frac{\partial u}{\partial \eta}$ with respect to η gives

$$u(\xi, \eta) = \int \Psi(\eta) d\eta = p(\eta) + q(\xi)$$

where $p(\eta)$ is an antiderivative of $\Psi(\eta)$ that involves only η.

Thus

$$u(x, t) = p(x + ct) + q(x - ct)$$

for any twice differentiable functions p and q. This is d'Alembert's formula.

If we have appropriate initial conditions, then we can specify p and q as we now show.

Suppose that

$$u(x,0)=f(x) \text{ and } u_t(x,0)=g(x).$$

So

$$u(x,0)=p(x+0)+q(x-0)=p(x)+q(x)=f(x) \tag{1}$$

and

$$u_t(x,t)=cp'(x+ct)-cq'(x-ct)$$

so

$$u_t(x,0)=cp'(x+0)-cq'(x-0)=cp'(x)-cq'(x)=g(x). \tag{2}$$

If we differentiate Eq. (1) and multiply by c, we get

$$cp'(x)+cq'(x)=cf'(x). \tag{3}$$

Adding Eqs. (2) and (3) gives

$$2cp'(x)=g(x)+cf'(x)$$

so that

$$p'(x)=\frac{1}{2c}g(x)+\frac{1}{2}f'(x).$$

Integrating, we get

$$p(x)=\frac{1}{2c}\int_0^x g(\tau)d\tau+K+\frac{1}{2}f(x).$$

Then

$$q(x)=f(x)-p(x)=f(x)-\left[\frac{1}{2c}\int_0^x g(\tau)d\tau+K+\frac{1}{2}f(x)\right]$$

$$=\frac{1}{2}f(x)-\frac{1}{2c}\int_0^x g(\tau)d\tau-K.$$

So

$$p(x+ct)=\frac{1}{2c}\int_0^{x+ct} g(\tau)d\tau+\frac{1}{2}f(x+ct)+K$$

and

$$q(x-ct)=\frac{1}{2}f(x-ct)-\frac{1}{2c}\int_0^{x-ct} g(\tau)d\tau-K$$

$$=\frac{1}{2}f(x-ct)+\frac{1}{2c}\int_{x-ct}^{0} g(\tau)d\tau-K.$$

Finally, we have

$$u(x,t) = p(x+ct) + q(x-ct) = \frac{1}{2}[f(x+ct) + f(x-ct)] + \frac{1}{2c}\int_{x-ct}^{x+ct} g(\tau)d\tau. \tag{4}$$

Eq. (4) is also called d'Alembert's formula.

We shall give the solution for the heat equation later in the text, but we give a cartoon of graphs below showing how the solutions of the wave equation and the heat equation evolve in time for the same initial condition of a Gaussian distribution. In Fig. 5.4.1, we show the evolution of the wave equation, and in Fig. 5.4.2 we show the evolution of the heat equation. Notice that for the wave equation, the distribution splits into two equal parts, each as the same shape but one half the size as the original distribution. The wave equation is the prototypical example of a hyperbolic PDE, and the solutions of such equations behave in this manner. For the heat equation, the distribution of the initial condition diffuses in time. (The heat equation is also called the diffusion equation.) The heat equation is the prototypical example of a parabolic PDE, and the solutions of such equations behave in this manner.

Graphs of the solution to the wave equation as it evolves from its initial condition through t=1, 2 and 3.

The graphs of the solution to the heat equation for $t = 0.2$, 0.5 and 1.

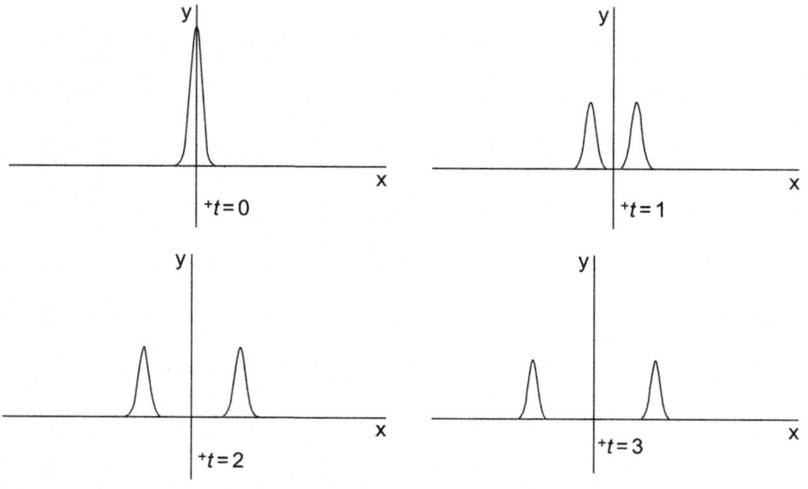

FIGURE 5.4.1

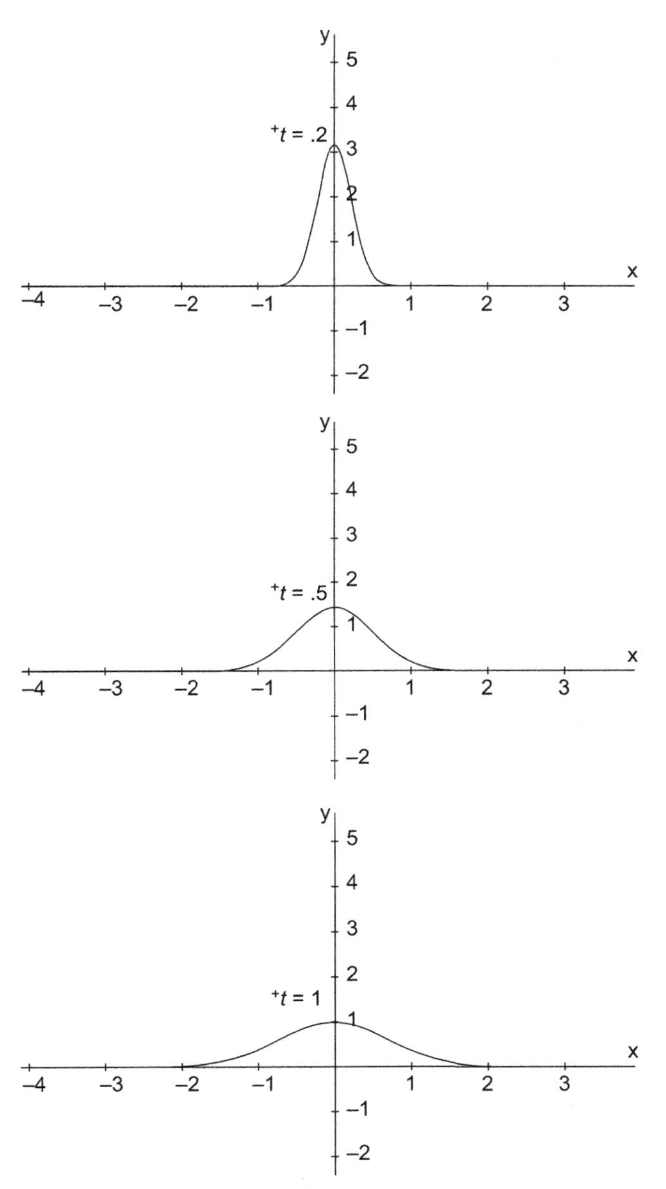
y
5
4
t = .2
3
2
1
x
–4
–3
–2
–1
1
2
3
–1
–2
y
5
4
3
2
t = .5
1
x
–4
–3
–2
–1
1
2
3
–1
–2
y
5
4
3
2
t = 1
1
x
–4
–3
–2
–1
1
2
3
–1
–2

FIGURE 5.4.2

EXERCISES

1. Solve the wave equation on $[0,\pi]$ if $u(x,0) = 0$ and $u_t(x,0) = \sin x$.
2. Solve the wave equation on $[0,\pi]$ if $u(x,0) = \sin x$ and $u_t(x,0) = 0$.
3. Solve the wave equation on $[0,1]$ if $u(x,0) = 1-x^2$ and $u_t(x,0) = 0$.
4. Solve the wave equation on $[0,1]$ if

$$u(x,0) = \begin{cases} x, 0 < x < \dfrac{1}{2} \\ \\ 1 - x, \dfrac{1}{2} \le x < 1 \end{cases}, u_t(x,0) = \sin \pi x.$$

5. Solve the wave equation on $[0,1]$ if $u(x,0) = 0$ and $u_t(x,0) = 1$.
6. Consider the PDE

$$u_{tt}(x,t) + 2au_t(x,t) + a^2u(x,t) = a^2u_{xx}(x,t) \;\; t > 0, -\infty < x < \infty; a > 0 \qquad (5)$$

which models voltage in a power line.

a. Show that if $y(x,t)$ satisfies $y_{tt}(x,t) = a^2 y_{xx}(x,t)$, then $u(x,t) = e^{-at}y(x,t)$ satisfies Eq. (5).

b. Find the solution for Eq. (5) with initial conditions $u(x,0) = 0$ and $u_t(x,0) = \sin x$.

c. Find the solution for Eq. (5) with initial conditions $u(x,0) = \sin x$ and $u_t(x,0) = 0$.

7. Solve the wave equation on $-\infty < x < \infty$ if

a. $u(x,0) = \frac{1}{1+x^2}, \quad u_t(x,0) = \frac{1}{1+x^2}$.

b. $u(x,0) = e^{-x^2}, \quad u_t(x,0) = xe^{-x^2}$.

5.6 CONVERTING SECOND-ORDER PARTIAL DIFFERENTIAL EQUATIONS TO STANDARD FORM

Next we demonstrate how to convert a second-order PDE of the form

$$A\frac{\partial^2 u(x,y)}{\partial x^2} + B\frac{\partial^2 u(x,y)}{\partial x \partial y} + C\frac{\partial^2 u(x,y)}{\partial x^2} = 0$$

into one of the three equations we studied earlier by a linear change of variables. The process is computationally similar to rotating axes in the plane to convert an equation of the form

$$Ax^2 + Bxy + Cy^2 + Dx + Ey + F = 0$$

into a parabola, ellipse, or a hyperbola in standard form. We now review that process.

Start with a vector $\widehat{v}$. We find the coordinates of $\widehat{v}$ with respect to two sets of axes. The first set of axes, (x,y), is in standard position, and the second set of axes, (x',y'), is

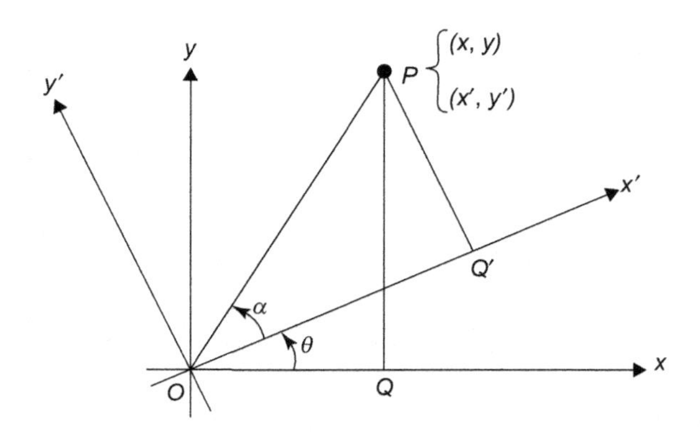

FIGURE 5.5.1

obtained by rotating the first set of axes counterclockwise through an angle θ. (See Fig. 5.5.1.)

Suppose in the (x',y') system, the vector $\widehat{v}$ makes an angle of α with the x' axis. Then in the (x,y) system, the vector $\widehat{v}$ makes an angle of $\alpha + \theta$ with the x axis. Then

$$x' = |\widehat{v}| \cos \alpha, \quad y' = |\widehat{v}| \sin \alpha$$

$$x = |\widehat{v}| \cos(\alpha + \theta), \quad y = |\widehat{v}| \sin(\alpha + \theta).$$

So

$$x = |\widehat{v}| \cos (\alpha + \theta) = |\widehat{v}|[\cos \alpha \cos \theta - \sin \alpha \sin \theta] = x' \cos \theta - y' \sin \theta$$

$$y = |\widehat{v}| \sin (\alpha + \theta) = |\widehat{v}|[\sin \alpha \cos \theta + \sin \theta \cos \alpha] = x' \sin \theta + y' \cos \theta.$$

We want to find $A', \ldots, F'$ so that

$$Ax^2 + Bxy + Cy^2 + Dx + Ey + F = A'x'^2 + B'x'y' + C'y'^2 + D'x' + E'y' + F'.$$

Now

$$\begin{aligned} Ax^2 + Bxy + Cy^2 + Dx + Ey + F &= A(x' \cos \theta - y' \sin \theta)^2 \\ &+ B(x' \cos \theta - y' \sin \theta)(x' \sin \theta + y' \cos \theta) + \cdots + F. \end{aligned} \tag{1}$$

If one expands the right-hand side of Eq. (1) and sets the result equal to

$$A'x'^2 + B'x'y' + C'y'^2 + D'x' + E'y' + F'$$

then one finds that

$$A' = A\cos^2\theta + B\sin\theta\cos\theta + C\sin^2\theta$$
$$B' = B\left(\cos^2\theta - \sin^2\theta\right) + 2(C - A)\sin\theta\cos\theta$$
$$C' = A\sin^2\theta - B\sin\theta\cos\theta + C\cos^2\theta$$
$$D' = D\cos\theta + E\sin\theta$$
$$E' = -D\sin\theta + E\cos\theta$$
$$F' = F.$$

We want to choose θ so there is no $x'y'$ term; i.e., we want $B' = 0$. We use the identities

$$\cos^2\theta - \sin^2\theta = \cos(2\theta) \text{ and } 2\sin\theta\cos\theta = \sin(2\theta)$$

to get

$$B' = B\cos(2\theta) + (C - A)\sin(2\theta)$$

so that $B' = 0$ if $B\cos(2\theta) = (A - C)\sin(2\theta)$ or

$$\frac{\sin(2\theta)}{\cos(2\theta)} = \tan(2\theta) = \frac{B}{A - C}$$

so

$$\theta = \frac{1}{2}\tan^{-1}\left(\frac{B}{A - C}\right).$$

Example:
We convert the equation

$$x^2 + 3xy + y^2 = 7$$

to standard form.

We have $A = C = 1$ and $B = 3$. Thus $2\theta = 90°$ so $\theta = 45°$ and

$$x = x'\cos\theta - y'\sin\theta = \frac{x' - y'}{\sqrt{2}}, \quad y = x'\sin\theta + y'\cos\theta = \frac{x' + y'}{\sqrt{2}}$$

so

$$x^2 + 3xy + y^2 = \left(\frac{x' - y'}{\sqrt{2}}\right)^2 + 3\left(\frac{x' - y'}{\sqrt{2}}\right)\left(\frac{x' + y'}{\sqrt{2}}\right) + \left(\frac{x' + y'}{\sqrt{2}}\right)^2$$
$$= \frac{5x'^2 - y'^2}{2} = 7$$

which is the equation of a hyperbola.

Repeating what we said earlier, recall that if $B^2 - 4AC > 0$ the graph is a hyperbola, if $B^2 - 4AC = 0$ the graph is an ellipse, if $B^2 - 4AC < 0$ the graph is a parabola. We follow the same nomenclature with second-order PDEs, designating them as hyperbolic, elliptic, or parabolic according to the sign of $B^2 - 4AC$.

We now demonstrate by an example that the same change of variables used to rotate the axes to convert a second-degree equation to standard form works to convert a second-order PDE to standard form.

Example:
We convert the equation

$$\frac{\partial^2 u(x,y)}{\partial x^2}+3\frac{\partial^2 u(x,y)}{\partial x\partial y}+\frac{\partial^2 u(x,y)}{\partial x^2}=7$$

to

$$A\frac{\partial^2 u(x',y')}{\partial x'^2}+B\frac{\partial^2 u(x',y')}{\partial y'^2}=C$$

using the change of variables

$$x=\frac{x'-y'}{\sqrt{2}},\quad y=\frac{x'+y'}{\sqrt{2}}.$$

Recall the chain rule for partial derivatives: If u is a function of x and y and x and y are each functions of x' and y', then

$$\frac{\partial u}{\partial x'}=\frac{\partial u}{\partial x}\frac{\partial x}{\partial x'}+\frac{\partial u}{\partial y}\frac{\partial y}{\partial x'}\quad\text{and}\quad\frac{\partial u}{\partial y'}=\frac{\partial u}{\partial x}\frac{\partial x}{\partial y'}+\frac{\partial u}{\partial y}\frac{\partial y}{\partial y'}.$$

We have

$$\frac{\partial x}{\partial x'}=\frac{1}{\sqrt{2}},\quad\frac{\partial x}{\partial y'}=\frac{-1}{\sqrt{2}},\quad\frac{\partial y}{\partial x'}=\frac{1}{\sqrt{2}},\quad\frac{\partial y}{\partial y'}=\frac{1}{\sqrt{2}}$$

so

$$\frac{\partial u}{\partial x'}=\frac{\partial u}{\partial x}\frac{1}{\sqrt{2}}+\frac{\partial u}{\partial y}\frac{1}{\sqrt{2}}=\frac{1}{\sqrt{2}}(u_x+u_y)\quad\text{and}$$

$$\frac{\partial u}{\partial y'}=\frac{\partial u}{\partial x}\left(\frac{-1}{\sqrt{2}}\right)+\frac{\partial u}{\partial y}\frac{1}{\sqrt{2}}=\frac{1}{\sqrt{2}}(u_y-u_x).$$

Then

$$\begin{aligned}\frac{\partial^2 u}{\partial x'^2}=\frac{\partial}{\partial x'}\left(\frac{\partial u}{\partial x'}\right)&=\frac{\partial}{\partial x'}\left(\frac{1}{\sqrt{2}}(u_x+u_y)\right)\\&=\frac{1}{\sqrt{2}}\frac{\partial}{\partial x}(u_x+u_y)\frac{\partial x}{\partial x'}+\frac{1}{\sqrt{2}}\frac{\partial}{\partial y}(u_x+u_y)\frac{\partial y}{\partial x'}\\&=\frac{1}{\sqrt{2}}(u_{xx}+u_{xy})\frac{1}{\sqrt{2}}+\frac{1}{\sqrt{2}}(u_{yx}+u_{yy})\frac{1}{\sqrt{2}}=\frac{1}{2}u_{xx}+u_{xy}+\frac{1}{2}u_{yy}.\end{aligned}$$

Likewise,

$$\frac{\partial^2 u}{\partial y'^2} = \frac{\partial}{\partial y'}\left(\frac{\partial u}{\partial y'}\right) = \frac{\partial}{\partial y'}\left(\frac{1}{\sqrt{2}}(u_y - u_x)\right)$$

$$= \frac{1}{\sqrt{2}}\left[\frac{\partial}{\partial x}(u_y - u_x)\frac{\partial x}{\partial y'} + \frac{\partial}{\partial y}(u_y - u_x)\frac{\partial y}{\partial y'}\right]$$

$$= \frac{1}{\sqrt{2}}\left\{[u_{xy} - u_{xx}]\left(\frac{-1}{\sqrt{2}}\right) + [u_{yy} - u_{yx}]\left(\frac{1}{\sqrt{2}}\right)\right\}$$

$$= \frac{1}{2}u_{xx} - u_{xy} + \frac{1}{2}u_{yy}.$$

We now show

$$\frac{\partial^2 u}{\partial x^2} + 3\frac{\partial^2 u}{\partial x \partial y} + \frac{\partial^2 u}{\partial y^2} = \frac{5\frac{\partial^2 u}{\partial x'^2} - \frac{\partial^2 u}{\partial y'^2}}{2}.$$

We have

$$\frac{5\frac{\partial^2 u}{\partial x'^2} - \frac{\partial^2 u}{\partial y'^2}}{2} = \frac{5}{2}\left(\frac{1}{2}u_{xx} + u_{xy} + \frac{1}{2}u_{yy}\right) - \frac{1}{2}\left(\frac{1}{2}u_{xx} - u_{xy} + \frac{1}{2}u_{yy}\right)$$

$$= u_{xx}\left(\frac{5}{4} - \frac{1}{4}\right) + u_{xy}\left(\frac{5}{4} + \frac{1}{4}\right) + u_{yy}\left(\frac{5}{4} - \frac{1}{4}\right) = u_{xx} + 3u_{xy} + u_{yy}.$$

EXERCISES

1. Classify the PDEs below as parabolic, hyperbolic, or elliptic and convert them to standard form.

   a. $u_{xx} - 2u_{xy} + 9u_{yy} - u_x + 5u_y = 0.$
   b. $u_{xx} + 4u_{xy} - 9u_{yy} - 3u_y = 0.$
   c. $u_{xx} - u_{xy} = 0.$
   d. $u_{xx} - 6u_{xy} + 4u_{yy} + u_x = 0.$

CHAPTER

Sturm–Liouville Theory

6

6.1 INTRODUCTION

Earlier, we studied Fourier series for trigonometric functions. One reason this is important is because the ordinary differential equation

$$y''(x) + \lambda y(x) = 0$$

with periodic boundary conditions arises when solving certain partial differential equations using separation of variables. Our approach was to recognize the problem of solving the differential equation as an eigenfunction/eigenvalue problem, the solutions to the equation being the eigenfunctions. It was necessary to determine when the Fourier expansion of a function in terms of the eigenfunctions converges to the function. In that problem, we considered pointwise convergence, uniform convergence, and L^2 convergence. For the equation above, the eigenfunctions are sines and cosines.

We shall see that other ordinary differential equations arise when the partial differential equations that we shall study are solved by separation of variables, including Bessel's equation and Legendre's equation. Each of these is a Sturm–Liouville differential equation. In this chapter, we present the problem of solving a Sturm–Liouville differential equation as an eigenfunction/eigenvalue problem and find conditions on a function to ensure that the Fourier expansion of the function in terms of eigenfunctions converges to the function. We shall consider only uniform convergence and L^2 convergence.

A differential equation of the form

$$[r(x)y'(x)]' - q(x)y(x) + \lambda\rho(x)y(x) = 0 \quad \text{for} \quad a < x < b$$

where $\rho(x)$ and $q(x)$ are continuous, and $r(x)$ has a continuous derivative on (a, b), and $\rho(x)$ and $r(x)$ are positive, and $q(x)$ is nonnegative on $[a, b]$ is called a Sturm–Liouville differential equation. We take $a = 0$ and $b = 1$ to simplify the notation. This equation can also be written as

$$[r(x)y'(x)]' - q(x)y(x) = -\lambda\rho(x)y(x)$$

Mathematical Physics with Partial Differential Equations. https://doi.org/10.1016/B978-0-12-814759-7.00006-5

or

$$-\frac{1}{\rho(x)}\left\{[r(x)y'(x)]' - q(x)y(x)\right\} = \lambda y(x).$$

We shall view this as the operator equation $L[y] = \lambda y$ where

$$L[y] = -\frac{1}{\rho(x)}\left\{[r(x)y'(x)]' - q(x)y(x)\right\}.$$

We restrict L to a set of functions on which it is self-adjoint. For example, one case that we shall see in the next section is the set of functions for which $y(0) = y(1) = 0$.

EXERCISES

1. Show that

$$a(x)y''(x) + b(x)y'(x) + c(x)y(x) + \lambda d(x)y(x) = 0; \quad a < x < b; \quad a(x) \neq 0$$

can be transformed to a Sturm–Liouville problem by dividing by $a(x)$ and multiplying by $\exp\left(\int \frac{b(x)}{a(x)} dx\right)$.

2. Use the idea of Problem 1 to convert the following to a Sturm–Liouville equation.

a. $(1 - x^2)y''(x) - 2xy'(x) + n(n+1)y(x) = 0, \ -1 < x < 1$. (Legendre differential equation)

b. $y''(x) - 2xy'(x) + 2ny(x) = 0, \ -\infty < x < \infty$. (Hermite differential equation)

c. $(1 - x^2)y''(x) - xy'(x) + n^2y(x) = 0, \ -1 < x < 1$. (Chebyshev differential equation)

3. Show that

$$L[y] = -\frac{1}{\rho(x)}\left\{[r(x)y'(x)]' - q(x)y(x)\right\}$$

is a linear operator.

6.2 THE SELF-ADJOINT PROPERTY OF A STURM–LIOUVILLE EQUATION

We first show there is a space of functions for which

$$L[y] = -\frac{1}{\rho(x)}\left\{[r(x)y'(x)]' - q(x)y(x)\right\}$$

is self-adjoint with respect to the inner product

$$\langle f, g\rangle_\rho = \int_0^1 f(x)\rho(x)g(x)dx.$$

So $\rho(x)$ is the weight function for the inner product, which is the reason we require $\rho(x) > 0$.

To show that L is self-adjoint on a class of functions with respect to this inner product, we must show that

$$\langle L[y_1], y_2\rangle_\rho = \langle y_1, L[y_2]\rangle_\rho$$

for every y_1, y_2 in the class of functions. We determine conditions for which this is the case.

Now

$$\langle L[y_1], y_2\rangle_\rho = -\int_0^1 \frac{1}{\rho(x)}\left\{\left[r(x)y_1'(x)\right]' - q(x)y_1(x)\right\}y_2(x)\rho(x)dx$$

and

$$\langle y_1, L[y_2]\rangle_\rho = -\int_0^1 \frac{1}{\rho(x)}\left\{\left[r(x)y_2'(x)\right]' - q(x)y_2(x)\right\}y_1(x)\rho(x)dx$$

so that

$$\langle L[y_1], y_2\rangle_\rho - \langle y_1, L[y_2]\rangle_\rho = -\int_0^1 \left\{\left[r(x)y_1'(x)\right]'y_2(x) - \left[r(x)y_2'(x)\right]'y_1(x)\right\}dx. \tag{1}$$

A key observation is that

$$\left[r(x)y_1'(x)\right]'y_2(x) - \left[r(x)y_2'(x)\right]'y_1(x) = \left[r(x)y_1'(x)y_2(x) - r(x)y_2'(x)y_1(x)\right]'$$

which is easily checked (Exercise 1). So

$$\begin{aligned}\langle L[y_1], y_2\rangle_\rho - \langle y_1, L[y_2]\rangle_\rho &= -\int_0^1 \left\{\left[r(x)y_1'(x)\right]'y_2(x) - \left[r(x)y_2'(x)\right]'y_1(x)\right\}dx \\ &= \left[r(x)y_1'(x)y_2(x) - r(x)y_2'(x)y_1(x)\right]\Big|_{x=0}^{x=1} \\ &= \left[r(1)y_1'(1)y_2(1) - r(1)y_2'(1)y_1(1)\right] \\ &\quad - \left[r(0)y_1'(0)y_2(0) - r(0)y_2'(0)y_1(0)\right].\end{aligned} \tag{2}$$

Thus L will be self-adjoint when the right-hand side of Eq. (2) is zero. We now give conditions that ensure this.

1. If the problem imposes homogeneous boundary conditions. That is, if every function $y(x)$ in the vector space satisfies $y(1) = y(0) = 0$.

2. If $r(1) = r(0) = 0$. Then the right-hand side of Eq. (2) is zero regardless of the boundary values of $y(x)$.
3. If $r(1) = 0$ but $r(0) \neq 0$, then for the right-hand side of Eq. (2) to be zero, we must have

$$r(0)y_1'(0)y_2(0) - r(0)y_2'(0)y_1(0) = 0. \tag{3}$$

This will be the case if $y_1'(0)y_2(0) - y_2'(0)y_1(0) = 0$.
We argue that this occurs if there are constants c_1 and c_2 with $c_2 \neq 0$ for which

$$c_1y_1(0) + c_2y_1'(0) = 0$$

and

$$c_1y_2(0) + c_2y_2'(0) = 0.$$

For, if this is the case, then

$$c_1y_1(0)y_2(0) + c_2y_1'(0)y_2(0) = 0$$

and

$$c_1y_1(0)y_2(0) + c_2y_1(0)y_2'(0) = 0$$

so that

$$c_2\left[y_1'(0)y_2(0) - y_1(0)y_2'(0)\right] = 0$$

which means that Eq. (3) holds.
4. If $r(0) = 0$ but $r(1) \neq 0$, then an argument similar to the case above shows that for the right-hand side of Eq. (2) to be zero, we must have constants d_1 and d_2 with $d_2 \neq 0$ for which

$$d_1y_1(1) + d_2y_1'(1) = 0$$

and

$$d_1y_2(1) + d_2y_2'(1) = 0.$$

5. If $r(0) \neq 0$ and $r(1) \neq 0$, then the conditions of cases 3 and 4 must both hold.

EXERCISES

1. Show that if $y_1(x)$, $y_2(x)$, and $r(x)$ are functions with second derivatives, then

$$\left[r(x)y_1'(x)\right]'y_2(x) - \left[r(x)y_2'(x)\right]'y_1(x) = \left[r(x)y_1'(x)y_2(x) - r(x)y_2'(x)y_1(x)\right]'.$$

2. Verify the orthogonality of the eigenfunctions for Legendre's equation

$$\left[(1 - x^2)y'(x)\right]' + \lambda y(x) = 0, \quad -1 < x < 1.$$

What is the weight function? Are appropriate boundary conditions necessary?

3. Verify the orthogonality of the eigenfunctions for Bessel's equation of order m

$$(xy'(x))' + \left(\lambda x - \frac{m^2}{x}\right)y(x) = 0.$$

What is the weight function? Are appropriate boundary conditions necessary?

4. Verify the orthogonality of the eigenfunctions for Hermite's equation

$$y''(x) - 2xy'(x) + 2ny(x) = 0, \quad -\infty < x < \infty.$$

What is the weight function? Are appropriate boundary conditions necessary?

6.3 COMPLETENESS OF EIGENFUNCTIONS FOR STURM–LIOUVILLE EQUATIONS

We continue to consider the problem

$$[r(x)y'(x)]' - q(x)y(x) = -\lambda \rho(x)y(x), \quad 0 < x < 1, \quad y(0) = y(1) = 0.$$

as an eigenvalue/eigenvector problem with

$$L[y] = -\frac{1}{\rho(x)}\left\{[r(x)y'(x)]' - q(x)y(x)\right\}.$$

Note that we have now imposed homogeneous boundary conditions, and these conditions ensure that L is self-adjoint.

In this section, we show that if $f(x)$ is an integrable function for which

$$\langle f,f \rangle_\rho = \int_0^1 |f(x)|^2 \rho(x)dx < \infty$$

then, $f(x)$ is the L^2 limit of the Fourier series of $f(x)$ in terms of the eigenfunctions for the Sturm–Liouville equation. That is,

$$\lim_{N\to\infty} \int_0^1 [S_N(x) - f(x)]^2 \rho(x)dx = 0$$

where

$$S_N(x) = \sum_{n=1}^{N} \frac{\langle \phi_n(x), f(x) \rangle_\rho}{\langle \phi_n(x), \phi_n(x) \rangle_\rho} \phi_n(x)$$

and $\phi_n(x)$ is the nth eigenfunction of the Sturm–Liouville equation.

This is an involved process, and the level of mathematics is more sophisticated than in the rest of the text. If one wishes to bypass the proof and accept the result, there is no loss of continuity.

Here are the major steps:

1. Show that each eigenvalue is positive.
2. Since 0 is not an eigenvalue, there is a Green's function $G(x,\xi)$ for the differential equation

$$[r(x)y'(x)]' - q(x)y(x) = F(x) \quad y(0) = y(1) = 0.$$

We show that if the eigenvalues are $\lambda_1, \lambda_2, \lambda_3, \ldots, \lambda_N$, then

$$\sum_{n=1}^{N} \frac{1}{\lambda_n^2} \le \int_0^1 \int_0^1 [G(x,\xi)]^2 \rho(x)\rho(\xi)\,dx\,d\xi.$$

If there are infinitely many eigenvalues, this means the series $\sum_{n=1}^{\infty}(1/\lambda_n^2)$ converges, and so $\lim_{n\to\infty}(1/\lambda_n) = 0$.

3. Show there are infinitely many eigenfunctions. This is the most involved step and is beyond the scope of the text
4. Show that

$$\int_0^1 [S_N(x) - f(x)]^2 \rho(x)\,dx \le \frac{1}{\lambda_N}\Psi$$

where Ψ is a bounded function.

Step 1.

Suppose that $y(x)$ is an eigenfunction of L with eigenvalue λ. That is,

$$L[y] = -\frac{1}{\rho(x)}\left\{[r(x)y'(x)]' - q(x)y(x)\right\} = \lambda y(x), \quad y(0) = y(1) = 0.$$

We show that $\lambda > 0$. Using integration by parts, with

$$u = y(x) \quad du = y'(x)$$

$$dv = [r(x)y'(x)]' \quad v = [r(x)y'(x)]$$

we have

$$\int_0^1 y(x)[r(x)y'(x)]'dx = [y(x)r(x)y'(x)]\Big|_{x=0}^{x=1} - \int_0^1 r(x)[y'(x)]^2 dx.$$

Since $y(0) = y(1) = 0$, we have

$$\int_0^1 y(x)[r(x)y'(x)]'dx = -\int_0^1 r(x)[y'(x)]^2 dx.$$

Now,

$$\lambda\langle y, y\rangle_\rho = \lambda\int_0^1 y(x)\rho(x)y(x)dx$$

and

$$\begin{aligned}\langle y, Ly\rangle_\rho &= -\int_0^1 y(x)\rho(x)\frac{1}{\rho(x)}\left\{[r(x)y'(x)]' - q(x)y(x)\right\}dx \\ &= -\int_0^1 y(x)\left\{[r(x)y'(x)]' - q(x)y(x)\right\}dx \\ &= \int_0^1 r(x)[y'(x)]^2dx + \int_0^1 y(x)q(x)y(x)dx.\end{aligned}$$

Since

$$\lambda\langle y, y\rangle_\rho = \langle y, Ly\rangle_\rho$$

we have

$$\lambda\int_0^1 y(x)\rho(x)y(x)dx = \int_0^1 r(x)[y'(x)]^2dx + \int_0^1 q(x)[y(x)]^2dx$$

so

$$\lambda = \frac{\displaystyle\int_0^1 r(x)[y'(x)]^2dx + \int_0^1 q(x)[y(x)]^2dx}{\langle y, y\rangle_\rho}.$$

Now $r(x)$ is positive and $q(x)$ is nonnegative, and $y'(x)$ cannot be zero everywhere. This is because if $y'(x) = 0$ everywhere, then $y(x)$ would be constant, the boundary condition $y(0) = 0$ would mean that $y(x)$ would be the zero function. This is impossible for an eigenvector. Thus $\lambda > 0$. This completes Step 1.

Step 2.

Recall that

$$[r(x)y'(x)]' - q(x)y(x) + \lambda\rho(x)y(x) = -F(x), \;\; y(0) = y(1) = 0 \tag{1}$$

has a unique solution if and only if

$$[r(x)y'(x)]' - q(x)y(x) + \lambda\rho(x)y(x) = 0, \;\; y(0) = y(1) = 0$$

has no solution other than $y(x) = 0$.

Since 0 is not an eigenvalue for Eq. (1), there is a Green's function for

$$[r(x)y'(x)]' - q(x)y(x) = -F(x), \;\; y(0) = y(1) = 0. \tag{2}$$

Let $G(x, \xi)$ be the Green's function for Eq. (2) and take $F(x) = \lambda\rho(x)u(x)$ to get that, if $u(x)$ is an eigenfunction of Eq. (1), with eigenvalue λ, then

$$u(x) = \lambda\int_0^1 G(x, \xi)\rho(\xi)u(\xi)d\xi.$$

Fix $x = \bar{x}$. Consider $G(\bar{x}, \xi)$. For any function $f(\xi)$, we can write the Fourier expansion of $f(\xi)$ in terms of the eigenfunctions of an operator (even if we do not know the expansion converges to the function). If the eigenfunctions are u_1, u_2, ... and the inner product is $\langle , \rangle_\rho$, the Fourier expansion is

$$\sum_{n=1}^{\infty} \frac{\langle u_n(\xi), f(\xi) \rangle_\rho}{\langle u_n(\xi), u_n(\xi) \rangle_\rho} u_n(\xi).$$

We are not claiming at this point that the Fourier series of the function is equal to the function.
We do this for $f(\xi) = G(\bar{x}, \xi)$ to get

$$\sum_{n=1}^{\infty} \frac{\langle u_n(\xi), G(\bar{x}, \xi) \rangle_\rho}{\langle u_n(\xi), u_n(\xi) \rangle_\rho} u_n(\xi) = \sum_{n=1}^{\infty} \frac{\left(\int_0^1 u_n(\xi) G(\bar{x}, \xi) \rho(x) dx \right)}{\langle u_n(\xi), u_n(\xi) \rangle_\rho} u_n(\xi).$$

Bessel's inequality states that if $\sum c_n u_n$ is the Fourier expansion for $f(\xi)$, then for any N

$$\sum_{n=1}^{N} c_n^2 \langle u_n(\xi), \ u_n(\xi) \rangle_\rho \leq \int_0^1 [f(\xi)]^2 \rho(\xi) d\xi.$$

For this problem,

$$f(\xi) = G(\bar{x}, \xi) \quad \text{and} \quad c_n = \frac{\displaystyle\int_0^1 u_n(\xi) G(\bar{x}, \xi) \rho(\xi) d\xi}{\displaystyle\int_0^1 [u_n(\xi)]^2 \rho(\xi) d\xi}$$

so we have, for any N

$$\sum_{n=1}^{N} \frac{\left(\displaystyle\int_0^1 u_n(\xi) G(\bar{x}, \xi) \rho(\xi) d\xi \right)^2}{\displaystyle\int_0^1 [u_n(\xi)]^2 \rho(\xi) d\xi} \leq \int_0^1 [G(\bar{x}, \xi)]^2 \rho(\xi) d\xi.$$

But

$$\int_0^1 u_n(\xi) G(\bar{x}, \xi) \rho(\xi) d\xi = \frac{u_n(\bar{x})}{\lambda_n}$$

so we have

$$\sum_{n=1}^{N} \frac{[u_n(\bar{x})]^2}{\lambda_n^2} \cdot \frac{1}{\displaystyle\int_0^1 [u_n(\xi)]^2 \rho(\xi) d\xi} \leq \int_0^1 [G(\bar{x}, \xi)]^2 \rho(\xi) d\xi. \tag{3}$$

Now multiply both sides of (3) by $\rho(\overline{x})$ then integrate with respect to $\overline{x}$ from 0 to 1 to get

$$\sum_{n=1}^{N} \frac{1}{\int_0^1 [u_n(\xi)]^2 \rho(\xi) d\xi} \frac{1}{\lambda_n^2} \int_{\overline{x}=0}^{1} [u_n(\overline{x})]^2 \rho(\overline{x}) d\overline{x}$$

$$\le \int_{\overline{x}=0}^{1} \left(\int_{\xi=0}^{1} [G(\overline{x}, \xi)]^2 \rho(\xi) d\xi \right) \rho(\overline{x}) d\overline{x}.$$

So

$$\sum_{n=1}^{N} \frac{1}{\lambda_n^2} \le \int_0^1 \left(\int_0^1 [G(\overline{x}, \xi)]^2 \rho(\xi) d\xi \right) \rho(\overline{x}) d\overline{x}. \tag{4}$$

Since $G(\overline{x}, \xi)$ is continuous on $[0, 1] \times [0, 1]$, it is bounded, and so the right-hand side of Eq. (4) is finite. Thus if there are infinitely many eigenvalues, then the series $\sum_{n=1}^{\infty} \frac{1}{\lambda_n^2}$ must converge and $\lim_{n\to\infty} (1/\lambda_n) = 0$.

This completes Step 2.

Step 3.

The proof of this step is beyond the scope of the text. A salient part of Step 3 is described in Step 4. A proof of Step 3 may be found in Courant and Hilbert, Vol. I.

Step 4.

In this part of the proof, we need a few facts to make the argument flow more smoothly. First, if $f(x)$ is continuously differentiable with $f(0) = f(1) = 0$, then

$$\int_0^1 r(x) f'(x) u_n'(x) dx = \int_0^1 f(x) \left[-r(x) u_n'(x) \right]' dx.$$

This is because if we integrate by parts with

$$w = r(x) u_n'(x) \quad dw = \left[r(x) u_n'(x) \right]'$$

$$dv = f'(x) \quad v = f(x)$$

we get

$$\int_0^1 r(x) f'(x) u_n'(x) dx = r(x) u_n'(x) f(x) \Big|_{x=0}^{x=1} - \int_0^1 f(x) \left[r(x) u_n'(x) \right]' dx$$

$$= \int_0^1 f(x) \left[-r(x) u_n'(x) \right]' dx$$

because $f(0) = f(1) = 0$.

From this we get

$$\int_0^1 [r(x)f'(x)u_n'(x) + q(x)f(x)u_n(x)]dx = \int_0^1 f(x)\{[-r(x)u_n'(x)]' + q(x)u_n(x)\}dx$$
$$= \int_0^1 f(x)\lambda_n\rho(x)u_n(x)dx = \lambda_n \int_0^1 f(x)\rho(x)u_n(x)dx = \lambda_n c_n \int_0^1 \rho(x)[u_n(x)]^2 dx. \quad (5)$$

The last step follows because

$$c_n = \frac{\int_0^1 f(x)\rho(x)u_n(x)dx}{\int_0^1 \rho(x)[u_n(x)]^2 dx},$$

so

$$\int_0^1 f(x)\rho(x)u_n(x)dx = c_n \int_0^1 \rho(x)[u_n(x)]^2 dx.$$

A second fact is that if we replace $f(x)$ by $u_m(x)$ in

$$\int_0^1 [r(x)f'(x)u_n'(x) + q(x)f(x)u_n(x)]dx = \int_0^1 f(x)\lambda_n\rho(x)u_n(x)dx$$

we get

$$\int_0^1 [r(x)u_m'(x)u_n'(x) + q(x)u_m(x)u_n(x)]dx = \int_0^1 u_m(x)\lambda_n\rho(x)u_n(x)dx$$
$$= \begin{cases} \lambda_n \int_0^1 [u_n(x)]^2\rho(x)dx & \text{if} \quad m = n \\ 0 & \text{if} \quad m \neq n \end{cases}. \quad (6)$$

We now apply these equalities. We continue to assume that $f(x)$ is a function for which $f(0) = f(1) = 0$ and whose Fourier expansion is $\sum_{n=1}^{\infty} c_n u_n$ with

$$c_n = \frac{\int_0^1 f(x)\rho(x)u_n(x)dx}{\int_0^1 \rho(x)[u_n(x)]^2 dx}.$$

We want to establish a bound on

$$\int_0^1 \rho(x)\left[f(x) - \sum_{n=1}^{N} c_n u_n\right]^2 dx$$

so that the bound will go to 0 as. $N \to \infty$.

A key part of the proof of Step 3 is to show that the minimum of the set

$$\left\{ \left. \frac{\int_0^1 \left[r(x)(\phi'(x))^2 + q(x)(\phi(x))^2\right]dx}{\int_0^1 \rho(x)(\phi(x))^2 dx} \right| \phi(x)\epsilon C^2[0,1] \quad \text{and} \quad \phi(0) = \phi(1) = 0 \right\}$$

is actually achieved, that the minimum value of this set is λ_1, and the minimizing function is a multiple of. u_1.

One then proceeds to take the minimum of the set above, except restricting $\phi(x)$ to be in the orthogonal complement of u_1. That is, we take the minimum of the set

$$\left\{ \left. \frac{\int_0^1 \left[r(x)(\phi'(x))^2 + q(x)(\phi(x))^2\right]dx}{\int_0^1 \rho(x)(\phi(x))^2 dx} \right| \phi(x)\epsilon C^2[0,1] \quad \text{and} \right.$$

$$\left. \phi(0) = \phi(1) = 0 \quad \text{and} \quad \int_0^1 \rho(x)u_1(x)\phi\ (x)dx = 0 \right\}.$$

The minimum value of this set is λ_2 and the minimizing function is a multiple of u_2. One then continues recursively to show that λ_k is the minimum value of the set

$$\left\{ \left. \frac{\int_0^1 \left[r(x)(\phi'(x))^2 + q(x)(\phi(x))^2\right]dx}{\int_0^1 \rho(x)(\phi(x))^2 dx} \right| \phi(x)\epsilon C^2[0,1] \quad \text{and} \right.$$

$$\left. \phi(0) = \phi(1) = 0 \quad \text{and} \quad \int_0^1 \rho(x)u_i(x)\phi(x)dx = 0 \quad \text{for} \quad i = 1, 2, \ldots, k-1 \right\}$$

and the minimizing function is a multiple of u_k.

Thus for any function $\phi(x)$ for which

$\int_0^1 \rho(x)u_i(x)\phi(x)dx = 0$ for $i = 1, 2, \ldots, k-1$, we have

$$\lambda_k \le \frac{\int_0^1 \left[r(x)(\phi'(x))^2 + q(x)(\phi(x))^2\right]dx}{\int_0^1 \rho(x)(\phi(x))^2 dx}. \tag{7}$$

We show in Exercise 1 that

$$\int_0^1 \rho(x)\left[f(x) - \sum_{n=1}^{k-1} c_n u_n(x)\right]u_i(x)dx = 0 \quad \text{for} \quad i \le k-1.$$

In Eq. (7) we take

$$\phi(x) = f(x) - \sum_{n=1}^{k-1} c_n u_n(x)$$

to get

$$\lambda_k \le \frac{\int_0^1 \left\{r(x)\left[f'(x) - \sum_{n=1}^{k-1} c_n u_n'(x)\right]^2 + q(x)\left[f(x) - \sum_{n=1}^{k-1} c_n u_n(x)\right]^2\right\}dx}{\int_0^1 \rho(x)\left[f(x) - \sum_{n=1}^{k-1} c_n u_n(x)\right]^2 dx}$$

so that

$$\lambda_k \int_0^1 \rho(x)\left[f(x) - \sum_{n=1}^{k-1} c_n u_n(x)\right]^2 dx$$
$$\le \int_0^1 \left\{r(x)\left[f'(x) - \sum_{n=1}^{k-1} c_n u_n'(x)\right]^2 + q(x)\left[f(x) - \sum_{n=1}^{k-1} c_n u_n(x)\right]^2\right\}dx$$

and thus

$$\int_0^1 \rho(x)\left[f(x) - \sum_{n=1}^{k-1} c_n u_n(x)\right]^2 dx$$
$$\le \frac{1}{\lambda_k}\int_0^1 \left\{r(x)\left[f'(x) - \sum_{n=1}^{k-1} c_n u_n'(x)\right]^2 + q(x)\left[f(x) - \sum_{n=1}^{k-1} c_n u_n(x)\right]^2\right\}dx. \tag{8}$$

Now

$$
\begin{aligned}
&\int_0^1 \left\{ r(x)\left[f'(x) - \sum_{n=1}^{k-1} c_n u_n'(x)\right]^2 + q(x)\left[f(x) - \sum_{n=1}^{k-1} c_n u_n(x)\right]^2 \right\} dx \\
&\quad = \int_0^1 r(x)[f'(x)]^2 dx - 2\sum_{n=1}^{k-1} c_n \int_0^1 r(x)f'(x)u_n'(x)dx \\
&\quad + \sum_{m=1}^{k-1}\sum_{n=1}^{k-1} c_n c_m \int_0^1 r(x)u_n'(x)u_m'(x)dx + \int_0^1 q(x)[f(x)]^2 dx \\
&\quad -2\sum_{n=1}^{k-1} c_n \int_0^1 q(x)f(x)u_n(x)dx \\
&\quad + \sum_{m=1}^{k-1}\sum_{n=1}^{k-1} c_n c_m \int_0^1 q(x)u_n(x)u_m(x)dx = \int_0^1 \left\{ r(x)[f'(x)]^2 + q(x)[f(x)]^2 \right\} dx \\
&\quad -2\sum_{n=1}^{k-1} c_n \int_0^1 \left[r(x)f'(x)u_n'(x) + q(x)f(x)u_n(x) \right] dx \\
&\quad + \sum_{m=1}^{k-1}\sum_{n=1}^{k-1} c_n c_m \int_0^1 \left[r(x)u_n'(x)u_m'(x) + q(x)u_n(x)u_m(x) \right] dx.
\end{aligned} \tag{9}
$$

By Eq. (5)

$$
\begin{aligned}
\int_0^1 \left[r(x)f'(x)u_n'(x) + q(x)f(x)u_n(x) \right] &= \lambda_n \int_0^1 f(x)\rho(x)u_n(x)dx \\
&= \lambda_n c_n \int_0^1 \rho(x)[u_n(x)]^2 dx
\end{aligned}
$$

and

$$
\begin{aligned}
\int_0^1 \left[r(x)u_m'(x)u_n'(x) + q(x)u_m(x)u_n(x) \right] dx &= \int_0^1 u_m(x)\lambda_n\rho(x)u_n(x)dx \\
&= \begin{cases} \lambda_n \int_0^1 [u_n(x)]^2 \rho(x)dx & \text{if} \quad m = n \\ 0 & \text{if} \quad m \neq n. \end{cases}
\end{aligned}
$$

Thus the right-hand side of Eq. (9) is

$$\int_0^1 \left\{r(x)[f'(x)]^2 + q(x)[f(x)]^2\right\}dx - 2\sum_{n=1}^{k-1} c_n\lambda_n \int_0^1 f(x)\rho(x)u_n(x)dx$$
$$+\sum_{n=1}^{k-1} c_n^2\lambda_n \int_0^1 [u_n(x)]^2\rho(x)dx = \int_0^1 \left\{r(x)[f'(x)]^2 + q(x)[f(x)]^2\right\}dx$$
$$-\sum_{n=1}^{k-1} c_n^2\lambda_n \int_0^1 [u_n(x)]^2\rho(x)dx.$$

So from Eq. (8) we have

$$\int_0^1 \rho(x)\left[f(x) - \sum_{n=1}^{k-1} c_n u_n(x)\right]^2 dx$$
$$\leq \frac{1}{\lambda_k}\left\{\int_0^1 \left\{r(x)[f'(x)]^2 + q(x)[f(x)]^2\right\}dx - \sum_{n=1}^{k-1} c_n^2\lambda_n \int_0^1 [u_n(x)]^2\rho(x)dx\right\}.$$

Since $\lim_{k\to\infty}(1/\lambda_k) = 0$, we have

$$\lim_{k\to\infty}\left\langle f(x) - \sum_{n=1}^{k-1} c_n u_n(x),\ f(x) - \sum_{n=1}^{k-1} c_n u_n(x)\right\rangle_\rho$$
$$= \lim_{k\to\infty}\int_0^1 \rho(x)\left[f(x) - \sum_{n=1}^{k-1} c_n u_n(x)\right]^2 dx = 0.$$

Thus if

$$\int_0^1 \rho(x)[f(x)]^2 dx < \infty$$

then the Fourier series of $f(x)$ converges to $f(x)$ in the L^2 sense (i.e., in the mean). Said another way, the eigenfunctions for the Sturm–Liouville problem are complete.

EXERCISES

1. Show that

$$\int_0^1 \rho(x)\left[f(x) - \sum_{n=1}^{k-1} c_n u_n(x)\right]u_i(x)dx = 0 \quad \text{for} \quad i \leq k-1.$$

2. Verify that the eigenvalues and eigenfunctions of

$$y''(x) + \lambda y(x) = 0; \quad y(0) = 0,\ y'(1) + y(1) = 0$$

are $\tan\sqrt{\lambda} = -\sqrt{\lambda}$; $y_n(x) = \sin\left(x\sqrt{\lambda_n}\right)$.

3. Find the eigenvalues and eigenfunctions for $y''(x) + \lambda y(x) = 0$ with the following boundary conditions:
a. $y(0) = 0, \ y'(1) = 0.$
b. $y'(0) = 0, \ y(1) = 0.$
c. $y(0) = 0, \ y'(1) - y(1) = 0.$

6.4 UNIFORM CONVERGENCE OF FOURIER SERIES

We want to determine conditions on $f(x)$ that will ensure that the Fourier series of $f(x)$ converges uniformly to $f(x)$. In this case, we mean Fourier series of eigenfunctions of an eigenvalue problem. One example of these we have seen previously—the sine and cosine functions.

Suppose that $\sum c_n u_n(x)$ is the Fourier series of $f(x)$. Since each function $u_n(x)$ is an eigenfunction of L, in our setting, it is differentiable and thus continuous. We shall determine conditions on $f(x)$ that ensure $\sum c_n u_n(x)$ is uniformly Cauchy. That will mean that $\sum c_n u_n(x)$ converges uniformly, and the limit is a continuous function that we call $g(x)$. It will follow that $f(x) = g(x)$.

We have shown that if $G(x, y)$ is the Green's function for the second-order linear differential operator L where the interval is $0 < x < 1$, and if $\{\phi_n\}$ is the set of normalized eigenfunctions for L and λ_m is the eigenvalue for ϕ_m and no eigenvalue is 0, then

$$G(x, y) = \sum_n \frac{\phi_n(x)\phi_n(y)}{\lambda_n}.$$

If the inner product is

$$\langle f(x), g(x) \rangle_\rho = \int_0^1 f(x)\rho(x)g(x)dx$$

we have for $\{u_n\}$, the set of eigenfunctions that are not necessarily normalized

$$G(x, y) = \sum_n \frac{u_n(x)u_n(y)}{\lambda_n \langle u_n, u_n \rangle_\rho} = \sum_n \frac{u_n(x)u_n(y)}{\lambda_n \int_0^1 \rho(w)[u_n(w)]^2 dw}.$$

Also recall Schwarz's inequality states that for a_i, b_i, $i = 1,\ldots,n$

$$\sum_{i=1}^n a_i b_i \le \left(\sum_{i=1}^n a_i^2\right)^{1/2} \left(\sum_{i=1}^n b_i^2\right)^{1/2}$$

so that

$$\left(\sum_{i=1}^n a_i b_i\right)^2 \le \left(\sum_{i=1}^n a_i^2\right)\left(\sum_{i=1}^n b_i^2\right).$$

Fix $\overline{x}$. Let

$$a_i = c_i\sqrt{\lambda_i\int_0^1 \rho(x)[u_n(x)]^2dx} \quad b_i = \frac{u_i(\overline{x})}{\sqrt{\lambda_i\int_0^1 \rho(x)[u_i(x)]^2dx}}.$$

Then

$$a_ib_i = c_iu_i(\overline{x}), \;\; a_i^2 = c_i^2\lambda_i\int_0^1 \rho(x)[u_i(x)]^2dx, \;\; b_i^2 = \frac{[u_i(\overline{x})]^2}{\lambda_i\int_0^1 \rho(x)[u_i(x)]^2dx}$$

and so

$$\left(\sum_{i=1}^n c_iu_i(\overline{x})\right)^2 \le \left(\sum_{i=1}^n c_i^2\lambda_i\int_0^1 \rho(x)[u_i(x)]^2dx\right)\left(\sum_{i=1}^n \frac{[u_i(\overline{x})]^2}{\lambda_i\int_0^1 \rho(x)[u_i(x)]^2dx}\right).$$

Now

$$G(\overline{x},\, \overline{x}) = \sum_i \frac{[u_i(\overline{x})]^2}{\lambda_i\int_0^1 \rho(x)[u_i(x)]^2dx}$$

so

$$\left(\sum_{i=1}^n c_iu_i(\overline{x})\right)^2 \le \left(\sum_{i=1}^n c_i^2\lambda_i\int_0^1 \rho(x)[u_i(x)]^2dx\right)G(\overline{x},\, \overline{x}).$$

We want to invoke the Cauchy criterion for uniform convergence, so we change the values over which we sum to get, for positive integers M and N with $M < N$,

$$\left(\sum_{i=M+1}^N c_iu_i(\overline{x})\right)^2 \le \left(\sum_{i=M+1}^N c_i^2\lambda_i\int_0^1 \rho(x)[u_i(x)]^2dx\right)G(\overline{x},\, \overline{x}).$$

Now $G(\overline{x},\, \overline{x})$ is uniformly bounded, so $\sum_{i=M+1}^N c_iu_i(\overline{x})$ will be uniformly Cauchy if $\sum_{i=1}^\infty c_i^2\lambda_i\int_0^1 \rho(x)[u_i(x)]^2dx$ converges. We now derive conditions for which that is the case.

Suppose $\phi(x) = \sum c_nu_n(x)$. We want to find the kth Fourier coefficient in the expansion of

$$\frac{1}{\rho(x)}\left\{[r(x)\phi'(x)]' - q(x)\phi(x)\right\}.$$

That is, we compute

$$\begin{aligned} a_k &= \int_0^1 \frac{1}{\rho(x)}\left\{[r(x)\phi'(x)]' - q(x)\phi(x)\right\}\rho(x)u_k(x)dx \\ &= \int_0^1 \left\{[r(x)\phi'(x)]' - q(x)\phi(x)\right\}u_k(x)dx. \end{aligned}$$

We integrate $\int_0^1 [r(x)\phi'(x)]' u_k(x)dx$ by parts with

$$w = u_k(x) \quad dw = u_k{}'(x)$$
$$dv = [r(x)\phi'(x)]' \quad v = r(x)\phi'(x)$$

to get

$$\int_0^1 [r(x)\phi'(x)]' u_k(x)dx = r(x)\phi'(x)u_k(x)|_{x=0}^{x=1} - \int_0^1 r(x)\phi'(x)u_k{}'(x)dx$$
$$= -\int_0^1 r(x)\phi'(x)u_k{}'(x)dx.$$

We integrate $\int_0^1 \phi(x)\left[r(x)u_k{}'(x)\right]' dx$ by parts with

$$w = \phi(x) \quad dw = \phi'(x)$$
$$dv = \left[r(x)u_k{}'(x)\right]' \quad v = r(x)u_k{}'(x)$$

to get

$$\int_0^1 \phi(x)\left[r(x)u_k{}'(x)\right]' dx = \phi(x)\, r(x)u_k{}'(x)|_{x=0}^{x=1} - \int_0^1 r(x)u_k{}'(x)\, \phi'(x)dx$$
$$= -\int_0^1 r(x)u_k{}'(x)\phi'(x)dx.$$

Thus

$$\int_0^1 [r(x)\phi'(x)]' u_k(x)dx = \int_0^1 \phi(x)\left[r(x)u_k{}'(x)\right]' dx$$

and so

$$\int_0^1 \left\{[r(x)\phi'(x)]' - q(x)\phi(x)\right\} u_k(x)dx = \int_0^1 \phi(x)\left\{\left[r(x)u_k{}'(x)\right]' - q(x)u_k(x)\right\}dx.$$

Now

$$\left[r(x)u_k{}'(x)\right]' - q(x)u_k(x) = -\lambda_k \rho(x)u_k(x)$$

so

$$\int_0^1 \phi(x)\left\{\left[r(x)u_k{}'(x)\right]' - q(x)u_k(x)\right\}dx = -\lambda_k \int_0^1 \phi(x)\rho(x)u_k(x)dx = -\lambda_k c_k.$$

Thus we have the Fourier expansion

$$\frac{1}{\rho(x)}\left\{[r(x)\phi'(x)]' - q(x)\phi(x)\right\} \sim -\sum \lambda_k c_k u_k.$$

Note that

$$\left\langle f, \frac{1}{\rho}\left[(rf')' - qf\right]\right\rangle_\rho = \int_0^1 f(x)\rho(x)\frac{1}{\rho(x)}\left\{[r(x)f'(x)]' - q(x)f(x)\right\}dx$$
$$= \int_0^1 f(x)\left\{[r(x)f'(x)]' - q(x)f(x)\right\}dx$$
$$= \int_0^1 f(x)[r(x)f'(x)]^{dx} - \int_0^1 q(x)[f(x)]^2 dx.$$

We have seen that integrating by parts gives

$$\int_0^1 f(x)\left\{[r(x)f'(x)]'\right\}dx = -\int_0^1 r(x)[f'(x)]^2 dx,$$

so

$$\left\langle f, \frac{1}{\rho}\left[(rf')' - qf\right]\right\rangle_\rho = -\left[\int_0^1 \left(r(x)[f'(x)]^2 + q(x)[f(x)]^2\right)dx\right].$$

We also have

$$\left\langle f(x), -\sum \lambda_k c_k u_k(x)\right\rangle_\rho = -\sum \lambda_k c_k \langle f(x),\ u_k(x)\rangle_\rho$$
$$= -\sum \lambda_k c_k \int_0^1 f(x)\rho(x)u_k(x)dx.$$

Now

$$c_k = \frac{\int_0^1 f(x)\rho(x)u_k(x)dx}{\int_0^1 \rho(x)[u_k(x)]^2 dx}$$

so

$$\int_0^1 f(x)\rho(x)u_k(x)dx = c_k \int_0^1 \rho(x)[u_k(x)]^2 dx$$

and thus

$$-\sum \lambda_k c_k \int_0^1 f(x)\rho(x)u_k(x)dx = -\sum \lambda_k c_k c_k \int_0^1 \rho(x)[u_k(x)]^2 dx$$
$$= -\sum \lambda_k c_k^2 \int_0^1 \rho(x)[u_k(x)]^2 dx.$$

Since the vectors $\{u_n\}$ form a complete set, and since

$$\frac{1}{\rho(x)}\left\{[r(x)\phi'(x)]' - q(x)\phi(x)\right\} \sim -\sum \lambda_k c_k u_k,$$

if $f(x)$ is a function that has a continuous second derivative and $f(0) = f(1) = 0$,we have

$$\left\langle f(x), \frac{1}{\rho(x)}\left\{[r(x)\phi'(x)]' - q(x)\phi(x)\right\}\right\rangle_\rho = \left\langle f(x), -\sum \lambda_k c_k u_k\right\rangle_\rho.$$

That is,

$$-\left[\int_0^1 \left(r(x)[f'(x)]^2 + q(x)[f(x)]^2\right)dx\right] = \left\langle f(x), \frac{1}{\rho(x)}\left\{[r(x)\phi'(x)]' - q(x)\phi(x)\right\}\right\rangle_\rho$$

$$= \left\langle f(x), -\sum \lambda_k c_k u_k\right\rangle_\rho$$

$$= -\sum \lambda_k c_k^2 \int_0^1 \rho(x)[u_k(x)]^2 dx.$$

Thus

$$\int_0^1 \left\{r(x)[f'(x)]^2 + q(x)[f(x)]^2\right\}dx = \sum \lambda_k c_k^2 \int_0^1 \rho(x)[u_k(x)]^2 dx,$$

so if $\int_0^1 \left(r(x)[f'(x)]^2 + q(x)[f(x)]^2\right)dx$ is finite, then the series $\sum \lambda_k c_k^2 \int_0^1 \rho(x)[u_k(x)]^2 dx$ converges. This means we can make $\sum_{i=M+1}^N c_i^2 \lambda_i \int_0^1 \rho(x)[u_i(x)]^2 dx$ as small as we like, by making M sufficiently large.

Summarizing, we have

If $f(x)$ has a continuous second derivative and $f(0) = f(1) = 0$ and

$$\int_0^1 \left\{r(x)[f'(x)]^2 + q(x)[f(x)]^2\right\}dx$$

is finite, then the Fourier series of $f(x)$ with respect to the eigenfunctions of L converges uniformly to $f(x)$.

CHAPTER

Using Generating Functions to Solve Specialized Differential Equations

7

7.1 INTRODUCTION

Our focus for the rest of the text will be solving Laplace's equation, the wave equation, and the heat equation using different techniques and in different scenarios. Some of these will involve curvilinear coordinates, and in curvilinear coordinates certain specialized differential equations will arise. In this chapter, we describe the solutions to these specialized equations through generating functions. We recommend investigating these specific equations when the need arises.

There are several types of generating functions, and their uses in mathematics abound. We will use generating functions to give explicit formulae for solutions to some differential equations that occur prominently in mathematical physics. In addition to providing solutions, the search for these functions will uncover relationships among them that may not be as transparent with other techniques.

The fundamental idea of a generating function is that we express a function as a power series, and the coefficients of the power series form a sequence.

Examples:

The function

$$f(x) = \frac{1}{1-x}$$

can be expressed as

$$f(x) = 1 + x + x^2 + x^3 + \cdots.$$

Each coefficient of x^n is 1, so $f(x)$ generates the sequence $\{1,1,1,\ldots\}$.

With generating functions, we are proceeding formally and will not be concerned about convergence of the power series.

The function

$$f(x) = \frac{x}{1-x-x^2}$$

Mathematical Physics with Partial Differential Equations. https://doi.org/10.1016/B978-0-12-814759-7.00007-7

can be expressed as

$$f(x) = x + x^2 + 2x^3 + 3x^4 + 5x^5 + \cdots$$

so $f(x)$ generates the sequence $\{1,1,2,3,5,\ldots\}$, which is the Fibonacci sequence.

In the setting of this chapter, a generating function $G(x,t)$ is a function of two variables, and we will express the function as a power series in t. That is, we will have

$$G(x,t) = P_0(x) + P_1(x)t + P_2(x)t^2 + P_3(x)t^3 + \cdots = \sum_{n=0}^{\infty} P_n(x)t^n.$$

We will be given the function $G(x,t)$ and be asked to determine the functions $P_n(x)$ and show that each function $P_n(x)$ satisfies a certain differential equation.

Four important differential equations in physics are Laguerre's equation, Hermite's equation, Legendre's equation, and Bessel's equation. Each of these has a solution that is a polynomial of order n. For Laguerre's equation, Hermite's equation, and Legendre's equation n is any nonnegative integer, and for Bessel's equation, any integer n has an associated solution.

One way to obtain an expression for these polynomials is through generating functions. The idea is that we hypothesize a generating function $G(x,t)$, expand $G(x,t)$ in a Maclaurin series, and group the terms in powers of t. We then have an expression of the form

$$G(x,t) = g_0(x) + g_1(x)t + g_2(x)t^2 + g_3(x)t^3 + \cdots.$$

The function $g_i(x)$ will turn out to be the ith Hermite polynomial in the case of the generator for Hermite polynomials and, similarly, for the other cases. Thus, we have a conceptually simple method of finding these functions.

To justify the claim that these are indeed the desired functions, we must show that they satisfy the appropriate differential equation. To do this, one computes

$$\frac{\partial}{\partial x}G(x,t) \text{ and } \frac{\partial}{\partial t}G(x,t)$$

and from these, obtains recurrence relations. With some manipulations, the functions $g_i(x)$ are shown to satisfy the pertinent differential equations, thus justifying their names as Hermite polynomials, Legendre polynomials, Laguerre polynomials, and Bessel functions. This technique is also helpful in determining relationships among different orders of the functions that might not be transparent.

The major difficulty is finding the generating function for a particular equation, and there is not an established algorithm or strategy for doing this.

The generating functions that we will use are exponential functions that use the Maclaurin expansion

$$e^{f(x,t)} = 1 + f(x,t) + \frac{(f(x,t))^2}{2!} + \frac{(f(x,t))^3}{3!} + \cdots.$$

7.2 GENERATING FUNCTION FOR LAGUERRE POLYNOMIALS

Laguerre's equation is a second-order differential equation of the form

$$xy'' + (1 - x)y' + ny = 0$$

where n is a positive integer.

Here, we develop the solutions to Laguerre's equation from a generating function.

The generating function for Laguerre polynomials is

$$G(x,t) = \frac{1}{1-t}\exp\left(-\frac{xt}{1-t}\right) = \sum_{n=0}^{\infty} \frac{L_n(x)}{n!}t^n$$

where $L_n(x)$ will be shown to be polynomials that satisfy Laguerre's equation for the value n.

We describe $\partial G/\partial t$ in two ways, and set the expressions equal to one another. This will yield a recursion relation. Doing the same thing for $\partial G/\partial x$ gives a second recursion relation. Manipulating the two relations will enable us to demonstrate that $L_n(x)$ is a solution for Laguerre's equation.

We will now determine the two recurrence relations. The first is obtained by noting

$$(1-t)^2\frac{\partial G}{\partial t} = (1 - t - x)G.$$

This says

$$(1-t)^2\sum_{n=0}^{\infty}\frac{L_n(x)}{n!}nt^{n-1} = (1-t-x)\sum_{n=0}^{\infty}\frac{L_n(x)}{n!}t^n.$$

Expanding gives

$$\sum_{n=0}^{\infty}\frac{L_n(x)}{n!}nt^{n-1} - 2\sum_{n=0}^{\infty}\frac{L_n(x)}{n!}nt^n + \sum_{n=0}^{\infty}\frac{L_n(x)}{n!}nt^{n+1} =$$

$$\sum_{n=0}^{\infty}\frac{L_n(x)}{n!}t^n - \sum_{n=0}^{\infty}\frac{L_n(x)}{n!}t^{n+1} - x\sum_{n=0}^{\infty}\frac{L_n(x)}{n!}t^n.$$

Reindexing yields

$$-\sum_{n=0}^{\infty}x\frac{L_n(x)}{n!}t^n = \sum_{n=0}^{\infty}\left[\frac{L_{n+1}(x)}{n!} - 2n\frac{L_n(x)}{n!} + n(n-1)\frac{L_{n-1}(x)}{n!} - \frac{L_n(x)}{n!} + \frac{L_{n-1}(x)}{n!}\right]t^n.$$

This lead to the first recurrence relation

$$(n+1)L_{n+1}(x) = (2n+1-x)L_n(x) - n^2 L_{n-1}(x). \tag{1}$$

Next, we use the fact that

$$\frac{\partial G}{\partial x} = \frac{-t}{1-t}G$$

so

$$(1-t)\frac{\partial G}{\partial x} = -tG.$$

Now

$$\frac{\partial G}{\partial x} = \frac{\partial}{\partial x}\sum_{n=0}^{\infty}\frac{L_n(x)}{n!}t^n = \sum_{n=0}^{\infty}\frac{L_n'(x)}{n!}t^n$$

so

$$(1-t)\frac{\partial G}{\partial x} = (1-t)\sum_{n=0}^{\infty}\frac{L_n'(x)}{n!}t^n = \sum_{n=0}^{\infty}\frac{L_n'(x)}{n!}t^n - \sum_{n=0}^{\infty}\frac{L_n'(x)}{n!}t^{n+1}.$$

Also

$$-tG = -t\sum_{n=0}^{\infty}\frac{L_n(x)}{n!}t^n = -\sum_{n=0}^{\infty}\frac{L_n(x)}{n!}t^{n+1}.$$

Now

$$\sum_{n=0}^{\infty}\frac{L_n(x)}{n!}t^{n+1} = \sum_{n=1}^{\infty}\frac{L_{n-1}(x)}{(n-1)!}t^n = \sum_{n=1}^{\infty}\frac{nL_{n-1}(x)}{n!}t^n = \sum_{n=0}^{\infty}\frac{nL_{n-1}(x)}{n!}t^n =$$

$$\sum_{n=0}^{\infty}\frac{L_n'(x)}{n!}t^{n+1} = \sum_{n=1}^{\infty}\frac{L_{n-1}'(x)}{(n-1)!}t^n = \sum_{n=1}^{\infty}\frac{nL_{n-1}'(x)}{n!}t^n = \sum_{n=0}^{\infty}\frac{nL_{n-1}'(x)}{n!}t^n$$

so

$$(1-t)\frac{\partial G}{\partial x} = \sum_{n=0}^{\infty}\frac{L_n'(x)}{n!}t^n - \sum_{n=0}^{\infty}\frac{L_n'(x)}{n!}t^{n+1} = \sum_{n=0}^{\infty}\frac{L_n'(x)}{n!}t^n - \sum_{n=0}^{\infty}\frac{nL_{n-1}'(x)}{n!}t^n$$

and

$$-tG = -t\sum_{n=0}^{\infty}\frac{L_n(x)}{n!}t^n = -\sum_{n=0}^{\infty}\frac{nL_{n-1}(x)}{n!}t^n.$$

Thus we have

$$\sum_{n=0}^{\infty} \frac{L_n'(x)}{n!} t^n - \sum_{n=0}^{\infty} \frac{nL_{n-1}'(x)}{n!} t^n = -\sum_{n=0}^{\infty} \frac{nL_{n-1}(x)}{n!} t^n$$

and so

$$L_n'(x) - nL_{n-1}'(x) = -nL_{n-1}(x) \tag{2}$$

which is the second recurrence relation.

We now manipulate these relations to show that $L_n(x)$ is a solution to Laguerre's equation.

The problem now is to manipulate the recurrence relations to show that $L_n(x)$ is a solution to Laguerre's equation.

In Eq. (2), replace n by $n+1$ to get

$$L_{n+1}'(x) = (n+1)L_n'(x) - (n+1)L_n(x). \tag{3}$$

Differentiate Eq. (3) to get

$$L_{n+1}''(x) = (n+1)L_n''(x) - (n+1)L_n'(x).$$

Replace n *by* $n+1$ to get

$$L_{n+2}''(x) = (n+2)L_{n+1}''(x) - (n+2)L_{n+1}'(x).$$

Use Eq. (3) to get

$$\begin{aligned} L_{n+2}''(x) &= (n+2)L_{n+1}''(x) - (n+2)\left[(n+1)L_n'(x) - (n+1)L_n(x)\right] \\ &= (n+1)(n+2)\left[L_{n+1}''(x) - 2L_n'(x) + L_n(x)\right]. \end{aligned}$$

The first recurrence relation is

$$L_{n+1}(x) + (x - 2n - 1)L_n(x) + n^2 L_{n-1}(x) = 0.$$

Differentiating (using the product rule on the second term) gives

$$L_{n+1}'(x) + (x - 2n - 1)L_n'(x) + L_n(x) + n^2 L_{n-1}'(x) = 0.$$

Differentiating a second time gives

$$L_{n+1}''(x) + (x - 2n - 1)L_n''(x) + L_n'(x) + L_n'(x) + n^2 L_{n-1}''(x) = 0.$$

Replace n *by* $n+1$ to get

$$L_{n+2}''(x) + xL_{n+1}''(x) + 2L_{n+1}'(x) - (2n+2)L_{n+1}''(x) + (n+1)^2 L_n''(x) = 0. \tag{4}$$

Substituting

$$L_{n+1}'(x) = (n+1)L_n'(x) - (n+1)L_n(x)$$

$$L_{n+1}''(x) = (n+1)L_n''(x) - (n+1)L_n'(x)$$

$$L''_{n+2}(x) = (n+1)(n+2)\left[L''_{n+1}(x) - 2L'_n(x) + L_n(x)\right]$$

into Eq. (4) gives

$$\begin{aligned}&(n+1)(n+2)\left[L''_{n+1}(x) - 2L'_n(x) + L_n(x)\right] + x\left[(n+1)L''_n(x) - (n+1)L'_n(x)\right]\\&+2\left[(n+1)L'_n(x) - (n+1)L_n(x)\right]\\&-(2n+2)\left[(n+1)L''_n(x) - (n+1)L'_n(x)\right] + (n+1)^2 L''_n(x) = 0.\end{aligned}$$

Collecting terms, we have

$$xL''_n(x) + (1-x)L'_n(x) + nL_n(x) = 0.$$

Thus, $L_n(x)$ is a solution to Laguerre's equation.

EXERCISES

1. Compute the first four Laguerre polynomials using the generating function.
2. Compute the first four Laguerre polynomials using Rodrigues formula

$$L_n(x) = e^x \frac{d^n}{dx^n}\left(x^n e^{-x}\right).$$

3. Using a computer algebra system (CAS) evaluate

$$\int_0^\infty e^{-x} L_2(x) L_3(x) dx \text{ and } \int_0^\infty e^{-x} L_3(x) L_3(x) dx.$$

7.3 HERMITE'S DIFFERENTIAL EQUATION

Hermite's differential equation is

$$y''(x) - 2xy'(x) + 2ny(x) = 0.$$

For Hermite polynomials, we will show that the generating function is

$$G(x,t) = e^{2tx-t^2}.$$

Expanding in a power series

$$G(x,t) = \sum_{n=0}^{\infty} \frac{\left(2tx - t^2\right)^n}{n!} = \sum_{n=0}^{\infty} \frac{t^n(2x-t)^n}{n!}.$$

Generating functions are convenient because they provide an easy way to compute the coefficients of a power series. Similar to Maclaurin series, if

$$G(x,t) = a_0(x) + a_1(x)t + a_2(x)t^2 + a_3(x)t^3 + \cdots$$

then

$$\frac{\partial^k}{\partial t^k} G(x,t)|_{t=0} = k! a_k(x).$$

In our case, this means

$$\frac{\partial^k}{\partial t^k} G(x,t)|_{t=0} = k! \frac{H_k(x)}{k!} = H_k(x)$$

where $H_k(x)$ is the kth Hermite polynomial.

We now compute the recurrence equations.

Since

$$G(x,t) = e^{2tx-t^2} = \sum_{n=0}^{\infty} \frac{H_n(x)}{n!} t^n$$

we have

$$\frac{\partial}{\partial x} G(x,t) = 2te^{2tx-t^2} = 2t \sum_{n=0}^{\infty} \frac{H_n(x)}{n!} t^n = \sum_{n=0}^{\infty} \frac{2t^{n+1}}{n!} H_n(x)$$

and we also have

$$\frac{\partial}{\partial x} G(x,t) = \frac{\partial}{\partial x} \sum_{n=0}^{\infty} \frac{H_n(x)}{n!} t^n = \sum_{n=0}^{\infty} \frac{t^n}{n!} H_n'(x).$$

Thus, we have computed $\frac{\partial}{\partial x} G(x,t)$ in two different ways. In the first method we differentiated the generating function, and in the second we differentiated the power series. We get the first recurrence relation by equating the coefficients of t^n in the two power series expansions. Typically, we will need to reindex some of the series.

We have

$$\sum_{n=0}^{\infty} \frac{2t^{n+1}}{n!} H_n(x) = \sum_{n=0}^{\infty} \frac{t^n}{n!} H_n'(x)$$

Reindexing the term on the left gives

$$\sum_{n=0}^{\infty} \frac{2t^{n+1}}{n!} H_n(x) = \sum_{n=1}^{\infty} \frac{2t^n}{(n-1)!} H_{n-1}(x).$$

Also,

$$\frac{d}{dx} H_0(x) = \frac{d}{dx}(1) = 0$$

so

$$\sum_{n=0}^{\infty} \frac{t^n}{n!} H_n'(x) = \sum_{n=1}^{\infty} \frac{t^n}{n!} H_n'(x)$$

and we have

$$\sum_{n=1}^{\infty} \frac{2t^n}{(n-1)!} H_{n-1}(x) = \sum_{n=1}^{\infty} \frac{t^n}{n!} H_n'(x).$$

Equating the coefficients of t^n gives,

$$\frac{2}{(n-1)!} H_{n-1}(x) = \frac{1}{n!} H_n'(x)$$

so

$$2nH_{n-1}(x) = H_n'(x). \tag{1}$$

This is the first recurrence equation that we will use later.

To get the second recurrence relation, we do a similar procedure, but using $\frac{\partial}{\partial t} G(x,t)$.

We have

$$\frac{\partial}{\partial t} G(x,t) = (2x-2t)e^{2tx-t^2} = (2x-2t) \sum_{n=0}^{\infty} \frac{H_n(x)}{n!} t^n = \sum_{n=0}^{\infty} \frac{(2x-2t)H_n(x)}{n!} t^n$$

and

$$\frac{\partial}{\partial t} G(x,t) = \frac{\partial}{\partial t} \sum_{n=0}^{\infty} \frac{H_n(x)}{n!} t^n = \sum_{n=0}^{\infty} n \frac{H_n(x)}{n!} t^{n-1} = \sum_{n=1}^{\infty} n \frac{H_n(x)}{n!} t^{n-1}$$

$$= \sum_{n=1}^{\infty} \frac{H_n(x)}{(n-1)!} t^{n-1} = \sum_{n=0}^{\infty} \frac{H_{n+1}(x)}{n!} t^n.$$

Thus,

$$\sum_{n=0}^{\infty} \frac{(2x-2t)H_n(x)}{n!} t^n = \sum_{n=0}^{\infty} \frac{H_{n+1}(x)}{n!} t^n.$$

Now

$$\sum_{n=0}^{\infty} \frac{(2x-2t)H_n(x)}{n!} t^n = \sum_{n=0}^{\infty} \frac{-2tH_n(x)}{n!} t^n + \sum_{n=0}^{\infty} \frac{2xH_n(x)}{n!} t^{n+1}$$

and

$$\sum_{n=0}^{\infty} \frac{-2tH_n(x)}{n!} t^n = \sum_{n=0}^{\infty} \frac{-2H_n(x)}{n!} t^{n+1} = \sum_{n=1}^{\infty} \frac{-2H_{n-1}(x)}{(n-1)!} t^n = \sum_{n=1}^{\infty} \frac{-2nH_{n-1}(x)}{n!} t^n.$$

Thus,

$$\sum_{n=1}^{\infty} \frac{-2nH_{n-1}(x)}{n!} t^n + \sum_{n=0}^{\infty} \frac{2xH_n(x)}{n!} t^n = \sum_{n=0}^{\infty} \frac{H_{n+1}(x)}{n!} t^n.$$

When $n = 0$

$$\frac{2xH_n(x)}{n!} t^n = 2xH_0(x) = 2x(1) = 2x$$

and

$$\frac{H_{n+1}(x)}{n!} t^n = H_1(x) = 2x$$

so

$$\sum_{n=1}^{\infty} \frac{-2nH_{n-1}(x)}{n!} t^n + \sum_{n=1}^{\infty} \frac{2xH_n(x)}{n!} t^n = \sum_{n=1}^{\infty} \frac{H_{n+1}(x)}{n!} t^n$$

and so

$$-2nH_{n-1}(x) + 2xH_n(x) = H_{n+1}(x) \quad n = 1, 2, \ldots \tag{2}$$

This is the second recurrence relation. The problem now is to manipulate the recurrence relations to show that $H_n(x)$ solves Hermite's equation.

Differentiating Eq. (2) gives

$$H'_{n+1}(x) = 2H_n(x) + 2xH'_n(x) - 2nH'_{n-1}(x). \tag{3}$$

We also found in Eq. (1) that $H_n(x)$ solves the differential equation

$$2nH_{n-1}(x) = H'_n(x) \qquad (1)$$

so

$$H''_n(x) = 2nH'_{n-1}(x).$$

Replacing n by $n + 1$ Eq. (1) gives

$$H'_{n+1}(x) = 2(n+1)H_n(x).$$

Thus

$$H'_{n+1}(x) = 2H_n(x) + 2xH'_n(x) - 2nH'_{n-1}(x)$$

is equivalent to

$$2(n+1)H_n(x) = 2H_n(x) + 2xH'_n(x) - H''_n(x)$$

$$\left(\text{since } H''_n(x) = 2nH'_{n-1}(x) \text{ and } H'_{n+1}(x) = 2(n+1)H_n(x)\right)$$

or

$$2nH_n(x) = 2xH_n'(x) - H_n''(x)$$

so

$$H_n''(x) - 2xH_n'(x) + 2nH_n(x) = 0$$

which is Hermite's equation.

Thus, we have justified the claim that $H_n(x)$ is an nth degree polynomial that satisfies Hermite's equation and so is a Hermite polynomial.

EXERCISES

1. Use the generating function for Hermite polynomials to find $H_n(x)$ for $n = 1,2,3$ and show

$$H_n(x) = (-1)^n e^{x^2} \frac{d^n}{dx^n} e^{-x^2} \quad \text{for } n = 1, 2, 3.$$

2. Using a CAS find

$$\int_{-\infty}^{\infty} e^{-x^2} H_2(x)H_3(x)dx \quad \text{and} \quad \int_{-\infty}^{\infty} e^{-x^2} H_3(x)H_3(x)dx.$$

7.4 GENERATING FUNCTION FOR LEGENDRE'S EQUATION

The differential equation

$$(1 - x^2)y''(x) - 2xy'(x) + n(n+1)y(x) = 0$$

is Legendre's equation.

We will show that the function

$$G(x, t) = \frac{1}{\sqrt{1 - 2tx + t^2}}$$

is the generating function for Legendre polynomials.

Denoting the Legendre polynomial of degree n by $P_n(x)$, we have

$$G(x, t) = \sum_{n=0}^{\infty} P_n(x)t^n.$$

We will get two recursion relations by computing

$$\frac{\partial}{\partial t}G(x,t) \text{ and } \frac{\partial}{\partial x}G(x,t).$$

Letting

$$G(x,t) = \frac{1}{\sqrt{1-2tx+t^2}}$$

gives

$$\frac{\partial}{\partial x}G(x,t) = \frac{t}{(1-2xt+t^2)^{3/2}} = \frac{t}{(1-2xt+t^2)}\frac{1}{(1-2xt+t^2)^{1/2}}$$
$$= \frac{t}{(1-2xt+t^2)}G(x,t) = \frac{t}{(1-2xt+t^2)}\sum_{n=0}^{\infty}P_n(x)t^n$$

and expressing

$$G(x,t) = \sum_{n=0}^{\infty}P_n(x)t^n$$

gives

$$\frac{\partial}{\partial x}G(x,t) = \sum_{n=0}^{\infty}P'_n(x)t^n.$$

Thus, we have

$$\frac{t}{(1-2xt+t^2)}\sum_{n=0}^{\infty}P_n(x)t^n = \sum_{n=0}^{\infty}P'_n(x)t^n.$$

So

$$t\sum_{n=0}^{\infty}P_n(x)t^n = (1-2xt+t^2)\sum_{n=0}^{\infty}P'_n(x)t^n.$$

Expanding gives

$$\sum_{n=0}^{\infty}P_n(x)t^{n+1} = \sum_{n=0}^{\infty}P'_n(x)t^n - \sum_{n=0}^{\infty}2xP'_n(x)t^{n+1} + \sum_{n=0}^{\infty}P'_n(x)t^{n+2}.$$

Reindex to get

$$\sum_{n=0}^{\infty}P_n(x)t^{n+1} = \sum_{n=-1}^{\infty}P'_{n+1}(x)t^{n+1} - \sum_{n=0}^{\infty}2xP'_n(x)t^{n+1} + \sum_{n=1}^{\infty}P'_{n-1}(x)t^{n+1}.$$

So for $n \geq 1$ *we* have

$$P_n(x) = P'_{n+1}(x) - 2xP'_n(x) + P'_{n-1} \tag{1}$$

which is our first recurrence equation.

Also

$$\frac{\partial}{\partial t}G(x,t) = \frac{x-t}{(1-2xt+t^2)^{\frac{3}{2}}} = \frac{x-t}{(1-2xt+t^2)}\frac{1}{(1-2xt+t^2)^{\frac{1}{2}}}$$

$$= \frac{x-t}{(1-2xt+t^2)}G(x,t) = \frac{x-t}{(1-2xt+t^2)}\sum_{n=0}^{\infty} P_n(x)t^n$$

and

$$\frac{\partial}{\partial t}G(x,t) = \frac{\partial}{\partial t}\sum_{n=0}^{\infty} P_n(x)t^n = \sum_{n=0}^{\infty} nP_n(x)t^{n-1}$$

so

$$\frac{x-t}{(1-2xt+t^2)}\sum_{n=0}^{\infty} P_n(x)t^n = \sum_{n=0}^{\infty} nP_n(x)t^{n-1}$$

or

$$(x-t)\sum_{n=0}^{\infty} P_n(x)t^n = (1-2xt+t^2)\sum_{n=0}^{\infty} nP_n(x)t^{n-1}.$$

Expanding gives

$$\sum_{n=0}^{\infty} xP_n(x)t^n - \sum_{n=0}^{\infty} P_n(x)t^{n+1} = \sum_{n=0}^{\infty} nP_n(x)t^{n-1} - \sum_{n=0}^{\infty} 2xnP_n(x)t^n + \sum_{n=0}^{\infty} nP_n(x)t^{n+1}.$$

Reindexing, we have

$$\sum_{n=0}^{\infty} xP_n(x)t^n - \sum_{n=1}^{\infty} P_{n-1}(x)t^n = \sum_{n=-1}^{\infty} (n+1)P_{n+1}(x)t^n - \sum_{n=0}^{\infty} 2xnP_n(x)t^n + \sum_{n=0}^{\infty} (n-1)P_{n-1}(x)t^n.$$

So

$$xP_n(x) - P_{n-1}(x) = (n+1)P_{n+1}(x) - 2xnP_n(x) + (n-1)P_{n-1}(x)$$

or

$$(n+1)P_{n+1}(x) - (2xn+x)P_n(x) + nP_{n-1}(x) = 0 \tag{2}$$

which is the second recurrence equation.

Our goal is to demonstrate

$$(1 - x^2)P_n''(x) + 2xP_n'(x) + n(n+1)P_n(x) = 0.$$

This will be done by manipulating

$$P_n(x) + 2xP_n'(x) = P_{n-1}'(x) + P_{n+1}'(x) \qquad (3)$$

and

$$P_n(x) + 2xP_n'(x) = P_{n-1}'(x) + P_{n+1}'(x). \qquad (4)$$

We have

$$(2n + 1)xP_n(x) = (n+1)P_{n+1}(x) + nP_{n-1}(x) \qquad (5)$$

$$P_{n+1}'(x) + P_{n-1}'(x) = 2xP_n'(x) + P_n(x). \qquad (6)$$

Adding 2 × Eq. (5) to (2n + 1) × Eq. (6) gives

$$(2n + 1)P_n(x) = P_{n+1}'(x) - P_{n-1}'(x). \qquad (7)$$

Computing [Eq. (6) + Eq. (7)]/2 gives

$$P_{n+1}'(x) = (n+1)P_n(x) + xP_n'(x). \qquad (8)$$

Similarly, [Eq. (6) – Eq. (7)]/2 gives

$$P_{n-1}'(x) = nP_n(x) + xP_n'(x). \qquad (9)$$

In Eq. (7), replace n by $n - 1$ to get

$$[2(n-1) + 1]P_{n-1}'(x) = P_n'(x) - P_{n-2}'(x)$$

or

$$(2n - 1)P_{n-1}'(x) = P_n'(x) - P_{n-2}'(x). \qquad (10)$$

Multiply Eq. (9) by x to get

$$xP_{n-1}'(x) = nxP_n(x) + x^2P_n'(x). \qquad (11)$$

Add Eq. (10) to Eq. (11) to get

$$(1 - x^2)P_n'(x) = nP_{n-1}(x) - nxP_n(x). \qquad (12)$$

Differentiate Eq. (12) to get

$$(1 - x^2)P_n''(x) - 2xP_n'(x) = nP_{n-1}'(x) - nP_n(x) - nxP_n'(x). \qquad (13)$$

Multiply Eq. (9) by n to get

$$nP'_{n-1}(x) = n^2 P_n(x) + nxP'_n(x). \tag{14}$$

Adding Eq. (13) to Eq. (14) gives

$$(1 - x^2)P''_n(x) - 2xP'_n(x) + n(n+1)P_n(x) = 0.$$

Thus, $P_n(x)$ is a polynomial of degree n, which satisfies Legendre's equation and is a Legendre polynomial.

EXERCISES

1. Use the generating function to find $P_2(x)$, $P_3(x)$, $P_4(x)$.
2. Use a CAS to evaluate
   a. $\int_{-1}^{1} P_3(x)P_4(x)dx.$
   b. $\int_{-1}^{1} P_3(x)P_3(x)dx.$
   c. $\int_{-1}^{1} P_4(x)P_4(x)dx.$
3. Show that

$$P_3(x) = \frac{1}{2^3 3!}\frac{d^3}{dx^3}\left(x^2 - 1\right)^3.$$

4. Show that

$$P_2(\cos\theta) = \frac{1}{4}(1 + 3\cos(2\theta))$$

7.5 GENERATOR FOR BESSEL FUNCTIONS OF THE FIRST KIND

Bessel's equation is

$$x^2 y''(x) + xy'(x) + \left(x^2 - n^2\right)y(x) = 0.$$

We will show that

$$G(x,t) = \exp\left[\frac{x}{2}\left(t - t^{-1}\right)\right]$$

is a generator for Bessel functions of the first kind.

We will show that $J_n(x)$ satisfies Bessel's equation.

We will get two recurrence relations by taking the partial derivatives of the generating function. We have

$$\begin{aligned}\frac{\partial G(x,t)}{\partial t} &= \frac{\partial}{\partial t}\exp\left[\frac{x}{2}\left(t-t^{-1}\right)\right] = \exp\left[\frac{x}{2}\left(t-t^{-1}\right)\right]\left(\frac{x}{2}+\frac{x}{2t^2}\right)\\ &= \left(\frac{x}{2}\right)\left(1+\frac{1}{t^2}\right)\exp\left[\frac{x}{2}\left(t-t^{-1}\right)\right] = \left(\frac{x}{2}\right)\left(1+\frac{1}{t^2}\right)\sum_{n=-\infty}^{\infty} J_n(x)t^n\\ &= \sum_{n=-\infty}^{\infty}\left(\frac{x}{2}\right)\left[J_n(x)t^n + J_n(x)t^{n-2}\right]\\ &= \sum_{n=-\infty}^{\infty}\left(\frac{x}{2}\right)J_n(x)t^n + \sum_{n=-\infty}^{\infty}\left(\frac{x}{2}\right)J_{n+2}(x)t^n\end{aligned}$$

We also have

$$\frac{\partial G(x,t)}{\partial t} = \frac{\partial}{\partial t}\sum_{n=-\infty}^{\infty} J_n(x)t^n = \sum_{n=-\infty}^{\infty} nJ_n(x)t^{n-1} = \sum_{n=-\infty}^{\infty}(n+1)J_{n+1}(x)t^n.$$

Thus,

$$\left(\frac{x}{2}\right)\left[J_n(x)+J_{n+2}\right] = (n+1)J_{n+1}(x)$$

or

$$J_{n-1}(x)+J_{n+1} = \frac{2n}{x}J_n(x) \tag{1}$$

which is the first recurrence relation.

Also,

$$\begin{aligned}\frac{\partial G(x,t)}{\partial x} &= \frac{\partial}{\partial x}\exp\left[\frac{x}{2}\left(t-t^{-1}\right)\right] = \exp\left[\frac{x}{2}\left(t-t^{-1}\right)\right]\left(\frac{1}{2}\right)\left(t-t^{-1}\right)\\ &= \left(\frac{1}{2}\right)\left(t-t^{-1}\right)\sum_{n=-\infty}^{\infty} J_n(x)t^n\\ &= \left(\frac{1}{2}\right)\left[\sum_{n=-\infty}^{\infty} J_n(x)t^{n+1} - \sum_{n=-\infty}^{\infty} J_n(x)t^{n-1}\right]\\ &= \left(\frac{1}{2}\right)\left[\sum_{n=-\infty}^{\infty} J_{n-1}(x)t^n - \sum_{n=-\infty}^{\infty} J_{n+1}(x)t^n\right].\end{aligned}$$

We also have

$$\frac{\partial G(x,t)}{\partial x} = \frac{\partial}{\partial x}\sum_{n=-\infty}^{\infty} J_n(x)t^n = \sum_{n=-\infty}^{\infty} J_n'(x)t^n.$$

Thus,

$$\left(\frac{1}{2}\right)[J_{n-1}(x) - J_{n+1}(x)] = J_n'(x)$$

or

$$J_{n-1}(x) - J_{n+1}(x) = 2J_n'(x) \tag{2}$$

which is the second recurrence relation.

Adding Eqs. (1) and (2) gives

$$J_{n-1}(x) + J_{n+1} = \frac{2n}{x}J_n(x) \qquad (1)$$

$$J_{n-1}(x) - J_{n+1}(x) = 2\,J_n'(x) \qquad (2)$$

$$J_{n-1}(x) = \frac{n}{x}J_n(x) + J_n'(x)$$

or

$$xJ_n'(x) = xJ_{n-1}(x) - nJ_n(x) \tag{3}$$

and so

$$nxJ_n'(x) = nxJ_{n-1}(x) - n^2J_n(x).$$

Thus,

$$nxJ_n'(x) - nxJ_{n-1}(x) + n^2J_n(x) = 0. \tag{3'}$$

Differentiating Eq. (3) gives

$$xJ_n''(x) + J_n'(x) = xJ_{n-1}'(x) + J_{n-1}(x) - nJ_n'(x)$$

or

$$xJ_n''(x) + (n+1)J_n'(x) - J_{n-1}(x) - xJ_{n-1}'(x) = 0. \tag{4}$$

Multiplying Eq. (4) by x gives

$$x^2J_n''(x) + (n+1)xJ_n'(x) - xJ_{n-1}(x) - x^2J_{n-1}'(x) = 0. \tag{4'}$$

Eq. (4′) – Eq. (3′) gives

$$\begin{aligned}&\left[x^2J_n''(x) + (n+1)xJ_n'(x) - xJ_{n-1}(x) - x^2J_{n-1}'(x)\right] \qquad \text{Eq. (4')}\\&-\left[nxJ_n'(x) - nxJ_{n-1}(x) + n^2J_n(x)\right] \qquad \text{Eq. (3')}\\&= x^2J_n''(x) + xJ_n'(x) - n^2J_n(x) + (n-1)xJ_{n-1}(x) - x^2J_{n-1}'(x) = 0.\end{aligned} \tag{5}$$

We will show in Exercise 1 that

$$(n-1)xJ_{n-1}(x) - x^2J'_{n-1}(x) = x^2J_n(x).$$

Knowing this, we have

$$\begin{aligned} &x^2J''_n(x) + xJ'_n(x) - n^2J_n(x) + (n-1)xJ_{n-1}(x) - x^2J'_{n-1}(x) \\ &= x^2J''_n(x) + xJ'_n(x) - n^2J_n(x) + x^2J_n(x) \\ &= x^2J''_n(x) + xJ'_n(x) + (x^2 - n^2)J_n(x) = 0 \end{aligned} \tag{6}$$

which is Bessel's equation.

EXERCISE 1

Show that

$$(n-1)xJ_{n-1}(x) - x^2J'_{n-1}(x) = x^2J_n(x).$$

CHAPTER

Separation of Variables in Cartesian Coordinates

8

8.1 INTRODUCTION

For the remainder of the text, we concentrate on solving partial differential equations that involve the Laplacian. We analyze the three prototypical equations—the heat equation, the wave equation, and the Laplace's equation—in significant detail. We shall consider four techniques of solving partial differential equations; separation of variables, the Fourier transform, the Laplace transform, and Green's functions. In this chapter we solve each of these equations in Cartesian coordinates by separation of variables.

The idea of separation of variables is to assume that the solution to the partial differential equation, $u(\alpha, \beta, \gamma)$, can be written as

$$u(\alpha, \beta, \gamma) = f(\alpha)g(\beta)h(\gamma)$$

and determine an ordinary differential equation (ODE) that each of $f(\alpha)$, $g(\beta)$, and $h(\gamma)$ must satisfy. Each of the ODEs is then solved, and the solutions are "pasted together" to give the solution to the partial differential equation. The validity of the solution should be verified because we began with the assumption that the variables could be separated.

In Section 8.2, we consider the case of Laplace's equation in two variables. We shall see that in all of the examples of this chapter the resulting ODEs are familiar and elementary to solve.

In Section 8.3, we analyze Laplace's equation in three variables.

In Section 8.4, we give a detailed description of the heat equation in one space dimension, and in Section 8.5 we study the wave equation in one space dimension.

8.2 SOLVING LAPLACE'S EQUATION ON A RECTANGLE

We begin with Laplace's equation in two variables

$$\Delta u(x, y) = \frac{\partial^2 u}{\partial x^2} + \frac{\partial^2 u}{\partial y^2} = 0$$

Mathematical Physics with Partial Differential Equations. https://doi.org/10.1016/B978-0-12-814759-7.00008-9

and hypothesize that

$$u(x,y) = X(x)\,Y(y).$$

Then

$$\frac{\partial^2 u}{\partial x^2} = X''(x)\,Y(y) \quad \text{and} \quad \frac{\partial^2 u}{\partial y^2} = X(x)\,Y''(y)$$

so

$$X''(x)\,Y(y) + X(x)\,Y''(y) = 0.$$

Hence

$$\frac{X''(x)\,Y(y)}{X(x)\,Y(y)} + \frac{X(x)\,Y''(y)}{X(x)\,Y(y)} = \frac{X''(x)}{X(x)} + \frac{Y''(y)}{Y(y)} = 0$$

and thus

$$\frac{X''(x)}{X(x)} = -\frac{Y''(y)}{Y(y)}. \tag{1}$$

The left-hand side of Eq. (1) is a function only of x and the right hand side a function only of y, so the common value must be a constant, which we denote λ. So we have

$$X(x) = \lambda X(x) \quad \text{and} \quad Y(y) = -\lambda Y(y)$$

or

$$X''(x) - \lambda X(x) = 0 \tag{2a}$$

$$Y''(y) + \lambda Y(y) = 0. \tag{2b}$$

There are three cases: $\lambda = 0$, $\lambda > 0$, and $\lambda < 0$.

If $\lambda = 0$, then $X''(x) = 0$ and $X(x) = Ax + B$ and likewise, $Y(y) = Cy + D$. The more interesting cases that allow for nontrivial boundary conditions are $\lambda > 0$ and $\lambda < 0$.

Suppose that $\lambda > 0$. Then $X''(x) + \lambda X(x) = 0$ has the solution

$$X(x) = A\cos\sqrt{\lambda}x + B\sin\sqrt{\lambda}x$$

and $Y''(y) - \lambda Y(y) = 0$ has the solution

$$Y(y) = C\cosh\sqrt{\lambda}y + D\sinh\sqrt{\lambda}y.$$

The case $\lambda < 0$ is nearly identical and is left as Exercise 1.

We continue to consider the case $\lambda > 0$ and assign boundary conditions to the rectangle $0 \le x \le a$, $0 \le y \le b$. See Fig. 8.2.1.

We assign the boundary conditions

$$u(x,0) = f_1(x),\ \ 0 \le x \le a; \quad u(x,b) = f_2(x),\ \ 0 \le x \le a;$$

$$u(0,y) = g_1(y),\ \ 0 \le y \le b; \quad u(a,y) = g_2(y),\ \ 0 \le y \le b.$$

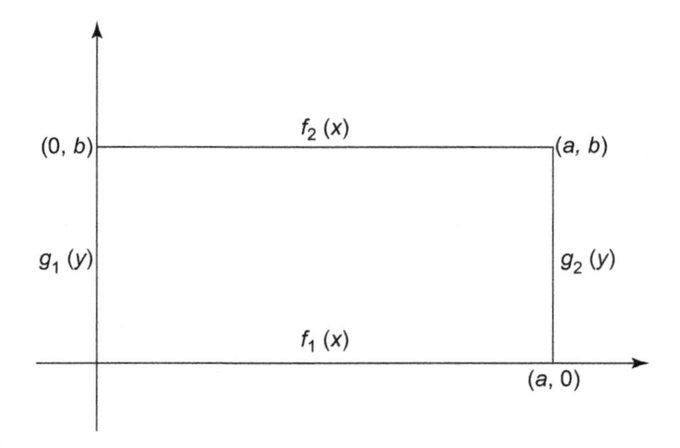

FIGURE 8.2.1

The simplest way to solve the problem is to consider four boundary value problems $\Delta u(x, y) = 0$ with the boundary values on three of the sides being zero and the value of the given function on the fourth side. Solve each of the four problems, and sum the solutions. The result will be $\Delta u(x, y) = 0$, and all four boundary conditions will be satisfied. One such boundary value problem, which we now consider, will be

$$\Delta u(x, y) = 0,$$

$$u(x,0) = f_1(x), \;\; 0 \le x \le a; \quad u(x,b) = 0, \;\; 0 \le x \le a;$$

$$u(0,y) = 0, \;\; 0 \le y \le b; \quad u(a,y) = 0, \;\; 0 \le y \le b.$$

We have

$$X(x) = A\cos\sqrt{\lambda}x + B\sin\sqrt{\lambda}x, \;\; X(0) = 0, \;\; X(a) = 0.$$

Now

$$X(0) = A,$$

so

$$A = 0; \quad X(a) = B\sin\sqrt{\lambda}a = 0.$$

To avoid having only the trivial solution, we must have $\sqrt{\lambda}a = n\pi$, so for each integer n

$$\lambda_n = \frac{n^2\pi^2}{a^2}$$

is an eigenvalue for the boundary value problem

$$X''(x) + \lambda X(x) = 0, \;\; X(0) = 0, \;\; X(a) = 0$$

and

$$X_n(x) = \sin\left(\frac{n\pi x}{a}\right)$$

is the corresponding eigenfunction.

Now consider

$$Y''(y) - \lambda_n Y(y) = 0; \quad Y(b) = 0. \tag{3}$$

We have determined that $\lambda_n = \dfrac{n^2\pi^2}{a^2}$ so Eq. (3) is

$$Y''(y) - \frac{n^2\pi^2}{a^2} Y(y) = 0; \quad Y(b) = 0.$$

The solution to

$$Y''(y) - \frac{n^2\pi^2}{a^2} Y(y) = 0$$

is

$$Y_n(y) = C\cosh\left(\frac{n\pi y}{a}\right) + D\sinh\left(\frac{n\pi y}{a}\right)$$

and the boundary condition $Y_n(b) = 0$ gives the solution

$$C\cosh\left(\frac{n\pi b}{a}\right) + D\sinh\left(\frac{n\pi b}{a}\right) = 0$$

so

$$D = \frac{-C\cosh\left(\dfrac{n\pi b}{a}\right)}{\sinh\left(\dfrac{n\pi b}{a}\right)} = -C\coth\left(\frac{n\pi b}{a}\right)$$

and

$$\begin{aligned} Y_n(y) &= C_n\cosh\left(\frac{n\pi y}{a}\right) + D_n\sinh\left(\frac{n\pi y}{a}\right) \\ &= C_n\left[\cosh\left(\frac{n\pi y}{a}\right) - \coth\left(\frac{n\pi b}{a}\right)\sinh\left(\frac{n\pi y}{a}\right)\right]. \end{aligned}$$

A more convenient way to write the solution is

$$Y_n(y) = F_n\sinh\left[\frac{n\pi(b-y)}{a}\right]$$

as the eigenfunction for

$$Y''(y) + \frac{n^2\pi^2}{a^2} Y(y) = 0; \quad Y(b) = 0$$

as we verify in Exercise 2.

Thus, we have

$$u_n(x,y) = X_n(x)Y_n(y) = \sin\left(\frac{n\pi x}{a}\right)\left[\cosh\left(\frac{n\pi y}{a}\right) - \coth\left(\frac{n\pi b}{a}\right)\sinh\left(\frac{n\pi y}{a}\right)\right]$$

or

$$\sin\left(\frac{n\pi x}{a}\right)\sinh\left[\frac{n\pi(b-y)}{a}\right],$$

and so

$$\begin{aligned} u(x,y) &= \sum_{n=1}^{\infty} c_n u_n(x,y) \\ &= \sum_{n=1}^{\infty} c_n \sin\left(\frac{n\pi x}{a}\right)\left[\cosh\left(\frac{n\pi y}{a}\right) - \coth\left(\frac{n\pi b}{a}\right)\sinh\left(\frac{n\pi y}{a}\right)\right] \qquad (4) \\ &= \sum_{n=1}^{\infty} c_n \sin\left(\frac{n\pi x}{a}\right)\sinh\left[\frac{n\pi(b-y)}{a}\right]. \end{aligned}$$

We now determine the constants c_n so that

$$f_1(x) = u(x,0) = \sum_{n=1}^{\infty} c_n u_n(x,0) = \sum_{n=1}^{\infty} c_n \sin\left(\frac{n\pi x}{a}\right)\sinh\left[\frac{n\pi(b-0)}{a}\right]$$

$$= \sum_{n=1}^{\infty} c_n \sin\left(\frac{n\pi x}{a}\right)\sinh\left[\frac{n\pi b}{a}\right].$$

Letting $d_n = c_n \sinh\left[\frac{n\pi b}{a}\right]$, we get

$$f_1(x) = u(x,0) = \sum_{n=1}^{\infty} d_n \sin\left(\frac{n\pi x}{a}\right)$$

which is the Fourier expansion of $f_1(x)$ in a sine series. The coefficients are given by

$$d_n = \frac{2}{a}\int_0^a f_1(x)\sin\left(\frac{n\pi x}{a}\right)dx.$$

Thus

$$c_n = \frac{d_n}{\sinh\left[\frac{n\pi b}{a}\right]} = \frac{\frac{2}{a}\int_0^a f_1(x)\sin\left(\frac{n\pi x}{a}\right)dx}{\sinh\left[\frac{n\pi b}{a}\right]}$$

and

$$u(x,y)=\sum_{n=1}^{\infty}c_n u_n(x,y)=\sum_{n=1}^{\infty}c_n\sin\left(\frac{n\pi x}{a}\right)\sinh\left[\frac{n\pi(b-y)}{a}\right]$$

$$=\sum_{n=1}^{\infty}\left[\frac{\frac{2}{a}\int_0^a f_1(x)\sin\left(\frac{n\pi x}{a}\right)dx}{\sinh\left[\frac{n\pi b}{a}\right]}\right]\sin\left(\frac{n\pi x}{a}\right)\sinh\left[\frac{n\pi(b-y)}{a}\right].$$

In Exercise 3 we show the solution

$$\Delta u(x,y)=0,$$

$$u(x,0)=0,\ 0<x<a;\quad u(a,y)=0,\ 0<x<a;$$

$$u(0,y)=g_1(y),\ 0<y<b;\quad u(a,y)=0,\ 0<y<b$$

is

$$u(x,y)=\sum_{n=1}^{\infty}\left\{\left[\frac{2}{b}\int_0^b g_1(y)\sin\frac{n\pi y}{b}dy\right]\cosh\frac{n\pi x}{b}\right.$$

$$\left.-\left[\frac{2}{b}\int_0^b g_1(y)\sin\frac{n\pi y}{b}dy\right]\coth\frac{n\pi a}{b}\sinh\frac{n\pi x}{b}\right\}\sin\frac{n\pi y}{b}.$$

Example:
We consider the case where $a=b=1, f_1(x)=x^2, g_1(y)=y$.
Two cases are pertinent to this example. For the first case,

$$\Delta u(x,y)=0,$$

$$u(x,0)=x^2,\ 0<x<1;\quad u(x,1)=0,\ 0<x<1;$$

$$u(0,y)=0,\ 0<y<1;\quad u(1,y)=0,\ 0<y<1$$

the solution is

$$u(x,y)=\sum_{n=1}^{\infty}\left[\frac{\frac{2}{1}\int_0^1 x^2\sin\left(\frac{n\pi x}{1}\right)dx}{\sinh\left[\frac{n\pi 1}{1}\right]}\right]\sin\left(\frac{n\pi x}{1}\right)\sinh\left[\frac{n\pi(1-y)}{1}\right].$$

For the second case,

$$\Delta u(x,y)=0,$$

$$u(x,0)=0,\ 0<x<1;\quad u(x,1)=0,\ 0<x<1;$$

$$u(0,y)=y,\ 0<y<1;\quad u(1,y)=0,\ 0<y<1$$

the solution is

$$u(x,y)=\sum_{n=1}^{\infty}\left\{\left[\frac{2}{1}\int_0^1 y\sin\frac{n\pi y}{1}dy\right]\cosh\frac{n\pi x}{1}\right.$$

$$\left.-\left[\frac{2}{1}\int_0^1 y\sin\frac{n\pi y}{1}dy\right]\coth\frac{n\pi 1}{1}\sinh\frac{n\pi x}{1}\right\}\sin\frac{n\pi y}{1}$$

and we would add the solutions.

EXERCISES

1. Describe the solutions to Eqs. (2a) and (2b) in the case $\lambda < 0$.
2. Verify that

$$Y_n(y)=\sinh\left[\frac{n\pi(b-y)}{a}\right]$$

is a solution to

$$Y''(y)+\frac{n^2\pi^2}{a^2}Y(y)=0;\quad Y(b)=0.$$

3. Show that the solution to

$$\Delta u(x,y)=0,$$
$$u(x,0)=0,\ 0<x<a;\quad u(a,y)=0,\ 0<x<a;$$
$$u(0,y)=g_1(y),\ 0<y<b;\quad u(a,y)=0,\ 0<y<b$$

is

$$u(x,y)=\sum_{n=1}^{\infty}\left\{\left[\frac{2}{b}\int_0^b g_1(y)\sin\frac{n\pi y}{b}dy\right]\cosh\frac{n\pi x}{b}\right.$$

$$\left.-\left[\frac{2}{b}\int_0^b g_1(y)\sin\frac{n\pi y}{b}dy\right]\coth\frac{n\pi a}{b}\sinh\frac{n\pi x}{b}\right\}\sin\frac{n\pi y}{b}.$$

4. Find the solution to

$$\Delta u(x,y)=0,$$
$$u(x,0)=0,\ \ 0<x<a;\quad u(x,b)=f_2(x),\ \ 0<x<a;$$
$$u(0,y)=0,\ \ 0<y<b;\quad u(a,y)=0,\ \ 0<y<b.$$

5. Find the solution to

$$\Delta u(x,y)=0,$$
$$u(x,0)=0,\ \ 0<x<a;\quad u(x,b)=0,\ \ 0<x<a;$$
$$u(0,y)=0,\ \ 0<y<b;\quad u(a,y)=g_2(y),\ \ 0<y<b.$$

6. Find the solution to

$$\Delta u(x,y)=0,$$
$$u(x,0)=0,\ \ 0<x<\pi;\quad u(x,\pi)=\sin x,\ \ 0<x<\pi;$$
$$u(0,y)=0,\ \ 0<y<\pi;\quad u(\pi,y)=0,\ \ 0<y<\pi.$$

7. Find the solution to

$$\Delta u(x,y)=0,$$
$$u(x,0)=0,\ \ 0<x<\pi;\quad u(x,\pi)=0,\ \ 0<x<\pi;$$
$$u(0,y)=\cos y,\ \ 0<y<\pi;\quad u(\pi,y)=0,\ \ 0<y<\pi.$$

8. Find the solution to

$$\Delta u(x,y)=0,$$
$$u(x,0)=\sin x-\sin 4\,x,\ \ 0<x<\pi;\quad u(x,\pi)=0,\ \ 0<x<\pi;$$
$$u(0,y)=0, u(\pi,y)=0.$$

9. Find the solution to

$$\Delta u(x,y)=0,$$
$$u(x,0)=0,\ \ 0<x<\pi;\quad u(x,\pi)=\pi,\ \ 0<x<\pi;\quad u(0,y)=y, u(\pi,y)=y.$$

10. Find the solution to Laplace's equation on the semiinfinite strip $0<x<1$, $0<y<\infty$ given by

$$\Delta u(x,y)=0,$$

$$u(x, 0) = f(x), \ 0 < x < 1;$$

$$u(0, y) = 0, \ 0 < y < \infty; \quad u(1, y) = 0, \ 0 < y < \infty; \quad \lim_{y \to \infty} u(x, y) = 0.$$

11. Use separation of variables to write the ODEs necessary to solve the following problems:

a. $\Delta u(x, y) = u(x, y)$

b. $\Delta u(x, y) + u_x(x, y) = u(x, y)$

c. $\Delta u(x, y) = u_x(x, y) = u_y(x, y)$.

12. We shall see that the steady state for the heat equation in two dimensions is given by

$$T_{xx}(x, y) + T_{yy}(x, y) = 0$$

Find the steady-state temperature distribution for the square $[0, \pi] \times [0, \pi]$ if the boundary conditions are

$$T_x(0, y) = T_x(\pi, y) = 0 \quad \text{(which is the case for insulated edges)} \quad \text{and}$$
$$T(x, 0) = 0, \ T(x, \pi) = T_0 \sin x.$$

8.3 LAPLACE'S EQUATION ON A CUBE

We now consider Laplace's equation on a cube. We solve the problem on a cube whose edges are each of length π instead of an arbitrary parallelepiped to make the computations somewhat less cumbersome. We model the solution on the two-dimensional case. Major differences are that the boundary consists of six faces rather than four edges, and the boundary conditions are functions of two variables. Accordingly, we will need to solve boundary value problems of the type

$$\Delta u(x, y, z) = \frac{\partial^2 u}{\partial x^2} + \frac{\partial^2 u}{\partial y^2} + \frac{\partial^2 u}{\partial z^2} = 0; \quad 0 < x < \pi, \ 0 < y < \pi, \ 0 < z < \pi; \quad 0 < z < \pi;$$

$$u(x, y, z) = 0 \quad \text{if} \quad x = 0, \ x = \pi, \ y = 0, \ y = \pi, \ z = \pi;$$

$$u(x, y, 0) = f(x, y).$$

The approach is identical in spirit to the two-dimensional case. We hypothesize that $u(x, y, z) = X(x)Y(y)Z(z)$ so that

$$\frac{\partial^2 u}{\partial x^2} = X''(x)Y(y)Z(z), \quad \frac{\partial^2 u}{\partial y^2} = X(x)Y''(y)Z(z), \quad \frac{\partial^2 u}{\partial z^2} = X(x)Y(y)Z''(z).$$

Then

$$\frac{\partial^2 u}{\partial x^2}+\frac{\partial^2 u}{\partial y^2}+\frac{\partial^2 u}{\partial z^2}=X''(x)Y(y)Z(z)+X(x)Y''(y)Z(z)+X(x)Y(y)Z''(z)=0.$$

Dividing by $X(x)Y(y)Z(z)$ gives

$$\frac{X''(x)}{X(x)}+\frac{Y''(y)}{Y(y)}+\frac{Z''(z)}{Z(z)}=0$$

so

$$\frac{X''(x)}{X(x)}+\frac{Y''(y)}{Y(y)}=-\frac{Z''(z)}{Z(z)}. \tag{1}$$

The left-hand side of Eq. (1) is a function of x and y, and the right-hand side is a function of z so each must be a constant that we denote α. Thus

$$\frac{X''(x)}{X(x)}+\frac{Y''(y)}{Y(y)}=\alpha \quad \text{and} \quad \frac{Z''(z)}{Z(z)}=-\alpha.$$

Since

$$\frac{X''(x)}{X(x)}+\frac{Y''(y)}{Y(y)}=\alpha,$$

then

$$\frac{X''(x)}{X(x)}=\alpha-\frac{Y''(y)}{Y(y)}. \tag{2}$$

Then, reasoning as before, each side of Eq. (2) must be a constant. Let $\beta=\frac{X''(x)}{X(x)}$. We then have

$$\frac{X''(x)}{X(x)}=\beta,$$

so

$$X''(x)-\beta X(x)=0.$$

Also,

$$\frac{Y''(y)}{Y(y)}=\alpha-\frac{X''(x)}{X(x)}=\alpha-\beta$$

so

$$Y''(y)-(\alpha-\beta)Y(y)=0.$$

Collecting the equations and noting the boundary conditions, we have

$$X''(x)-\beta X(x)=0; \quad X(0)=X(\pi)=0 \tag{3}$$

$$Y''(y) - (\alpha - \beta)Y(y) = 0; \quad Y(0) = Y(\pi) = 0 \tag{4}$$

$$Z''(z) + \alpha Z(z) = 0; \quad Z(\pi) = 0. \tag{5}$$

We determine appropriate values for the constants. The equation

$$X''(x) - \beta X(x) = 0; \quad X(0) = X(\pi) = 0$$

because of the boundary conditions must have $-\beta > 0$. Experience with the two-dimensional case prompts us to take $\beta = -n^2$. We thus have Eq. (3) is

$$X''(x) + n^2 X(x) = 0; \quad X(0) = X(\pi) = 0.$$

We saw in the two-dimensional case, the solution for this equation is

$$X_n(x) = \sin nx.$$

Likewise, for the equation

$$Y''(y) - (\alpha - \beta)Y(y) = 0; \quad Y(0) = Y(\pi) = 0$$

we must have $-(\alpha - \beta) > 0$, and we take $\alpha - \beta = -m^2$. We then have for Eq. (4)

$$Y''(y) + m^2 Y(y) = 0; \quad Y(0) = Y(\pi) = 0$$

for which the solution is

$$Y_m(y) = \sin my.$$

Note that $\alpha = \beta - m^2 = -m^2 - n^2$ so the equation

$$Z''(z) + \alpha Z(z) = 0; \quad Z(\pi) = 0$$

is

$$Z''(z) - \left(m^2 + n^2\right)Z(z) = 0; \quad Z(\pi) = 0$$

and we saw in the two-dimensional case the solution for this equation is

$$Z_{mn}(z) = \sinh\sqrt{m^2 + n^2}(\pi - z).$$

Thus, the solution to

$$\Delta u(x, y, z) = \frac{\partial^2 u}{\partial x^2} + \frac{\partial^2 u}{\partial y^2} + \frac{\partial^2 u}{\partial z^2} = 0; \ 0 < x < \pi, \ 0 < y < \pi, \ 0 < z < \pi; \ 0 < z < \pi;$$

$$u(x, y, z) = 0 \quad \text{if} \quad x = 0, \ x = \pi, \ y = 0, \ y = \pi, \ z = \pi$$

is

$$u(x, y, z) = \sum_{m=1}^{\infty} \sum_{n=1}^{\infty} a_{mn} \sin nx \sin my \sinh\sqrt{m^2 + n^2}(\pi - z).$$

We now determine the constants a_{mn} so that $u(x, y, 0) = f(x, y)$. We have

$$u(x,y,0) = \sum_{m=1}^{\infty}\sum_{n=1}^{\infty} a_{mn} \sin nx \sin my \sinh\sqrt{m^2+n^2}(\pi - 0)$$

$$= \sum_{m=1}^{\infty}\sum_{n=1}^{\infty} a_{mn} \sin nx \sin my \sinh \pi\sqrt{m^2+n^2}.$$

Let $c_{mn} = a_{mn} \sinh \pi\sqrt{m^2+n^2}$. Then

$$f(x,y) = u(x,y,0) = \sum_{m=1}^{\infty}\sum_{n=1}^{\infty} c_{mn} \sin nx \sin my.$$

To satisfy the boundary condition $u(x, y, 0) = f(x, y)$ we must choose

$$c_{mn} = \frac{4}{\pi^2}\int_0^{\pi}\left(\int_0^{\pi} f(x,y)\sin nx\, dx\right)\sin my\, dy.$$

Then

$$a_{mn} = \frac{c_{mn}}{\sinh \pi\sqrt{m^2+n^2}} = \frac{\dfrac{4}{\pi^2}\displaystyle\int_0^{\pi}\left(\int_0^{\pi} f(x,y)\sin nx\, dx\right)\sin my\, dy}{\sinh \pi\sqrt{m^2+n^2}}.$$

Thus,

$$u(x,y,z) = \sum_{m=1}^{\infty}\sum_{n=1}^{\infty} a_{mn} \sin nx \sin my \sinh\sqrt{m^2+n^2}(\pi - z)$$

$$= \sum_{m=1}^{\infty}\sum_{n=1}^{\infty}\left[\frac{\dfrac{4}{\pi^2}\displaystyle\int_0^{\pi}\left(\int_0^{\pi} f(x,y)\sin nx\, dx\right)\sin my\, dy}{\sinh \pi\sqrt{m^2+n^2}}\right]$$

$$\times \sin nx \sin my \sinh\sqrt{m^2+n^2}(\pi - z).$$

EXERCISES

1. Solve the initial value problem

$$\Delta u(x,y,z) = \frac{\partial^2 u}{\partial x^2} + \frac{\partial^2 u}{\partial y^2} + \frac{\partial^2 u}{\partial z^2} = 0; \quad 0 < x < \pi,\ 0 < y < \pi,\ 0 < z < \pi; \quad 0 < z < \pi;$$

$$u(x,y,z) = 0 \quad \text{if} \quad x = 0, \; x = \pi, \; y = 0, \; y = \pi, \; z = \pi; u(x,y,0) = f(x,y)$$

for the following values of $f(x, y)$.

a. $f(x, y) = xy$.

b. $f(x, y) = y^2$.

2. Use separation of variables to solve the initial value problem

$$\Delta u(x,y,z) = \frac{\partial^2 u}{\partial x^2} + \frac{\partial^2 u}{\partial y^2} + \frac{\partial^2 u}{\partial z^2} = 0; \quad 0 < x < \pi, \; 0 < y < \pi, \; 0 < z < \pi; \quad 0 < z < \pi;$$

$$\frac{\partial u}{\partial x} = 0, \quad x = 0, \; x = \pi; \quad u = 0, \; y = 0, \; y = \pi, \; z = \pi.$$

3. Use separation of variables to solve

$$\frac{\partial^2 u}{\partial x^2} + \frac{\partial^2 u}{\partial y^2} + \frac{\partial^2 u}{\partial z^2} = u, \; 0 < x < \pi, \; 0 < y < \pi, \; 0 < z < \pi; \quad 0 < z < \pi;$$

$$u = 0 \quad \text{if} \quad x = \pi, \; y = 0, \; z = 0; \quad u_y = 0 \quad \text{if} \quad y = \pi; \quad u_z = 0$$
$$\text{if} \quad z = \pi; \quad u_x(x,y,0) = f(x,y).$$

8.4 SOLVING THE WAVE EQUATION IN ONE DIMENSION BY SEPARATION OF VARIABLES

Consider the wave equation

$$u_{tt}(x,t) - c^2 u_{xx}(x,t) = 0; \quad u(0,t) = u(L,t) = 0; \quad u(x,0) = f(x), \; u_t(x,0) = g(x).$$

Note that in the wave equation, we need two initial conditions because of the $u_{tt}(x, t)$ term and two boundary conditions because of the $u_{xx}(x, t)$ term.

Suppose that

$$u(x,t) = X(x)T(t).$$

Then

$$u_{tt}(x,t) = X(x)T''(t) \quad \text{and} \quad u_{xx}(x,t) = X''(x)T(t).$$

Thus

$$u_{tt}(x,t) - c^2 u_{xx}(x,t) = X(x)T''(t) - c^2 X''(x)T(t) = 0, \text{ or}$$
$$X(x)T''(t) = c^2 X''(x)T(t).$$

Dividing by $X(x)\,T(t)$ gives

$$\frac{T''(t)}{T(t)} = c^2\frac{X''(x)}{X(x)}. \tag{1}$$

The left-hand side of Eq. (1) is a function only of t, and the right-hand side is a function only of x, so the common value is a constant that we denote $-\alpha^2$. (We show the constant must be negative in Exercise 1.)

So

$$\frac{T''(t)}{T(t)} = -\alpha^2 \quad \text{or} \quad T''(t) + \alpha^2 T(t) = 0$$

the solution for which is

$$T(t) = A\sin\alpha t + B\cos\alpha t.$$

Also,

$$c^2\frac{X''(x)}{X(x)} = -\alpha^2 \quad \text{or} \quad X''(x) + \frac{\alpha^2}{c^2}X(x) = 0$$

the solution for which is

$$X(x) = D\sin\frac{\alpha x}{c} + E\cos\frac{\alpha x}{c}.$$

The boundary values will determine the value of $-\alpha^2$. Since

$$u(0,t) = X(0)T(t) = 0$$

we must have

$$X(0) = E = 0$$

so

$$X(x) = D\sin\frac{\alpha x}{c}.$$

Then

$$u(L,t) = X(L)T(t) = 0 \quad \text{forces} \quad X(L) = D\sin\frac{\alpha L}{c} = 0.$$

To avoid having only the trivial solution, we must have

$$\frac{\alpha_n L}{c} = n\pi$$

where n is an integer.

Thus

$$\alpha_n = \frac{n\pi c}{L}$$

where n is an integer.

Many derivations of the wave equation in one dimension will substitute $c = 1/v$. We now do that so as to conform to the more common expression of the solution. Thus, we have

$$\alpha_n = \frac{n\pi}{vL}$$

where n is an integer.

So

$$T_n(t) = A_n \sin\frac{n\pi t}{vL} + B_n \cos\frac{n\pi t}{vL}$$

and

$$X_n(x) = D_n \sin\frac{n\pi x}{L}$$

and thus, by superposition,

$$u(x,t) = \sum_n u_n(x,t) = \sum_n X_n(x)T_n(t) = \sum_n D_n \sin\frac{n\pi x}{L}\left[A_n \sin\frac{n\pi t}{vL} + B_n \cos\frac{n\pi t}{vL}\right].$$

We use the initial conditions to determine the constants. We have

$$f(x) = u(x,0) = \sum_n B_n D_n \sin\frac{n\pi x}{L}.$$

Let $\beta_n = B_n D_n$. Then

$$f(x) = u(x,0) = \sum_n \beta_n \sin\frac{n\pi x}{L}$$

and β_n is determined by

$$\beta_n = \frac{2}{L}\int_0^L f(x)\sin\frac{n\pi x}{L}\,dx.$$

We also have

$$u_t(x,t) = \sum_n \left[\left(\frac{n\pi}{vL}\right)A_n \cos\left(\frac{n\pi t}{vL}\right) - \left(\frac{n\pi}{vL}\right)B_n \sin\left(\frac{n\pi t}{vL}\right)\right]D_n \sin\left(\frac{n\pi x}{L}\right)$$

so

$$g(x) = u_t(x,0) = \sum_n \left(\frac{n\pi}{vL}\right)A_n D_n \sin\left(\frac{n\pi x}{L}\right).$$

Let

$$\gamma_n = \left(\frac{n\pi}{vL}\right)A_n D_n \quad \text{so} \quad A_n D_n = \frac{\gamma_n vL}{n\pi}.$$

Then

$$g(x) = \sum_n \gamma_n \sin\left(\frac{n\pi x}{L}\right) \quad \text{and} \quad \gamma_n = \frac{2}{L}\int_0^L g(x)\sin\frac{n\pi x}{L}\,dx$$

and so

$$\begin{aligned}
u(x,t) &= \sum_n D_n \sin\frac{n\pi x}{L}\left[A_n \sin\frac{n\pi t}{vL} + B_n \cos\frac{n\pi t}{vL}\right] \\
&= \sum_n A_n D_n \sin\left(\frac{n\pi x}{L}\right)\sin\left(\frac{n\pi t}{vL}\right) + B_n D_n \sin\left(\frac{n\pi x}{L}\right)\cos\left(\frac{n\pi t}{vL}\right) \\
&= \sum_n \left[\left(\frac{\gamma_n vL}{n\pi}\right)\sin\left(\frac{n\pi x}{L}\right)\sin\left(\frac{n\pi t}{vL}\right) + \beta_n \sin\left(\frac{n\pi x}{L}\right)\cos\left(\frac{n\pi t}{vL}\right)\right] \\
&= \sum_n \left[\left(\frac{vL}{n\pi}\right)\left(\frac{2}{L}\int_0^L g(x)\sin\frac{n\pi x}{L}\,dx\right)\sin\left(\frac{n\pi x}{L}\right)\sin\left(\frac{n\pi t}{vL}\right)\right. \\
&\quad \left. + \left(\frac{2}{L}\int_0^L f(x)\sin\frac{n\pi x}{L}\,dx\right)\sin\left(\frac{n\pi x}{L}\right)\cos\left(\frac{n\pi t}{vL}\right)\right].
\end{aligned}$$

We should still check that the series converges, and the series is a solution.

Example:

Suppose that we initially distort the string by lifting it at the center of the interval by lifting it by the amount α. See Fig. 8.4.1.

The boundary conditions are then

$$u(0,t) = u(L,t) = 0$$

and one initial condition (because the string is at rest immediately before it is released) is

$$u_t(x,0) = 0.$$

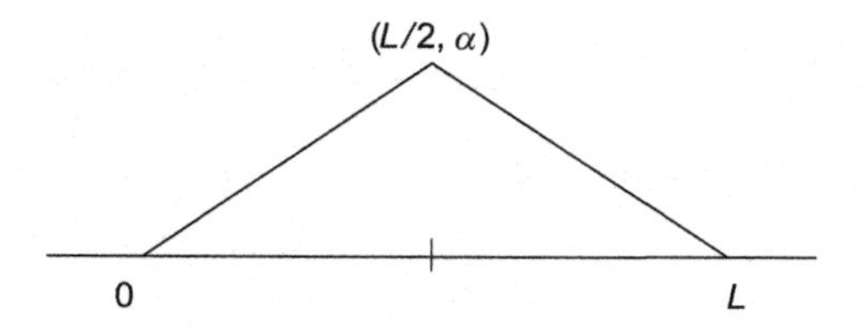

FIGURE 8.4.1

The other initial condition, $u(x, 0)$, is the equation of the graph in Fig. 8.4.1. In Exercise 2 we show that this is

$$u(x,0) = \begin{cases} \dfrac{2\alpha x}{L} & 0 \le x \le \dfrac{L}{2} \\ \dfrac{2\alpha}{L}(L-x) & \dfrac{L}{2} \le x \le L \end{cases}. \tag{2}$$

Note that $u(x, 0)$ is $f(x)$ in the derivation above. Thus, in the formula we derived

$$\beta_n = \frac{2}{L}\int_0^L f(x)\sin\frac{n\pi x}{L}dx = \frac{2}{L}\left[\int_0^{L/2}\frac{2\alpha x}{L}\sin\frac{n\pi x}{L}dx + \int_{L/2}^L \frac{2\alpha}{L}(L-x)\sin\frac{n\pi x}{L}dx\right].$$

One can check with a computer algebra system that the value of this integral is

$$\beta_n = \frac{8\alpha}{n^2\pi^2}\sin\left(\frac{n\pi}{2}\right).$$

Since $g(x) = u_t(x, 0) = 0$, we have

$$\gamma_n = \frac{2}{L}\int_0^L g(x)\sin\frac{n\pi x}{L}dx = 0.$$

Substituting into

$$u(x,t) = \sum_n\left[\left(\frac{\gamma_n v L}{n\pi}\right)\sin\left(\frac{n\pi x}{L}\right)\sin\left(\frac{n\pi t}{vL}\right) + \beta_n\sin\left(\frac{n\pi x}{L}\right)\cos\left(\frac{n\pi t}{vL}\right)\right]$$

gives

$$u(x,t) = \sum_{n=1}^{\infty}\left(\frac{8\alpha}{n^2\pi^2}\sin\left(\frac{n\pi}{2}\right)\right)\sin\left(\frac{n\pi x}{L}\right)\cos\left(\frac{n\pi t}{vL}\right).$$

Note that when n is even, $\sin\left(\frac{n\pi}{2}\right) = 0$.

Example:
We compute the kinetic and potential energy of a vibrating string.
Let $u(x, t)$ denote the vertical distance of the string at point x at time t. Kinetic energy K is computed according to

$$K = \frac{1}{2}mv^2.$$

Divide the interval $[0, L]$ into n equal subintervals of length Δx by inserting $x_0, x_1, x_2, \ldots, x_n$ with $0 = x_0 < x_1 < x_2 < \cdots < x_n = L$ and $\Delta x = x_i - x_{i-1}$. The mass of the string between x_{i-1} and x_i is $\rho\Delta x$ and the velocity at x_i is

$$\left.\frac{\partial u(x,t)}{\partial t}\right|_{x=x_i,}.$$

Thus the kinetic energy of the string is approximately given by the Riemann sum

$$\sum_{i=1}^{n} \frac{1}{2}\rho\Delta x \left(\left. \frac{\partial u(x,t)}{\partial t} \right|_{x=x_i,} \right)^2$$

so, in the limit as $\Delta x \to 0$ we get the exact value of the kinetic energy is

$$K = \frac{\rho}{2} \int_0^L \left(\frac{\partial u(x,t)}{\partial t} \right)^2 dx.$$

To compute the potential energy, we appeal to Hooke's law. According to Hooke's the force a spring exerts is proportional to the distance the spring is distorted from equilibrium. If the spring constant is k and the distance the spring is distorted is x, we have $f = -kx$. The work performed in moving a unit mass from a displaced distance a to a displaced distance b is

$$W = \int_a^b f(x)dx = \int_a^b -kx\, dx = -\frac{1}{2}k\left[b^2 - a^2\right]$$

which is the difference in potential energy between the points.

If T is the tension of the nonstretched string then we have that the potential energy is approximately

$$U \approx \frac{1}{2}\frac{T}{\Delta x}\sum_{i=1}^{n}(u(x_i,t) - u(x_{i-1},t))^2 = \frac{T}{2}\sum_{i=1}^{n}\left(\frac{u(x_i,t) - u(x_{i-1},t)}{\Delta x}\right)^2 \Delta x.$$

Taking the limit as $\Delta x \to 0$, we get

$$U = \frac{T}{2}\int_0^L \left(\frac{\partial u}{\partial x}\right)^2 dx.$$

EXERCISES

1. In Eq. (1), we derived

$$\frac{T''(t)}{T(t)} = c^2 \frac{X''(x)}{X(x)}.$$

Since $T''(t)/T(t)$ is a function only of t and $c^2 \frac{X''(x)}{X(x)}$ is a function only of x, it must be that this is a constant. Show that this constant is negative.

2. Show that the equation given for $u(x, 0)$ in the example is valid.
3. Find the solution for the wave equation if we have the same conditions as in the example, except instead of plucking the string at the point $x = L/2$ we pluck it at the point $x = L/3$.

4. Solve the wave equation for the following initial conditions on $0 < x < L$:

a. $u(x, 0) = 0$, $u_t(x, 0) = 2$.

b. $u(x, 0) = \sin \pi x/L$, $u_t(x, 0) = 0$.

c. $u(x, 0) = x(L-x)$, $u_t(x, 0) = 2$.

d. $u(x, 0) = \sin \pi x/L$, $u_t(x, 0) = \sin \pi x/L$.

e. $u(x, 0) = 0$, $u_t(x, 0) = x$.

5. Verify that the wave equation

$$c^2 u_{xx} - u_{tt} = h(x,t), \text{ with initial conditions } u(x,0) = f(x), \ u_t(x,0) = g(x)$$

has as its solution

$$u(x,t) = \frac{f(x+ct) + f(x-ct)}{2} + \frac{1}{2c}\int_{x-ct}^{x+ct} g(\theta)d\theta + \frac{1}{2c}\int_0^t \left[\int_{x-c(t-\tau)}^{x+c(t-\tau)} h(\theta,\tau)d\theta\right] d\tau.$$

6. Consider an infinitely long string that is released from rest with displacement $f(x) = e^{-x^2}$. Show that a solution to the wave equation with these initial conditions is

$$u(x,t) = \frac{1}{2}\left[e^{-(x-ct)^2} + e^{-(x+ct)^2}\right].$$

7. Show that energy is conserved in a vibrating string. To do this, show the time derivative of the energy we derived for the vibrating string in the last example is zero. Take $T = \rho c^2$.

8. Show that the solution to

$$u_{tt} = c^2 u_{xx} - g, \ 0 < x < \pi, \quad t > 0, \ u(0,t) = 0,$$
$$u(\pi,t) = 0, \ u(x,0) = 0, \ u_t(x,0) = 0$$

is

$$u(x,t) = \frac{4g}{c^2}\left[\sum_{n=1}^{\infty} \frac{\sin(2n-1)x}{(2n-1)^3}\cos(2n-1)ct - \frac{1}{8}x(\pi - x)\right].$$

This models a string that is initially at rest and at equilibrium that when released falls due to the force of gravity.

9. In this exercise we solve

$$u_{tt}(x,t) = u_{xx}(x,t) - 2au_t(x,t); \quad u(0,t) = u(\pi,t) = 0; \quad u(x,0) = 0, \ u_t(x,0) = b.$$

a. Show that separation of variables leads to the ODEs

$$X''(x) + \lambda X(x) = 0, \;\; X(0) = 0, \;\; X(\pi) = 0$$
$$T''(t) + 2aT'(t) + \lambda T(t) = 0, \;\; T(0) = 0.$$

b. Show that $\lambda_n = n^2$ and that the solutions to the resulting ODEs are

$$X_n(x) = A_n \sin nx$$
$$T_n(t) = B_n e^{-at} \sin\left(\sqrt{n^2 - a^2}\, t\right).$$

c. Conclude that

$$u(x,t) = \sum_{n=1}^{\infty} C_n e^{-at} \sin\left(\sqrt{n^2 - a^2}\, t\right) \sin nx.$$

d. Use the initial condition $u_t(x, 0) = b$ to show

$$C_n = \frac{2b}{\pi} \cdot \frac{1 - (-1)^n}{n\sqrt{n^2 - a^2}}.$$

10. The telegraph equation is

$$u_{tt}(x,t) + au_t(x,t) = bu_{xx}(x,t) - cu(x,t)$$

where a, b, $c > 0$. Show that the substitution $u(x,t) = \exp\left(-\frac{at}{2}\right) y(x,t)$ yields the equation

$$y_{tt}(x,t) = by_{xx}(x,t) + \left(\frac{1}{4}a^2 - c\right) y(x,t).$$

8.5 SOLVING THE WAVE EQUATION IN TWO DIMENSIONS IN CARTESIAN COORDINATES BY SEPARATION OF VARIABLES

In this section we solve the wave equation

$$\Delta u(x,y,t) = \left(\frac{\partial^2 u}{\partial x^2} + \frac{\partial^2 u}{\partial x^2}\right) = \frac{1}{c^2} \frac{\partial^2 u}{\partial t^2} \quad 0 < x < a, \;\; 0 < y < b. \qquad (1)$$

The method combines the ideas of what we did in solving Laplace's equation in two dimensions and what we did for the wave equation in one dimension. As we did with Laplace's equation, we take $a = b = \pi$ to make the computations less cumbersome.

Our boundary conditions will be

$$u(x,0) = u(x,\pi) = u(0,y) = u(\pi,y) = 0 \quad \text{for all } t$$

and the initial conditions will be

$$u(x,y,0) = f(x,y), \quad u_t(x,y,0) = g(x,y).$$

We hypothesize

$$u(x,y,t) = X(x)Y(y)T(t).$$

Then Eq. (1) is

$$X''(x)Y(y)T(t) + X(x)Y''(y)T(t) = \frac{1}{c^2}X(x)Y(y)T''(t).$$

Dividing by $X(x)\ Y(y)\ T(t)$ gives

$$\left(\frac{X''(x)}{X(x)} + \frac{Y''(y)}{Y(y)}\right) = \frac{1}{c^2}\frac{T''(t)}{T(t)}. \tag{2}$$

The left-hand side of Eq. (2) is a function of x and y, and the right-hand side is a function of t, so each must be a constant λ. We show in Exercise 1 that this constant must be negative, so we let $-\alpha^2 = \lambda$. We then have

$$\frac{1}{c^2}\frac{T''(t)}{T(t)} = -\alpha^2 \quad \text{so} \quad T''(t) + c^2\alpha^2 T(t) = 0. \tag{3}$$

We also have

$$\left(\frac{X''(x)}{X(x)} + \frac{Y''(y)}{Y(y)}\right) = -\alpha^2 \quad \text{so} \quad \frac{X''(x)}{X(x)} = -\alpha^2 - \frac{Y''(y)}{Y(y)}. \tag{4}$$

The left-hand side of Eq. (4) is a function of x, and the right-hand side is a function of y, so each must be a constant. As in the case of Laplace's equation on a square, we must have

$$\frac{X''(x)}{X(x)} = -m^2 \quad \text{and} \quad \frac{Y''(y)}{Y(y)} = -n^2$$

where $m^2 + n^2 = \alpha^2$.

Thus we have

$$X(x) + m^2 X(x) = 0 \quad \text{and} \quad Y(y) + n^2\, Y(y) = 0.$$

We also have the boundary conditions

$$X(0) = X(\pi) = 0 \quad \text{and} \quad Y(0) = Y(\pi) = 0.$$

As we have seen on several occasions, the solutions are

$$X_m(x) = \sin mx, \quad Y_n(y) = \sin ny.$$

Returning to

$$T''(t) + c^2\alpha^2 T(t) = T''(t) + c^2(m^2 + n^2)T(t) = 0$$

we have

$$T_{mn}(t) = a_{mn} \cos\left[\left(c\sqrt{m^2+n^2}\right)t\right] + b_{mn} \sin\left[\left(c\sqrt{m^2+n^2}\right)t\right].$$

Thus,

$$u(x,y,t) = \sum_{m,n=1}^{\infty} u_{m.n}(x,y,t) = \sum_{m,n=1}^{\infty} \left\{a_{mn} \cos\left[\left(c\sqrt{m^2+n^2}\right)t\right] + b_{mn} \sin\left[\left(c\sqrt{m^2+n^2}\right)t\right]\right\} \sin mx \sin ny. \tag{5}$$

We use the initial conditions to determine a_{mn} and b_{mn}. We have

$$f(x,y) = u(x,y,0) = \sum_{m,n=1}^{\infty} a_{mn} \sin mx \cos ny$$

so

$$a_{mn} = \frac{4}{\pi^2} \int_0^{\pi} \left[\int_0^{\pi} f(x,y) \sin ny\, dy\right] \sin mx\, dx.$$

We also have

$$u_t(x,y,0) = \sum_{m,n=1}^{\infty} \left\{b_{mn}\left[\left(c\sqrt{m^2+n^2}\right)\right]\right\} \sin mx \sin ny.$$

Letting $c_{mn} = b_{mn}\left[\left(c\sqrt{m^2+n^2}\right)\right]$ we have

$$g(x,y) = u_t(x,y,0) = \sum_{m,n=1}^{\infty} c_{mn} \sin mx \sin ny$$

and

$$c_{mn} = \frac{4}{\pi^2} \int_0^{\pi} \left[\int_0^{\pi} g(x,y) \sin ny\, dy\right] \sin mx\, dx$$

so

$$b_{mn} = \frac{c_{mn}}{c\sqrt{m^2+n^2}} = \frac{\frac{4}{\pi^2} \int_0^{\pi} \left[\int_0^{\pi} g(x,y) \sin ny\, dy\right] \sin mx\, dx}{c\sqrt{m^2+n^2}}.$$

Substituting the values of a_{mn} and b_{mn} into Eq. (5) gives the solution.

Example:
Solve the wave equation on the rectangle $0 \le x \le \pi$, $0 \le y \le \pi$ given that

$$f(x,y) = u(x,y,0) = x(\pi - x)y(\pi - y) \quad \text{and} \quad g(x,y) = u_t(x,y,0) = 0.$$

Since $g(x, y) = 0$, then $b_{mn} = 0$. Also

$$a_{mn} = \frac{4}{\pi^2}\int_0^{\pi}\left[\int_0^{\pi} f(x,y)\sin ny\, dy\right]\sin mx\, dx$$

$$= \frac{4}{\pi^2}\int_0^{\pi}\left[\int_0^{\pi} x(\pi - x)y(\pi - y)\sin ny\, dy\right]\sin mx\, dx.$$

EXERCISES

Solve the wave equation on the rectangle $0 < x < \pi$, $0 < y < \pi$ given the following initial data (do not attempt to simplify the expressions for a_{mn} and b_{mn}):

1. $f(x, y) = u(x, y, 0) = 1 - x^2 - y^2$; $g(x, y) = u_t(x, y, 0) = \sin x \cos y$.
2. $f(x, y) = u(x, y, 0) = 1 - x \sin y$; $g(x, y) = u_t(x, y, 0) = x - 2y$.
3. $f(x, y) = u(x, y, 0) = 1 - x - y^2$; $g(x, y) = u_t(x, y, 0) = \sin x$.
4. $f(x, y) = u(x, y, 0) = \sin x \sin y$; $g(x, y) = u_t(x, y, 0) = 0$. (It is possible to simplify this problem with reasonable effort.)
5. Solve $u_{tt}(x, y, t) = u_{xx}(x, y, t) + u_{yy}(x, y, t)$, $0 \le x \le \pi$, $0 \le y \le \pi$, with boundary conditions $u(x, 0, t) = u(x, \pi, t) = u(0, y, t) = u(\pi, y, t) = 0$ and initial conditions $u(x, y, 0) = x(x - \pi)y(y - \pi)$, $u_t(x, y, 0) = 0$.

8.6 SOLVING THE HEAT EQUATION IN ONE DIMENSION USING SEPARATION OF VARIABLES

NO HEAT SOURCE

In one dimension the heat equation with no heat source is

$$\frac{\partial^2 u(x,t)}{\partial x^2} = \frac{1}{\alpha^2}\frac{\partial u(x,t)}{\partial t}. \tag{1}$$

Note: By using the change of variables $\tau = \alpha t$, the $1/\alpha^2$ term in Eq. (1) can be eliminated. In some examples we make this substitution.

To solve the equation, we need two boundary conditions $\left(\text{because of } \frac{\partial^2 u(x,t)}{\partial x^2}\right)$ and one initial condition $\left(\text{because of } \frac{\partial u(x,t)}{\partial t}\right)$. For our initial considerations we let these be

$$u(0,t) = 0, \quad u(0,L) = 0, \quad u(x,0) = f(x).$$

To apply separation of variables, we assume that the solution can be written as the product of functions of a single variable. In this problem we assume

$$u(x,t) = X(x)T(t). \tag{2}$$

From Eqs. (1) and (2), we get

$$X''(x)T(t) = \frac{1}{\alpha^2}X(x)T'(t). \tag{3}$$

Dividing Eq. (3) by $X(x)\ T(t)$ gives

$$\frac{X''(x)}{X(x)} = \frac{1}{\alpha^2}\frac{T'(t)}{T(t)}. \tag{4}$$

Eq. (4) again illustrates the salient point of the technique of separation of variables. We have one side of the equation as a function of one set of variables (in this case, x) and the other side as a totally different set of variables (in this case, t). Thus, each side must be the same constant.

Suppose

$$\frac{X''(x)}{X(x)} = \beta \quad \text{and} \quad \frac{1}{\alpha^2}\frac{T'(t)}{T(t)} = \beta.$$

Then we must solve

$$X''(x) - \beta X(x) = 0 \quad \text{and} \quad \frac{T'(t)}{T(t)} = \alpha^2\beta.$$

The solution to the second equation is

$$\ln(T(t)) = \alpha^2\beta t + C \quad \text{or} \quad T(t) = T(0)e^{\alpha^2\beta t}.$$

We now determine the sign of β. If $\beta = 0$, then $T(t)$ is constant, which is impossible unless $u(x, 0) = f(x) = 0$. This yields the trivial solution $u(x, t) = 0$. If $\beta > 0$ then $T(t)$ grows exponentially, which is impossible with no heat source. Thus, $\beta < 0$, and we set $\beta = -\lambda^2$. We must now solve

$$X''(x) + \lambda^2 X(x) = 0 \tag{5a}$$

and

$$T'(t) + \alpha^2\lambda^2 T(t) = 0. \tag{5b}$$

The solutions to Eqs. (5a) and (5b) are

$$X(x) = a_1 \sin(\lambda x) + a_2 \cos(\lambda x) \tag{6a}$$

and

$$T(t) = a_3 e^{-\alpha^2\lambda^2 t} \tag{6b}$$

respectively. We use the boundary conditions and initial condition to determine a_1, a_2, and a_3.

We have

$$u(0,t) = X(0)T(t) = 0$$
$$u(L,t) = X(L)T(t) = 0.$$

Since $T(t) = 0$ yields the trivial solution, we must have $X(0) = X(L) = 0$.

In Eq. (6a), when $x = 0$ we have

$$X(0) = a_1 \sin(\lambda 0) + a_2 \cos(\lambda 0) = a_2 = 0.$$

So

$$X(x) = a_1 \sin(\lambda x)$$

and thus when $x = L$ we have

$$X(L) = a_1 \sin(\lambda L) = 0.$$

One possibility is that $a_1 = 0$, but that gives the trivial solution $u(x, t) = 0$. The only other possibility is $\lambda L = n\pi$ where n is an integer. Thus, we must have $\lambda_n = n\pi/L$ where n is an integer, and so

$$X_n(x) = \sin\left(\frac{n\pi x}{L}\right).$$

Thus

$$T_n(t) = e^{-\alpha^2\lambda_n^2 t} = \exp\left[-\left(\frac{\alpha n\pi}{L}\right)^2 t\right]$$

and for any positive integer n we have

$$u_n(x,t) = X_n(x)\,T_n(t) = \sin\left(\frac{n\pi x}{L}\right)\exp\left[-\left(\frac{\alpha n\pi}{L}\right)^2 t\right]$$

is a solution to

$$\frac{\partial^2 u(x,t)}{\partial x^2} = \frac{1}{\alpha^2}\frac{\partial u(x,t)}{\partial t} \qquad u(0,t) = u(L,t) = 0. \tag{7}$$

By superposition

$$\sum_{n=1}^{\infty} c_n u_n(x,t)$$

is also a solution to Eq. (7).

We use the initial condition $u(x, 0) = f(x)$ to determine the c_n's. Now

$$\begin{aligned} u(x,0) &= \sum_{n=1}^{\infty} c_n u_n(x,0) \\ &= \sum_{n=1}^{\infty} c_n \sin\left(\frac{n\pi x}{L}\right)\exp\left[-\left(\frac{\alpha n\pi}{L}\right)^2 0\right] = \sum_{n=1}^{\infty} c_n \sin\left(\frac{n\pi x}{L}\right) = f(x). \end{aligned}$$

We need to express $f(x)$ as a sine series. We have

$$f(x)=\sum_{n=1}^{\infty} c_n \sin\left(\frac{n\pi x}{L}\right)$$

so

$$c_n=\frac{2}{L}\int_0^L f(x)\sin\left(\frac{n\pi x}{L}\right)dx.$$

Thus,

$$u(x,t)=\sum_{n=1}^{\infty} c_n \sin\left(\frac{n\pi x}{L}\right)\exp\left[-\left(\frac{\alpha n\pi}{L}\right)^2 t\right]$$

$$=\sum_{n=1}^{\infty}\left[\frac{2}{L}\int_0^L f(x)\sin\left(\frac{n\pi x}{L}\right)dx\right]\sin\left(\frac{n\pi x}{L}\right)\exp\left[-\left(\frac{\alpha n\pi}{L}\right)^2 t\right]$$

is the solution to the equation

$$\frac{\partial^2 u(x,t)}{\partial x^2}=\frac{1}{\alpha^2}\frac{\partial u(x,t)}{\partial t}\quad u(0,t)=u(L,t)=0\quad u(x,0)=f(x)$$

if indeed $u(x,t)$ can be written as $X(x)\,T(t)$. We leave it as Exercise 1 to show the solution we have asserted is valid.

THE INITIAL CONDITION IS THE DIRAC-DELTA FUNCTION

We now consider the equation

$$\frac{\partial^2 u(x,t)}{\partial x^2}=\frac{1}{\alpha^2}\frac{\partial u(x,t)}{\partial t}\quad u(-\pi,t)=u(\pi,t)=0\quad u(x,0)=\delta_0(x).$$

This equation would be appropriate to model a sudden release of pollution where the boundary of the region is absorbing. (The heat equation is also models diffusion.)

To do this problem, we use the Fourier expansion of the Dirac-delta function, $\delta_{x_0}(x)$. We note that

$$a_n=\frac{1}{\pi}\int_{-\pi}^{\pi}\delta_{x_0}(x)\cos(nx)dx=\frac{1}{\pi}\cos(nx_0)$$

$$b_n=\frac{1}{\pi}\int_{-\pi}^{\pi}\delta_{x_0}(x)\sin(nx)dx=\frac{1}{\pi}\sin(nx_0)$$

so

$$\delta_{x_0}(x)=\frac{1}{2\pi}+\frac{1}{\pi}\sum_{n=1}^{\infty}[\cos(nx_0)\cos(nx)+\sin(nx_0)\sin(nx)].$$

Now

$$\cos(nx_0)\cos(nx) + \sin(nx_0)\sin(nx) = \cos(nx - nx_0) = \cos[n(x - x_0)]$$

so the Fourier expansion of $\delta_{x_0}(x)$ is

$$\frac{1}{2\pi} + \frac{1}{\pi}\sum_{n=1}^{\infty}\cos[n(x - x_0)]$$

and the Fourier expansion of $\delta_0(x)$ is

$$\frac{1}{2\pi} + \frac{1}{\pi}\sum_{n=1}^{\infty}\cos(nx).$$

As in the previous example, we assume

$$u(x,t) = X(x)T(t).$$

Since we know $T(0) = 1$, we have from previous examples, $u(x, 0) = X(x)$. But

$$u(x,0) = \delta_0(x) = \frac{1}{2\pi} + \frac{1}{\pi}\sum_{n=1}^{\infty}\cos(nx).$$

As before,

$$T_n(t) = e^{-\alpha^2 n^2 t}$$

so that

$$u(x,t) = \sum_{n=0}^{\infty} X_n(x)T_n(t) = X_0(x)T_0(t) + \sum_{n=1}^{\infty} X_n(x)T_n(t)$$

$$= \frac{1}{2\pi} + \frac{1}{\pi}\sum_{n=1}^{\infty}\cos(nx)e^{-\alpha^2 n^2 t}$$

where we use the fact that the coefficients have been determined by the initial condition.

We note that another approach to the problem is to use the fundamental solution, which we derive in Chapter 11, to get

$$u(x,t) = \frac{1}{\sqrt{4\pi kt}}\exp\left(-\frac{x^2}{kt}\right).$$

EXERCISES

1. Solve the heat equation

$$\frac{\partial u(x,t)}{\partial t} = \alpha^2\frac{\partial^2 u(x,t)}{\partial x^2} \qquad 0 < x < \pi, \ \ t > 0$$

for the following initial and boundary conditions:

a. $u(0, t) = u(\pi, t) = 0$; $u(x, 0) = 1 - \cos x$.
b. $u(0, t) = u(\pi, t) = 0$; $u(x, 0) = \sin^2(x)$.
c. $u(0, t) = u(\pi, t) = 0$; $u(x, 0) = x(\pi - x)$.
d. $u_x(0, t) = u_x(\pi, t) = 0$; $u(x, 0) = x(\pi - x)$.

2. Solve

$$\frac{\partial u(x,t)}{\partial t} = \frac{\partial^2 u(x,t)}{\partial x^2} - u(x,t) \quad 0 < x < \pi, \ t > 0$$

$$u(0,t) = 0, \quad \frac{\partial u(\pi,t)}{\partial x} = 0; \quad u(x,0) = \sin x.$$

3. Solve the heat equation

$$\frac{\partial u(x,t)}{\partial t} = \frac{\partial^2 u(x,t)}{\partial x^2} \quad 0 < x < \pi, \ t > 0$$

$$\frac{\partial u(0,t)}{\partial x} = 0, \quad \frac{\partial u(\pi,t)}{\partial x} = 0; \quad u(x,0) = x.$$

4. Solve the heat equation

$$\frac{\partial u(x,t)}{\partial t} = \frac{\partial^2 u(x,t)}{\partial x^2} \quad -\pi < x < \pi, \ t > 0; \quad u(-\pi,t) = u(\pi,t);$$

$$\frac{\partial u(-\pi,t)}{\partial x} = \frac{\partial u(\pi,t)}{\partial x}; \quad u(x,0) = c\delta_0(x).$$

5. In this exercise we solve by separation of variables the equation

$$\frac{\partial u(x,t)}{\partial t} - \frac{\partial^2 u(x,t)}{\partial x^2} + au(x,t) = 0 \quad 0 < x < \pi, \ t > 0,$$

$$u(0,t) = 0, \quad \frac{\partial u(\pi,t)}{\partial x} = 0, \quad u(x,0) = x(\pi - x).$$

a. Show that separation of variables yields the equations

$$X''(x) + \lambda^2 X(x) = 0$$

$$T'(t) + \left(a + \lambda^2\right) T(t) = 0.$$

b. Show that the eigenvalues are $\lambda_n = n + \frac{1}{2}$.
c. Show that

$$T_n(t) = e^{-\left[(2n-1)^2 + 4a\right]t/4}; \quad X_n(x) = \sin\left(n - \frac{1}{2}\right)x$$

and thus

$$u(x,t) = \sum_{n=1}^{\infty} c_n e^{-\left[(2n-1)^2 + 4a\right]t/4} \sin\left(n - \frac{1}{2}\right)x.$$

d. Let $k = n - 1/2$ so that

$$u(x,0) = \sum_{k=1}^{\infty} b_k \sin kx = x(\pi - x).$$

Find b_k and show that

$$u(x,t) = \sum_{n=1}^{\infty} \left[\frac{32}{\pi(2n-1)^3} + \frac{8\cos(n\pi)}{(2n-1)^2}\right] e^{-\left[(2n-1)^2 + 4a\right]t/4} \sin\left(n - \frac{1}{2}\right)x.$$

8.7 STEADY STATE OF THE HEAT EQUATION

The heat equation has a steady (equilibrium) state exactly when

$$\lim_{t\to\infty} \frac{\partial u(x,t)}{\partial t} = 0.$$

If we have a rod of length L, then having a steady state means for each value of x, $0 \le x \le L$, there is a number (temperature) T_x, for which $\lim_{t\to\infty} u(x,t) = T_x$.

One of the simplest cases is constant boundary conditions

$$u(0,t) = T_0, \quad u(L,t) = T_L.$$

In the event that the heat equation has a heat source, $g(x)$, which depends only on x, the heat equation is

$$\frac{\partial u(x,t)}{\partial t} = \frac{\partial^2 u(x,t)}{\partial x^2} + g(x)$$

where we have rescaled the model so that $\alpha^2 = 1$.

We consider the heat equation with a source

$$\frac{\partial u(x,t)}{\partial t} = \frac{\partial^2 u(x,t)}{\partial x^2} + g(x), \quad u(0,t) = T_0, \quad u(L,t) = T_L, \quad u(x,0) = f(x).$$

(If $g(x) > 0$ then $g(x)$ is said to be a heat source; if $g(x) < 0$ then $g(x)$ is said to be a heat sink.) If there is a steady state, then

$$\lim_{t\to\infty}\frac{\partial u(x,t)}{\partial t}=0=\lim_{t\to\infty}\left[\frac{\partial^2 u(x,t)}{\partial x^2}+g(x)\right].$$

Note that if $\lim_{t\to\infty}\dfrac{\partial u(x,t)}{\partial t}=0$, then $\lim_{t\to\infty}\dfrac{\partial^2 u(x,t)}{\partial x^2}$ must exist.

Let $u_S(x)$ be the steady state; i.e., $u_S(x)=\lim_{t\to\infty}u(x,t)=T_x$. Then $u_S(x)$ must satisfy

$$\frac{d^2u_S(x)}{dx^2}+g(x)=0,\;\; u_S(0)=T_0,\;\; u_S(L)=T_L.$$

From $u(x, t)$ we subtract the steady state $u_S(x)$ to obtain

$$v(x,t)=u(x,t)-u_S(x).$$

We show that $v(x, t)$ satisfies the heat equation. We have

$$\frac{\partial v(x,t)}{\partial t}=\frac{\partial u(x,t)}{\partial t},\;\;\frac{\partial^2 v(x,t)}{\partial x^2}=\frac{\partial^2 u(x,t)}{\partial x^2}-\frac{d^2u_S(x)}{dx^2}=\frac{\partial^2 u(x,t)}{\partial x^2}+g(x)$$

so

$$\frac{\partial v(x,t)}{\partial t}=\frac{\partial u(x,t)}{\partial t}=\frac{\partial^2 u(x,t)}{\partial x^2}+g(x)=\frac{\partial^2 v(x,t)}{\partial x^2}$$

and

$$\begin{aligned}v(0,t)&=u(0,t)-u_S(0)=T_0-T_0=0\\ v(L,t)&=u(L,t)-u_S(L)=T_L-T_L=0\\ v(x,0)&=f(x)-u_S(x)\quad 0\le x\le L.\end{aligned}$$

We have seen how to solve

$$\frac{\partial v(x,t)}{\partial t}=\frac{\partial^2 v(x,t)}{\partial x^2},\;\; v(0,t)=0,\;\; v(0,t)=0,\;\; v(x,0)=f(x)-u_S(x)\quad 0\le x\le L$$

in the previous section. The function $u_S(x)$ must satisfy

$$\frac{d^2u_S(x)}{dx^2}=-g(x),\;\; u_S(0)=T_0,\;\; u_S(L)=T_L.$$

In the next two examples, we apply this method to solve forms of the heat equation.

Example:
Suppose there is no heat source, but the body reaches a steady state.
In this problem, we continue to rescale so that $\alpha^2=1$.
Find the steady state and the solution for the heat equation that is given by

$$\frac{\partial u(x,t)}{\partial t}=\frac{\partial^2 u(x,t)}{\partial x^2}\;\; u(0,t)=T_1,\;\; u(L,t)=T_2,\;\; u(x,0)=f(x).$$

Since we are given the temperature of the body approaches a steady state, we also have

$$\lim_{t\to\infty}\frac{\partial u(x,t)}{\partial t}=0.$$

Because $u(0, t) = T_1$ and $u(L, t) = T_2$, we must have

$$u_S(0)=\lim_{t\to\infty}u(0,t)=T_1 \quad \text{and} \quad u_S(L)=\lim_{t\to\infty}u(L,t)=T_2.$$

Now

$$\lim_{t\to\infty}\frac{\partial^2 u(x,t)}{\partial x^2}=\lim_{t\to\infty}\frac{\partial u(x,t)}{\partial t}=0$$

and

$$\lim_{t\to\infty}\frac{\partial^2 u(x,t)}{\partial x^2}=\frac{d^2 u_S(x)}{dx^2}$$

so

$$\frac{d^2 u_S(x)}{dx^2}=0.$$

Thus

$$u_S(x)=Ax+B.$$

We have $T_1 = u_S(0) = B$ and $T_2 = u_S(L) = AL + B = AL + T_1$, so $A = (T_2 - T_1)/L$. Thus,

$$u_S(x)=\frac{(T_2-T_1)}{L}x+T_1.$$

If, as before, we let

$$v(x,t)=u(x,t)-u_S(x)$$

then we have seen

$$\frac{\partial v(x,t)}{\partial t}=\frac{\partial^2 v(x,t)}{\partial x^2},\quad v(0,t)=0,\quad v(0,t)=0,\quad v(x,0)=f(x)-u_S(x)\quad 0\le x\le L.$$

As in the previous section, (although here we have set $\alpha^2 = 1$)

$$v(x,t)=\sum_{n=1}^{\infty}c_n\exp\left(\frac{-\alpha^2n^2\pi^2t}{L}\right)\sin\left(\frac{n\pi x}{L}\right).$$

To determine c_n we must use

$$v(x,0)=f(x)-u_S(x)=f(x)-\left[\left(\frac{T_2-T_1}{L}\right)x+T_1\right].$$

Thus

$$c_n = \frac{2}{L}\int_0^L \left[f(x) - \left(T_1 + \frac{T_2 - T_1}{L}x\right)\right]\sin\left(\frac{n\pi x}{L}\right)dx.$$

Thus,

$$v(x,t) = \sum_{n=1}^{\infty} c_n \exp\left(\frac{-\alpha^2 n^2 \pi^2 t}{L}\right)\sin\left(\frac{n\pi x}{L}\right)$$

$$= \sum_{n=1}^{\infty}\left[\frac{2}{L}\int_0^L \left[f(x) - \left(T_1 + \frac{T_2 - T_1}{L}x\right)\right]\sin\left(\frac{n\pi x}{L}\right)dx\right]$$

$$\times \exp\left(\frac{-\alpha^2 n^2 \pi^2 t}{L}\right)\sin\left(\frac{n\pi x}{L}\right).$$

Next we consider the case where there is a constant heat source and an equilibrium state.

Example:
Consider the equation

$$\frac{\partial u(x,t)}{\partial t} = \frac{\partial^2 u(x,t)}{\partial x^2} + \beta,\ u(0,t) = T_1,\ u(0,L) = T_2,\ u(x,0) = f(x) \qquad (1)$$

where β is a constant.

We first find the equilibrium state $u_S(x)$ that satisfies the equation

$$\frac{d^2 u_S(x)}{dx^2} + \beta = 0 \quad u_S(0) = T_1,\ u_S(L) = T_2.$$

Then

$$\frac{d^2 u_S(x)}{dx^2} = -\beta$$

so

$$u_S(x) = -\frac{\beta}{2}x^2 + Ax + B.$$

Since $u_S(0) = T_1$, we have $B = T_1$ and

$$u_S(L) = T_2 = -\frac{\beta}{2}L^2 + AL + T_1$$

so

$$AL = T_2 - T_1 + \frac{\beta}{2}L^2 \quad \text{and} \quad A = \frac{T_2 - T_1}{L} + \frac{\beta L}{2}.$$

Therefore

$$u_S(x) = -\frac{\beta}{2}x^2 + \left(\frac{T_2 - T_1}{L} + \frac{\beta L}{2}\right)x + T_1.$$

We let

$$v(x,t) = u(x,t) - u_S(x).$$

If we can find $v(x, t)$, then we know $u(x, t)$. Again

$$\frac{\partial v(x,t)}{\partial t} = \frac{\partial^2 v(x,t)}{\partial x^2}, \quad v(0,t) = 0, \quad v(0,t) = 0, \quad v(x,0) = f(x) - u_S(x) \quad 0 \le x \le L.$$

We proceed as before to find

$$v(x,t) = \sum_{n=1}^{\infty} c_n \exp\left(\frac{-\alpha^2 n^2 \pi^2 t}{L}\right) \sin\left(\frac{n\pi x}{L}\right).$$

To determine c_n we must use

$$v(x,0) = f(x) - u_S(x) = f(x) - \left[-\frac{\beta}{2}x^2 + \left(\frac{T_2 - T_1}{L} + \frac{\beta L}{2}\right)x + T_1\right].$$

Thus

$$c_n = \frac{2}{L}\int_0^L \left\{f(x) - \left[-\frac{\beta}{2}x^2 + \left(\frac{T_2 - T_1}{L} + \frac{\beta L}{2}\right)x + T_1\right]\right\} \sin\left(\frac{n\pi x}{L}\right) dx.$$

Thus,

$$\begin{aligned} v(x,t) &= \sum_{n=1}^{\infty} c_n \exp\left(\frac{-\alpha^2 n^2 \pi^2 t}{L}\right) \sin\left(\frac{n\pi x}{L}\right) \\ &= \sum_{n=1}^{\infty} \left[\frac{2}{L}\int_0^L \left\{f(x) - \left[-\frac{\beta}{2}x^2 + \left(\frac{T_2 - T_1}{L} + \frac{\beta L}{2}\right)x + T_1\right]\right\} \right. \\ &\quad \left. \times \sin\left(\frac{n\pi x}{L}\right) dx\right] \exp\left(\frac{-\alpha^2 n^2 \pi^2 t}{L}\right) \sin\left(\frac{n\pi x}{L}\right). \end{aligned}$$

Finally, we have

$$\begin{aligned} u(x,t) &= v(x,t) + u_S(x) \\ &= \sum_{n=1}^{\infty} \left[\frac{2}{L}\int_0^L \left\{f(x) - \left[-\frac{\beta}{2}x^2 + \left(\frac{T_2 - T_1}{L} + \frac{\beta L}{2}\right)x + T_1\right]\right\} \right. \\ &\quad \left. \times \sin\left(\frac{n\pi x}{L}\right) dx\right] \exp\left(\frac{-\alpha^2 n^2 \pi^2 t}{L}\right) \sin\left(\frac{n\pi x}{L}\right) \\ &\quad + \left[-\frac{\beta}{2}x^2 + \left(\frac{T_2 - T_1}{L} + \frac{\beta L}{2}\right)x + T_1\right]. \end{aligned}$$

EXERCISES

1. Find the steady state and the solution to the heat equation

$$\frac{\partial u(x,t)}{\partial t} = \frac{\partial^2 u(x,t)}{\partial x^2} + g(x) \quad 0 < x < L, \ t > 0, \ u(0,t) = T_0, \ u(L,t) = T_L,$$
$$u(x,0) = f(x)$$

for the following functions $g(x)$.

a. $g(x) = 2$.
b. $g(x) = \sin x$.
c. $g(x) = x^2$.
d. $g(x) = 3e^{-x}$.
e. $g(x) = 1 + 3x$.

2. Find the solutions of problem 1 for $f(x) = T_0 + (T_L - T_0)\sin\left(\frac{\pi x}{2L}\right)$.

3. Solve the heat equation

$$\frac{\partial u(x,t)}{\partial t} = \frac{\partial^2 u(x,t)}{\partial x^2} \quad 0 < x < \pi; \quad u(0,t) = T_0, \ u(\pi,t) = T_L, \ u(x,0) = 1 - x$$

assuming there is an equilibrium state.

4. Solve the heat equation

$$\frac{\partial u(x,t)}{\partial t} = \frac{\partial^2 u(x,t)}{\partial x^2}; \quad 0 < x < 1; \quad u(0,t) = T_0, \ u(1,t) = T_1,$$
$$u(x,0) = (1-x)T_0 + xT_1$$

assuming there is an equilibrium state.

5. In this problem we solve the heat equation

$$\frac{\partial u(x,t)}{\partial t} = \frac{\partial^2 u(x,t)}{\partial x^2} + f(x,t), \ 0 < x < \pi, \ t > 0; \quad u(0,t) = 0, \ u(\pi,t) = 0,$$
$$u(x,0) = g(x).$$

a. Show that the solution to

$$\frac{\partial u(x,t)}{\partial t} = \frac{\partial^2 u(x,t)}{\partial x^2}, \ 0 < x < \pi, \ t > 0; \quad u(0,t) = 0, \ u(\pi,t) = 0, \ u(x,0) = g(x)$$

is of the form

$$\sum_{n=1}^{\infty} A_n e^{-n^2 t} \sin(nx)$$

where the A_n's are determined using $g(x)$.

b. Assume that $f(x, t)$ can be expanded in a sine series

$$f(x,t) = \sum_{n=1}^{\infty} f_n(t)\sin(nx).$$

Show that

$$f_n(t) = \frac{2}{\pi}\int_0^{\pi} f(x,t)\sin(nx)dx$$

and thus

$$f(x,t) = \sum_{n=1}^{\infty}\left[\frac{2}{\pi}\int_0^{\pi} f(x,t)\sin(nx)dx\right]\sin(nx).$$

c. Let

$$u(x,t) = \sum_{n=1}^{\infty} B_n(t)\sin(nx).$$

Show $u(x, 0) = u(x, \pi) = 0$. Also show that $\frac{\partial u(x,t)}{\partial t} = \frac{\partial^2 u(x,t)}{\partial x^2} + f(x,t)$ with $u(x, t)$ as defined above gives the equation

$$\sum_{n=1}^{\infty} \frac{d}{dt}B_n(t)\sin(nx) = -\sum_{n=1}^{\infty} n^2 B_n(t)\sin(nx) + \sum_{n=1}^{\infty} f_n(t)\sin(nx).$$

This means

$$\frac{d}{dt}B_n(t) + n^2 B_n(t) = f_n(t). \tag{2}$$

d. Show that the solution to Eq. (2) is

$$B_n(t) = \frac{\int f_n(t)e^{n^2 t}dt + C_n}{e^{n^2 t}}$$

where the C_n's are to be determined.

e. Use that $u(x,0) = g(x) = \sum_{n=1}^{\infty} B_n(0)\sin(nx)$ to conclude

$$B_n(0) = \frac{2}{\pi}\int_0^{\pi} g(x)\sin(nx)dx.$$

f. Show $C_n = B_n(0)$ and thus

$$B_n(t) = e^{-n^2 t}\int_0^t f_n(\tau)e^{n^2\tau}d\tau + B_n(0)e^{-n^2 t}.$$

g. Show that

$$u(x,t) = \sum_{n=1}^{\infty} B_n(0)e^{-n^2 t}\sin(nx) + \sum_{n=1}^{\infty}\left(\int_0^t e^{-n^2(t-\tau)}f_n(\tau)d\tau\right)\sin(nx).$$

6. Use the result of problem 5 to solve

$$\frac{\partial u(x,t)}{\partial t} = \frac{\partial^2 u(x,t)}{\partial x^2} + xq(t), \quad 0 < x < \pi, \ t > 0; \quad u(0,t) = 0,$$
$$u(\pi,t) = 0, \ u(x,0) = 0.$$

It may be helpful to use the Fourier expansion

$$x = 2\sum_{n=1}^{\infty}\frac{(-1)^{n+1}}{n}\sin nx.$$

8.8 CHECKING THE VALIDITY OF THE SOLUTION

In this chapter we have assumed that the solution to the initial value problem could be written as the product of functions of one variable. With that assumption, we found the solution. Along the way, we have cautioned that the result should be verified. In this section we show one method for checking the solution. Some facts about series of functions that we shall use are

1. The Weierstrass M-test.
2. If $\sum a_n x^n$ is a power series with radius of convergence R, then
   a. $\sum a_n x^n$ converges uniformly and absolutely on the interval $[-R+\varepsilon, R-\varepsilon]$ for any $\varepsilon > 0$.
   b. $\frac{d}{dx}\sum a_n x^n = \sum n a_n x^{n-1}$.
   c. The series $\sum a_n n^k x^n$ has the same radius of convergence as $\sum a_n x^n$ for any positive integer k.

Consider the heat equation

$$u_{xx}(x,t) = \frac{1}{\alpha^2}u_t(x,t); \quad u(0,t) = u(\pi,t) = 0; \quad u(x,0) = f(x).$$

We have found that the solution using separation of variables is

$$u(x,t) = \sum_{n=1}^{\infty} u_n(x,t)$$

where

$$u_n(x,t) = \left(\frac{2}{\pi}\int_0^{\pi} f(x)\sin nx\ dx\right)\sin nxe^{-(\alpha n)^2 t}.$$

Now

$$\left|\frac{2}{\pi}\int_0^{\pi} f(x)\sin nx\ dx\right| \le \frac{2}{\pi}\int_0^{\pi} |f(x)|dx.$$

If $f(x)$ is piecewise continuous, then it is bounded. Thus there is a number M for which

$$\frac{2}{\pi}\int_0^{\pi} |f(x)|dx \le M.$$

So for every positive integer n, we have

$$|u_n(x,t)| \le \left|M \sin nxe^{-(\alpha n)^2 t}\right| = Me^{-(\alpha n)^2 t}.$$

Note that

$$\frac{d^2}{dx^2}(u_n(x,t)) = -n^2 u_n(x,t) \quad \text{and} \quad \frac{d}{dt}(u_n(x,t)) = -(\alpha n)^2 u_n(x,t)$$

so that

$$\left|\frac{d^2}{dx^2}(u_n(x,t))\right| \le n^2 Me^{-(\alpha n)^2 t} \quad \text{and} \quad \left|\frac{d}{dt}(u_n(x,t))\right| \le n^2\alpha^2 Me^{-(\alpha n)^2 t}.$$

Since for any $\varepsilon > 0$ the series of numbers

$$\sum_{n=1}^{\infty} n^2 e^{-n^2\varepsilon}$$

converges, each of the series

$$\sum_{n=1}^{\infty} u_n(x,t), \quad \sum_{n=1}^{\infty} \frac{\partial^2}{\partial x^2}(u_n(x,t)) \quad \text{and} \quad \sum_{n=1}^{\infty} \frac{\partial}{\partial t}(u_n(x,t))$$

converges uniformly for $t \ge \varepsilon$, where ε is an arbitrary positive number. Thus, if we let

$$u(x,t) = \sum_{n=1}^{\infty} u_n(x,t)$$

then

$$\frac{\partial^2}{\partial x^2} u(x,t) = \sum_{n=1}^{\infty} \frac{\partial^2}{\partial x^2} u_n(x,t) \quad \text{and} \quad \frac{\partial}{\partial t} u(x,t) = \sum_{n=1}^{\infty} \frac{\partial}{\partial t} u_n(x,t)$$

if $t \geq \varepsilon$. Since

$$\frac{\partial^2}{\partial x^2}u_n(x,t) - \frac{1}{\alpha^2}\frac{\partial}{\partial t}u_n(x,t) = 0 \quad \text{and} \quad u_n(0,t) = u_n(\pi,t) = 0 \quad \text{for every } n$$

we have

$$\frac{\partial^2}{\partial x^2}u(x,t) - \frac{1}{\alpha^2}\frac{\partial}{\partial t}u(x,t) = 0 \quad \text{and} \quad u(0,t) = u(\pi,t) = 0 \quad \text{for } t \geq \varepsilon.$$

Finally, we must show

$$\lim_{t \downarrow 0} u(x,t) = f(x).$$

To proceed, we assume that $f(x)$ is continuous on $[0, \pi]$ and $f(0) = f(\pi) = 0$ and that

$$\int_0^\pi [f'(x)]^2 dx < \infty.$$

The last assumption ensures that the Fourier series for $f(x)$ converges uniformly to $f(x)$.

Let $s_N(x, t)$ be the Nth partial sum of

$$\sum_{n=1}^{\infty} b_n e^{-(\alpha n)^2 t} \sin nx$$

i.e.,

$$s_N(x,t) = \sum_{n=1}^{N} b_n e^{-(\alpha n)^2 t} \sin nx.$$

Then

$$s_N(x,0) = \sum_{n=1}^{N} b_n \sin nx.$$

Since the Fourier series of $f(x)$ converges uniformly to $f(x)$, given $\varepsilon > 0$ there is a number $N(\varepsilon)$ so that if $n > N(\varepsilon)$ then

$$|s_N(x,0) - f(x)| = \left| \sum_{n=1}^{N} b_n \sin nx - f(x) \right| < \frac{\varepsilon}{2}$$

so

$$|s_N(x,0) - s_M(x,0)| = |s_N(x,0) - f(x) + f(x) - s_M(x,0)| \leq |s_N(x,0) - f(x)|$$
$$+ |f(x) - s_M(x,0)| < \frac{\varepsilon}{2} + \frac{\varepsilon}{2} = \varepsilon$$

if M, $N > N(\varepsilon)$. Thus $\{s_n(x,0)\}$ is a uniformly Cauchy sequence of continuous functions that converges uniformly to a continuous function, and this function must be $f(x)$.

A fact that we do not prove (see Weinberger, pp. 58–60) is that in our setting we must then have

$$|s_N(x,t) - s_M(x,t)| \leq \varepsilon \quad \text{if} \quad 0 \leq x \leq \pi, \ \ t \geq 0 \quad \text{if} \quad M, \ \ N > N(\varepsilon).$$

This shows

$$\lim_{t \downarrow 0} u(x,t) = f(x).$$

We note that in verifying the solution for the heat equation we were aided substantially by the fact that we had an exponentially decreasing factor. This is not true with the wave equation. To prove validity of the solution for the wave equation, we need some properties of the wave equation that we have not yet developed. A proof of the validity of the solution of the wave equation may be found in Brown and Churchill, pp. 338–341.

CHAPTER

Solving Partial Differential Equations in Cylindrical Coordinates Using Separation of Variables

9

9.1 INTRODUCTION

In Chapter 8, we solved Laplace's equation, the wave equation, and the heat equation in Cartesian coordinates using separation of variables. In this chapter, we solve the same equations in polar or cylindrical coordinates. In Cartesian coordinates, the ordinary differential equations (ODEs) that arose were simple to solve. We shall see that, in cylindrical and spherical coordinates, not all the ODEs are as agreeable. The solutions to these more difficult ODEs go by the names Bessel functions and Legendre polynomials. In cylindrical coordinates, we need only Bessel functions. We begin this section by showing how these equations arise.

AN EXAMPLE WHERE BESSEL FUNCTIONS ARISE

Consider the heat equation

$$u_t = K\Delta u.$$

The reason the equations that arise from separation of variables in cylindrical coordinates are not as simple as in Cartesian coordinates is the form of the Laplacian. In cylindrical coordinates, the Laplacian is given by

$$\Delta u = u_{rr} + \frac{1}{r}u_r + \frac{1}{r^2}u_{\theta\theta} + u_{zz}.$$

It will simplify our computations and still allow us to demonstrate how Bessel functions arise if we assume that u is a function of r, θ and t, but not a function of z.

We suppose

$$u(r,\theta,t) = R(r)\Theta(\theta)T(t)$$

so that

$$u_t = K\Delta u$$

Mathematical Physics with Partial Differential Equations. https://doi.org/10.1016/B978-0-12-814759-7.00009-0

can be expressed as

$$R(r)\Theta(\theta)T'(t) = K\left[R''(r)\Theta(\theta)T(t) + \frac{1}{r}R'(r)\Theta(\theta)T(t) + \frac{1}{r^2}R(r)\Theta''(\theta)T(t)\right].$$

Dividing by $KR(r)\Theta(\theta)T(t)$ gives

$$\frac{1}{K}\frac{T'(t)}{T(t)} = \frac{R''(r)}{R(r)} + \frac{1}{r}\frac{R'(r)}{R(r)} + \frac{1}{r^2}\frac{\Theta''(\theta)}{\Theta(\theta)}. \tag{1}$$

The left-hand side of Eq. (1) is a function only of t, and the right-hand side is a function of r and θ, so it must be that each is a constant. In Exercise 1, we show that this is a negative number that we denote $-\lambda$. Thus

$$\frac{1}{K}\frac{T'(t)}{T(t)} = -\lambda \text{ or } T'(t) + \lambda K T(t) = 0. \tag{2}$$

Also,

$$\frac{R''(r)}{R(r)} + \frac{1}{r}\frac{R'(r)}{R(r)} + \frac{1}{r^2}\frac{\Theta''(\theta)}{\Theta(\theta)} = -\lambda.$$

So

$$\frac{R''(r)}{R(r)} + \frac{1}{r}\frac{R'(r)}{R(r)} + \lambda = -\frac{1}{r^2}\frac{\Theta''(\theta)}{\Theta(\theta)}$$

or

$$r^2\left[\frac{R''(r)}{R(r)} + \frac{1}{r}\frac{R'(r)}{R(r)} + \lambda\right] = -\frac{\Theta''(\theta)}{\Theta(\theta)}. \tag{3}$$

The left-hand side of Eq. (3) is a function of r and the right-hand side is a function of θ, so each must be a constant that we denote μ. Thus we have

$$\Theta''(\theta) + \mu\,\Theta(\theta) = 0 \tag{4}$$

and

$$r^2\left[\frac{R''(r)}{R(r)} + \frac{1}{r}\frac{R'(r)}{R(r)} + \lambda\right] = \mu \text{ or } \frac{R''(r)}{R(r)} + \frac{1}{r}\frac{R'(r)}{R(r)} + \lambda = \frac{\mu}{r^2}$$

so

$$R''(r) + \frac{1}{r}R'(r) + \left(\lambda - \frac{\mu}{r^2}\right)R(r) = 0. \tag{5}$$

Thus, to solve the heat equation in polar coordinates, we need to solve Eqs. (2), (4) and (5). Of these, only Eq. (5) requires additional attention. It is (like) a Bessel equation, and we construct its solution in the next section. This equation will arise when we use the Laplacian in polar or cylindrical coordinates in the wave equation or the heat equation. In Laplace's equation, we shall see the equation is of the form

$$R''(r) + \frac{1}{r}R'(r) - \left(m^2 + \frac{n^2}{r^2}\right)R(r) = 0$$

and will have to be handled differently (although very similarly).

Had we assumed that the function u also depended on z and that

$$u(r, \theta, z, t) = R(r)\Theta(\theta)T(t)Z(z),$$

Eq. (5) would still have been the only complicated ODE that would have arisen (see Exercise 3.)

Eq. (5) is a "Bessel-like" equation. We next define a Bessel equation, demonstrate one solution to such equations, and then make a transformation that will enable us to solve the equation above. (Because this is a second-order differential equation, there are two solutions, but one is unbounded at $r = 0$. Because of physical considerations, this will be an inadmissible solution for our problems.)

A Bessel equation is an equation of the form

$$x^2y''(x) + xy'(x) + (x^2 - \nu^2)y(x) = 0, \quad 0 \leq x < \infty.$$

The method of solution that we use is power series:

Step 1: Hypothesize a solution of the form

$$y = \sum_{n=0}^{\infty} a_n x^{n+\alpha}.$$

For the solution to be bounded at $x = 0$, we require that $\alpha \geq 0$.

Step 2: Differentiate and collect terms. We have

$$y' = \sum_{n=0}^{\infty} a_n(n + \alpha)x^{n+\alpha-1}$$

so

$$xy' = \sum_{n=0}^{\infty} a_n(n + \alpha)x^{n+\alpha}$$

$$y'' = \sum_{n=0}^{\infty} a_n(n + \alpha)(n + \alpha - 1)x^{n+\alpha-2}$$

so

$$x^2y'' = \sum_{n=0}^{\infty} a_n(n + \alpha)(n + \alpha - 1)x^{n+\alpha}$$

then

$$x^2y'' + xy' + (x^2 - \nu^2) = \sum_{n=0}^{\infty}\left[a_n(n+\alpha)(n+\alpha-1)x^{n+\alpha}\right] + \left[a_n(n+\alpha)x^{n+\alpha}\right]$$
$$+ \left[\left(a_n x^{n+\alpha+2}\right) - \nu^2 a_n x^{n+\alpha}\right] = 0.$$

This can be written as

$$a_0\left[\alpha(\alpha-1) + \alpha - \nu^2\right]x^\alpha + a_1\left[(\alpha+1)\alpha + (\alpha+1) - \nu^2\right]x^{\alpha+1}$$
$$+\sum_{n=2}^{\infty}\left\{a_n\left[((n+\alpha))((n+\alpha)-1) + (n+\alpha) - \nu^2\right] + a_{n-2}\right\}x^{\alpha+n}$$
$$= a_0\left(\alpha^2 - \nu^2\right)x^\alpha + a_1\left[(\alpha+1)^2 - \nu^2\right]$$
$$+\sum_{n=2}^{\infty}\left\{\left[(n+\alpha)^2 - \nu^2\right]a_n + a_{n-2}\right\}x^{\alpha+n} = 0.$$

The coefficient of each power of x must be 0. The coefficient of x^α must be 0 and this gives the indicial equation, from which we determine the value of α. If $a_0 \neq 0$, we have

$$\alpha^2 - \nu^2 = 0$$

so $\alpha^2 = \nu^2$. Also,

$$a_1\left[(\alpha+1)^2 - \nu^2\right] = a_1\left[(\alpha+1)^2 - \alpha^2\right] = a_1[2\alpha+1] = 0.$$

If the solution is bounded, then α is nonnegative and $a_1 = 0$.

Step 3: The recurrence relation is

$$a_n\left[(n+\alpha)((n+\alpha)-1) + (n+\alpha) - \nu^2\right] + a_{n-2} = 0$$

or

$$a_n\left[(n+\alpha)((n+\alpha)-1) + (n+\alpha) - \nu^2\right] = a_n\left[(n+\alpha)^2 - \nu^2\right] + a_{n-2} = 0.$$

Substituting ν for α gives

$$a_n\left[(n+\nu)^2 - \nu^2\right] = a_n n(n+2\nu) = -a_{n-2} \text{ or } a_n = \frac{-a_{n-2}}{n(n+2\nu)}.$$

Because $a_1 = 0$, then $a_k = 0$ for every odd integer k.

Step 4: We now determine a general description of a_{2k}. We have

$$a_2 = -\frac{a_0}{2(2+2\nu)}$$

$$a_4 = -\frac{a_2}{4(4+2\nu)} = \frac{(-1)}{4(4+2\nu)}\frac{(-1)a_0}{2(2+2\nu)}$$

$$a_6 = -\frac{a_4}{6(6+2\nu)} = -\frac{a_0}{2^3(1\cdot 2\cdot 3)(3+\nu)(2+\nu)(1+\nu)}$$

and

$$a_{2k} = \frac{(-1)^k a_0}{2^{2k}(k!)(k+\nu)(k-1+\nu)\cdots(1+\nu)}.$$

Thus one of the solutions of Bessel's equation is

$$y_1(x) = a_0 x^\nu \left[1 + \sum_{k=1}^{\infty} \frac{(-1)^k x^{2k}}{2^{2k}(k!)(k+\nu)(k-1+\nu)\cdots(1+\nu)}\right].$$

This is a solution for any value of a_0. (Notice that we have not imposed any boundary conditions.) This is Bessel's function of the first kind of order ν, denoted $J_\nu(x)$. If we let $a_0 = 1/\nu!2^\nu$. We can express $J_\nu(x)$ as

$$J_\nu(x) = \sum_{k=0}^{\infty} \frac{(-1)^k x^{2k+\nu}}{2^{2k+\nu}k!(k+\nu)!} = \sum_{n=0}^{\infty} \frac{(-1)^k \left(\frac{x}{2}\right)^{2k+\nu}}{k!(k+\nu)!}. \tag{6}$$

By the ratio test, this series converges for all values of x, and in Exercise 5 we show that it is a solution for $x > 0$.

In the case that the equation is of the form

$$R''(r) + \frac{1}{r}R'(r) - \left(m^2 + \frac{n^2}{r^2}\right)R(r) = 0,$$

as will be the case for Laplace's equation, the solution is a modified Bessel function (sometimes called a Bessel function with imaginary argument) denoted by $I_\nu(x)$, which is defined by

$$I_\nu(x) = \frac{x^\nu}{2^\nu \nu!}\left[1 + \sum_{n=1}^{\infty} \frac{x^{2n}}{2^{2n}n!(1+\nu)\cdots(n+\nu)}\right].$$

(See Pinsky, 1998, p. 187.) The modified Bessel's function is obtained by replacing x with ix in Bessel's equation.

Bessel's equation is a second-order equation, so there will be two linearly independent solutions. For our purposes, we shall not be concerned with the second solution because it diverges at $x = 0$. The discussion and derivation of the second solution can be found in many differential equation texts, including Boyce and DiPrima (2008), but the second solution is of the form

$$y_2(x) = y_1(x)\ln|x| + |x|^\nu \sum_{n=1}^{\infty} b_n x^n.$$

The two equations given at the beginning of this section are of the form

$$y''(r) + \frac{(d-1)}{r}y'(r) + \left(\lambda - \frac{\mu}{r^2}\right)y(r) = 0. \qquad (1)$$

In cylindrical coordinates, $d = 2$ and μ will typically be m^2. In spherical coordinates, $d = 3$ and μ will typically be $k(k+1)$.

In our later work we consider this to be an eigenvector—eigenvalue problem.

EXERCISES

1. Show that the constant that arises in Eq. (1) is negative.
2. Repeat the separation of variables argument in the case that

$$u(r, \theta, z, t) = R(r)\Theta(\theta)T(t)Z(z).$$

3. Repeat the separation of variables argument for the wave equation.
4. Show that $J_\nu(x)$ that arises in Eq. (6) converges for $x \geq 0$ and solves

$$x^2y''(x) + xy'(x) + (x^2 - \nu^2)y(x) = 0, \quad 0 \leq x < \infty.$$

5. Give a bounded solution for the following equations:
   a. $x^2y''(x) + xy'(x) + (x^2 - 4)y(x) = 0.$
   b. $x^2y''(x) + xy'(x) + x^2y(x) = 0.$
   c. $y''(x) + (1/x)y'(x) + \left(1 - \frac{9}{x^2}\right)y(x) = 0.$

9.2 THE SOLUTION TO BESSEL'S EQUATION IN CYLINDRICAL COORDINATES

In our analysis of the heat equation in cylindrical coordinates using separation of variables, we arrived at an equation of the form

$$r^2\frac{d^2R(r)}{dr^2} + r\frac{dR(r)}{dr} + \left[\lambda r^2 - \mu\right]R(r) = 0.$$

For our applications, $\mu = n^2$ where n is an integer, so we want to solve

$$r^2\frac{d^2R(r)}{dr^2} + r\frac{dR(r)}{dr} + \left[\lambda r^2 - n^2\right]R(r) = 0. \qquad (1)$$

Eq. (1) is not a Bessel equation but can be transformed into a Bessel equation by the change of variables $x = r\sqrt{\lambda}$, as we now demonstrate. We have

$$\frac{dR(r)}{dr} = \frac{dR(x)}{dx}\frac{dx}{dr} = \frac{dR(x)}{dx}\sqrt{\lambda} \text{ and}$$

$$\frac{d^2R(r)}{dr^2} = \frac{d^2R(x)}{dx^2}\lambda$$

so

$$r^2\frac{d^2R(r)}{dr^2} + r\frac{dR(r)}{dr} + \left[\lambda r^2 - n^2\right]R(r) = 0$$

is transformed to

$$\begin{aligned}&\left(\frac{x}{\sqrt{\lambda}}\right)^2\lambda\frac{d^2R(x)}{dx^2} + \left(\frac{x}{\sqrt{\lambda}}\right)\sqrt{\lambda}\,\frac{dR(x)}{dx} + \left[\lambda\frac{x^2}{\lambda} - n^2\right]R(x)\\ &= x^2\frac{d^2R(x)}{dx^2} + x\frac{dR(x)}{dx} + \left[x^2 - n^2\right]R(x) = 0.\end{aligned} \tag{2}$$

Eq. (2) is a Bessel equation that has two solutions. The solution that is bounded at $x = 0$ is typically the only one of interest to us. That solution is given by

$$R(x) = J_n(x) = J_n\left(r\sqrt{\lambda}\right).$$

Any Bessel function of the first type has infinitely many positive roots. The graphs of several Bessel functions of the first type are shown in Fig. 9.2.1.

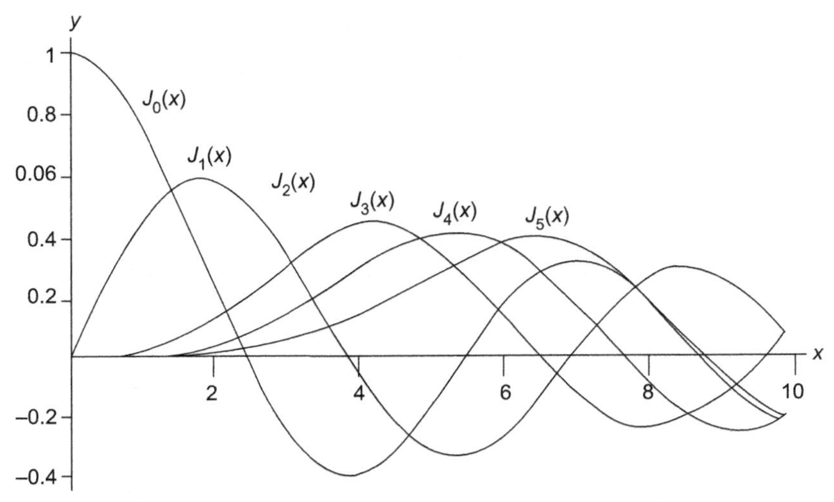

FIGURE 9.2.1

Let $\{x_{n,m}\}$ denote the positive roots of $J_n(x)$; i.e., $J_n(x_{n,m}) = 0$. In many equations involving a cylinder, one of the boundary conditions is that the value of $R(r)$ at the surface of the cylinder is constant. Suppose that the radius of the cylinder is a and $R(a) = 0$. (If $R(a)$ is a different constant, this boundary condition can be obtained by rescaling the temperature. We could also rescale the length to take $a = 1$, which is common.) Thus, in cylindrical coordinates, an initial value problem of interest is

$$x^2\frac{d^2R(x)}{dx^2} + x\frac{dR(x)}{dx} + \left[x^2 - n^2\right]R(x) = 0, \quad R(a) = 0. \tag{3}$$

The solution to Eq. (3) that is continuous at $x = 0$ is $R(x) = J_n(x) = J_n(r\sqrt{\lambda})$ and the boundary condition requires that $J_n(a) = 0$. Since we have $J_n(x_{n,m}) = 0$, if we let

$$R_m(r) = J_n\left(x_{n,m}\frac{r}{a}\right)$$

then

$$R_m(a) = J_n\left(x_{n,m}\frac{a}{a}\right) = J_n\left(x_{n,m}\right) = 0.$$

Thus,

$$R_m(r) = J_n\left(x_{n,m}\frac{r}{a}\right)$$

is an eigenfunction for the initial value problem given by Eq. (3). The eigenvalue for $R_m(r)$ is $\left(\frac{x_{n,m}}{a}\right)^2$ as we verify in Exercise 1.

In Exercise 2 we show that

$$x^2\frac{d^2R(x)}{dx^2} + x\frac{dR(x)}{dx} + \left[x^2 - n^2\right]R(x) = 0, \quad R(a) = 0$$

is a Sturm–Liouville problem with weight function $w(x) = x$.

By the Sturm–Liouville theory, the eigenfunctions are complete. That is, if $f(x)$ is a function on $[0, a]$ for which

$$\int_0^a x[f(x)]^2dx < \infty,$$

then

$$f(x) = \sum_{m=0}^{\infty} b_m J_n\left(x_{n,m}\frac{x}{a}\right)$$

where

$$b_m = \frac{\int_0^a f(x)x\, J_n\left(x_{n,m}\frac{x}{a}\right)dx}{\int_0^a x\left[J_n\left(x_{n,m}\frac{x}{a}\right)\right]^2 dx}.$$

The final points in this section are reiterated with some intuitive explanation in Section 9.5 when we discuss the heat equation on a disk.

EXERCISES

1. Show that

$$J_n\left(x_{n,m}\frac{x}{a}\right)$$

is an eigenfunction for

$$x^2\frac{d^2R(x)}{dx^2}+x\frac{dR(x)}{dx}+\left[x^2-n^2\right]R(x)=0,\quad R(a)=0,\ R(x)\ \text{is bounded at}\ x=0$$

with eigenvalue

$$\left(\frac{x_{n,m}}{a}\right)^2.$$

2. Show that

$$x^2\frac{d^2R(x)}{dx^2}+x\frac{dR(x)}{dx}+\left[x^2-n^2\right]R(x)=0,\quad R(a)=0$$

is a Sturm–Liouville problem with weight function $w(x) = x$.

9.3 SOLVING LAPLACE'S EQUATION IN CYLINDRICAL COORDINATES USING SEPARATION OF VARIABLES

In cylindrical coordinates, Laplace's equation is

$$\Delta u(r,\theta,z)=u_{rr}+\frac{1}{r}u_r+\frac{1}{r^2}u_{\theta\theta}+u_{zz}=0.$$

We solve the boundary value problem

$$u_{rr}+\frac{1}{r}u_r+\frac{1}{r^2}u_{\theta\theta}+u_{zz}=0;\quad 0<r<a,\quad 0<z<b;$$

$$u(r,\theta,0)=u(r,\theta,b)=0,\quad u(a,\theta,z)=f(\theta,z).$$

We hypothesize that

$$u(r,\theta,z)=R(r)\Theta(\theta)Z(z)$$

so

$$\begin{aligned}\Delta u(r,\theta,z)= {} & R''(r)\Theta(\theta)Z(z)+\frac{1}{r}R'(r)\Theta(\theta)Z(z)\\ & +\frac{1}{r^2}R(r)\Theta''(\theta)Z(z)+R(r)\Theta(\theta)Z''(z)=0.\end{aligned}$$

Dividing by $R(r)\,\Theta(\theta)\,Z(z)$ gives

$$\frac{R''(r)}{R(r)}+\frac{1}{r}\frac{R'(r)}{R(r)}+\frac{1}{r^2}\frac{\Theta''(\theta)}{\Theta(\theta)}+\frac{Z''(z)}{Z(z)}=0$$

or

$$\frac{R''(r)}{R(r)}+\frac{1}{r}\frac{R'(r)}{R(r)}+\frac{1}{r^2}\frac{\Theta''(\theta)}{\Theta(\theta)}=-\frac{Z''(z)}{Z(z)}. \tag{1}$$

The left-hand side of Eq. (1) is a function of r and θ, and the right-hand side is a function of z, so each is a positive constant that we call C. Thus we have

$$-\frac{Z''(z)}{Z(z)}=C$$

so

$$Z''(z)+CZ(z)=0.$$

Then

$$Z(z)=A\cos\left(\sqrt{C}\,z\right)+B\sin\left(\sqrt{C}\,z\right)$$

and

$$Z(0)=0,$$

so

$$A=0;\qquad Z(b)=0$$

so

$$\sqrt{C}=\frac{n\pi}{b}\text{ and }C=\left(\frac{n\pi}{b}\right)^2.$$

Thus

$$Z_n(z)=\sin\left(\frac{n\pi z}{b}\right).$$

We also have

$$\frac{R''(r)}{R(r)}+\frac{1}{r}\frac{R'(r)}{R(r)}+\frac{1}{r^2}\frac{\Theta''(\theta)}{\Theta(\theta)}=C$$

so

$$r^2\left(\frac{R''(r)}{R(r)}+\frac{1}{r}\frac{R'(r)}{R(r)}-C\right)=-\frac{\Theta''(\theta)}{\Theta(\theta)}$$

and there is a constant D with

$$-\frac{\Theta''(\theta)}{\Theta(\theta)}=D$$

so

$$\Theta''(\theta) + D\Theta(\theta) = 0$$

and

$$r^2\left[\frac{R''(r)}{R(r)} + \frac{1}{r}\frac{R'(r)}{R(r)} - C\right] = D. \tag{2}$$

The periodicity conditions $\Theta(\pi) = \Theta(-\pi)$ and $\Theta'(\pi) = \Theta'(-\pi)$ require that $D > 0$. We let $D = m^2$ and so we have

$$r^2\left[\frac{R''(r)}{R(r)} + \frac{1}{r}\frac{R'(r)}{R(r)} - C\right] = D \text{ or } \frac{R''(r)}{R(r)} + \frac{1}{r}\frac{R'(r)}{R(r)} - C - \frac{D}{r^2} = 0$$

so

$$R(r) + \frac{1}{r}R'(r) - \left(C + \frac{D}{r^2}\right)R(r) = R(r) + \frac{1}{r}R'(r) - \left(\left(\frac{n\pi}{b}\right)^2 + \frac{m^2}{r^2}\right)R(r) = 0.$$

It will simplify the notation if we let $b = \pi$, and we now make that substitution. We then have

$$R''(r) + \frac{1}{r}R'(r) - \left(n^2 + \frac{m^2}{r^2}\right)R(r) = 0.$$

Recapping, there are three equations we must solve:

$$Z''(z) + n^2 Z(z) = 0, \quad Z(0) = Z(\pi) = 0$$

$$\Theta''(\theta) + m^2\Theta(\theta) = 0, \quad \Theta(\pi) = \Theta(-\pi) \text{ and } \Theta'(\pi) = \Theta'(-\pi)$$

$$R''(r) + \frac{1}{r}R'(r) - \left(n^2 + \frac{m^2}{r^2}\right)R(r) = 0. \tag{3}$$

Eq. (3) has no boundary condition. Instead, we require that the solution is bounded at $r = 0$.

Note that to solve Eq. (3), we will need to use a modified Bessel function.

We have shown that

$$Z_n(z) = a_n \sin nz.$$

The equation

$$\Theta''(\theta) + m^2\Theta(\theta) = 0$$

has only the continuity conditions $\Theta(\pi) = \Theta(-\pi)$ and $\Theta'(\pi) = \Theta'(-\pi)$ so

$$\Theta_m(\theta) = b_m \cos m\theta + c_m \sin m\theta.$$

The equation

$$R''(r) + \frac{1}{r}R'(r) - \left(n^2 + \frac{m^2}{r^2}\right)R(r) = 0$$

has $A_{mn}I_m(nr)$ as its solution. We choose $A_{mn} = 1/I_m(na)$ so that

$$R_{mn}(r) = A_{mn}I_m(nr) = \frac{I_m(nr)}{I_m(na)}$$

and so

$$R_{mn}(a) = \frac{I_m(na)}{I_m(na)} = 1.$$

Thus, we have

$$\begin{aligned} u_{mn}(r,\theta,z) &= R_{mn}(r)\Theta_m(\theta)Z_n(z) \\ &= A_{mn}I_m(nr)(b_m \cos m\theta + c_m \sin m\theta)a_n \sin nz \\ &= \frac{1}{I_m(na)}I_m(nr)(b_m \cos m\theta + c_m \sin m\theta)a_n \sin nz. \end{aligned}$$

The constants can be combined to write

$$u_{nm}(r,\theta,z) = \frac{1}{I_m(na)}I_m(nr)\sin nz(d_{nm} \cos m\theta + e_{nm} \sin m\theta).$$

Thus,

$$\begin{aligned} u(r,\theta,z) = {} & \frac{1}{2}\sum_{n=1}^{\infty} d_{n0}\frac{1}{I_0(na)}I_0(nr) \\ & + \sum_{m,\, n=1}^{\infty} \frac{1}{I_m(na)}I_m(nr)\sin nz(d_{nm} \cos m\theta + e_{nm} \sin m\theta). \end{aligned}$$

We apply the boundary condition $u(a,\theta, z) = f(\theta, z)$ to get

$$\begin{aligned} u(a,\theta,z) = f(\theta,z) = {} & \frac{1}{2}\sum_{n=1}^{\infty} d_{n0}\frac{1}{I_0(na)}I_0(na) \\ & + \sum_{m,\, n=1}^{\infty} \frac{1}{I_m(na)}I_m(na)\sin nz(d_{nm} \cos m\theta + e_{nm} \sin m\theta) \\ = {} & \frac{1}{2}\sum_{n=1}^{\infty} d_{n0} + \sum_{m,\, n=1}^{\infty} \sin nz(d_{nm} \cos m\theta + e_{nm} \sin m\theta). \end{aligned}$$

Then

$$d_{nm} = \frac{1}{\pi}\frac{2}{\pi}\int_{z=0}^{\pi}\left[\int_{\theta=0}^{2\pi} f(\theta,z)\cos m\theta d\theta\right]\sin nzdz$$

and

$$e_{nm} = \frac{1}{\pi}\frac{2}{\pi}\int_{z=0}^{\pi}\left[\int_{\theta=0}^{2\pi} f(\theta,z)\sin m\theta d\theta\right]\sin nzdz.$$

We note that the z−coordinate can be problematic. Some sources, such as Pinsky, *Partial Differential Equations and Boundary-Value Problems with Applications*, Third Edition, do the problem in polar coordinates. If no boundary conditions on $Z(z)$ are given, the sign of C cannot be assigned. We have followed the problem and solution as given in Weinberger, *A First Course in Partial Differential Equations*.

EXERCISES

1. Solve

$$\begin{aligned}\Delta u = 0, \quad 0 < r < 1, \ 0 < z < \pi; \ u(r,\theta,0) = u(r,\theta,\pi) = 0;\\ u(1,\theta,z) = \ z(1-z).\end{aligned}$$

2. Solve

$$\begin{aligned}\Delta u = 0, \quad 0 < r < 1, \ 0 < z < \pi; \ u(r,\theta,0) = u(r,\theta,\pi) = 0;\\ u(1,\theta,z) = \ \sin\theta.\end{aligned}$$

3. a. Solve Laplace's equation on a cylinder in the case that the solution is independent of z and θ.

b. Use the result in part a to solve Laplace's equation in the cylinder $r_1 < r < r_2$ with boundary conditions $u(r_1) = A$ and $u(r_2) = B$.

c. Recall that the solution to Laplace's equation provides the steady state of the heat equation. Use the result in part b to find the steady state of the temperature between two concentric cylinders, where the inner cylinder is kept at temperature A and the outer cylinder is kept at temperature B.

4. Determine the steady-state temperature of a cylinder of radius 2 and height π where the boundary conditions, which are the top and bottom temperatures, are kept at $T = 0$ and the temperature on the curved surface is $T(z) = \sin z$. Assume the solution is independent of θ.

5. Solve Laplace's equation on a cylinder of radius 1 and height π where the solution is independent of z with boundary condition $u(1,\theta) = \sin\theta$.

6. a. Solve Laplace's equation on concentric circles of radius 1 and 2 with boundary conditions $u(1,\theta) = \sin\theta$; $u(2,\theta) = 4\sin\theta$.

b. Find an expression the coefficients must satisfy to solve Laplace's equation on concentric circles of radius R_1 and R_2 with boundary conditions $u(R_1,\theta) = f(\theta)$; $u(R_2,\theta) = g(\theta)$.

7. In this problem we solve Laplace's equation on a circle. Solve

$$\Delta u(r,\theta) = u_{rr} + \frac{1}{r}u_r + \frac{1}{r^2}u_{\theta\theta} = 0; \quad 0 < r < a, \; u(a,\theta) = f(\theta).$$

Show the solution is of the form

$$u(r,\theta) = \frac{a_0}{2} + \sum_{n=1}^{\infty}(a_n \cos n\theta + b_n \sin n\theta)$$

and find an expression for a_n and b_n.

9.4 THE WAVE EQUATION ON A DISK (THE DRUMHEAD PROBLEM)

The wave equation on a cylinder is well illustrated by the drumhead problem, even though the z-variable does not appear in the problem. This problem was solved by Euler in the mid-18th century. The problem states that we have a membrane stretched over a circular form whose radius is a. We let $u(r,\theta,t)$ denote the vertical deviation from equilibrium at the point (r,θ) at time t. We want to solve the equation

$$u_{tt}(r,\theta,t) = c^2\Delta u(r,\theta,t) = c^2\left(u_{rr} + \frac{1}{r}u_r + \frac{1}{r^2}u_{\theta\theta}\right)$$

with boundary condition

$$u(a,\ \theta,t) = 0$$

and initial conditions

$$u(r,\theta,0) = f_1(r,\theta)$$

$$u_t(r,\theta,0) = f_2(r,\theta).$$

If we proceed as we have with separation of variables, letting

$$u(r,\theta,t) = R(r)\Theta(\theta)T(t)$$

we arrive at three ODEs:

$$T''(t) + c^2\lambda^2 T(t) = 0 \tag{1a}$$

$$\Theta''(\theta) + n^2\Theta(\theta) = 0 \tag{1b}$$

$$r^2R''(r) + rR'(r) + \left(\lambda^2r^2 - n^2\right)R(r) = 0 \tag{1c}$$

as we show in Exercise 1. We note the periodicity of $\Theta(\theta)$ gives n^2 in Eq. (1b).

Note that

$$u(a, \theta, t) = 0$$

forces $R(a) = 0$.

The solutions to these equations are

$$T(t) = A\cos(c\lambda t) + B\sin(c\lambda t)$$
$$\Theta_n(\theta) = C_n\cos n\theta + D_n\sin n\theta$$
$$R_n(r) = E_n J_n(\lambda r).$$

The boundary condition $R(a) = 0$ forces $J_n(\lambda a) = 0$. We let x_{mn} denote the mth positive root of $J_n(x)$, and define

$$\lambda_{mn} = \frac{x_{mn}}{a}; \quad n = 0, 1, 2, \ldots; \quad m = 1, 2, 3, \ldots.$$

We now have

$$T_{mn}(t) = A_{mn}\cos(c\lambda_{mn}t) + B_{mn}\sin(c\lambda_{mn}t)$$
$$J_n(\lambda_{mn}r) = J_n\left(\frac{x_{mn}}{a}r\right)$$

and so

$$u_{mn}(r, \theta, t) = J_n\left(\frac{x_{mn}}{a}r\right)[C_n\cos n\theta + D_n\sin n\theta][A_{mn}\cos(c\lambda_{mn}t) + B_{mn}\sin(c\lambda_{mn}t)].$$

Thus

$$\begin{aligned}
u(r, \theta, t) &= \sum_{n=0}^{\infty}\sum_{m=1}^{\infty} J_n\left(\frac{x_{mn}}{a}r\right)[C_n\cos n\theta + D_n\sin n\theta][A_{mn}\cos(c\lambda_{mn}t) \\
&\quad + B_{mn}\sin(c\lambda_{mn}t)] \\
&= \sum_{n=0}^{\infty}\sum_{m=1}^{\infty} J_n\left(\frac{x_{mn}}{a}r\right)[C_n\cos n\theta + D_n\sin n\theta][A_{mn}\cos(c\lambda_{mn}t)] \\
&\quad + \sum_{n=0}^{\infty}\sum_{m=1}^{\infty} J_n\left(\frac{x_{mn}}{a}r\right)[C_n\cos n\theta + D_n\sin n\theta][B_{mn}\sin(c\lambda_{mn}t)] \\
&= \sum_{n=0}^{\infty}\sum_{m=1}^{\infty} J_n\left(\frac{x_{mn}}{a}r\right)[E_{mn}\cos n\theta + F_{mn}\sin n\theta][\cos(c\lambda_{mn}t)] \\
&\quad + \sum_{n=0}^{\infty}\sum_{m=1}^{\infty} J_n\left(\frac{x_{mn}}{a}r\right)[G_{mn}\cos n\theta + H_{mn}\sin n\theta][\sin(c\lambda_{mn}t)]
\end{aligned}$$

where $E_{mn} = C_nA_{mn}$, $F_{mn} = D_nA_{mn}$, $G_{mn} = C_nB_{mn}$, $H_{mn} = D_nB_{mn}$.

We used the boundary condition $R(a) = u(a,\theta,t) = 0$ to get $J_n(\lambda a) = 0$ and thereby determine the values of λ_{mn}. We now use the initial conditions

$$u(r,\theta,0) = f_1(r,\theta)$$
$$u_t(r,\theta,0) = f_2(r,\theta)$$

to determine the constants. We have

$$u(r,\theta,0) = \sum_{n=0}^{\infty}\sum_{m=1}^{\infty} J_n\left(\frac{x_{mn}}{a}r\right)[E_{mn}\cos n\theta + F_{mn}\sin n\theta] = f_1(r,\theta)$$

so

$$\begin{aligned}\int_0^{2\pi} f_1(r,\theta)\cos k\theta d\theta &= \int_0^{2\pi}\sum_{n=0}^{\infty}\sum_{m=1}^{\infty} J_n\left(\frac{x_{mn}}{a}r\right)\\ &\quad\times[E_{mn}\cos n\theta + F_{mn}\sin n\theta]\cos k\theta d\theta\\ &= \int_0^{2\pi}\sum_{n=0}^{\infty}\sum_{m=1}^{\infty} J_n\left(\frac{x_{mn}}{a}r\right)[E_{mn}\cos n\theta]\cos k\theta d\theta\\ &\quad+\int_0^{2\pi}\sum_{n=0}^{\infty}\sum_{m=1}^{\infty} J_n\left(\frac{x_{mn}}{a}r\right)[F_{mn}\sin n\theta]\cos k\theta d\theta\\ &= \int_0^{2\pi}\sum_{n=0}^{\infty}\sum_{m=1}^{\infty} J_n\left(\frac{x_{mn}}{a}r\right)[E_{mn}\cos n\theta]\cos k\theta d\theta\\ &= \sum_{m=1}^{\infty}\int_0^{2\pi} J_k\left(\frac{x_{mk}}{a}r\right)[E_{mk}\cos k\theta]\cos k\theta d\theta\\ &= \pi\sum_{m=1}^{\infty} E_{mk}J_k\left(\frac{x_{mk}}{a}r\right).\end{aligned}$$

Then

$$\int_0^a\left(\int_0^{2\pi} f_1(r,\theta)\cos k\theta d\theta\right)J_k\left(\frac{x_{jk}}{a}r\right)rdr = \pi\sum_{m=1}^{\infty} E_{mk}\int_0^a J_k\left(\frac{x_{mk}}{a}r\right)J_k\left(\frac{x_{jk}}{a}r\right)rdr$$

$$= \pi E_{jk}\int_0^a J_k\left(\frac{x_{jk}}{a}r\right)J_k\left(\frac{x_{jk}}{a}r\right)rdr \quad\text{since}\quad \int_0^a J_k\left(\frac{x_{mk}}{a}r\right)J_k\left(\frac{x_{jk}}{a}r\right)rdr = 0 \quad\text{if } m\neq j.$$

Thus,

$$E_{jk} = \frac{\frac{1}{\pi}\int_0^a \left(\int_0^{2\pi} f_1(r,\theta)\cos k\theta d\theta\right) J_k\left(\frac{x_{jk}}{a}r\right) rdr}{\int_0^a J_k\left(\frac{x_{jk}}{a}r\right) J_k\left(\frac{x_{jk}}{a}r\right) rdr}.$$

By a similar analysis,

$$F_{jk} = \frac{\frac{1}{\pi}\int_0^a \left(\int_0^{2\pi} f_1(r,\theta)\sin k\theta d\theta\right) J_k\left(\frac{x_{jk}}{a}r\right) rdr}{\int_0^a J_k\left(\frac{x_{jk}}{a}r\right) J_k\left(\frac{x_{jk}}{a}r\right) rdr}.$$

Now,

$$u_t(r,\theta,t) = \sum_{n=0}^{\infty}\sum_{m=1}^{\infty} J_n\left(\frac{x_{mn}}{a}r\right) c\lambda_{mn}[E_{mn}\cos n\theta + F_{mn}\sin n\theta][-\sin(c\lambda_{mn}t)]$$
$$+\sum_{n=0}^{\infty}\sum_{m=1}^{\infty} J_n\left(\frac{x_{mn}}{a}r\right) c\lambda_{mn}[G_{mn}\cos n\theta + H_{mn}\sin n\theta][\cos(c\lambda_{mn}t)]$$

so

$$u_t(r,\theta,0) = \sum_{n=0}^{\infty}\sum_{m=1}^{\infty} J_n\left(\frac{x_{mn}}{a}r\right) c\lambda_{mn}[G_{mn}\cos n\theta + H_{mn}\sin n\theta].$$

With an analysis very similar to the one above, we get

$$c\lambda_{jk}G_{jk} = \frac{\frac{1}{\pi}\int_0^a \left(\int_0^{2\pi} f_2(r,\theta)\cos k\theta d\theta\right) J_k\left(\frac{x_{jk}}{a}r\right) rdr}{\int_0^a J_k\left(\frac{x_{jk}}{a}r\right) J_k\left(\frac{x_{jk}}{a}r\right) rdr}$$

so

$$G_{jk} = \frac{\frac{1}{\pi}\int_0^a \left(\int_0^{2\pi} f_2(r,\theta)\cos k\theta d\theta\right) J_k\left(\frac{x_{jk}}{a}r\right) rdr}{c\lambda_{jk}\int_0^a J_k\left(\frac{x_{jk}}{a}r\right) J_k\left(\frac{x_{jk}}{a}r\right) rdr}.$$

Likewise,

$$H_{jk} = \frac{\frac{1}{\pi}\int_0^a \left(\int_0^{2\pi} f_2(r,\theta)\sin k\theta d\theta\right) J_k\left(\frac{x_{jk}}{a}r\right) rdr}{c\lambda_{jk}\int_0^a J_k\left(\frac{x_{jk}}{a}r\right) J_k\left(\frac{x_{jk}}{a}r\right) rdr}.$$

EXERCISES

1. Show that separation of variables leads to Eqs. (1a), (1b), and (1c). What does $\Theta(\pi) = \Theta(-\pi)$ and $\Theta'(\pi) = \Theta'(-\pi)$ say about the values of n in Eq. (1b)?
In problems 2–6, give the formulas for the Fourier coefficients.
2. Solve the drumhead problem in the case $a = \pi$, $u_t(r,\theta,0) = 0$, $u(r,\theta,0) = \sin\theta$.
3. Solve the drumhead problem in the case $a = \pi$, $u_t(r,\theta,0) = \sin\theta$, $u(r,\theta,0) = 0$.
4. Solve the drumhead problem in the case $a = 1$, $u_t(r,\theta,0) = 0$, $u(r,\theta,0) = 1 - r$.
5. Solve the drumhead problem in the case $a = 1$, $u_t(r,\theta,0) = \sin \pi r$, $u(r,\theta,0) = 1 - r$.

9.5 THE HEAT EQUATION ON A DISK

In analyzing the heat equation on a disk by separation of variables, our focus will be to better understand Bessel functions. We approach this by constructing the problem so that the only solution of Bessel's equation will be the Bessel function of order zero of the first kind. We want to understand our problem as an eigenvalue/eigenfunction problem, and we examine how in some ways our solution parallels a familiar problem that uses sine and cosine functions.

Consider a disk of radius a centered at the origin. The heat equation in polar coordinates is

$$u_t(r,\theta,t) = K\left(u_{rr}(r,\theta,t) + \frac{1}{r}u_r(r,\theta,t) + \frac{1}{r^2}u_{\theta\theta}(r,\theta,t)\right). \tag{1}$$

For convenience, we set $K = 1$. We set the initial condition to be $u(r,\theta,0) = f(r,\theta)$ and to achieve the desired form of the problem, we set the boundary condition $u(a,\theta,t) = 0$, $t > 0$. As usual, we hypothesize

$$u(r,\theta,t) = R(r)\Theta(\theta)T(t)$$

so that Eq. (1) becomes

$$R(r)\Theta(\theta)T'(t) = R''(r)\Theta(\theta)T(t) + \frac{1}{r}R'(r)\Theta(\theta)T(t) + \frac{1}{r^2}R(r)\Theta''(\theta)T(t)$$

and dividing by $R(r)\ \Theta(\theta)\ T(t)$ gives

$$\frac{T'(t)}{T(t)} = \frac{R''(r)}{R(r)} + \frac{1}{r}\frac{R'(r)}{R(r)} + \frac{1}{r^2}\frac{\Theta''(\theta)}{\Theta(\theta)}. \tag{2}$$

The left-hand side of Eq. (2) is a function of t, and the right-hand side is a function of r and θ, so each must be a constant that we designate k. We now determine the sign of k. We have

$$\frac{T'(t)}{T(t)} = k$$

so

$$T(t) = e^{kt}.$$

As there are no heat sources, $u(r,\theta,t)$ cannot grow unboundedly. Thus k must be negative. We set $k = -\lambda^2$. Thus we have

$$T'(t) = -\lambda^2 T(t).$$

Setting

$$\frac{R''(r)}{R(r)} + \frac{1}{r}\frac{R'(r)}{R(r)} + \frac{1}{r^2}\frac{\Theta''(\theta)}{\Theta(\theta)} = -\lambda^2$$

and rearranging gives

$$r^2\frac{R''(r)}{R(r)} + r\frac{R'(r)}{R(r)} + \lambda^2 r^2 = -\frac{\Theta''(\theta)}{\Theta(\theta)}.$$

Again, this means each side of the equation must be a constant that we denote j. We now determine the sign of j. We have

$$-\frac{\Theta''(\theta)}{\Theta(\theta)} = j$$

so

$$\Theta(\theta) = -j\Theta(\theta) \quad \text{and} \quad \Theta(\theta) + j\Theta(\theta) = 0.$$

The periodicity conditions $\Theta(\pi) = \Theta(-\pi)$, and $\Theta'(\pi) = \Theta'(-\pi)$ force $j \geq 0$. We set $j = \mu^2$.

We thus have three equations

$$T'(t) = -\lambda^2 T(t)$$

$$\Theta''(\theta) + \mu^2\Theta(\theta) = 0$$

$$r^2\frac{R''(r)}{R(r)} + r\frac{R'(r)}{R(r)} + \lambda^2 r^2 = \mu^2.$$

The last equation can be rewritten as

$$r^2R''(r) + rR'(r) + (\lambda^2r^2 - \mu^2)R(r) = 0.$$

We want to analyze a Bessel equation of order 0. If we assume that $u(r,\theta,t)$ is independent of θ, then $\mu^2 = 0$ and we get

$$r^2R''(r) + rR'(r) + \lambda^2r^2R(r) = 0.$$

This is a Bessel equation of order 0. It has two solutions, but only $J_0(r)$ is bounded at $r = 0$ so that is the only admissible solution. Thus we have

$$r^2R''(r) + rR'(r) = -\lambda^2r^2R(r)$$

or

$$R''(r) + \frac{1}{r}R'(r) = -\lambda^2R(r). \tag{3}$$

If we take

$$L[R] = R''(r) + \frac{1}{r}R'(r)$$

then Eq. (3) can be recognized as an eigenvalue/eigenvector problem

$$L[R] = -\lambda^2R(r)$$

for which $J_0(r)$ is an eigenfunction.

We have imposed the boundary condition $u(a,\theta,t) = 0$, $t > 0$, so we must have $J_0(a) = 0$.

We want to emphasize the similarities between this problem and the familiar problem

$$L[\Theta] = \Theta''.$$

This is an eigenvalue/eigenvector problem with two eigenfunctions, sin $\alpha\theta$ and cos $\alpha\theta$. If we add the boundary condition $\Theta(a) = 0$, then sin $\alpha\theta$ is the only admissible eigenfunction, and $\alpha = \frac{n\pi}{a}$. Thus, $\sin\left(\frac{n\pi}{a}x\right)$ are the eigenfunctions and $-\left(\frac{n\pi}{a}\right)^2$ are the eigenvalues.

Returning to the Bessel function case, we know that $J_0(r)$ (like sin θ) has infinitely many values where the function is 0. Previously, we have designated these as $x_{01}, x_{02}, x_{03}, \ldots$. Then

$$J_0\left(\frac{x_{0n}}{a}r\right)$$

are the eigenfunctions that satisfy the boundary conditions. Thus, the numbers

$$-\left(\frac{x_{0n}}{a}\right)^2$$

are the eigenvalues for

$$L[R] = -\lambda^2 R(r), \quad R(a) = 0.$$

Going back to the problem of Fourier series, recall that if $\{\phi_1(x), \phi_2(x), \ldots\}$ is a complete orthogonal set of functions and $f(x)$ is a suitably well-behaved function, then

$$f(x) = \sum_{n=1}^{\infty} a_n \phi_n(x)$$

where

$$a_n = \frac{\langle f, \phi_n \rangle}{\langle \phi_n, \phi_n \rangle}.$$

The Sturm–Liouville theory tells us that $\left\{J_0\left(\frac{x_{0n}}{a} r\right)\right\}$ is a complete orthogonal set of functions. The inner product for this problem is

$$\langle f, g \rangle_\rho = \int_0^a f(x) x g(x) dx$$

as we verified in Exercise 2, Section 6.2.

In our particular problem, we set the initial condition to be $u(r,\theta,0) = f(r,\theta)$. Since we later assumed independence of θ, we amend this so that the initial condition is $u(r,\theta,0) = f(r)$.

Because of the independence of θ, our solution is

$$u(r,t) = \sum_{n=1}^{\infty} a_n R_n(r) T(t) = \sum_{n=1}^{\infty} a_n J_0\left(\frac{x_{0n}}{a} r\right) e^{-kt}$$

with

$$f(r) = u(r,0) = \sum_{n=1}^{\infty} a_n J_0\left(\frac{x_{0n}}{a} r\right)$$

so that

$$a_n = \frac{\int_0^a f(r) r J_0\left(\frac{x_{0n}}{a} r\right) dr}{\int_0^a J_0\left(\frac{x_{0n}}{a} r\right) r J_0\left(\frac{x_{0n}}{a} r\right) dr}.$$

Thus we have

$$u(r,t) = \sum_{n=1}^{\infty} \left[\frac{\int_0^a f(r) r J_0\left(\frac{x_{0n}}{a} r\right) dr}{\int_0^a J_0\left(\frac{x_{0n}}{a} r\right) r J_0\left(\frac{x_{0n}}{a} r\right) dr} \right] J_0\left(\frac{x_{0n}}{a} r\right).$$

EXERCISES

1. Describe (do not work out the calculus) the solution to the heat equation on a disk of radius a with boundary condition $u(a,\theta,t) = 0$ if $u(r,0) = \sin\left(\frac{\pi r}{a}\right)$.
2. Describe (do not work out the calculus) the solution to the heat equation on a disk of radius a with boundary condition $u(a,\theta,t) = 0$ if $u(r,0) = a - r$.
3. Describe (do not work out the calculus) the solution to the heat equation on a disk of radius a with boundary condition $u(a,\theta,t) = 0$ if $u(r,0) = 1 - \cos\left(\frac{\pi r}{a}\right)$.

CHAPTER

Solving Partial Differential Equations in Spherical Coordinates Using Separation of Variables

10

10.1 AN EXAMPLE WHERE LEGENDRE EQUATIONS ARISE

In Chapter 9 we studied solving partial differential equations (PDEs) in which the Laplacian appeared in cylindrical coordinates using separation of variables. We saw that among the differential equations that arose was a Bessel (or, at least a "Bessel-like") equation. In this chapter, we follow a similar approach except we work in spherical coordinates. We shall see that in addition to a Bessel equation we encounter a differential equation called Legendre's equation.

Legendre equations arise when solving a PDE in spherical coordinates that uses the Laplacian. We demonstrate this with the wave equation.

In spherical coordinates, the Laplacian is

$$\Delta u(r,\theta,\vartheta)=\frac{1}{r^2}\frac{\partial}{\partial r}\left(r^2u_r\right)+\frac{1}{r^2\sin\theta}\frac{\partial}{\partial\theta}(\sin\theta\ u_\theta)+\frac{1}{r^2\sin^2\theta}u_{\vartheta\vartheta}$$

so the wave equation in spherical coordinates is

$$u_{tt}=K\Delta u=K\left[\frac{1}{r^2}\frac{\partial}{\partial r}\left(r^2u_r\right)+\frac{1}{r^2\sin\theta}\frac{\partial}{\partial\theta}(\sin\theta\ u_\theta)+\frac{1}{r^2\sin^2\theta}u_{\vartheta\vartheta}\right].$$

We note that

$$\frac{\partial}{\partial\theta}(\sin\theta u_\theta)=\cos\theta\ u_\theta+\sin\theta\ u_{\theta\theta}$$

and

$$\frac{1}{r^2}\frac{\partial}{\partial r}\left(r^2u_r\right)=\frac{1}{r^2}\left(2ru_r+r^2u_{rr}\right)=\frac{2}{r}u_r+u_{rr}$$

Mathematical Physics with Partial Differential Equations. https://doi.org/10.1016/B978-0-12-814759-7.00010-7

so

$$\begin{aligned}\Delta u(r,\theta,\vartheta) &= \frac{1}{r^2}\frac{\partial}{\partial r}(r^2 u_r) + \frac{1}{r^2 \sin\theta}\frac{\partial}{\partial \theta}(\sin\theta u_\theta) + \frac{1}{r^2\sin^2\theta}u_{\vartheta\vartheta} \\ &= \left(\frac{2}{r}u_r + u_{rr}\right) + \frac{1}{r^2\sin\theta}(\cos\theta u_\theta + \sin\theta u_{\theta\theta}) + \frac{1}{r^2\sin^2\theta}u_{\vartheta\vartheta} \qquad (1) \\ &= \left(\frac{2}{r}u_r + u_{rr}\right) + \frac{1}{r^2}\left(\cot\theta u_\theta + u_{\theta\theta} + \frac{u_{\vartheta\vartheta}}{\sin^2\theta}\right).\end{aligned}$$

We assume $u(r,\theta,\vartheta,t) = R(r)\Theta(\theta)\Phi(\vartheta)T(t)$. Then Eq. (1) can be written as

$$\begin{aligned}\Delta u(r,\theta,\vartheta) &= \frac{2}{r}(R'\Theta\Phi T + R''\Theta\Phi T) \\ &\quad + \frac{1}{r^2}\left(\cot\theta R\Theta'\Phi T + R\Theta''\Phi T + \frac{1}{\sin^2\theta}R\Theta\Phi'' T\right)\end{aligned}$$

and

$$\frac{\Delta u}{u} = \frac{2}{r}\frac{R'}{R} + \frac{R''}{R} + \frac{1}{r^2}\left(\cot\theta\frac{\Theta'}{\Theta} + \frac{\Theta''}{\Theta} + \frac{1}{\sin^2\theta}\frac{\Phi''}{\Phi}\right).$$

Also

$$\frac{1}{K}\frac{u_{tt}}{u} = \frac{1}{K}\frac{T''}{T}.$$

We rewrite the wave equation as

$$\frac{\Delta u}{u} = \frac{1}{K}\frac{T''}{T}$$

or

$$\frac{2}{r}\frac{R'}{R} + \frac{R''}{R} + \frac{1}{r^2}\left(\cot\theta\frac{\Theta'}{\Theta} + \frac{\Theta''}{\Theta} + \frac{1}{\sin^2\theta}\frac{\Phi''}{\Phi}\right) = \frac{1}{K}\frac{T''}{T}. \qquad (2)$$

In what follows, we argue that Eq. (2) can be solved by solving four ordinary differential equations (ODEs), one for each of the functions T, Θ, Φ, and R. The algebra is somewhat tedious, but we are doing what we have done before. That is, we show, one step at a time, that independence of an expression of certain variables forces the expression to be constant.

The expression on the right-hand side of Eq. (2) depends only on t, and the expression on the left-hand side is independent of t. Therefore each side is constant. Thus we have

$$\frac{1}{K}\frac{T''}{T} = -\lambda \text{ or } T'' + \lambda K T = 0 \qquad (3)$$

where $\lambda > 0$.

(As before, the sign of the constants will be justified in the exercises.)

Multiply Eq. (2) by r^2 to get

$$2r\frac{R'}{R}+r^2\frac{R''}{R}+\left(\cot\theta\frac{\Theta'}{\Theta}+\frac{\Theta''}{\Theta}+\frac{1}{\sin^2\theta}\frac{\Phi''}{\Phi}\right)=r^2\frac{1}{K}\frac{T''}{T}=-\lambda r^2$$

so

$$\cot\theta\frac{\Theta'}{\Theta}+\frac{\Theta''}{\Theta}+\frac{1}{\sin^2\theta}\frac{\Phi''}{\Phi}=-\left(2r\frac{R'}{R}+r^2\frac{R''}{R}+\lambda r^2\right). \tag{4}$$

The left-hand side of Eq. (4) is independent of r and the right-hand side depends only on r, so each side must be a constant. We set

$$\cot\theta\frac{\Theta'}{\Theta}+\frac{\Theta''}{\Theta}+\frac{1}{\sin^2\theta}\frac{\Phi''}{\Phi}=-\mu. \tag{5}$$

Multiply Eq. (5) by $\sin^2\theta$ to get

$$\sin^2\theta\cot\theta\frac{\Theta'}{\Theta}+\sin^2\theta\frac{\Theta''}{\Theta}+\frac{\Phi''}{\Phi}=-\mu\sin^2\theta$$

so

$$\sin^2\theta\cot\theta\frac{\Theta'}{\Theta}+\sin^2\theta\frac{\Theta''}{\Theta}+\mu\sin^2\theta=-\frac{\Phi''}{\Phi}. \tag{6}$$

The terms on the left-hand side of Eq. (6) depend only on θ, and the right-hand side depends only on φ, so Φ''/Φ is another constant. We set

$$\frac{\Phi''}{\Phi}=-\nu \tag{7}$$

so

$$\Phi''+\nu\Phi=0.$$

Thus Eq. (6) is

$$\sin^2\theta\cot\theta\frac{\Theta'}{\Theta}+\sin^2\theta\frac{\Theta''}{\Theta}+\mu\sin^2\theta=\nu.$$

Multiply by $\Theta/\sin^2\theta$ to get

$$\cot\theta\,\Theta'+\Theta''+\mu\Theta=\nu\frac{\Theta}{\sin^2\theta}$$

so

$$\cot\theta\,\Theta'+\Theta''+\left(\mu-\frac{\nu}{\sin^2\theta}\right)\Theta=0. \tag{8}$$

Finally, we determine the ODE that $R(r)$ must satisfy. We began with Eq. (2)

$$\frac{2}{r}\frac{R'}{R}+\frac{R''}{R}+\frac{1}{r^2}\left(\cot\theta\frac{\Theta'}{\Theta}+\frac{\Theta''}{\Theta}+\frac{1}{\sin^2\theta}\frac{\Phi''}{\Phi}\right)=\frac{1}{K}\frac{T''}{T}$$

and have set

$$\frac{1}{K}\frac{T''}{T} = -\lambda \text{ and } \frac{\Phi''}{\Phi} = -\nu.$$

The equation

$$\frac{\Phi''}{\Phi} = -\nu$$

can be written $\Phi'' + \nu\Phi = 0$. Since Φ is periodic, $\nu > 0$, so we take $\nu = m^2$.

So Eq. (2) could now be written

$$\frac{1}{R}\left(R'' + \frac{2}{r}R'\right) + \frac{1}{r^2}\left(\frac{\nu}{\sin^2\theta} - \mu\right) + \frac{1}{r^2\sin^2\theta}(-\nu) + \lambda = 0$$

or

$$\left(R'' + \frac{2}{r}R'\right) + \left(\lambda - \frac{\mu}{r^2}\right)R = 0.$$

If we look back at what we have done, we see that to solve the wave equation in spherical coordinates using separation of variables, we need to solve the following ODEs:

$$T'' + \lambda KT = 0 \tag{9}$$

$$\Phi'' + m^2\Phi = 0 \tag{10}$$

$$\cot\theta\,\Theta' + \Theta'' + \left(\mu - \frac{\nu}{\sin^2\theta}\right)\Theta = 0 \tag{11}$$

$$\left(R'' + \frac{2}{r}R'\right) + \left(\lambda - \frac{\mu}{r^2}\right)R = 0. \tag{12}$$

The first two of these equations are familiar. The fourth is a "Bessel-like" equation. It is similar to what we considered in Chapter 9, but different enough so that we will give it a separate analysis. The third equation can be converted into a "Legendre-like" equation that we shall study. In the next sections, we discuss the solutions of these equations.

10.2 THE SOLUTION TO BESSEL'S EQUATION IN SPHERICAL COORDINATES

In spherical coordinates, the "Bessel-like" equation that must be solved when analyzing equations that involve the Laplacian using separation of variables is

$$\frac{d^2R(r)}{dr^2} + \frac{2}{r}\frac{dR(r)}{dr} + \left[\lambda - \frac{\mu}{r^2}\right]R(r) = 0$$

or

$$r^2\frac{d^2R(r)}{dr^2}+2r\frac{dR(r)}{dr}+\left[\lambda r^2-\mu\right]R(r)=0.$$

In practice, it is most common that $\mu = n(n+1)$ and $\lambda = k^2$ where n and k are positive integers. Thus we seek a solution to

$$\frac{d^2R(r)}{dr^2}+\frac{2}{r}\frac{dR(r)}{dr}+\left[k^2-\frac{n(n+1)}{r^2}\right]R(r)=0$$

or

$$r^2\frac{d^2R(r)}{dr^2}+2r\frac{dR(r)}{dr}+\left[k^2r^2-n(n+1)\right]R(r)=0 \tag{1}$$

that is bounded at $r = 0$. We convert Eq. (1) to a Bessel equation by first making a change of variables and then transforming a function. The change of variables is $x = kr$.

Then

$$\frac{dR(r)}{dr}=\frac{dR(x)}{dx}\frac{dx}{dr}=\frac{dR(x)}{dx}k \text{ and } \frac{d^2R(r)}{dr^2}=\frac{d^2R(x)}{dx^2}k^2$$

so

$$r\frac{dR(r)}{dr}=\frac{x}{k}\frac{dR(x)}{dx}k=x\frac{dR(x)}{dx} \text{ and}$$

$$r^2\frac{d^2R(r)}{dr^2}=\left(\frac{x}{k}\right)^2\frac{d^2R(x)}{dx^2}k^2=x^2\frac{d^2R(x)}{dx^2}.$$

Thus

$$r^2\frac{d^2R(r)}{dr^2}+2r\frac{dR(r)}{dr}+\left[r^2k^2-n(n+1)\right]R(r)$$

$$=x^2\frac{d^2R(x)}{dx^2}+2x\frac{dR(x)}{dx}+\left[x^2-n(n+1)\right]R(x)=0.$$

Now let

$$Y(x)=\sqrt{x}R(x) \text{ or } R(x)=x^{-\frac{1}{2}}Y(x).$$

Then

$$\frac{dR(x)}{dx}=x^{-\frac{1}{2}}\frac{dY(x)}{dx}-\frac{1}{2}x^{-\frac{3}{2}}Y(x)$$

$$\frac{d^2R(x)}{dx^2} = x^{-\frac{1}{2}}\frac{d^2Y(x)}{dx^2} - \frac{dY(x)}{dx}\frac{1}{2}x^{-\frac{3}{2}} - \frac{1}{2}\left[x^{-\frac{3}{2}}\frac{dY(x)}{dx} - \frac{3}{2}x^{-\frac{5}{2}}Y(x)\right]$$

$$= x^{-\frac{1}{2}}\frac{d^2Y(x)}{dx^2} - x^{-\frac{3}{2}}\frac{dY(x)}{dx} + \frac{3}{4}x^{-\frac{5}{2}}Y(x).$$

So

$$x^2\frac{d^2R(x)}{dx^2} + 2x\frac{dR(x)}{dx} + \left[x^2 - n(n+1)\right]R(x)$$

$$= x^2\left[x^{-\frac{1}{2}}\frac{d^2Y(x)}{dx^2} - x^{-\frac{3}{2}}\frac{dY(x)}{dx} + \frac{3}{4}x^{-\frac{5}{2}}Y(x)\right] + 2x\left[x^{-\frac{1}{2}}\frac{dY(x)}{dx} - \frac{1}{2}x^{-\frac{3}{2}}Y(x)\right]$$

$$+\left[x^2 - n(n+1)\right]x^{-\frac{1}{2}}Y(x) = 0.$$

Multiplying by $x^{1/2}$ gives

$$x^2\left[\frac{d^2Y(x)}{dx^2} - x^{-1}\frac{dY(x)}{dx} + \frac{3}{4}x^{-2}Y(x)\right] + 2x\left[\frac{dY(x)}{dx} - \frac{1}{2}x^{-1}Y(x)\right]$$
$$+\left[x^2 - n(n+1)\right]Y(x) = 0$$

or

$$x^2\frac{d^2Y(x)}{dx^2} + (-x+2x)\frac{dY(x)}{dx} + \left[\frac{3}{4} - 1 + x^2 - n(n+1)\right]Y(x)$$

$$= x^2\frac{d^2Y(x)}{dx^2} + x\frac{dY(x)}{dx} + \left[x^2 - \left(n^2 + n + \frac{1}{4}\right)\right]Y(x)$$

$$= x^2\frac{d^2Y(x)}{dx^2} + x\frac{dY(x)}{dx} + \left[x^2 - \left(n + \frac{1}{2}\right)^2\right]Y(x) = 0.$$

The equation

$$x^2\frac{d^2Y(x)}{dx^2} + x\frac{dY(x)}{dx} + \left[x^2 - \left(n + \frac{1}{2}\right)^2\right]Y(x) = 0 \qquad (2)$$

is a Bessel equation of half-integer order. The requirement that the solution be bounded at $x = 0$ means the solution is of the form

$$Y(x) = AJ_{n+\frac{1}{2}}(x).$$

Now

$$R(x) = \frac{Y(x)}{\sqrt{x}} = A\frac{J_{n+\frac{1}{2}}(x)}{\sqrt{x}} \text{ and } R(r) = A\frac{J_{n+\frac{1}{2}}(kr)}{\sqrt{kr}}.$$

It is common to take the constant A to be $\sqrt{\pi/2}$ and denote the spherical solution of the first kind (the solution that is bounded when $r = 0$) by

$$j_n(r) = \left(\frac{\pi}{2}\right)^{1/2} \frac{J_{n+\frac{1}{2}}(r)}{\sqrt{r}}$$

so that

$$R(r) = \left(\frac{\pi}{2}\right)^{1/2} \frac{J_{n+\frac{1}{2}}(kr)}{\sqrt{kr}} \equiv j_n(kr).$$

Next we find the solution to

$$y(r) + \frac{2}{r}y'(r) + \left(\lambda - \frac{k(k+1)}{r^2}\right)y(r) = 0$$

or

$$r^2 y(r) + 2ry'(r) + \left[\lambda r^2 - k(k+1)\right]y(r) = 0.$$

We consider the case $\lambda > 0$. We let

$$y(r) = \sum_{n=0}^{\infty} a_n r^{n+\alpha}$$

and proceed as we did in Chapter 9. We have

$$y'(r) = \sum_{n=0}^{\infty} a_n(n+\alpha)r^{n+\alpha-1} \quad y''(r) = \sum_{n=0}^{\infty} a_n(n+\alpha)(n+\alpha-1)r^{n+\alpha-2}$$

so

$$ry'(r) = \sum_{n=0}^{\infty} a_n(n+\alpha)r^{n+\alpha} \quad r^2 y''(r) = \sum_{n=0}^{\infty} a_n(n+\alpha)(n+\alpha-1)r^{n+\alpha}.$$

Then

$$
\begin{aligned}
&r^2y(r) + 2ry'(r) + \left[\lambda r^2 - k(k+1)\right]y(r) \\
&= \sum_{n=0}^{\infty} a_n[(n+\alpha)(n+\alpha-1) + 2(n+\alpha) - k(k+1)]r^{n+\alpha} + \lambda\sum_{n=0}^{\infty} a_n r^{n+\alpha+2} \\
&= \sum_{n=0}^{\infty} a_n[(n+\alpha)(n+\alpha+1) - k(k+1)]r^{n+\alpha} + \lambda\sum_{n=0}^{\infty} a_n r^{n+\alpha+2} \\
&= \sum_{n=0}^{\infty} a_n[(n+\alpha)(n+\alpha+1) - k(k+1)]r^{n+\alpha} + \lambda\sum_{n=2}^{\infty} a_{n-2} r^{n+\alpha} \\
&= a_0[(\alpha)(\alpha+1) - k(k+1)]r^{\alpha} + a_1[(1+\alpha)(2+\alpha) - k(k+1)]r^{\alpha+1} \\
&\quad + \sum_{n=2}^{\infty}\{a_n[(n+\alpha)(n+\alpha+1) - k(k+1)] + \lambda a_{n-2}\}r^{n+\alpha} = 0.
\end{aligned}
$$

Each coefficient must be 0, so $a_0[(\alpha)(\alpha+1) - k(k+1)] = 0$, and if $a_0 \neq 0$, then $\alpha = k$. If $\alpha = k$, since $a_1[(1+\alpha)(2+\alpha) - k(k+1)] = 0$, then $a_1 = 0$. Since

$$
\begin{aligned}
&a_n[(n+\alpha)(n+\alpha+1) - k(k+1)] + \lambda a_{n-2} \\
&= a_n[(n+k)(n+k+1) - k(k+1)] + \lambda a_{n-2} = 0
\end{aligned}
$$

we have the recurrence relation

$$
a_n = \frac{-\lambda a_{n-2}}{(n+k)(n+k+1) - k(k+1)} = \frac{-\lambda a_{n-2}}{n(n+2k+1)}.
$$

If $a_1 = 0$, then $a_{2n+1} = 0$ for the very positive integer n.
By induction, one can show

$$
\begin{aligned}
y(r) &= a_0 r^k\left[1 + \sum_{n=1}^{\infty}\frac{(-\lambda)^n r^{2n}}{2^n n!(2k+3)(2k+5)\cdots(2k+2n+1)}\right] \\
&= a_0 r^k\left[1 + \sum_{n=1}^{\infty}\frac{(-\lambda)^n r^{2n}}{2^{2n} n!\left(k+\frac{3}{2}\right)\left(k+\frac{5}{2}\right)\cdots\left(k+n+\frac{1}{2}\right)}\right].
\end{aligned} \tag{3}
$$

Let $l = k + 1/2$. Then the right-hand side of Eq. (3) is

$$
a_0 r^{l-\frac{1}{2}}\left[1 + \sum_{n=1}^{\infty}\frac{(-\lambda)^n r^{2n}}{2^{2n} n!(l+1)(l+2)\cdots(l+n)}\right].
$$

If we set $\lambda = 1$ and denote the resulting expression by $j_k(r)$, we have

$$j_k(r) = \frac{a_0}{\sqrt{r}} J_l(r) = \frac{a_0}{\sqrt{r}} J_{k+1/2}(r).$$

The function $j_k(r)$ is called the kth spherical Bessel function.

We next show that $j_k\left(\sqrt{\lambda}\, r\right)$ satisfies

$$y''(r) + \frac{2}{r} y'(r) - \frac{\mu}{r^2} y(r) = \lambda y(r).$$

We let $z = \sqrt{\lambda} r$ so that $\frac{d}{dr} y(z) = \frac{d}{dr} y\left(\sqrt{\lambda}\, r\right) = \sqrt{\lambda}\, \frac{d}{dz} y(z)$.
Also

$$\frac{d^2}{dr^2} y\left(\sqrt{\lambda}\, r\right) = \lambda \frac{d^2}{dz^2} y(z).$$

Then

$$\begin{aligned}
&\frac{d^2}{dr^2} y\left(\sqrt{\lambda}\, r\right) + \frac{2}{r} \frac{d}{dr} y\left(\sqrt{\lambda}\, r\right) - \frac{\mu}{r^2} y\left(\sqrt{\lambda}\, r\right) \\
&= \lambda \frac{d^2}{dz^2} y(z) + \frac{2\sqrt{\lambda}}{r} \frac{d}{dz} y(z) - \frac{\mu}{r^2} y(z) \\
&= \lambda \left[\frac{d^2}{dz^2} y(z) + \frac{2}{\sqrt{\lambda} r} \frac{d}{dz} y(z) - \frac{\mu}{\left(\sqrt{\lambda}\, r\right)^2} y(z) \right] \\
&= \lambda \left[\frac{d^2}{dz^2} y(z) + \frac{2}{z} \frac{d}{dz} y(z) - \frac{\mu}{z^2} y(z) \right] = \lambda y(z)
\end{aligned}$$

since

$$\frac{d^2}{dz^2} y(z) + \frac{2}{z} \frac{d}{dz} y(z) - \frac{\mu}{z^2} y(z) = y(z).$$

What we have done is shown that Bessel functions—the solutions to Bessel's equation—are eigenfunctions for a particular linear operator. In the next section we show that Legendre polynomials—the solutions to Legendre's equation—are also eigenfunctions of a linear operator. These problems are one case of a general theory, where the differential operator is self-adjoint so that eigenfunctions belonging to different eigenvalues are orthogonal with respect to an appropriate weight function.

10.3 LEGENDRE'S EQUATION AND ITS SOLUTIONS

We have seen that in solving a PDE that uses the Laplacian by separation of variables in spherical coordinates, it is necessary to solve

$$\cot\theta\, \Theta' + \Theta'' + \left(\mu - \frac{\nu}{\sin^2\theta}\right)\Theta = 0. \tag{1}$$

Our procedure will be similar to what we did in solving "Bessel-like" equations. We begin by solving Legendre's equation. With a change of variables, Eq. (1) can be transformed to a "Legendre-like" equation. We then solve the "Legendre-like" equation.

Legendre's equation is

$$(1-x^2)y''(x) - 2xy'(x) + \mu y(x) = 0 \quad -1 < x < 1. \tag{2}$$

We solve the equation by power series; that is, we assume

$$y(x) = \sum_{n=0}^{\infty} a_n x^n$$

and determine a recurrence relation for the a_n's. We have

$$y'(x) = \sum_{n=0}^{\infty} a_n n x^{n-1} \text{ and } y''(x) = \sum_{n=0}^{\infty} a_n n(n-1)x^{n-2}$$

so

$$(1-x^2)y''(x) - 2xy'(x) + \mu y(x)$$
$$= \sum_{n=0}^{\infty} a_n n(n-1)x^{n-2} - \sum_{n=0}^{\infty} a_n n(n-1)x^n - 2\sum_{n=0}^{\infty} a_n n x^n + \mu \sum_{n=0}^{\infty} a_n x^n = 0. \tag{3}$$

Now

$$\sum_{n=0}^{\infty} a_n n(n-1)x^{n-2} = \sum_{n=2}^{\infty} a_n n(n-1)x^{n-2} = \sum_{n=0}^{\infty} a_{n+2}(n+2)(n+1)x^n$$

so Eq. (3) can be written as

$$\sum_{n=0}^{\infty} a_{n+2}(n+2)(n+1)x^n - \sum_{n=0}^{\infty} a_n[n(n-1)+2n-\mu]x^n$$
$$= \sum_{n=0}^{\infty} a_{n+2}(n+2)(n+1)x^n - \sum_{n=0}^{\infty} a_n\left[n^2+n-\mu\right]x^n = 0.$$

Thus we have

$$a_{n+2}(n+2)(n+1) = a_n\left(n^2+n-\mu\right) \text{ or } a_{n+2} = \frac{(n^2+n-\mu)}{(n+2)(n+1)}a_n.$$

So we have two linearly independent solutions

$$y_0(x) = a_0 + a_2x^2 + a_4x^4 + \cdots$$
$$y_1(x) = a_1x + a_3x^3 + a_5x^5 + \cdots.$$

In many problems, the value of μ is $k(k+1)$, where k is a positive integer. We show that if this is the case, then there is a solution to Legendre's equation that is a polynomial of degree k.

First, note that in the case $\mu = k(k+1)$ we have

$$a_{n+2} = \frac{(n^2 + n - k^2 - k)}{(n+2)(n+1)} a_n$$

so

$$a_{k+2} = \frac{(k^2 + k - k^2 - k)}{(k+2)(k+1)} a_k = 0$$

and if $a_0 = 0$, then $a_{k+2} = a_{k+4} = a_{k+6} = \ldots = 0$.

To get the polynomial solution when k is even, take $a_1 = 0$ and $a_0 \neq 0$. The solution is

$$y(x) = a_0 + a_2 x^2 + a_4 x^4 + \cdots + a_k x^k.$$

To get the polynomial solution when k is odd, take $a_0 = 0$ and $a_1 \neq 0$. The solution is

$$y(x) = a_1 x + a_3 x^3 + \cdots + a_k x^k.$$

In these cases, the solution is called the Legendre polynomial of degree k.

We have seen how the differential equation

$$\cot\theta\, \Theta' + \Theta'' + \left(\mu - \frac{\nu}{\sin^2\theta}\right)\Theta = 0$$

arises in problems that involve the Laplacian which have spherical symmetry. We now convert this to a Legendre equation. We let $x = \cos\theta$. Then

$$\Theta' = \frac{d\Theta}{d\theta} = \frac{d\Theta}{dx}\frac{dx}{d\theta} = -\sin\theta \frac{d\Theta}{dx}.$$

To compute Θ'', we have

$$\Theta'' = \frac{d}{d\theta}\left(\frac{d\Theta}{d\theta}\right) = \frac{d}{d\theta}\left(-\sin\theta\frac{d\Theta}{dx}\right) = -\sin\theta\frac{d}{d\theta}\left(\frac{d\Theta}{dx}\right) + \left(\frac{d\Theta}{dx}\right)\frac{d}{d\theta}(-\sin\theta).$$

Now

$$\frac{d}{d\theta}\left(\frac{d\Theta}{dx}\right) = \frac{d}{dx}\left(\frac{d\Theta}{dx}\right)\frac{dx}{d\theta} = \frac{d^2\Theta}{dx^2}(-\sin\theta)$$

so

$$\begin{aligned}\Theta'' &= -\sin\theta \frac{d}{d\theta}\left(\frac{d\Theta}{dx}\right) + \left(\frac{d\Theta}{dx}\right)\frac{d}{d\theta}(-\sin\theta)\\ &= (-\sin\theta)\frac{d^2\Theta}{dx^2}(-\sin\theta) + \left(\frac{d\Theta}{dx}\right)(-\cos\theta)\\ &= \sin^2\theta\frac{d^2\Theta}{dx^2} - \cos\theta\frac{d\Theta}{dx}.\end{aligned}$$

Thus

$$\begin{aligned}\cot\theta\,\Theta' + \Theta'' + \left(\mu - \frac{\nu}{\sin^2\theta}\right)\Theta &= \cot\theta\left(-\sin\theta\frac{d\Theta}{dx}\right)\\ &+ \left[\sin^2\theta\frac{d^2\Theta}{dx^2} - \cos\theta\frac{d\Theta}{dx}\right] + \left(\mu - \frac{\nu}{\sin^2\theta}\right)\Theta\\ &= -\cos\theta\frac{d\Theta}{dx} + \sin^2\theta\frac{d^2\Theta}{dx^2} - \cos\theta\frac{d\Theta}{dx} + \left(\mu - \frac{\nu}{\sin^2\theta}\right)\Theta\\ &= \sin^2\theta\frac{d^2\Theta}{dx^2} - 2\cos\theta\frac{d\Theta}{dx} + \left(\mu - \frac{\nu}{\sin^2\theta}\right)\Theta = 0.\end{aligned}$$

Now use

$$\cos\theta = x, \quad \sin^2\theta = 1 - x^2$$

to get

$$(1-x^2)\frac{d^2\Theta}{dx^2} - 2x\frac{d\Theta}{dx} + \left(\mu - \frac{\nu}{1-x^2}\right)\Theta = 0. \tag{4}$$

This is not exactly of the form of Legendre's equation, but if we take $\mu = l(l+1)$ and $\nu = m^2$, then Eq. (4) is

$$(1-x^2)\frac{d^2\Theta}{dx^2} - 2x\frac{d\Theta}{dx} + \left(l(l+1) - \frac{m^2}{1-x^2}\right)\Theta = 0. \tag{5}$$

Eq. (5) often is the equation that needs to be solved in physical problems. In such problems, m and l are integers with $m \le l$. The solution is called an associated Legendre function, which we study in the next section.

EXERCISES

1. Legendre's polynomial of degree n, denoted $P_n(x)$, is a solution (there are two) to the differential equation

$$(1-x^2)y''(x) - 2xy'(x) + n(n+1)y(x) = 0,$$
$$-1 < x < 1$$

where n is a nonnegative integer.

a. Verify that $P_0(x) = 1$ and $P_1(x) = x$ are Legendre polynomials.
b. Given that Legendre polynomials satisfy the recursion relation

$$(n+1)P_{n+1}(x) - (2n+1)xP_n(x) + nP_{n-1}(x) = 0, \quad n \geq 1,$$

find $P_2(x)$, $P_3(x)$, and $P_4(x)$.

2. Rodrigues' formula can be used to generate Legendre polynomials. This formula is

$$P_n(x) = \frac{(-1)^n}{2^n n!} \frac{d^n}{dx^n}(1-x^2)^n.$$

Verify that Rodrigues' formula is valid for $n = 0, 1, 2, 3$.

3. Verify for $n = 0, 1, 2, 3$ that

$$\int_{-1}^{1} [P_n(x)]^2 dx = \frac{2}{2n+1}.$$

(Thus Legendre's polynomials are not normalized.)

4. If

$$f(x) = \sum_{n=0}^{\infty} a_n P_n(x)$$

show that

$$a_k = \frac{1}{2}\left[(2k+1)\int_{-1}^{1} f(x)P_k(x)dx\right].$$

5. Use the result of Exercise 4 to find the first two nonzero terms for the expansion of $f(x) = x^2$ in terms of Legendre polynomials.

10.4 ASSOCIATED LEGENDRE FUNCTIONS

We relate the solution of

$$(1-x^2)y''(x) - 2xy'(x) + \left(l(l+1) - \frac{m^2}{1-x^2}\right)y(x) = 0 \tag{1}$$

to Legendre polynomials using the substitution

$$y(x) = (1-x^2)^{\frac{m}{2}} u(x)$$

where $u(x)$ is a solution to

$$(1-x^2)u''(x) - 2x(m+1)u'(x) + [l(l+1) + m(m+1)]u(x) = 0.$$

With this substitution, we have

$$y' = (1-x^2)^{\frac{m}{2}}u' - mx(1-x^2)^{\frac{m}{2}-1}u$$

so that

$$-2xy' = -2x(1-x^2)^{\frac{m}{2}}u' + 2mx^2(1-x^2)^{\frac{m}{2}-1}u.$$

Also

$$y'' = (1-x^2)^{\frac{m}{2}}u'' - mx(1-x^2)^{\frac{m}{2}-1}u' - \left[mx(1-x^2)^{\frac{m}{2}-1}u' + mu\frac{d}{dx}x(1-x^2)^{\frac{m}{2}-1}\right].$$

Now

$$\begin{aligned}\frac{d}{dx}\left[x(1-x^2)^{\frac{m}{2}-1}\right] &= (1-x^2)^{\frac{m}{2}-1} + \left[x\left(\frac{m}{2}-1\right)(1-x^2)^{\frac{m}{2}-2}(-2x)\right] \\ &= (1-x^2)^{\frac{m}{2}-1} - 2x^2\left(\frac{m}{2}-1\right)(1-x^2)^{\frac{m}{2}-2}.\end{aligned}$$

Thus

$$\begin{aligned}y'' = {} & (1-x^2)^{\frac{m}{2}}u'' - 2mx(1-x^2)^{\frac{m}{2}-1}u' - mu(1-x^2)^{\frac{m}{2}-1} \\ & +2mux^2\left(\frac{m}{2}-1\right)(1-x^2)^{\frac{m}{2}-2}\end{aligned}$$

so

$$\begin{aligned}(1-x^2)y'' = {} & (1-x^2)^{\frac{m}{2}+1}u'' - 2mx(1-x^2)^{\frac{m}{2}}u' - mu(1-x^2)^{\frac{m}{2}} \\ & +2mux^2\left(\frac{m}{2}-1\right)(1-x^2)^{\frac{m}{2}-1}.\end{aligned}$$

Since

$$-2xy' = 2mx^2(1-x^2)^{\frac{m}{2}-1}u - 2x(1-x^2)^{\frac{m}{2}}u'$$

we have

$$\begin{aligned}(1-x^2)y'' - 2xy' = {} & (1-x^2)^{\frac{m}{2}+1}u'' - 2x(m+1)(1-x^2)^{\frac{m}{2}}u' \\ & + \left[2mx^2\left(\frac{m}{2}-1\right)(1-x^2)^{\frac{m}{2}-1} - m(1-x^2)^{\frac{m}{2}} + 2mx^2(1-x^2)^{\frac{m}{2}-1}\right]u.\end{aligned}$$

Thus the coefficient of the u term in

$$(1-x^2)y''-2xy'+\left(l(l+1)-\frac{m^2}{1-x^2}\right)y$$

$$=(1-x^2)^{\frac{m}{2}+1}u''-2x(m+1)(1-x^2)^{\frac{m}{2}}u'$$

$$+\left[2mx^2\left(\frac{m}{2}-1\right)(1-x^2)^{\frac{m}{2}-1}-m(1-x^2)^{\frac{m}{2}}+2mx^2(1-x^2)^{\frac{m}{2}-1}\right.$$

$$\left.+\left(l(l+1)-\frac{m^2}{1-x^2}\right)(1-x^2)^{\frac{m}{2}}\right]u=0$$

is

$$2mx^2\left(\frac{m}{2}-1\right)(1-x^2)^{\frac{m}{2}-1}-m(1-x^2)^{\frac{m}{2}}+2mx^2(1-x^2)^{\frac{m}{2}-1}$$

$$+\left(l(l+1)-\frac{m^2}{1-x^2}\right)(1-x^2)^{\frac{m}{2}}$$

$$=(1-x^2)^{\frac{m}{2}-1}\left[2mx^2\left(\frac{m}{2}-1\right)+2mx^2-m^2\right]+(1-x^2)^{\frac{m}{2}}[l(l+1)-m]$$

$$=(1-x^2)^{\frac{m}{2}-1}(m^2x^2-m^2)+(1-x^2)^{\frac{m}{2}}[l(l+1)-m]$$

$$=-m^2(1-x^2)^{\frac{m}{2}}+(1-x^2)^{\frac{m}{2}}[l(l+1)-m]$$

$$=(1-x^2)^{\frac{m}{2}}[l(l+1)-m(m+1)].$$

Hence

$$(1-x^2)y-2xy'+\left(l(l+1)-\frac{m^2}{1-x^2}\right)y$$

$$=(1-x^2)^{\frac{m}{2}+1}u''-2x(m+1)(1-x^2)^{\frac{m}{2}}u'+(1-x^2)^{\frac{m}{2}}[l(l+1)-m(m+1)]u$$

$$=(1-x^2)^{\frac{m}{2}}\{(1-x^2)u''-2x(m+1)u'+[l(l+1)-m(m+1)]u\}.$$

So

$$(1-x^2)y''-2xy'+\left(l(l+1)-\frac{m^2}{1-x^2}\right)y=0$$

if and only if

$$(1-x^2)u'' - 2x(m+1)u' + [l(l+1) - m(m+1)]u = 0. \tag{2}$$

Note that if $m = 0$, then Eq. (2) is Legendre's equation. Recall that Legendre's equation has two linearly independent solutions, and if l is a positive integer, then one of the solutions is a polynomial that we denote $P_l(x)$.

We now demonstrate that $P_l'(x)$ is a solution to Eq. (2) when $m = 1$.

Differentiating Eq. (2) gives

$$\begin{aligned}(1-x^2)u''' &- 2xu'' - 2(m+1)u' - 2x(m+1)u'' + [l(l+1) - m(m+1)]u' \\ &= (1-x^2)u''' - 2x(m+2)xu'' - [l(l+1) - m(m+1)]u' = 0.\end{aligned}$$

This means that if $P_l(x)$ is a solution to Eq. (2) for $m = 0$, that is, if

$$(1-x^2)P_l''(x) - 2xP_l'(x) + l(l+1)P_l(x) = 0$$

then

$$(1-x^2)P_l'''(x) - 2xP_l''(x) + [l(l+1) - 1(1+1)]P_l'(x) = 0.$$

That is, $P_l'(x)$ solves Eq. (2) when $m = 1$. One can follow this procedure to show inductively that $\frac{d^m}{dx^m}P_l(x)$ is a solution to Eq. (2) for any positive integer m. This means

$$y = (1-x^2)^{\frac{m}{2}}\frac{d^m}{dx^m}P_l(x)$$

is a solution to Eq. (1). These solutions are denoted $P_l^m(x)$ and

$$P_l^m(x) = (1-x^2)^{\frac{m}{2}}\frac{d^m}{dx^m}P_l(x)$$

are called associated Legendre functions.

We return to the original problem of solving

$$\cot\theta\,\Theta' + \Theta'' + \left(\mu - \frac{\nu}{\sin^2\theta}\right)\Theta = 0$$

in the case that $\mu = l(l+1)$ and $\nu = m^2$ where l and m are integers with $m \le l$. Thus we want to solve

$$\cot\theta\,\Theta' + \Theta'' + \left(l(l+1) - \frac{m^2}{\sin^2\theta}\right)\Theta = 0. \tag{3}$$

With the substitution $x = \cos\theta$, Eq. (3) became

$$(1-x^2)\frac{d^2\Theta}{dx^2} - 2x\frac{d\Theta}{dx} + \left(l(l+1) - \frac{m^2}{1-x^2}\right)\Theta = 0. \tag{4}$$

We have shown that the solution to Eq. (4) is

$$P_l^m(x) = (1-x^2)^{\frac{m}{2}} \frac{d^m}{dx^m} P_l(x)$$

and so the solution to Eq. (3) is $P_l^m(\cos\theta)$.

EXERCISES

1. Use the formula

$$P_l^m(x) = (1-x^2)^{\frac{m}{2}} \frac{d^m}{dx^m} P_l(x)$$

to compute $P_l^m(\cos\theta)$ for

a. $l = 0, \quad m = 0.$
b. $l = 0, \quad m = 1.$
c. $l = 1, \quad m = 1.$
d. $l = 2, \quad m = 1.$
e. $l = 3, \quad m = 3.$

10.5 LAPLACE'S EQUATION IN SPHERICAL COORDINATES

In this section, we solve the boundary value problem

$$\Delta u(r,\theta,\vartheta) = 0, \quad 0 < r < a; \quad u(a,\theta,\vartheta) = f(\theta,\vartheta)$$

using separation of variables.

The approach is the one we have been using; namely, we hypothesize that

$$u(r,\theta,\vartheta) = R(r)\Theta(\Theta)\Phi(\varphi)$$

and find an ODE of each R, Θ, and Φ. We solve each of the ODEs, and each solution will involve arbitrary constants. We then use the boundary condition $u(a,\theta,\vartheta) = f(\theta,\vartheta)$ to determine the constants.

We have already done most of the preliminary work, but we now repeat it. In Section 10.1, we found that in spherical coordinates

$$\Delta u(r,\theta,\vartheta) = \left(\frac{2}{r}u_r + u_{rr}\right) + \frac{1}{r^2}\left(\cot\theta u_\theta + u_{\theta\theta} + \frac{u_{\vartheta\vartheta}}{\sin^2\theta}\right).$$

Following what we did in Section 9.1, we get

$$\frac{\Delta u}{u} = \frac{2}{r}\frac{R'}{R} + \frac{R''}{R} + \frac{1}{r^2}\left(\cot\theta\frac{\Theta'}{\Theta} + \frac{\Theta''}{\Theta} + \frac{1}{\sin^2\theta}\frac{\Phi''}{\Phi}\right) = 0.$$

We then get

$$\frac{2}{r}\frac{R'}{R}+\frac{R''}{R}=-\frac{1}{r^2}\left(\cot\theta\frac{\Theta'}{\Theta}+\frac{\Theta''}{\Theta}+\frac{1}{\sin^2\theta}\frac{\Phi''}{\Phi}\right)$$

so

$$-r^2\left(\frac{2}{r}\frac{R'}{R}+\frac{R''}{R}\right)=\cot\theta\frac{\Theta'}{\Theta}+\frac{\Theta''}{\Theta}+\frac{1}{\sin^2\theta}\frac{\Phi''}{\Phi}. \tag{1}$$

We repeat the familiar argument that the left-hand side of Eq. (1) is a function of r and the right-hand side is a function of θ and φ, so each is a constant (that will turn out to be negative) that we denote $-\mu$.

We then have

$$\cot\theta\frac{\Theta'}{\Theta}+\frac{\Theta''}{\Theta}+\frac{1}{\sin^2\theta}\frac{\Phi''}{\Phi}=-\mu \tag{2}$$

and

$$r^2\left(\frac{2}{r}\frac{R'}{R}+\frac{R''}{R}\right)=\mu. \tag{3}$$

Multiplying Eq. (2) by $\sin^2\theta$ gives

$$\sin^2\theta\cot\theta\frac{\Theta'}{\Theta}+\sin^2\theta\frac{\Theta''}{\Theta}+\frac{\Phi''}{\Phi}=-\mu\sin^2\theta$$

so

$$\sin^2\theta\cot\theta\frac{\Theta'}{\Theta}+\sin^2\theta\frac{\Theta''}{\Theta}+\mu\sin^2\theta=-\frac{\Phi''}{\Phi}. \tag{4}$$

The left-hand side of Eq. (4) is a function of θ and the right-hand side a φ, so each is a constant that we denote ν. We now show that $\nu>0$. We have

$$-\frac{\Phi''}{\Phi}=\Phi$$

so

$$\Phi+\nu\Phi=0.$$

The periodicity of Φ forces $\nu>0$. We let $m^2=\nu$. Thus we have

$$\Phi''+m^2\Phi=0$$

and so

$$\Phi_m(\varphi)=a_m\cos(m\varphi)+b_m\sin(m\varphi). \tag{5}$$

We also have

$$\sin^2\theta\cot\theta\frac{\Theta'}{\Theta}+\sin^2\theta\frac{\Theta''}{\Theta}+\mu\sin^2\theta=m^2$$

and multiplying by $\Theta/\sin^2\theta$ gives

$$\cot\theta\Theta'+\Theta''+\mu\Theta=\frac{m^2\Theta}{\sin^2\theta}$$

or

$$\Theta''+\cot\theta\Theta'+\left(\mu-\frac{m^2}{\sin^2\theta}\right)\Theta=0. \tag{6}$$

In problems that concern us $\mu=l(l+1)$, so Eq. (6) becomes

$$\Theta''+\cot\theta\Theta'+\left(l(l+1)-\frac{m^2}{\sin^2\theta}\right)\Theta=0. \tag{7}$$

In Section 10.4, we found that the solution to Eq. (7) is $P_l{}^m(\cos\theta)$. Eq. (3) states

$$r^2\left(\frac{2}{r}\frac{R'}{R}+\frac{R''}{R}\right)=\mu$$

so

$$\frac{1}{R}\left(R''+\frac{2}{r}R'\right)=\frac{\mu}{r^2}$$

or

$$R''+\frac{2}{r}R'-\frac{\mu}{r^2}R=0.$$

If $\mu=l(l+1)$, then we have

$$R''+\frac{2}{r}R'-\frac{l(l+1)}{r^2}=0$$

or

$$r^2R''+2rR'-l(l+1)R=0. \tag{8}$$

As we show in Exercise 1, the solution to Eq. (8) is

$$R(r)=\alpha r^l.$$

If we want $R(a)=1$, we take $\alpha=1/a^l$.

Thus

$$\Delta u(r, \theta, \vartheta) = 0, \quad 0 < r < a$$

has the solution

$$\begin{aligned} u(r, \theta, \vartheta) &= \sum_{m,n=0}^{\infty} u_{m,n}(r, \theta, \vartheta) \\ &= \sum_{m,n=0}^{\infty} \Phi_m(\varphi) R(r) P_n{}^m(\cos\theta) \\ &= \sum_{m,n=0}^{\infty} r^l (a_{mn}\cos(m\varphi) + b_{mn}\sin(m\varphi)) P_n{}^m(\cos\theta). \end{aligned}$$

The constants are determined by expanding the initial condition $u(a, \theta, \vartheta) = f(\theta, \vartheta)$ in a Fourier series.

To expand in the Fourier series, we need that

$$\langle P_n{}^m(\cos\theta), P_n{}^m(\cos\theta)\rangle_{\sin\theta} = \int_0^{\pi} [P_n{}^m(\cos\theta)]^2 \sin\theta d\,\theta = \frac{2}{(2n+1)} \frac{(n+m)!}{(n-m)!}$$

so that expanding $f(\theta,\varphi)$ in a Fourier series

$$f(\theta, \varphi) = \sum_{n=0}^{\infty} \frac{1}{2} a_{n0} P_n(\cos\theta) + \sum_{m,n=1}^{\infty} (a_{mn}\cos(m\varphi) + b_{mn}\sin(m\varphi)) P_n{}^m(\cos\theta)$$

where

$$\begin{aligned} a_{mn} &= \frac{2n+1}{2\pi} \frac{(n-m)!}{(n+m)!} \int_0^{2\pi} \int_0^{\pi} f(\theta, \varphi) P_n{}^m(\cos\theta) \cos m\varphi \sin\theta d\,\theta d\varphi \\ b_{mn} &= \frac{2n+1}{2\pi} \frac{(n-m)!}{(n+m)!} \int_0^{2\pi} \int_0^{\pi} f(\theta, \varphi) P_n{}^m(\cos\theta) \sin m\varphi \sin\theta d\,\theta d\varphi. \end{aligned}$$

EXERCISES

1. For $f(\theta,\varphi) = 1$, compute a_{mn} and b_{mn} for m, $n = 0, 1, 2$.

10.6 RIGID ROTOR

In this section and the next two sections, we give some classical applications of some earlier work.

In the rigid rotor problem, we have two bodies separated by a rigid bar with no mass of length r_0. Let

$$m_1 = \text{mass of body } 1$$
$$m_2 = \text{mass of body } 2$$
$$r_1 = \text{distance from the center of mass to body } 1$$
$$r_2 = \text{distance from the center of mass to body } 2.$$

See Fig. 10.6.1.

We have

$$r_1 + r_2 = r_0 \quad m_1 + m_2 = M \quad \mu = \frac{m_1 m_2}{m_1 + m_2} = \text{reduced mass.}$$

The effect of the reduced mass is to reduce a two-body problem to a one-body problem.

$m_1 r_1 = m_2 r_2$ defines coordinates of atoms relative to center of mass

$$r_1 = \frac{\frac{m_1 m_2}{m_1 + m_2} r_0}{m_1} \quad r_2 = \frac{\frac{m_1 m_2}{m_1 + m_2} r_0}{m_2}$$

$$\text{Kinetic Energy } = \frac{1}{2} m_1 {r_1}^2 \omega^2 + \frac{1}{2} m_2 {r_2}^2 \omega^2$$

$$\text{Hamiltonion } H = \text{Kinetic Energy } + \text{Potential Energy} = T + V.$$

For this problem

$$V = \begin{cases} 0 & \text{if } r = r_0 \\ \infty & \text{if } r \neq r_0 \end{cases}$$

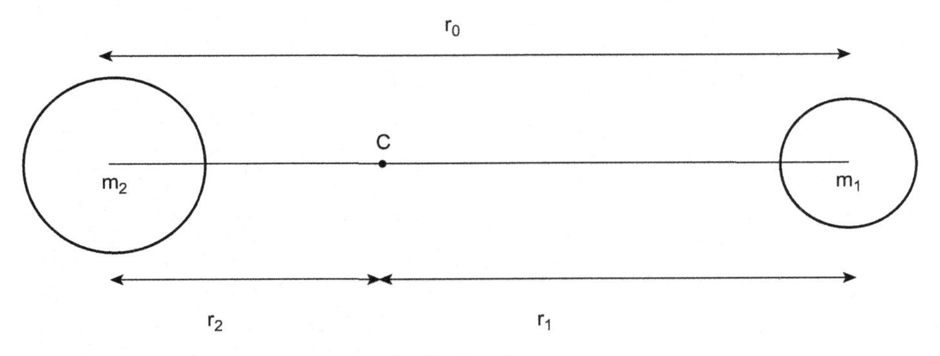

FIGURE 10.6.1

so

$$H = T + V = T = -\frac{h^2}{2\mu}\nabla^2.$$

In spherical coordinates

$$\nabla^2 = \frac{1}{r^2}\frac{\partial}{\partial r}\left(r^2\frac{\partial}{\partial r}\right) + \frac{1}{r^2 \sin\theta}\frac{\partial}{\partial\theta}\left(\sin\theta\frac{\partial}{\partial\theta}\right) + \frac{1}{r^2\sin^2\theta}\frac{\partial^2}{\partial\varphi^2}.$$

In this situation, r has the constant value r_0 so terms involving $\partial/\partial r$ do not appear and other places r is replaced by r_0, so

$$\nabla^2 = \frac{1}{{r_0}^2 \sin\theta}\frac{\partial}{\partial\theta}\left(\sin\theta\frac{\partial}{\partial\theta}\right) + \frac{1}{{r_0}^2\sin^2\theta}\frac{\partial^2}{\partial\varphi^2}.$$

Schrodinger's equation for this model is

$$-\frac{h^2}{2\mu}\nabla^2 Y(\theta,\phi) = EY(\theta,\phi)$$

or

$$-\frac{h^2}{2\mu {r_0}^2}\left[\frac{1}{\sin\theta}\frac{\partial}{\partial\theta}\left(\sin\theta\frac{\partial}{\partial\theta}\right) + \frac{1}{\sin^2\theta}\frac{\partial^2}{\partial\varphi^2}\right]Y(\theta,\phi) = EY(\theta,\phi)$$

which we rearrange to give

$$-\frac{h^2}{2\mu {r_0}^2}\frac{1}{\sin\theta}\frac{\partial}{\partial\theta}\left(\sin\theta\frac{\partial}{\partial\theta}\right)Y(\theta,\phi) - EY(r,\theta,\phi) = \frac{h^2}{2\mu {r_0}^2}\frac{1}{\sin^2\theta}\frac{\partial^2}{\partial\varphi^2}Y(\theta,\phi).$$

Multiplying by

$$\frac{2\mu {r_0}^2}{h^2}$$

gives

$$-\frac{1}{\sin\theta}\frac{\partial}{\partial\theta}\left(\sin\theta\frac{\partial}{\partial\theta}\right)Y(\theta,\phi) - \frac{2\mu {r_0}^2}{h^2}EY(\theta,\phi) = \frac{1}{\sin^2\theta}\frac{\partial^2}{\partial\varphi^2}Y(\theta,\phi).$$

Setting

$$\beta = \frac{2\mu {r_0}^2 E}{h^2}$$

gives

$$-\frac{1}{\sin\theta}\frac{\partial}{\partial\theta}\left(\sin\theta\frac{\partial}{\partial\theta}\right)Y(\theta,\phi) - \beta Y(\theta,\phi) = \frac{1}{\sin^2\theta}\frac{\partial^2}{\partial\varphi^2}Y(\theta,\phi)$$

and multiplying by $-\sin^2\theta$ gives

$$\sin\theta\frac{\partial}{\partial\theta}\left(\sin\theta\frac{\partial}{\partial\theta}\right)Y(\theta,\phi)+\beta\sin^2\theta Y(\theta,\phi)=\frac{\partial^2}{\partial\varphi^2}Y(\theta,\phi).$$

We apply separation of variables to find a solution. Let

$$Y(\theta,\phi)=\Theta(\theta)\Psi(\phi).$$

Then we have

$$\sin\theta\frac{\partial}{\partial\theta}\left(\sin\theta\frac{\partial}{\partial\theta}\right)\Theta(\theta)\Psi(\phi)+\beta\sin^2\theta\Theta(\theta)\Psi(\phi)=-\frac{\partial^2}{\partial\varphi^2}\Theta(\theta)\Psi(\phi)$$

or

$$\Psi(\phi)\sin\theta\frac{\partial}{\partial\theta}\left(\sin\theta\frac{\partial}{\partial\theta}\right)\Theta(\theta)+\beta\sin^2\theta\Theta(\theta)\Psi(\phi)=-\Theta(\theta)\frac{\partial^2}{\partial\varphi^2}\Psi(\phi).$$

Dividing by $\Theta(\theta)\Psi(\phi)$ gives

$$\frac{\sin\theta}{\Theta(\theta)}\frac{\partial}{\partial\theta}\left(\sin\theta\frac{\partial}{\partial\theta}\right)\Theta(\theta)+\beta\sin^2\theta=-\frac{1}{\Psi(\phi)}\frac{\partial^2}{\partial\varphi^2}\Psi(\phi). \tag{1}$$

The left-hand side of Eq. (1) is a function of only θ, and the right-hand side is a function of only ϕ so each side must be a constant that we denote m^2. Then

$$-\frac{1}{\Psi(\phi)}\frac{d^2}{d\varphi^2}\Psi(\phi)=m^2\quad\text{or}\quad\frac{1}{\Psi(\phi)}\frac{d^2}{d\varphi^2}\Psi(\phi)=-m^2.$$

Thus we have

$$\frac{d^2}{d\varphi^2}\Psi(\phi)+m^2\Psi(\phi)=0$$

for which the solution is

$$\Psi(\phi)=A\sin m\phi+B\cos m\phi.$$

The equation,

$$\frac{\sin\theta}{\Theta(\theta)}\frac{d}{d\theta}\left(\sin\theta\frac{d}{d\theta}\right)\Theta(\theta)+\beta\sin^2\theta=m^2$$

as we have seen before (Section 9.3), turns into Legendre's equation. In particular, with the substitution

$$x=\cos\theta,\ \ 1-x^2=\sin^2\theta,\ \ P(x)=\Theta(\theta)$$

the equation

$$\frac{\sin\theta}{\Theta(\theta)}\frac{d}{d\theta}\left(\sin\theta\frac{d}{d\theta}\right)\Theta(\theta)+\beta\sin^2\theta=m^2$$

becomes

$$(1-x^2)P''(x) - 2xP'(x) + \left(\beta - \frac{m^2}{1-x^2}\right)P(x) = 0.$$

It is beyond the scope of our work, but it can be shown that the energy levels are quantized according to

$$E_k = \frac{k(k+1)h^2}{2I}$$

where

$$I = m_1 {r_1}^2 + m_2 {r_2}^2.$$

10.7 ONE DIMENSION QUANTUM MECHANICAL OSCILLATOR

In going from classical mechanics to quantum mechanics, position x and momentum p are replaced by operators $\widehat{x}$ and $\widehat{p}$, where $\widehat{x}$ and $\widehat{p}$ operate on functions according to

$$\widehat{x}\psi(x) = x\psi(x) \text{ and } \widehat{p}\psi(x) = -ih\frac{\partial \psi(x)}{\partial x}$$

where $\psi(x)$ satisfies

$$\int_{-\infty}^{\infty} |\psi(x)|^2 dx < \infty.$$

In the classical case

$$\textit{Total energy} = \textit{kinetic energy} + \textit{potential energy}$$

and for the classical harmonic oscillator

$$\textit{kinetic energy} = \frac{1}{2m}p^2$$

and

$$\textit{potential energy} = \frac{1}{2}\, m\omega^2 x^2.$$

Replacing p by $\widehat{p}$ and x by $\widehat{x}$, we get that in the quantum mechanical case, the Hamiltonian H is given by

$$H = -\frac{h^2}{2m}\frac{d^2}{dx^2} + \frac{1}{2}\, m\omega^2 x^2.$$

In this case, the time-independent Schrodinger's equation is

$$-\frac{h^2}{2m}\frac{d^2\psi}{dx^2}+\frac{1}{2}m\omega^2x^2\psi=E\psi$$

where E is a real number that is the time-independent energy level. It will turn out that there are only certain acceptable values for E, and each acceptable value gives rise to an eigenvalue problem.

The acceptable values of the energy will be shown to be

$$E_n=(2n+1)\frac{h\omega}{2}, n=0,1,2,\ \ldots$$

and the solutions will be given by

$$\psi_n=\frac{1}{\sqrt{n!2^n}}\left(\frac{m\omega}{\pi h}\right)^{1/4}e^{-(m\omega^2/\pi h)}H_n\left(\sqrt{\frac{m\omega}{h}}x\right)$$

where H_n is the nth order Hermite polynomial that we will define.

Temporarily replacing ψ by u in the equation

$$-\frac{h^2}{2m}\frac{d^2\psi}{dx^2}+\frac{1}{2}m\omega^2x^2\psi=E\psi$$

and multiplying by

$$-\frac{2m}{h^2}$$

we get

$$\frac{d^2u}{dx^2}-\frac{2m}{h^2}\left(\frac{1}{2}m\omega^2x^2u\right)=-\frac{2m}{h^2}Eu$$

so

$$\frac{d^2u}{dx^2}-\frac{2m}{h^2}\left(\frac{1}{2}m\omega^2x^2u\right)+\frac{2m}{h^2}Eu=0$$

or

$$\frac{d^2u}{dx^2}+\frac{2m}{h^2}\left(E-\frac{1}{2}m\omega^2x^2\right)u=0.$$

Let

$$y=\sqrt{\frac{m\omega}{h}}x$$

then

$$x^2=\frac{h}{m\omega}y^2$$

so

$$\frac{du}{dx} = \frac{du}{dy}\frac{dy}{dx} = \frac{du}{dy}\sqrt{\frac{m\omega}{h}}$$

$$\frac{d^2u}{dx^2} = \frac{d}{dx}\left(\frac{du}{dx}\right) = \frac{d\left(\frac{du}{dx}\right)}{dy}\frac{dy}{dx} = \frac{d^2u}{dy^2}\sqrt{\frac{m\omega}{h}}\sqrt{\frac{m\omega}{h}} = \frac{m\omega}{h}\frac{d^2u}{dy^2}.$$

Thus

$$\frac{d^2u}{dx^2} + \frac{2m}{h^2}\left(E - \frac{1}{2}m\omega^2x^2\right)u = \frac{m\omega}{h}\frac{d^2u}{dy^2} + \frac{2m}{h^2}\left(E - \frac{1}{2}m\omega^2\frac{h}{m\omega}y^2\right)$$

$$= \frac{m\omega}{h}\frac{d^2u}{dy^2} + \frac{2m}{h^2}\left(E - \frac{1}{2}\omega h y^2\right) = 0.$$

Then

$$\frac{h}{m\omega}\left[\frac{m\omega}{h}\frac{d^2u}{dy^2} + \frac{2m}{h^2}\left(E - \frac{1}{2}\omega h y^2\right)u\right] = \frac{d^2u}{dy^2} + \frac{h}{m\omega}\frac{2m}{h^2}\left(E - \frac{1}{2}\omega h y^2\right)u$$

$$= \frac{d^2u}{dy^2} + \frac{2}{h\omega}\left(E - \frac{1}{2}\omega h y^2\right)u = \frac{d^2u}{dy^2} + \left(\frac{2E}{h\omega} - y^2\right)u = 0.$$

Letting

$$\epsilon = \frac{2E}{h\omega}$$

the equation is

$$\frac{d^2u}{dy^2} + (\epsilon - y^2)u = 0. \tag{1}$$

We solve Eq. (1) and find the allowable values of ϵ.

We first note that as $y \to \pm\infty$ the term $\epsilon - y^2$ is dominated by y^2. We thus make the approximation that for y large

$$\frac{d^2u}{dy^2} + (\epsilon - y^2)u \approx \frac{d^2u}{dy^2} - y^2u$$

and we can solve

$$\frac{d^2u}{dy^2} - y^2u = 0$$

to get

$$u(y) = Ae^{-\frac{y^2}{2}} + Be^{\frac{y^2}{2}}.$$

The term

$$Be^{\frac{y^2}{2}} \to \infty \text{ as } y \to \pm\infty$$

which precludes the function being normalizable. Thus $B = 0$ and for y large, the solution is approximated by

$$u(y) = Ae^{-\frac{y^2}{2}}.$$

For smaller values of y, we conjecture that the solution is of the form

$$u(y) = h(y)e^{-\frac{y^2}{2}}.$$

We want to find conditions that will determine $h(y)$.
We have

$$u'(y) = h(y)e^{-\frac{y^2}{2}}(-y) + h'(y)e^{-\frac{y^2}{2}} = e^{-\frac{y^2}{2}}[h'(y) - yh(y)]$$

$$\begin{aligned} u''(y) &= e^{-\frac{y^2}{2}}(-y)[h'(y) - yh(y)] + e^{-\frac{y^2}{2}}[h''(y) - h(y) - yh(y)] \\ &= e^{-\frac{y^2}{2}}\left[-yh'(y) + y^2h(y) + h''(y) - h(y) - yh'(y)\right] \\ &= e^{-\frac{y^2}{2}}\left[y^2h(y) - 2yh'(y) + h''(y)\right]. \end{aligned}$$

Then

$$\begin{aligned} \frac{d^2u}{dy^2} + (\epsilon - y^2)u &= e^{-\frac{y^2}{2}}\left[h''(y) - 2yh'(y) + h(y)\left(y^2 - 1 + \epsilon - y^2\right)\right] \\ &= e^{-\frac{y^2}{2}}[h''(y) - 2yh'(y) + (\epsilon - 1)h(y)] = 0. \end{aligned}$$

So we must solve

$$h''(y) - 2yh'(y) + (\epsilon - 1)h(y) = 0. \quad (2)$$

We will find $h(y)$ using a power series expansion.
If

$$h(y) = \sum_{n=0}^{\infty} a_n y^n$$

then

$$h'(y) = \sum_{n=0}^{\infty} a_n n y^{n-1} \text{ and } h''(y) = \sum_{n=0}^{\infty} a_n n(n-1)y^{n-2}$$

so

$$h''(y) - 2yh'(y) + (\epsilon - 1)h(y) = \sum_{n=0}^{\infty} a_n n(n-1)y^{n-2} - 2\sum_{n=0}^{\infty} a_n n y^n + (\epsilon - 1)\sum_{n=0}^{\infty} a_n y^n.$$

Now

$$\sum_{n=0}^{\infty} a_n n(n-1)y^{n-2} = \sum_{n=2}^{\infty} a_n n(n-1)y^{n-2} = \sum_{n=0}^{\infty} a_{n+2}(n+2)(n+1)y^n$$

so

$$\begin{aligned}\sum_{n=0}^{\infty} a_n n(n-1)y^{n-2} - 2\sum_{n=0}^{\infty} a_n n y^n + (\epsilon - 1)\sum_{n=0}^{\infty} a_n y^n \\ = \sum_{n=0}^{\infty} a_{n+2}(n+2)(n+1)y^n - 2\sum_{n=0}^{\infty} a_n n y^n + (\epsilon - 1)\sum_{n=0}^{\infty} a_n y^n \\ = \sum_{n=0}^{\infty} [a_{n+2}(n+2)(n+1) - 2na_n + (\epsilon - 1)a_n]y^n = 0.\end{aligned}$$

Then

$$a_{n+2}(n+2)(n+1) - 2n\, a_n + (\epsilon - 1)a_n = 0$$

or

$$a_{n+2} = \frac{1 - \epsilon + 2n}{(n+2)(n+1)} a_n.$$

We have

$$\lim_{n \to \infty} \frac{1 - \epsilon + 2n}{(n+2)(n+1)} = \frac{2}{n}.$$

So for large n, we have

$$a_{n+2} \approx \frac{2}{n}\, a_n.$$

For n even, we claim a solution to

$$a_{n+2} = \frac{2}{n}a_n$$

is

$$a_n = \frac{C}{\left[\left(\frac{n}{2}\right) - 1\right]!}.$$

If this is the case, then

$$a_{n+2} = \frac{C}{\left[\left(\frac{n+2}{2}\right) - 1\right]!} = \frac{C}{\left(\frac{n}{2}\right)!}.$$

Thus

$$h(y) \approx C\sum \frac{y^n}{\left[\left(\frac{n}{2}\right) - 1\right]!} \approx C\sum \frac{y^{2n}}{(n-1)!} = C\sum \frac{y^2 y^{2(n-1)}}{(n-1)!} = Cy^2 \sum \frac{\left(y^2\right)^{n-1}}{(n-1)!}$$
$$= Cy^2 e^{y^2}.$$

But then

$$u(y) = h(y)e^{-\frac{y^2}{2}} \approx e^{\frac{y^2}{2}}$$

for large values of y. This means the function cannot be normalized. The only possibility to have an acceptable solution is if the series terminates; i.e., there is a value of n for which $a_n = 0$.

The equation

$$a_{n+2} = \frac{1 - \epsilon + 2n}{(n+2)(n+1)}a_n$$

will terminate if and only if

$$\epsilon = 2n + 1$$

for some nonnegative integer n.

We seek functions $y_n(x) = H_n(x)$ that satisfy

$$y'' - 2xy' + 2ny = 0. \tag{3}$$

Eq. (3) is called the Hermite equation of order n, and $H_n(x)$ is called the Hermite polynomial of degree n. $H_n(x)$ is the nth degree polynomial where the coefficient of x^n is 2^n.

Example:
We compute $H_4(x)$.
Set

$$y = H_4(x) = a + bx + cx^2 + dx^3 + 2^4x^4$$

and we want $H_4(x)$ to solve

$$y'' - 2xy' + 2(4)y = y'' - 2xy' + 8y = 0.$$

Now

$$y' = b + 2cx + 3dx^2 + 4(2^4x^3)$$

so

$$2xy' = 2bx + 4cx^2 + 6dx^3 + 8(2^4x^4).$$

Also

$$y'' = 2c + 6dx + 12(2^4x^2)$$

so

$$\begin{aligned} y'' - 2xy' + 8y \\ &= [2c + 6dx + 12(2^4x^2)] - [2bx + 4cx^2 + 6dx^3 + 8(2^4x^4)] \\ &\quad + 8[a + bx + cx^2 + dx^3 + 2^4x^4] \\ &= (2c + 8a) + (6d - 2b + 8b)x + [12((2^4) - 4c + 8c]x^2 \\ &\quad + (-6d + 8d)x^3 + [-8(2^4) + 8(2^4)]x^4 = 0. \end{aligned}$$

So we must have

$$2c + 8a = 0 \tag{4}$$

$$6d - 2b + 8b = 0 \tag{5}$$

$$12(2^4) - 4c + 8c = 0 \tag{6}$$

$$-6d + 8d = 0. \tag{7}$$

Eq. (7) gives $d = 0$, so by Eq. (5), $b = 0$

Eq. (6) says $4c = -12(2^4)$ so $c = -48$
and from Eq. (4) we get $a = -\frac{1}{4}c = 12$.

Thus

$$H_4(x) = 12 - 48x^2 + 2^4x^4 = 16x^4 - 48x^2 + 12.$$

The first six Hermite polynomials are

$$\begin{aligned} H_0(x) &= 1 & H_1(x) &= 2x \\ H_2(x) &= 4x^2 - 2 & H_3(x) &= 8x^3 - 12x \\ H_4(x) &= 16x^4 - 48x^2 + 12 & H_5(x) &= 32x^5 - 160x^3 + 120x. \end{aligned}$$

We have found the solutions to

$$-\frac{h^2}{2m}\frac{d^2}{dx^2}\psi_n + \frac{1}{2}m\omega^2 x^2\psi_n = \frac{h\omega}{2}(2n+1)\psi_n$$

are

$$\psi_n(x) = N_n H_n(x)e^{-x^2/2}$$

where N_n is a constant chosen so that

$$\int_{-\infty}^{\infty} |\psi_n(x)|^2 dx = 1.$$

In fact,

$$\psi_n(x) = \left(\frac{m\omega}{\pi h}\right)^{1/4} \frac{1}{\sqrt{2^n n!}} H_n(x)e^{-x^2/2}.$$

Looking back at what we have done, we started with the energy of the classical harmonic oscillator

$$E = \frac{1}{2m}p^2 + \frac{1}{2}m\omega^2 x^2$$

and converted to the quantum mechanical case to get the Hamiltonian

$$H = -\frac{h^2}{2m}\frac{d^2}{dx^2} + \frac{1}{2}m\omega^2 x^2.$$

In this case, the time-independent Schrodinger's equation is

$$-\frac{h^2}{2m}\frac{d^2\Psi}{dx^2} + \frac{1}{2}m\omega^2 x^2\Psi = E\Psi$$

where E is a real number that is the time-independent energy level. It turned out that there are only certain acceptable values for E and each acceptable value gives rise to an eigenvalue problem.

The acceptable values of the energy are

$$E_n = (2n+1)\frac{h\omega}{2}, n = 0, 1, 2, \ldots$$

and the solutions are given by

$$\psi_n = \frac{1}{\sqrt{n!2^n}}\left(\frac{m\omega}{\pi h}\right)^{1/4} e^{-(m\omega^2/\pi h)} H_n\left(\sqrt{\frac{m\omega}{h}}x\right)$$

where H_n is the nth-order Hermite polynomial.

10.8 THE HYDROGEN ATOM

Schrodinger's equation for the Hydrogen atom is

$$-\frac{h^2}{2\mu}\left[\frac{1}{r^2}\frac{\partial}{\partial r}\left(r^2\frac{\partial\psi}{\partial r}\right)+\frac{1}{r^2\sin\theta}\frac{\partial}{\partial\theta}\left(\sin\theta\frac{\partial\psi}{\partial\theta}\right)+\frac{1}{r^2\sin^2\theta}\frac{\partial^2\psi}{\partial\phi^2}\right]-\frac{Ze^2}{4\pi\epsilon_0 r}\psi=E\psi.$$

Multiply by $-\frac{2\mu}{h^2}$ to get

$$\frac{1}{r^2}\frac{\partial}{\partial r}\left(r^2\frac{\partial\psi}{\partial r}\right)+\frac{1}{r^2\sin\theta}\frac{\partial}{\partial\theta}\left(\sin\theta\frac{\partial\psi}{\partial\theta}\right)+\frac{1}{r^2\sin^2\theta}\frac{\partial^2\psi}{\partial\phi^2}+\frac{2\mu}{h^2}\left(\frac{Ze^2}{4\pi\epsilon_0 r}+E\right)\psi=0.$$

Let

$$\psi(r,\theta,\phi)=R(r)Y(\theta,\phi).$$

Then

$$\frac{\partial\psi}{\partial r}=\frac{dR}{dr}Y,\ \frac{\partial\psi}{\partial\theta}=R\frac{\partial Y}{\partial\theta},\ \frac{\partial^2\psi}{\partial\phi^2}=R\frac{\partial^2 Y}{\partial\phi^2}.$$

So the equation becomes

$$\frac{1}{r^2}\frac{\partial}{\partial r}\left(r^2\frac{dR}{dr}Y\right)+\frac{1}{r^2\sin\theta}\frac{\partial}{\partial\theta}\left(\sin\theta R\frac{\partial Y}{\partial\theta}\right)+\frac{1}{r^2\sin^2\theta}R\frac{\partial^2 Y}{\partial\phi^2}$$
$$+\frac{2\mu}{h^2}\left(\frac{Ze^2}{4\pi\epsilon_0 r}+E\right)RY=0.$$

Multiply by $\frac{r^2}{RY}$ to get

$$\frac{1}{R}\frac{\partial}{\partial r}\left(r^2\frac{dR}{dr}\right)+\frac{1}{Y}\frac{1}{\sin\theta}\frac{\partial}{\partial\theta}\left(\sin\theta\frac{\partial Y}{\partial\theta}\right)+\frac{1}{Y}\frac{1}{\sin^2\theta}\frac{\partial^2 Y}{\partial\phi^2}+\frac{2\mu r^2}{h^2}\left(\frac{Ze^2}{4\pi\epsilon_0 r}+E\right)=0.$$

Thus

$$\frac{1}{R}\frac{\partial}{\partial r}\left(r^2\frac{dR}{dr}\right)+\frac{2\mu r^2}{h^2}\left(\frac{Ze^2}{4\pi\epsilon_0 r}+E\right)=-\left[\frac{1}{Y}\frac{1}{\sin\theta}\frac{\partial}{\partial\theta}\left(\sin\theta\frac{\partial Y}{\partial\theta}\right)+\frac{1}{Y}\frac{1}{\sin^2\theta}\frac{\partial^2 Y}{\partial\phi^2}\right]. \tag{1}$$

The expression on the left-hand side of (1) is a function only of r, and the expression on the right-hand side of (1) is a function only of θ and ϕ, so each must be a constant.

Let

$$\frac{1}{R}\frac{\partial}{\partial r}\left(r^2\frac{dR}{dr}\right)+\frac{2\mu r^2}{h^2}\left(\frac{Ze^2}{4\pi\epsilon_0 r}+E\right)=A.$$

Then

$$\frac{\partial}{\partial r}\left(r^2\frac{dR}{dr}\right)+\frac{2\mu r^2}{h^2}\left(\frac{Ze^2}{4\pi\epsilon_0 r}+E\right)R=AR$$

so

$$\frac{\partial}{\partial r}\left(r^2\frac{dR}{dr}\right)+\frac{2\mu r^2}{h^2}\left(\frac{Ze^2}{4\pi\epsilon_0 r}+E\right)R-AR=0. \quad (2)$$

Since

$$A=\frac{1}{R}\frac{\partial}{\partial r}\left(r^2\frac{dR}{dr}\right)+\frac{2\mu r^2}{h^2}\left(\frac{Ze^2}{4\pi\epsilon_0 r}+E\right)$$
$$=-\left[\frac{1}{Y}\frac{1}{\sin\theta}\frac{\partial}{\partial\theta}\left(\sin\theta\frac{\partial Y}{\partial\theta}\right)+\frac{1}{Y}\frac{1}{\sin^2\theta}\frac{\partial^2 Y}{\partial\phi^2}\right]$$

then

$$\frac{1}{Y}\frac{1}{\sin\theta}\frac{\partial}{\partial\theta}\left(\sin\theta\frac{\partial Y}{\partial\theta}\right)+\frac{1}{Y}\frac{1}{\sin^2\theta}\frac{\partial^2 Y}{\partial\phi^2}=-A$$

so

$$\frac{1}{\sin\theta}\frac{\partial}{\partial\theta}\left(\sin\theta\frac{\partial Y}{\partial\theta}\right)+\frac{1}{\sin^2\theta}\frac{\partial^2 Y}{\partial\phi^2}=-AY$$

or

$$\frac{1}{\sin\theta}\frac{\partial}{\partial\theta}\left(\sin\theta\frac{\partial Y}{\partial\theta}\right)+\frac{1}{\sin^2\theta}\frac{\partial^2 Y}{\partial\phi^2}+AY=0. \quad (3)$$

Let

$$Y(\theta,\phi)=\Theta(\theta)\Phi(\phi).$$

Then

$$\frac{\partial Y}{\partial\theta}=\Phi\frac{d\Theta}{d\theta} \text{ and } \frac{\partial^2 Y}{\partial\phi^2}=\Theta\frac{d^2\Phi}{d\phi^2}$$

so Eq. (3) is

$$\frac{1}{\sin\theta}\frac{d}{d\theta}\left(\sin\theta\Phi\frac{d\Theta}{d\theta}\right)+\frac{1}{\sin^2\theta}\Theta\frac{d^2\Phi}{d\phi^2}+A\Theta\Phi=0.$$

Dividing by $\Theta\Phi$ gives

$$\frac{1}{\sin\theta}\frac{1}{\Theta}\frac{d}{d\theta}\left(\sin\theta\frac{d\Theta}{d\theta}\right)+\frac{1}{\Phi}\frac{1}{\sin^2\theta}\frac{d^2\Phi}{d\phi^2}+A=0.$$

Multiply by $\sin^2\theta$ to get

$$\sin\theta\frac{1}{\Theta}\frac{d}{d\theta}\left(\sin\theta\frac{d\Theta}{d\theta}\right)+\frac{1}{\Phi}\frac{d^2\Phi}{d\phi^2}+A\sin^2\theta=0.$$

So

$$\sin\theta\frac{1}{\Theta}\frac{d}{d\theta}\left(\sin\theta\frac{d\Theta}{d\theta}\right)+A\sin^2\theta=-\frac{1}{\Phi}\frac{d^2\Phi}{d\phi^2}. \tag{4}$$

The expression on the left-hand side of (4) is a function only of θ, and the expression on the right-hand side of (4) is a function only of ϕ, so each must be a constant.

Let

$$\sin\theta\frac{1}{\Theta}\frac{d}{d\theta}\left(\sin\theta\frac{d\Theta}{d\theta}\right)+A\sin^2\theta=B$$

or

$$\sin\theta\frac{1}{\Theta}\frac{d}{d\theta}\left(\sin\theta\frac{d\Theta}{d\theta}\right)+A\sin^2\theta-B=0 \tag{5}$$

then

$$\frac{1}{\Phi}\frac{d^2\Phi}{d\phi^2}=-B$$

or

$$\frac{d^2\Phi}{d\phi^2}+B\Phi=0. \tag{6}$$

In Eq. (6), if we set $B=m^2$, then we have

$$\frac{d^2\Phi}{d\phi^2}+m^2\Phi=0. \tag{7}$$

One can express the solution to Eq. (7) as

$$\Phi(\phi)=c_1\cos(m\phi)+c_2\sin(m\phi)\ \text{ or }\ d_1e^{im\phi}+d_2e^{-im\phi}.$$

The first expression perhaps makes it more transparent that m must be an integer because of the continuity condition that $\Phi(0)=\Phi(2\pi)$. Because one is free to choose the "baseline" angle as one chooses, it is possible (and convenient) to express the solution as

$$\Phi(\phi)=d_1e^{im\phi}$$

which is the common choice in the literature.

We now consider Eq. (5) with $B = m^2$:

$$\sin\theta \frac{1}{\Theta}\frac{d}{d\theta}\left(\sin\theta\frac{d\Theta}{d\theta}\right) + A\sin^2\theta - B = \sin\theta\frac{d}{d\theta}\left(\sin\theta\frac{d\Theta}{d\theta}\right) + (A\sin^2\theta - m^2)\Theta$$

$$= \frac{1}{\sin\theta}\frac{d}{d\theta}\left(\sin\theta\frac{d\Theta}{d\theta}\right) + \left(A - \frac{m^2}{\sin^2\theta}\right) = 0.$$

This is similar to Legendre's equation. If we make the substitution

$$x = \cos\theta, P(\cos\theta) = \Theta(\theta)$$

then

$$\frac{d}{d\theta} = \frac{dx}{d\theta}\frac{d}{dx} = -\sin\theta\frac{d}{dx}$$

so

$$\frac{1}{\sin\theta}\frac{d}{d\theta}\left(\sin\theta\frac{d\Theta}{d\theta}\right) + \left(A - \frac{m^2}{\sin^2\theta}\right)$$

$$= \frac{1}{\sin\theta}\left(- \sin\theta\frac{d}{dx}\left(\sin\theta(-\sin\theta)\frac{dP}{dx}\right)\right) + \left(A - \frac{m^2}{\sin^2\theta}\right)$$

$$= \frac{d}{dx}\left[(1-x^2)\frac{dP}{dx}\right] + \left(A - \frac{m^2}{(1-x^2)}\right)$$

$$= (1-x^2)\frac{d^2P}{dx^2} - 2x\frac{dP}{dx} + \left(A - \frac{m^2}{(1-x^2)}\right) = 0.$$

The equation

$$(1-x^2)\frac{d^2P}{dx^2} - 2x\frac{dP}{dx} + \left(A - \frac{m^2}{(1-x^2)}\right) = 0$$

is similar to Legendre's equation; in fact, it is an associated Legendre's equation which is discussed in Section 10.4. The solutions are associated Legendre polynomials which are complicated. There are two solutions and these involve power series of the form

$$\sum_{n=0}^{\infty} a_{2n}x^{2n} \text{ and } \sum_{n=0}^{\infty} a_{2n+1}x^{2n+1}$$

where the coefficients satisfy

$$a_{n+2} = \frac{(n+m)(n+m+1) - A}{(n+1)(n+2)}a_n.$$

where m is fixed. For n large

$$\frac{(n+m)(n+m+1)-A}{(n+1)(n+2)} \approx \frac{(n+m)(n+m+1)}{(n+1)(n+2)} \approx 1$$

so the series converge only if there are finitely many nonzero terms. This forces

$$A = l(l+1)$$

for some positive integer l.

Associated Legendre polynomials are related to Legendre polynomials according to

$$P_l{}^m(x) = (-1)^m (1-x^2)^{\frac{m}{2}} \frac{d^m}{dx^m} P_l(x)$$

where $P_l(x)$ is the Legendre polynomial of order l.

We now solve Eq. (2) with $A = l(l+1)$. We have

$$\frac{\partial}{\partial r}\left(r^2 \frac{dR}{dr}\right) + \frac{2\mu}{h^2}\left(\frac{Ze^2}{4\pi\epsilon_0 r} + E\right)R - l(l+1)R = 0. \tag{2'}$$

Now

$$\frac{\partial}{\partial r}\left(r^2 \frac{dR}{dr}\right) = r^2 \frac{d^2R}{dr^2} + 2r\frac{dR}{dr}.$$

So Eq. (2′) can be written

$$\begin{aligned} r^2 \frac{d^2R}{dr^2} + 2r\frac{dR}{dr} + \frac{2\mu r^2}{h^2}\left(\frac{Ze^2}{4\pi\epsilon_0 r} + E\right)R - l(l+1)R \\ = r^2 \frac{d^2R}{dr^2} + 2r\frac{dR}{dr} + \frac{2\mu r^2}{h^2}\left(\frac{Ze^2}{4\pi\epsilon_0 r} + E\right)R - l(l+1)R = 0. \end{aligned}$$

Dividing by r^2 gives

$$\frac{d^2R}{dr^2} + \frac{2}{r}\frac{dR}{dr} + \frac{2\mu}{h^2}\left(\frac{Ze^2}{4\pi\epsilon_0 r} + E\right)R - \frac{l(l+1)}{r^2} = 0. \tag{8}$$

We investigate what happens as $r \to \infty$ in Eq. (8). In that case, the equation reduces to

$$\frac{d^2R}{dr^2} + \frac{2\mu E}{h^2} = 0. \tag{9}$$

The solution to Eq. (9) can be expressed either as sine and cosine functions or as exponential functions. In this case, the exponential form is more useful. So we have

$$R^*(r) = c_1 \exp\left(i\sqrt{\frac{2\mu E}{h^2}}r\right) + c_2 \exp\left(-i\sqrt{\frac{2\mu E}{h^2}}r\right).$$

Now $E < 0$, so we have

$$R^*(r) = c_1 \exp\left(-\sqrt{\frac{-2\mu E}{h^2}}r\right) + c_2 \exp\left(\sqrt{\frac{-2\mu E}{h^2}}r\right).$$

Since we want $R^*(r) \to 0$ as $r \to \infty$, we have

$$R^*(r) = c_1 \exp\left(-\sqrt{\frac{-2\mu E}{h^2}}r\right).$$

We now write $R(r)$ as the product of its asymptotic part and a part that reflects what happens for $r \approx 0$. We call this second part R^{**} so that $R = R^*R^{**}$. We solve for R^{**} as a power series so that

$$R = R^* R^{**} = c_1 \exp\left(-\sqrt{\frac{-2\mu E}{h^2}}r\right)\sum_{n=0}^{\infty} a_n r^n.$$

It is beyond the scope of the text to determine the a_n's.

CHAPTER

The Fourier Transform

11

11.1 INTRODUCTION

Note: The article "Fourier Transform" by Terrence Tao that appears in *The Princeton Companion to Mathematics* is a short, readable, and outstanding overview of the Fourier transform.

As Professor Tao notes, the Fourier transform is a technique that decomposes an arbitrary function into a superposition of functions that are, in some sense, symmetric. We have seen that periodic functions of a fairly general nature can be expressed as a superposition of sine and cosine functions by Fourier series. Thus we have decomposed a function into sine and cosine functions. The coefficients of these sine and cosine functions describe the extent to which various frequencies are present in the original function.

This choice of base or component functions is particularly advantageous because they are eigenfunctions of the prominent operator d_2/dx_2.

In this chapter we define the Fourier transform, describe how it decomposes certain functions, how it can be constructed as the limit of Fourier series as the period goes to infinity, and how it can be used to solve partial differential equations (PDEs).

The Fourier transform is perhaps the most widely used of the integral transforms. If $f : \mathbb{R} \rightarrow \mathbb{C}$ is a function for which $\int_{-\infty}^{\infty} |f(x)| dx < \infty$, then the Fourier transform of $f(x)$ is defined by

$$\mathcal{F}(f)(\xi) = \widehat{f}(\xi) = \frac{1}{\sqrt{2\pi}} \int_{-\infty}^{\infty} f(x) e^{-ix\xi} d\xi.$$

This definition is not universal. Variations have a multiplication or division factor of 2π and 2π may appear in the exponent of e. Also, the letter k is often used instead of the letter ξ.

The Fourier transform is closely related to Fourier series in its construction and interpretation. We shall see that the Fourier transform also decomposes a function according to frequencies, and $\widehat{f}(\xi)$ describes the extent to which the frequency ξ is present in $f(x)$.

Mathematical Physics with Partial Differential Equations. https://doi.org/10.1016/B978-0-12-814759-7.00011-9

11.2 THE FOURIER TRANSFORM AS A DECOMPOSITION

The most important application of the Fourier transform for us will be in solving PDEs. In the context of PDEs and mathematical physics, the most advantageous choice of the component functions is *plane waves*. A plane wave is a function $f: \mathbb{R}^n \to \mathbb{C}$ of the form

$$f(x) = c(\xi)e^{ix\cdot\xi}.$$

The physical interpretation of the variables is

$$x = \text{position}, \quad \xi = \text{frequency}, \quad |c(\xi)| = \text{amplitude}$$

of the plane wave.

We need to know what functions can be represented as the superposition of plane waves. We do not give a complete answer to this question, but Gaussian functions (functions of the form $f(x) = ae^{-bx^2}$) and other infinitely differentiable functions that "decrease rapidly" are included. This class of functions is sufficiently rich for our needs. With Fourier series, the superposition of the basic functions was expressed as a sum, but in the setting now under consideration, one must use an integral. Under some fairly general conditions, it is possible to take the inverse of the Fourier transform, which is given by

$$\mathcal{F}^{-1}(f(x)) = \frac{1}{\sqrt{2\pi}} \int_{\mathbb{R}^n} \widehat{f}(\xi)e^{ix\cdot\xi}d\xi.$$

where $\widehat{f}(\xi)$ is determined according to

$$\widehat{f}(\xi) = \frac{1}{\sqrt{2\pi}} \int_{\mathbb{R}^n} f(x)e^{-ix\cdot\xi}dx. \tag{1}$$

As we stated in the introduction, the function $\widehat{f}(\xi)$ given by Eq. (1) is called the Fourier transform of $f(x)$.

The physical interpretation of $\widehat{f}(\xi)$ is the prominence in $f(x)$ of a component that oscillates with frequency ξ.

One reason plane waves are the optimal choice for the component functions in the setting mentioned is that plane waves are eigenfunctions of the Laplacian. In particular,

$$\Delta\, c(\xi)e^{ix\cdot\xi} = -|\xi|^2 c(\xi)e^{2\pi ix\cdot\xi}.$$

We now describe how the Fourier transform allows us to view the Laplacian as a multiplication operator in Fourier space.

We express the Fourier representation of $\Delta f(x)$ in two ways. Since

$$f(x) = \frac{1}{\sqrt{2\pi}} \int_{-\infty}^{\infty} \widehat{f}(\xi)e^{ix\xi}d\xi$$

then

$$(\Delta f)(x) = \frac{1}{\sqrt{2\pi}} \int_{-\infty}^{\infty} \left(\widehat{\Delta f}\right)(\xi) e^{ix\xi}\, d\xi.$$

But we also have

$$(\Delta f)(x) = \Delta\left(\frac{1}{\sqrt{2\pi}} \int_{-\infty}^{\infty} \widehat{f}(\xi) e^{ix\xi} d\xi\right) = \frac{1}{\sqrt{2\pi}} \int_{-\infty}^{\infty} \Delta\left(\widehat{f}(\xi) e^{ix\xi}\right) d\xi$$

$$= \frac{1}{\sqrt{2\pi}} \int_{-\infty}^{\infty} \left(\widehat{f}(\xi) \Delta e^{ix\xi}\right) d\xi = \frac{1}{\sqrt{2\pi}} \int_{-\infty}^{\infty} \widehat{f}(\xi)(|\xi|^{\,2}) e^{ix\xi} d\xi.$$

Since $(\Delta f)(x)$ has a unique representation in terms of plane waves, we have $\widehat{\Delta f}(\xi) = -4\pi^2|\xi|^{\,2}\widehat{f}(\xi)$. One can continue this process to show

$$\widehat{\Delta^n f}(\xi) = (|\xi|^{\,2})^n \widehat{f}(\xi).$$

From this, one can define functions of Δ when the function is defined as a power series. For example, since

$$e^x = 1 + x + \frac{x^2}{2!} + \frac{x^3}{3!} + \ldots$$

one can define

$$\widehat{e^{\Delta} f}(\xi) = e^{-\xi^2} \widehat{f}(\xi).$$

One way of describing what we have demonstrated above is that in Fourier space, the Fourier transform diagonalizes the Laplacian.

11.3 THE FOURIER TRANSFORM FROM FOURIER SERIES

In this section we derive the formula for the Fourier transform of a function and show how is comes from Fourier series. The idea is to consider what happens if we begin with a periodic function and let the period go to infinity. Our arguments will freely interchange infinite sums and integrals and will not be rigorous, but the steps are legitimate if $f(x)$ is sufficiently well behaved. A more rigorous statement of the main result of this section is given in the next section.

Suppose $f(x)$ is periodic on $[-L, L]$ and is equal to its Fourier series. We use the exponential form of the Fourier series, so that

$$f(x) = \sum_{n=-\infty}^{\infty} c_n \exp\left(-\pi^2 k^2/a\right)$$

with

$$c_n = \frac{1}{2L}\int_{-L}^{L} f(x)e^{\frac{-in\pi x}{L}}dx.$$

Then

$$f(x) = \sum_{n=-\infty}^{\infty} c_n e^{\frac{in\pi x}{L}} = \sum_{n=-\infty}^{\infty}\left(\frac{1}{2L}\int_{-L}^{L} f(x)e^{\frac{-in\pi x}{L}}dx\right)e^{\frac{in\pi x}{L}}.$$

Let

$$\mu_n = \frac{n\pi}{L} \quad \text{and} \quad \Delta\mu = \frac{\pi}{L}.$$

Then

$$\sum_{n=-\infty}^{\infty} e^{\frac{in\pi x}{L}}\frac{\pi}{L} = \sum_{n=-\infty}^{\infty} e^{i\mu_n x}\Delta\mu$$

is a Riemann sum for

$$\int_{-\infty}^{\infty} e^{i\mu x}d\mu$$

and this Riemann sum converges to the integral as $L \to \infty$.

So we have

$$f(x) = \sum_{n=-\infty}^{\infty}\left(\frac{1}{2L}\int_{-L}^{L} f(x)e^{\frac{-in\pi x}{L}}dx\right)e^{\frac{in\pi x}{L}}$$

$$= \sum_{n=-\infty}^{\infty}\frac{1}{2\pi}\left(\int_{-L}^{L} f(x)e^{-i\mu_n x}dx\right)e^{i\mu_n x}\frac{\pi}{L}$$

and as $L \to \infty$

$$\sum_{n=-\infty}^{\infty}\frac{1}{2\pi}\left(\int_{-L}^{L} f(x)e^{-i\mu_n x}dx\right)e^{i\mu_n x}\frac{\pi}{L} \to \frac{1}{2\pi}\int_{-\infty}^{\infty}\left(\int_{-\infty}^{\infty} f(x)e^{-i\mu x}dx\right)e^{i\mu x}d\mu.$$

Also

$$\frac{1}{2\pi}\int_{-\infty}^{\infty}\left(\int_{-\infty}^{\infty} f(x)e^{-i\mu x}dx\right)e^{i\mu x}d\mu = \frac{1}{\sqrt{2\pi}}\int_{-\infty}^{\infty} F(\mu)e^{i\mu x}d\mu$$

where

$$F(\mu) = \frac{1}{\sqrt{2\pi}} \int_{-\infty}^{\infty} f(x) e^{-i\mu x} dx.$$

Thus we have the relations

$$f(x) = \frac{1}{\sqrt{2\pi}} \int_{-\infty}^{\infty} F(\mu) e^{i\mu x} d\mu \tag{1}$$

$$F(\mu) = \frac{1}{\sqrt{2\pi}} \int_{-\infty}^{\infty} f(x) e^{-i\mu x} dx. \tag{2}$$

Eqs. (1) and (2) provide the Fourier inversion formulas.

11.4 SOME PROPERTIES OF THE FOURIER TRANSFORM

In Chapter 1 we found that the Fourier transform of a Gaussian function is a Gaussian function. In fact, $\mathcal{F}\left(e^{-ax^2}\right)(k) = \sqrt{\frac{\pi}{a}} e^{-\pi^2 k^2/a}$. This suggests that there is an inverse Fourier transform for a class of functions that includes the Gaussian functions. Before stating a theorem that validates this, we note a relationship between a Gaussian function and its Fourier transform; namely, the wider the dispersion of the function, the narrower the dispersion of the Fourier transform of the function. (See Figs. 11.4.1A and 11.4.1B.)

Theorem: (Fourier Inversion Theorem)

Suppose that $f(x)$ and $\widehat{f}(\xi)$ are integrable functions and

$$g(x) = \frac{1}{\sqrt{2\pi}} \int_{-\infty}^{\infty} \widehat{f}(\xi) e^{ix\xi} d\xi.$$

Then $g(x) = f(x)$ almost everywhere.

(*Almost everywhere* is added for accuracy. It will not affect our work.)

Thus we have the relations

$$\widehat{f}(\xi) = \frac{1}{\sqrt{2\pi}} \int_{-\infty}^{\infty} f(x) e^{-ix\xi} dx \text{ and } f(x) = \frac{1}{\sqrt{2\pi}} \int_{-\infty}^{\infty} \widehat{f}(\xi) e^{ix\xi} d\xi$$

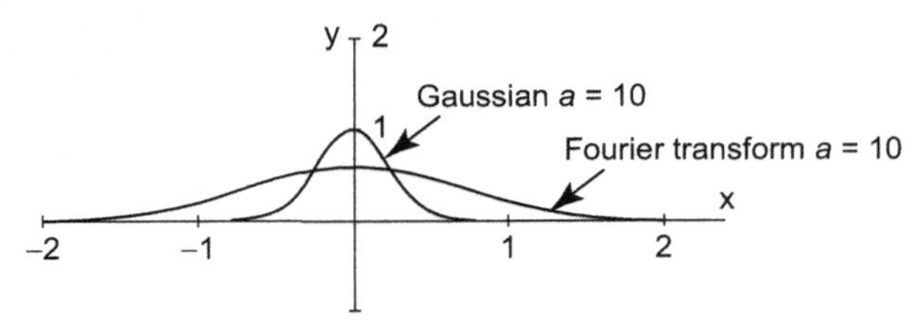

FIGURE 11.4.1A

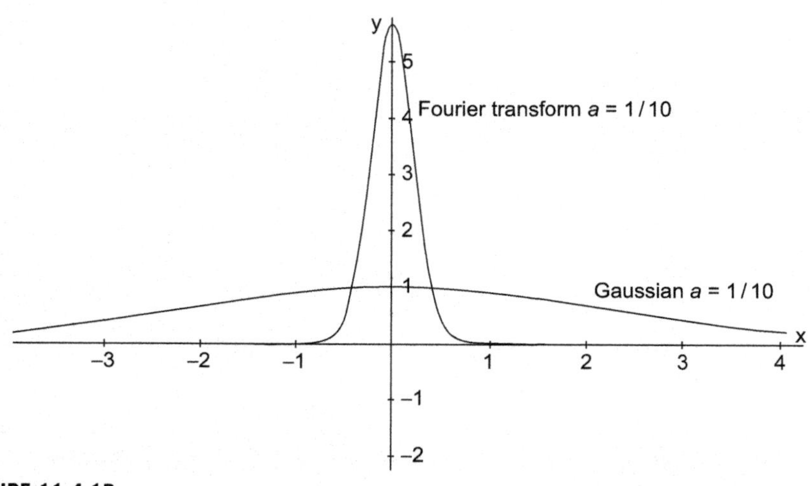

FIGURE 11.4.1B

if $f(x)$ is sufficiently well behaved. The latter equation defines the inverse Fourier transform.

The proof of the Fourier inversion theorem is beyond the scope of this text. (For a proof see Rudin, Real and Complex Analysis, Chapter 9.)

Next we describe the Fourier transform of the convolution of functions, and later we shall see that this is very useful in solving differential equations.

The convolution of the functions $f(x)$ and $g(x)$, denoted $(f * g)(x)$, is defined by

$$(f * g)(x) = \int_{-\infty}^{\infty} f(u)g(x-u)du.$$

We show in Exercise 1 that $(f * g)(x) = (g * f)(x)$; that is,

$$\int_{-\infty}^{\infty} f(u)g(x-u)du = \int_{-\infty}^{\infty} f(x-u)g(u)du.$$

Theorem: (Convolution Theorem)

If $f(x)$ and $g(x)$ are integrable, then

$$\left(\widehat{f * g}\right)(\xi) = \sqrt{2\pi}\hat{f}(\xi)\hat{g}(\xi).$$

Proof:
We have

$$\left(\widehat{f * g}\right)(\xi) = \frac{1}{\sqrt{2\pi}} \int_{x=-\infty}^{\infty} e^{-ix\xi}(f * g)(x)dx$$

$$= \frac{1}{\sqrt{2\pi}} \int_{x=-\infty}^{\infty} e^{-ix\xi} \left(\int_{u=-\infty}^{\infty} f(u)g(x-u)du \right) dx$$

$$= \frac{1}{\sqrt{2\pi}} \int_{x=-\infty}^{\infty} e^{-ix\xi} \left(\int_{u=-\infty}^{\infty} f(x-u)g(u)du \right) dx$$

$$= \frac{1}{\sqrt{2\pi}} \int_{u=-\infty}^{\infty} g(u) \left(\int_{x=-\infty}^{\infty} e^{-ix\xi} f(x-u)dx \right) du.$$

Let

$$z = x - u. \text{ Then } dz = dx, \quad x = z + u, \quad e^{-ix\xi} = e^{-iz\xi}e^{-iu\xi}.$$

Thus

$$\frac{1}{\sqrt{2\pi}} \int_{u=-\infty}^{\infty} g(u) \left(\int_{x=-\infty}^{\infty} e^{-ix\xi} f(x-u)dx \right) du$$

$$= \frac{1}{\sqrt{2\pi}} \int_{u=-\infty}^{\infty} g(u) \left(\int_{z=-\infty}^{\infty} e^{-iz\xi}e^{-iu\xi} f(z)dz \right) du$$

$$= \int_{u=-\infty}^{\infty} g(u)e^{-iu\xi}du \left(\frac{1}{\sqrt{2\pi}} \int_{z=-\infty}^{\infty} e^{-iz\xi} f(z)dz \right) du = \sqrt{2\pi}\widehat{f}(\xi)\widehat{g}(\xi).$$

We use alternate notation for the following corollary. The corollary is the form of the theorem that we shall use in solving differential equations.

Corollary:
If $F(\xi)$ and $G(\xi)$ are the Fourier transforms of $f(x)$ and $g(x)$, respectively, then

$$\mathcal{F}^{-1}(F(\xi)G(\xi)) = \frac{1}{\sqrt{2\pi}}(f * g)(x).$$

Theorem: (Parseval's Theorem)
If $\int_{-\infty}^{\infty} [f(x)]^2 dx$ is finite, then

$$\int_{-\infty}^{\infty} [f(x)]^2 dx = \int_{-\infty}^{\infty} [F(k)]^2 dk.$$

In the language of operator theory, this says that the Fourier transform is a unitary transformation.

Proof:

The main equations that we use in the proof are

$$f(x)=\frac{1}{\sqrt{2\pi}}\int_{-\infty}^{\infty}F(k)e^{-ikx}dk \quad \frac{1}{2\pi}\int_{-\infty}^{\infty}e^{i(x-x_0)}dx=\delta_{x_0}(x).$$

The second equation is developed in Exercise 7.

We have

$$\int_{-\infty}^{\infty}|f(x)|^2dx=\int_{-\infty}^{\infty}f(x)\overline{f}(x)dx$$

$$=\int_{x=-\infty}^{\infty}\left\{\left[\frac{1}{\sqrt{2\pi}}\int_{k=-\infty}^{\infty}F(k)e^{-ikx}dk\right]\left[\frac{1}{\sqrt{2\pi}}\int_{j=-\infty}^{\infty}\overline{F}(j)e^{ijx}dj\right]dx\right\}$$

$$=\int_{k=-\infty}^{\infty}F(k)\left\{\int_{j=-\infty}^{\infty}\overline{F}(j)\left[\frac{1}{2\pi}\int_{x=-\infty}^{\infty}e^{i(j-k)x}dx\right]dj\right\}dk$$

$$=\int_{k=-\infty}^{\infty}F(k)\left\{\int_{j=-\infty}^{\infty}\overline{F}(j)\delta_k(j)dj\right\}dk$$

$$=\int_{-\infty}^{\infty}F(k)\overline{F}(k)dk=\int_{-\infty}^{\infty}|F(k)|^2dk.$$

EXERCISES

1. Show that if $f(x)$ and $g(x)$ are integrable functions, then $(f*g)(x)=(g*f)(x)$; that is,

$$\int_{-\infty}^{\infty}f(u)g(x-u)du=\int_{-\infty}^{\infty}f(x-u)g(u)du.$$

2. Find the Fourier transform for

a. $f(x)=\begin{cases}a(1-a|x|) & \text{if } |x|<\frac{1}{a} \\ 0 & \text{otherwise}\end{cases}.$

b. $f(x)=\begin{cases}1 & a\le x\le b \\ 0 & \text{otherwise}\end{cases}.$

c. $f(x) = e^{-x^2}$.

d. $f(x) = e^{-a|x|}, \quad a > 0$.

e. $f(x) = \delta_a(x)$.

3. If $F(k)$ is the Fourier transform of $f(x)$, show the Fourier transform of $f'(x)$ is $ikF(k)$.
4. If $F(k)$ is the Fourier transform of $f(x)$, find
   a. The Fourier transform of $f(ax)$ in terms of $F(k)$ for $a > 0$.
   b. The Fourier transform of $f(x - a)$ in terms of $F(k)$ for $a > 0$.
5. Let $g(x,a)$ be the Gaussian function

$$g(x, a) = \frac{1}{\sqrt{2\pi a^2}} e^{-\frac{x^2}{2a^2}}.$$

The Fourier transform of $g(x,a)$ is

$$G(k, a) = e^{-\frac{k^2 a^2}{2}}.$$

Use the convolution theorem to show

$$g(x, a) * g(x, b) = g\left(x, \sqrt{a^2 + b^2}\right).$$

6. Show that

$$\int_{-\infty}^{\infty} |f(x) + g(x)|^2 dx = \int_{-\infty}^{\infty} |F(k) + G(k)|^2 dk.$$

11.5 SOLVING PARTIAL DIFFERENTIAL EQUATIONS USING THE FOURIER TRANSFORM

We now show how the Fourier transform is used to solve PDEs when the space variable is unbounded. In the PDEs that we consider, there will be two variables, time and space. We can take the Fourier transform with respect to either variable, but in most cases we take the transform with respect to the space variable. The technique for solution of PDEs consists of three steps: First, apply the Fourier transform with respect to the space variable, x, to the PDE and the initial conditions. This yields an ordinary differential equation and its boundary conditions. Second, solve the ordinary differential equation. Third, apply the inverse Fourier transform to the solution of the ordinary differential equation to obtain the solution to the original PDE. We demonstrate this technique with two examples.

Example:
Solving the heat equation using the Fourier transform.
We solve the heat equation in one dimension using the Fourier transform

$$u_t(x,t) = Ku_{xx}(x,t) \quad t>0, \quad -\infty < x < \infty \tag{1}$$

$$u(x,0) = f(x) \quad -\infty < x < \infty. \tag{2}$$

Let $U(\xi,t)$ be the Fourier transform of $u(x,t)$ and $F(\xi)$ be the Fourier transform of $f(x)$. Two key relations in this method are if

$$u(x,t) = \frac{1}{\sqrt{2\pi}} \int_{-\infty}^{\infty} U(\xi,t)e^{i\xi x}d\xi$$

then

$$\frac{\partial^n}{\partial t^n}u(x,t) = \frac{1}{\sqrt{2\pi}} \int_{-\infty}^{\infty} \left(\frac{\partial^n}{\partial t^n}U(\xi,t)\right)e^{i\xi x}d\xi \tag{3a}$$

and

$$\frac{\partial^n}{\partial x^n}u(x,t) = \frac{1}{\sqrt{2\pi}} \int_{-\infty}^{\infty} U(\xi,t)(i\xi)^n e^{i\xi x}d\xi. \tag{3b}$$

From Eq. (1), we have

$$0 = u_t(x,t) - Ku_{xx}(x,t) = \frac{1}{\sqrt{2\pi}} \int_{-\infty}^{\infty} \left[\frac{\partial}{\partial t}U(\xi,t) + K\xi^2 U(\xi,t)\right] e^{i\xi x}d\xi.$$

Holding ξ fixed, we solve the ordinary differential equation

$$\frac{d}{dt}U(\xi,t) + K\xi^2 U(\xi,t) = 0. \tag{4}$$

The initial condition for Eq. (4) is

$$U(\xi,0) = \mathcal{F}(u(x,0)) = \mathcal{F}(f(x)) = F(\xi).$$

From Eq. (4) we get

$$\int \frac{dU}{U} = -\int K\xi^2 dt$$

so

$$U(\xi,t) = U(\xi,0)e^{-K\xi^2 t} = F(\xi)e^{-K\xi^2 t}.$$

To find $u(x,t)$ we use

$$u(x,t) = \frac{1}{\sqrt{2\pi}} \int_{-\infty}^{\infty} U(\xi,t)e^{i\xi x}d\xi = \frac{1}{\sqrt{2\pi}} \int_{-\infty}^{\infty} F(\xi)e^{-K\xi^2 t}e^{i\xi x}d\xi. \tag{5}$$

To obtain a more explicit form of the solution, we would need to evaluate the integral in Eq. (5). In the case that $f(x)$ is a Gaussian function, this is reasonable to do (because $F(\xi)$ is then also Gaussian). In Exercise 2, we give such an example.

FUNDAMENTAL SOLUTION OF THE HEAT EQUATION

For the fundamental solution of the heat equation, we need to solve

$$u_t(x,t) = Ku_{xx}(x,t) \quad t > 0, \quad -\infty < x < \infty;$$
$$u(x,0) = \delta_0(x).$$

In this case,

$$U(\xi,0) = F(\xi) = \mathcal{F}\big(u(x,0)\big) = \mathcal{F}\big(\delta_0(x)\big) = \frac{1}{\sqrt{2\pi}}\int_{-\infty}^{\infty} \delta_0(x)e^{-i\xi x}d\xi = \frac{1}{\sqrt{2\pi}}$$

and Eq. (5) becomes

$$u(x,t) = \frac{1}{2\pi}\int_{-\infty}^{\infty} U(\xi,t)e^{i\xi x}d\xi = \frac{1}{2\pi}\int_{-\infty}^{\infty} e^{-K\xi^2 t}e^{i\xi x}d\xi.$$

In Section 1.5 we found

$$\frac{1}{2\pi}\int_{-\infty}^{\infty} e^{-ikx}e^{-\alpha t k^2}dk = \frac{1}{\sqrt{4\pi\alpha t}}e^{-x^2/4\alpha t}$$

so (taking the complex conjugate of the integral) the fundamental solution is

$$u(x,t) = \frac{1}{\sqrt{4\pi Kt}}e^{-x^2/4Kt}.$$

From this, the solution to the heat equation on an infinite bar with initial condition $u(x,0) = f(x)$ is

$$u(x,t) = \frac{1}{\sqrt{4\pi Kt}}\int_{-\infty}^{\infty} e^{-\frac{(x-y)^2}{4Kt}}f(y)dy.$$

Note that if we let

$$g(x) = \exp\left(-x^2/4Kt\right)$$

then

$$u(x,t) = \frac{1}{\sqrt{4\pi Kt}}(f * g)(x).$$

This is handy because if $f(x)$ and $g(x)$ are both Gaussian functions, say

$$f(x) = \exp\left(-x^2/2b^2\right) \text{ and } g(x) = \exp\left(-x^2/2b^2\right)$$

then

$$(f * g)(x) = \frac{1}{\sqrt{2\pi(a^2+b^2)}} \exp(-x^2/2(a^2+b^2)).$$

Example:
Solving the wave equation using the Fourier transform.
We solve the wave equation using the Fourier transform. Consider

$$u_{tt}(x,t) - c^2 u_{xx}(x,t) = 0, \quad t > 0, \quad -\infty < x < \infty \tag{6}$$

$$u(x,0) = f(x), \quad u_t(x,0) = g(x). \tag{7}$$

Let $U(\xi, t)$ denote the Fourier transform of (x,t), $F(\xi)$ the Fourier transform of $f(x)$, and $G(\xi)$ the Fourier transform of $g(x)$. We again use

$$u(x,t) = \frac{1}{\sqrt{2\pi}} \int_{-\infty}^{\infty} U(\xi,t) e^{i\xi x} d\xi$$

so

$$\frac{\partial^n}{\partial t^n} u(x,t) = \frac{1}{\sqrt{2\pi}} \int_{-\infty}^{\infty} \left(\frac{\partial^n}{\partial t^n} U(\xi,t) \right) e^{i\xi x} d\xi$$

and

$$\frac{\partial^n}{\partial x^n} u(x,t) = \frac{1}{\sqrt{2\pi}} \int_{-\infty}^{\infty} U(\xi,t)(i\xi)^n \, e^{i\xi x} d\xi.$$

From Eq. (6), we have

$$0 = u_{tt}(x,t) - c^2 u_{xx}(x,t) = \frac{1}{\sqrt{2\pi}} \int_{-\infty}^{\infty} \left[\frac{\partial^2}{\partial t^2} U(\xi,t) + c^2 \xi^2 U(\xi,t) \right] e^{i\xi x} d\xi.$$

So, holding ξ fixed, we need to solve the ordinary differential equation

$$\frac{d^2}{dt^2} U(\xi,t) + c^2 \xi^2 U(\xi,t) = 0 \tag{8}$$

with initial conditions

$$\mathcal{F}(u(x,0)) = \mathcal{F}(f(x)) = F(\xi); \quad \mathcal{F}(u_t(x,0)) = \mathcal{F}(g(x)) = G(\xi);$$

that is,

$$U(\xi,0) = F(\xi); \quad U_t(\xi,0) = G(\xi).$$

We get

$$U(\xi,t) = A(\xi)\cos(c\xi t) + B(\xi)\sin(c\xi t)$$

and so

$$F(\xi) = U(\xi,0) = A(\xi); \quad G(\xi) = U_t(\xi,0) = c\xi\, B(\xi) \quad \text{or } B(\xi) = \frac{G(\xi)}{c\xi}.$$

Thus

$$U(\xi,t) = F(\xi)\cos(c\xi t) + \frac{G(\xi)}{c\xi}\sin(c\xi t)$$

and so

$$u(x,t) = \frac{1}{\sqrt{2\pi}} \int_{-\infty}^{\infty} \left[F(\xi)\cos(c\xi t) + \frac{G(\xi)}{c\xi}\sin(c\xi t) \right] e^{i\xi x} d\xi.$$

With some manipulation, we can show this gives d'Alembert's formula. We have

$$\cos(c\xi t) = \frac{e^{ic\xi t} + e^{-ic\xi t}}{2} \quad \text{and} \quad \sin(c\xi t) = \frac{e^{ic\xi t} - e^{-ic\xi t}}{2i}.$$

So

$$\begin{aligned}
\int_{-\infty}^{\infty} F(\xi)\cos(c\xi t) e^{i\xi x} d\xi &= \frac{1}{2} \int_{-\infty}^{\infty} F(\xi) \left[e^{ic\xi t} + e^{-ic\xi t} \right] e^{i\xi x} d\xi \\
&= \frac{1}{2} \int_{-\infty}^{\infty} F(\xi) \left[e^{i\xi(x+ct)} + e^{i\xi(x-ct)} \right] d\xi \\
&= \frac{1}{\sqrt{2\pi}} \frac{1}{2} [f(x+ct) + f(x-ct)]
\end{aligned}$$

and

$$\int_{-\infty}^{\infty} \left[\frac{G(\xi)}{c\xi} \sin(c\xi t) \right] e^{i\xi x} d\xi = \frac{1}{2} \int_{-\infty}^{\infty} G(\xi) \left[\frac{e^{ic\xi t} - e^{-ic\xi t}}{ic\xi} \right] e^{i\xi x} d\xi$$

$$= \frac{1}{2} \int_{-\infty}^{\infty} G(\xi) \left[\frac{e^{i\xi(x+ct)} - e^{i\xi(x-ct)}}{ic\xi} \right] d\xi = \frac{1}{2c} \int_{-\infty}^{\infty} G(\xi) \left(\int_{x-ct}^{x+ct} e^{i\xi k} dk \right) d\xi$$

$$= \frac{1}{2c} \int_{x-ct}^{x+ct} \left(\int_{-\infty}^{\infty} G(\xi) e^{i\xi k} d\xi \right) dk = \frac{1}{\sqrt{2\pi}} \frac{1}{2c} \int_{x-ct}^{x+ct} g(k) dk.$$

Thus

$$\begin{aligned}
u(x,t) &= \frac{1}{\sqrt{2\pi}} \int_{-\infty}^{\infty} \left[F(\xi)\cos(c\xi t) + \frac{G(\xi)}{c\xi}\sin(c\xi t) \right] e^{i\xi x} d\xi \\
&= \frac{1}{2}[f(x+ct) + f(x-ct)] + \frac{1}{2c} \int_{x-ct}^{x+ct} g(k) dk.
\end{aligned}$$

We use Laplace's equation as an example to demonstrate another application of the convolution theorem.

Example:

Solving Laplace's equation using the Fourier transform.

Laplace's equation is

$$u_{xx}(x,y) + u_{yy}(x,y) = 0.$$

Suppose we have Laplace's equation on the half-plane $-\infty < x < \infty$, $y > 0$ with the boundary condition

$$u(x,0) = f(x) \text{ and suppose that } u(x,y) \text{ is bounded.}$$

We take the Fourier transform with respect to x to get

$$\mathcal{F}\{u(x,y)\} = \frac{1}{\sqrt{2\pi}} \int_{-\infty}^{\infty} u(x,y)e^{-ikx}dx \equiv U(k,y).$$

Then

$$\mathcal{F}\{u_{xx}(x,y)\} = -k^2 U(k,y), \quad \mathcal{F}\{u_{yy}(x,y)\} = \frac{d^2}{dy^2} U(k,y),$$
$$\mathcal{F}\{u(x,0)\} = U(k,0) = \mathcal{F}\{f(x)\} = F(k).$$

Taking the Fourier transform of Laplace's equation and the boundary condition above, we get the ordinary differential equation in y,

$$\frac{d^2}{dy^2} U(k,y) - k^2 U(k,y) = 0, \quad U(k,0) = F(k).$$

The solution to this equation is

$$U(k,y) = Ae^{|k|y} + Be^{-|k|y}.$$

If we assume that $U(k,y)$ is bounded as $y \to \infty$, then

$$U(k,y) = Be^{-|k|y}.$$

The initial condition gives $B = U(k,0) = F(k)$, so

$$U(k,y) = F(k)e^{-|k|y}.$$

If we take the inverse Fourier transform of $U(k,y)$, we obtain the solution $u(x,y)$.

In taking the inverse transform of $F(k)e^{-|k|y}$, we are taking the inverse of the product of two transforms, and this can be accomplished by using the convolution theorem. To apply the convolution theorem to this problem, we need to find $\mathcal{F}^{-1}\{e^{-|k|y}\}$.

Now

$$\mathcal{F}^{-1}\left\{e^{-|k|y}\right\} = \frac{1}{\sqrt{2\pi}} \int_{k=-\infty}^{\infty} e^{-|k|y} e^{ikx} dk$$
$$= \frac{1}{\sqrt{2\pi}} \int_{-\infty}^{0} e^{ky}[\cos(kx) + i\sin(kx)]dk$$
$$+ \frac{1}{\sqrt{2\pi}} \int_{0}^{\infty} e^{-ky}[\cos(kx) + i\sin(kx)]dk.$$

In

$$\int_{-\infty}^{0} e^{ky}[\cos(kx) + i\sin(kx)]dk$$

let $m = -k$. Then

$$\int_{-\infty}^{0} e^{ky}[\cos(kx) + i\sin(kx)]dk = \int_{\infty}^{0} e^{-my}[\cos(-mx) + i\sin(-mx)](-dm)$$
$$= \int_{0}^{\infty} e^{-my}[\cos(mx) - i\sin(mx)]dm.$$

Thus

$$\int_{k=-\infty}^{\infty} e^{-|k|y} e^{ikx} dk = 2\int_{0}^{\infty} e^{-ky}\cos(kx)dk.$$

Integrating by parts twice gives

$$\int_{0}^{\infty} e^{-ky}\cos(kx)dk = \frac{y}{x^2+y^2} \quad \text{if } y > 0.$$

Thus

$$\mathcal{F}^{-1}\left\{e^{-|k|y}\right\} = \frac{1}{\sqrt{2\pi}}\frac{2y}{x^2+y^2}.$$

If we let $\widehat{g}(k) = e^{-|k|y}$, we have

$$g(x) = \frac{1}{\sqrt{2\pi}}\frac{2y}{x^2+y^2}$$

and

$$g(x-z) = \frac{2y}{(x-z)^2+y^2}.$$

So

$$u(x,y) = \mathcal{F}^{-1}\{U(k,y)\} = \mathcal{F}^{-1}\{F(k)G(k)\} = (f*g)(x)$$
$$= \frac{1}{\pi}\int_{-\infty}^{\infty} \frac{yf(z)}{(x-z)^2+y^2}dz.$$

This is one form of Poisson's integral formula.

In Section 12.3 we use Poisson's integral in polar coordinates, which we developed in Exercise 8 of Section 4.3. This is the solution to the boundary value problem on a circle

$$\nabla u(r,\theta) = u_{rr} + \frac{1}{r}u_r + \frac{1}{r^2}u_{\theta\theta} = 0 \quad \text{for } r < R, u(R,\theta) = f(\theta)$$

which is given by

$$u(r,\theta) = \frac{R^2 - r^2}{2\pi}\int_{-\pi}^{\pi} \frac{f(\varphi)}{R^2 + r^2 - 2rR\cos(\theta - \varphi)}d\varphi.$$

EXERCISES

Solve the following PDEs using the Fourier transform.

1. $u_t(x,t) = Ku_{xx}(x,t), \quad u(x,0) = \frac{1}{1+x^2}, \quad -\infty < x < \infty,$ $t \geq 0$. Assume $\lim_{x\to\pm\infty} u(x,t) = 0$.

2. $u_t(x,t) = Ku_{xx}(x,t), \quad u(x,0) = ce^{-x^2}, \quad -\infty < x < \infty,$ $t \geq 0$. Assume $\lim_{x\to\pm\infty} u(x,t) = 0$.

3. Use Euler's formula

$$e^{i\theta} = \cos\theta + i\sin\theta$$

to show

$$\cos(c\xi t) = \frac{e^{ic\xi t} + e^{-ic\xi t}}{2} \text{ and } \sin(c\xi t) = \frac{e^{ic\xi t} - e^{-ic\xi t}}{2i}.$$

4. Solve the wave equation

$$u_{tt}(x,t) - c^2u_{xx}(x,t) = 0, \quad t > 0, \quad -\infty < x < \infty \tag{1}$$

$$u(x,0) = \frac{1}{1+x^2}, \quad u_t(x,0) = 0.$$

5. In digital processing, the function

$$sinc(x) = \frac{\sin x}{x}$$

is often used. Given that

$$\mathcal{F}(sinc(x)) = \widehat{f}(k) = \begin{cases} 1 & \text{if } |k| < 1 \\ 0 & \text{otherwise} \end{cases}.$$

Solve the wave equation

$$u_{tt}(x,t) - c^2 u_{xx}(x,t) = 0, \quad t > 0, \quad -\infty < x < \infty \tag{2}$$
$$u(x,0) = sinc(x), \quad u_t(x,0) = 0.$$

6. Solve Laplace's equation on the half-plane $-\infty < x < \infty,\ y > 0$ with the boundary condition

$$u(x,0) = e^{-|x|}, \text{ assuming that } u(x,y) \text{ is bounded.}$$

7. Solve Laplace's equation on the half-plane $-\infty < x < \infty,\ y > 0$ with the boundary condition

$$u(x,0) = \begin{cases} 1, & 0 \le x \le 2 \\ 0 & \text{otherwise} \end{cases}.$$

8. Justify, giving necessary hypotheses on $U(x,t)$, the formulas if

$$u(x,t) = \int_{-\infty}^{\infty} U(\xi,t) e^{i\xi x} d\xi$$

then

$$\frac{\partial}{\partial t} u(x,t) = \int_{-\infty}^{\infty} \left(\frac{\partial}{\partial t} U(\xi,t) \right) e^{i\xi x} d\xi$$

and

$$\frac{\partial}{\partial x} u(x,t) = \int_{-\infty}^{\infty} U(\xi,t)(i\xi)\, e^{i\xi x} d\xi.$$

11.6 THE SPECTRUM OF THE NEGATIVE LAPLACIAN IN ONE DIMENSION

If V is an n-dimensional vector space, and $A{:}V \to V$ is a linear operator, then A can be represented by an $n \times n$ matrix. (The entries of A depend on the basis of V.) The spectrum of A is the eigenvalues of A, which are the values of λ for which there is a nonzero vector $\widehat{x}$ with $A\widehat{x} = \lambda\widehat{x}$. These are the same numbers for which $(A - \lambda I)^{-1}$ does not exist.

If V is an infinite dimensional vector space, the spectrum is more complicated. The *resolvent* of the linear operator A is the set of complex numbers z for which $(A - zI)^{-1}: V \to V$ is a continuous (bounded) operator. The *spectrum* of A is the complement of the resolvent of A. The eigenvalues of A are part of the spectrum of A.

Our aim in this section is to show that for $A = -d^2/dx^2$ and $c > 0$ we have

$$(A + c)^{-1} f(x) = \frac{1}{\sqrt{2\pi}} \frac{\pi}{\sqrt{c}} \int_{-\infty}^{\infty} \exp(-\sqrt{c}\,|x - y|) f(x) dx$$

if $f(x)$ is sufficiently well behaved.

We let $m^2 = c$ and consider

$$-u''(x) + m^2 u(x) = f(x) \tag{1}$$

where $f(x)$ is a rapidly decreasing, infinitely differentiable function.

Proceeding formally, we rewrite Eq. (1) as

$$\left(-\frac{d^2}{dx^2} + m^2\right) u(x) = f(x)$$

and

$$u(x) = \left(-\frac{d^2}{dx^2} + m^2\right)^{-1} f(x).$$

Applying the Fourier transform to Eq. (1), we get

$$k^2 \widehat{U}(k) + m^2 \widehat{U}(k) = \widehat{F}(k)$$

so

$$\widehat{U}(k) = \left(k^2 + m^2\right)^{-1} \widehat{F}(k).$$

If there is a function $g(x)$ for which $\widehat{G}(k) = \left(k^2 + m^2\right)^{-1}$, then

$$\widehat{U}(k) = \left(k^2 + m^2\right)^{-1} \widehat{F}(k) = \widehat{G}(k)\widehat{F}(k) = \left(\widehat{g * f}\right)(k)$$

and

$$u(x) = (g * f)(x).$$

To find $g(x)$, we take the inverse Fourier transform of $(k^2 + m^2)^{-1}$. That is,

$$g(x) = \frac{1}{\sqrt{2\pi}} \int_{-\infty}^{\infty} \frac{e^{ikx}}{k^2 + m^2} dk. \tag{2}$$

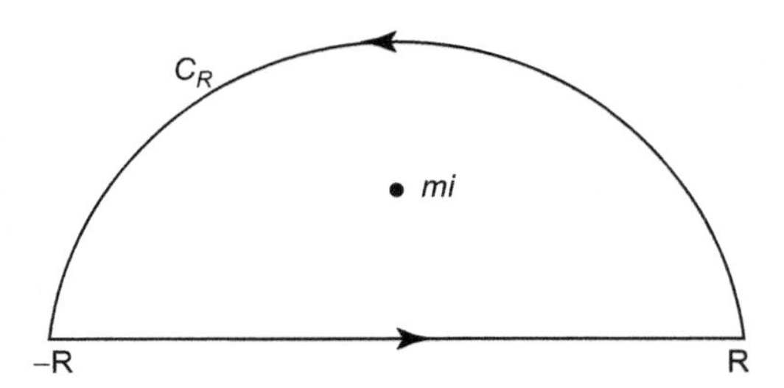

FIGURE 11.6.1

We shall show that

$$\int_{-\infty}^{\infty} \frac{e^{ikx}}{k^2+m^2}\,dk = \frac{\pi}{m} e^{-m|x|} \quad \text{for} \quad m > 0.$$

We do this by using the residue theorem for two cases. The first case is for $x > 0$. Consider the contour shown in Fig. 11.6.1.

The only pole of

$$f(k) = \frac{e^{ikx}}{k^2+m^2} = \frac{e^{ikx}}{(k-mi)(k+mi)}$$

that occurs within the contour is a simple pole at $k = mi$. The residue of $f(k)$ at $k = mi$ is

$$\lim_{k\to mi} (k-mi)\frac{e^{ikx}}{(k-mi)(k+mi)} = \frac{e^{i(mi)x}}{(mi+mi)} = \frac{e^{-mx}}{2mi}.$$

Thus

$$\int_{-R}^{R} \frac{e^{ikx}}{k^2+m^2}\,dk = 2\pi i\left(\frac{e^{-mx}}{2mi}\right) - \int_{C_R} \frac{e^{ikx}}{k^2+m^2}\,dk = \frac{\pi}{m}e^{-mx} - \int_{C_R} \frac{e^{ikx}}{k^2+m^2}\,dk.$$

If k is a point on C_R, then, because the imaginary part of $k > 0$ and $x > 0$, we have $\left|e^{ikx}\right| < 1$. Also for k a point on C_R,

$$\left|\frac{1}{k^2+m^2}\right| \le \frac{1}{(R-m)^2}.$$

Since the length of C_R is πR, we have

$$\left|\int_{C_R} \frac{e^{ikx}}{k^2+m^2}\,dk\right| \le \int_{C_R} \left|\frac{e^{ikx}}{k^2+m^2}\right| dk \le \frac{1}{(R-m)^2}\,\pi R$$

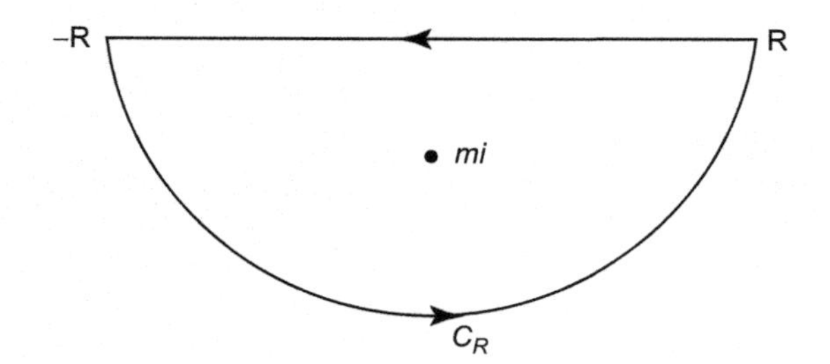

FIGURE 11.6.2

and so

$$\lim_{R\to\infty}\left|\int_{C_R}\frac{e^{ikx}}{k^2+m^2}dk\right| = 0.$$

Thus for $x > 0$ we have

$$\int_{-\infty}^{\infty}\frac{e^{ikx}}{k^2+m^2}dk = \frac{\pi}{m}e^{-mx} \quad \text{for } m > 0.$$

The second case is for $x < 0$. Consider the contour in Fig. 11.6.2. In this case, the only pole of

$$f(k) = \frac{e^{ikx}}{k^2+m^2} = \frac{e^{ikx}}{(k-mi)(k+mi)}$$

that occurs within the contour is a simple pole at $k = -mi$. The residue of $f(k)$ at $k = -mi$ is

$$\lim_{k\to -mi}(k+mi)\frac{e^{ikx}}{(k-mi)(k+mi)} = \frac{e^{i(-mi)x}}{(-mi-mi)} = \frac{e^{mx}}{-2mi}.$$

Thus for $x < 0$

$$\int_{-R}^{R}\frac{e^{ikx}}{k^2+m^2}dk = -\int_{R}^{-R}\frac{e^{ikx}}{k^2+m^2}dk = -\left[2\pi i\left(\frac{e^{mx}}{-2mi}\right) - \int_{C_R}\frac{e^{ikx}}{k^2+m^2}dk\right]$$

$$= -2\pi i\left(\frac{e^{mx}}{-2mi}\right) + \int_{C_R}\frac{e^{ikx}}{k^2+m^2}dk = \frac{\pi}{m}e^{mx} + \int_{C_R}\frac{e^{ikx}}{k^2+m^2}dk.$$

If k is a point on C_R, then, because the imaginary part of $k < 0$ and $x < 0$, we have $\left|e^{ikx}\right| < 1$. Also for k is a point on C_R,

$$\left|\frac{1}{k^2+m^2}\right| \le \frac{1}{(R-m)^2}.$$

Since the length of C_R is πR, we again have

$$\left|\int_{C_R}\frac{e^{ikx}}{k^2+m^2}dk\right| \leq \int_{C_R}\left|\frac{e^{ikx}}{k^2+m^2}\right|dk \leq \frac{1}{(R-m)^2}\pi R$$

and so

$$\lim_{R\to\infty}\left|\int_{C_R}\frac{e^{ikx}}{k^2+m^2}dk\right| = 0.$$

Thus for $x < 0$ we have

$$\int_{-\infty}^{\infty}\frac{e^{ikx}}{k^2+m^2}dk = \frac{\pi}{m}e^{mx} \quad \text{for } m > 0.$$

From the two cases we conclude

$$\int_{-\infty}^{\infty}\frac{e^{ikx}}{k^2+m^2}dk = \frac{\pi}{m}e^{-m|x|} \quad \text{for } m > 0.$$

Thus

$$u(x) = \widehat{f * g}(x) = \frac{1}{\sqrt{2\pi}}\int_{-\infty}^{\infty} g(x-y)f(y)dy = \frac{1}{\sqrt{2\pi}}\frac{\pi}{m}\int_{-\infty}^{\infty} e^{-m|x-y|}f(y)dy.$$

So we have formally written

$$\left(-\frac{d^2}{dx^2}+m^2\right)^{-1} f(x) = \frac{1}{\sqrt{2\pi}}\frac{\pi}{m}\int_{-\infty}^{\infty} e^{-m|x-y|}f(y)dy, \tag{3}$$

where $m > 0$.

Thus $\left(-\frac{d^2}{dx^2}+m^2\right)^{-1}$ is given by Eq. (3), and at least part of the spectrum of $-d^2/dx^2$ is the positive real axis.

In fact, the positive real axis is the entire spectrum of $-\Delta$. This is because $\left(-\frac{d^2}{dx^2}-m^2\right)^{-1}$ is defined and bounded, and because $-d^2/dx^2$ is self-adjoint, the spectrum is a subset of the real numbers.

11.7 THE FOURIER TRANSFORM IN THREE DIMENSIONS

The Fourier transform in three dimensions in rectangular coordinates is given by

$$\widehat{F}\left(\widehat{k}\right) = \frac{1}{\left(\sqrt{2\pi}\right)^3}\iiint_{R^3} f(x,y,z)e^{-i(k_x x+k_y y+k_z z)}dxdydz$$

where $\widehat{k} = (k_x, k_y, k_z)$. The inverse Fourier transform is determined by

$$f(x,y,z) = \frac{1}{(\sqrt{2\pi})^3} \iiint_{R^3} \widehat{F}\left(\widehat{k}\right) e^{i(k_x x + k_y y + k_z z)} dk_x dk_y dk_z$$

Some of the most important applications in three dimensions occur when $f(x,y,z)$ depends only on $r = |\widehat{r}| = \sqrt{x^2+y^2+z^2}$. In these instances, it is often advantageous to use spherical coordinates. Recall that in spherical coordinates $dV = r^2 \sin\theta dr d\theta d\varphi$, so that

$$\widehat{F}\left(\widehat{k}\right) = \frac{1}{(\sqrt{2\pi})^3} \int_{r=0}^{\infty} \int_{\theta=0}^{\pi} \int_{\varphi=0}^{2\pi} f(r) e^{-i\widehat{k}\cdot\widehat{r}} r^2 \sin\theta \, d\varphi \, d\theta \, dr.$$

We take $\widehat{k}$ so that it is parallel to the z-axis and then $\widehat{k}\cdot\widehat{r} = kr\cos\theta$ where $k = \left|\widehat{k}\right|$. Then

$$\widehat{F}\left(\widehat{k}\right) = \frac{1}{(\sqrt{2\pi})^3} \int_{r=0}^{\infty} \int_{\theta=0}^{\pi} \int_{\varphi=0}^{2\pi} f(r) e^{-ikr\cos\theta} r^2 \sin\theta \, d\varphi \, d\theta \, dr.$$

Now

$$\int_{\theta=0}^{\pi} \left(\int_{\varphi=0}^{2\pi} d\varphi \right) e^{-ikr\cos\theta} \sin\theta \, d\theta = 2\pi \int_{\theta=0}^{\pi} e^{-ikr\cos\theta} \sin\theta \, d\theta$$

$$= \frac{2\pi}{ikr} e^{-ikr\cos\theta}\Big|_{\theta=0}^{\pi} = \frac{2\pi}{ikr}\left(e^{-ikr(-1)} - e^{-ikr(1)}\right)$$

$$= \frac{2\pi}{ikr} 2i\sin(kr) = \frac{4\pi\sin(kr)}{kr}.$$

Thus

$$\widehat{F}(k) = \frac{1}{(\sqrt{2\pi})^3} \int_{r=0}^{\infty} f(r) \, r^2 \frac{4\pi\sin(kr)}{kr} dr = \sqrt{\frac{2}{\pi}} \int_{r=0}^{\infty} f(r) \, r \frac{\sin(kr)}{k} dr.$$

Similarly,

$$f(r) = \sqrt{\frac{2}{\pi}} \int_{r=0}^{\infty} \widehat{F}(k) \frac{k\sin(kr)}{r} dk.$$

We consider the example

$$-\Delta u(\widehat{x}) + m^2 u(\widehat{x}) = f(r)$$

in three dimensions. As in the one-dimensional case, we take the Fourier transform and find

$$\widehat{U}(k) = \left(k^2 + m^2\right)^{-1} \widehat{F}(k).$$

We then take the inverse Fourier transform of $(k^2 + m^2)^{-1}$; that is, we compute

$$g(r) = \sqrt{\frac{2}{\pi}} \int_0^\infty \frac{k \sin(kr)}{r(k^2 + m^2)} dk.$$

To compute the integral, we again use the residue theorem, but the estimates are more delicate than in the one-dimensional case. We begin by computing

$$\int_{-\infty}^{\infty} \frac{x \sin x}{x^2 + m^2} dx$$

using the residue theorem.

Let

$$f(z) = \frac{z}{z^2 + m^2}.$$

We compute

$$\int_C f(z) e^{iz} dz$$

where C is the contour shown in Fig. 11.7.1.

The function $f(z)e^{iz}$ has a pole of order 1 at $z = mi$. The residue of $f(z)e^{iz}$ at $z = mi$ is

$$\lim_{z \to mi} (z - mi) \frac{z e^{iz}}{(z - mi)(z + mi)} = \frac{mi e^{imi}}{mi + mi} = \frac{e^{-m}}{2}$$

so

$$\int_{-R}^{R} \frac{x e^{ix}}{x^2 + m^2} dx = 2\pi i \left(\frac{e^{-m}}{2} \right) - \int_{C_R} f(z) e^{iz} dz.$$

Thus

$$\int_{-R}^{R} \frac{x \sin x}{x^2 + m^2} dx = \operatorname{Im}\left(\pi i e^{-m}\right) - \operatorname{Im} \int_{C_R} f(z) e^{iz} dz.$$

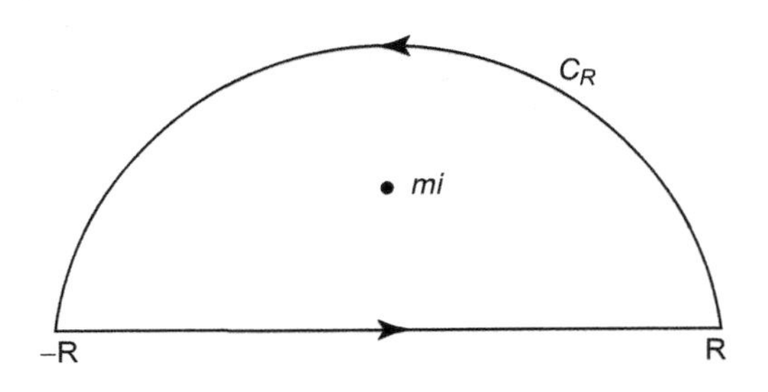

FIGURE 11.7.1

Consider

$$\left|\int_{C_R} f(z)e^{iz}dz\right|.$$

If z is a point on C_R, then

$$|f(z)| \le \frac{R}{(R-m)^2}$$

and the length of C_R is πR. So we can say

$$\left|\int_{C_R} f(z)e^{iz}dz\right| \le \frac{R}{(R-m)^2}\pi R$$

but

$$\lim_{R\to\infty} \frac{R}{(R-m)^2}\pi R \neq 0.$$

Thus more is needed to conclude that

$$\lim_{R\to\infty}\left|\int_{C_R} f(z)e^{iz}dz\right| = 0.$$

The pertinent result is Jordan's lemma. (See Brown and Churchill, p. 272.) It states that if $f(z)$ is analytic at all points in the upper half plane that are exterior to a circle $|z| = R_0$ and C_R denotes a semicircle $z = \mathrm{Re}^{i\theta}$, $0 \le \theta \le \pi$, where $R > R_0$ and $|f(z)| \le M_R$ for z on C_R and

$$\lim_{R\to\infty} M_R = 0$$

then for every positive real number a,

$$\lim_{R\to\infty} \int_{C_R} f(z)e^{iaz}dz = 0.$$

Thus

$$\int_{-\infty}^{\infty} \frac{x\sin x}{x^2+m^2}dx = \lim_{R\to\infty}\int_{-R}^{R} \frac{x\sin x}{x^2+m^2}dx$$

$$= \mathrm{Im}\left(\pi i e^{-m}\right) - \mathrm{Im}\lim_{R\to\infty}\int_{C_R} f(z)e^{iz}dz = \mathrm{Im}\left(\pi i e^{-m}\right) = \pi e^{-m}.$$

Now consider

$$\sqrt{\frac{2}{\pi}}\int_{-\infty}^{\infty}\frac{k\sin(kr)}{r(k^2+m^2)}dk = \sqrt{\frac{2}{\pi}}\,\frac{1}{r^3}\int_{-\infty}^{\infty}\frac{rk\sin(kr)}{(k^2+m^2)}rdk$$
$$= \sqrt{\frac{2}{\pi}}\,\frac{r^2}{r^3}\int_{-\infty}^{\infty}\frac{rk\sin(kr)}{r^2(k^2+m^2)}rdk$$
$$= \sqrt{\frac{2}{\pi}}\,\frac{1}{r}\int_{-\infty}^{\infty}\frac{rk\sin(kr)}{(r^2k^2+r^2m^2)}rdk$$
$$= \sqrt{\frac{2}{\pi}}\,\frac{1}{r}\int_{-\infty}^{\infty}\frac{u\sin(u)}{(u^2+r^2m^2)}du = \sqrt{2\pi}\,\frac{1}{r}e^{-mr}.$$

Since $\frac{k\sin(kr)}{r(k^2+m^2)}$ is an even function in k, and $r \geq 0$

$$\int_{0}^{\infty}\frac{k\sin(kr)}{r(k^2+m^2)}dk = \sqrt{\frac{2}{\pi}}\,\frac{1}{r}e^{-mr}.$$

Returning to

$$\widehat{U}(k) = (k^2+m^2)^{-1}\widehat{F}(k)$$

we now know

$$u(\widehat{x}) = (g*f)(\widehat{x}) = \frac{1}{\sqrt{2\pi}^3}\iiint_{\hat{y}}\frac{1}{|\widehat{x}-\widehat{y}|}e^{-m|\hat{x}-\hat{y}|}f(\widehat{y})d\widehat{y}.$$

EXERCISES

1. In three dimensions, find the Fourier transform of

$$f(r) = \frac{1}{(4\pi r)^{3/2}}e^{-r^2/4}.$$

CHAPTER

The Laplace Transform 12

12.1 INTRODUCTION

A second integral transform that plays a prominent role in the solution of differential equations is the Laplace transform. If $f(x)$ is a piecewise continuous function for which

$$f(x) = 0 \quad \text{if } x < 0$$

then the Laplace transform of $f(x)$, denoted $(\mathcal{L}f)(s)$, is defined to be

$$(\mathcal{L}f)(s) = \int_0^\infty e^{-sx} f(x) dx$$

for functions for which the integral converges.

We note a relationship between the Laplace transform and the Fourier transform. We have

$$(\mathcal{F}f)(s) = \frac{1}{\sqrt{2\pi}} \int_{-\infty}^{\infty} f(x) e^{-ixs} ds$$

so

$$(\mathcal{F}f)(-is) = \frac{1}{\sqrt{2\pi}} \int_{-\infty}^{\infty} f(x) e^{-ix(-is)} dx = \frac{1}{\sqrt{2\pi}} \int_{-\infty}^{\infty} f(x) e^{-xs} dx.$$

If $f(x) = 0$ for $x < 0$, then

$$(\mathcal{F}f)(-is) = \frac{1}{\sqrt{2\pi}} \int_{-\infty}^{\infty} f(x) e^{-xs} dx = \frac{1}{\sqrt{2\pi}} \int_0^{\infty} f(x) e^{-xs} dx = \frac{1}{\sqrt{2\pi}} (\mathcal{L}f)(s). \quad (1)$$

(This is one place where our definition of the Fourier transform makes things a little messier. There is a definition of the Fourier transform for which $(\mathcal{F}f)(-is) = (\mathcal{L}f)(s)$.)

Mathematical Physics with Partial Differential Equations. https://doi.org/10.1016/B978-0-12-814759-7.00012-0

EXERCISES

1. Show that

$$\mathcal{L}[t^n] = \frac{n!}{s^{n+1}} \quad s > 0.$$

2. Use the Maclaurin series for $\sin t$ and assume that the series can be integrated term by term to show

$$\mathcal{L}\{\sin t\} = \frac{1}{s^2 + 1}, \quad s > 1.$$

3. Use the Maclaurin series for

$$f(t) = \begin{cases} \dfrac{\sin t}{t}, & t \neq 0 \\ 1, & t = 0 \end{cases}$$

and assume that the series can be integrated term by term to show

$$\mathcal{L}\{f(t)\} = \tan^{-1}(1/s), \quad s > 1.$$

Note that the Taylor series for $\tan^{-1} x$ is $x - \frac{x^3}{3} + \frac{x^5}{5} - \cdots \quad x < 1$.

4. Use the Maclaurin series expansion for the order zero Bessel function

$$J_0(t) = \sum_{n=0}^{\infty} \frac{(-1)^n t^{2n}}{2^{2n}(n!)^2}$$

and assume that the series can be integrated term by term to show

$$\mathcal{L}\{J_0(t)\} = \left(s^2 + 1\right)^{-1/2}, \quad s > 1.$$

12.2 PROPERTIES OF THE LAPLACE TRANSFORM

The relationship between the Laplace and Fourier transforms suggests that certain properties of the two transforms are shared. For example, the Laplace transform can be viewed as a method to decompose a function. Also, as we now show, the analogous result about the Laplace transform of the convolution of two functions is the product of the Laplace transforms.

Theorem:

Let $F(s)$ and $G(s)$ denote the Laplace transforms of $f(x)$ and $g(x)$, respectively. Then

$$(\mathcal{L}(f * g))(s) = F(s)G(s).$$

Proof:
We have

$$(\mathcal{L}(f*g))(s) = \int_{t=0}^{\infty}(f*g)(t)e^{-st}dt = \int_{t=0}^{\infty}\left(\int_{z=0}^{t}f(z)g(t-z)dz\right)e^{-st}dt. \quad (1)$$

The reason the limits of integration are as they are in $\int_{z=0}^{t}f(z)g(t-z)dz$ is because $f(z)=0$ if $z<0$, and $g(t-z)=0$ if $z>t$. This is why the convolution of functions $f(x)$ and $g(x)$ that have a Laplace transform is usually given by

$$(f*g)(x) = \int_{z=0}^{t}f(z)g(x-z)dz$$

rather than

$$(f*g)(x) = \int_{z=-\infty}^{\infty}f(z)g(x-z)dz.$$

We change the order of integration in the double integral in Eq. (1). The region of integration is shown in Fig. 12.2.1.

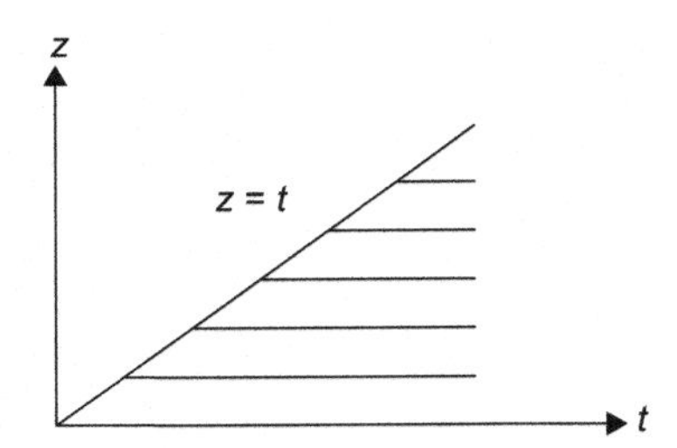

FIGURE 12.2.1

Changing the order of integration gives

$$\int_{t=0}^{\infty}\left(\int_{z=0}^{t}f(z)g(t-z)dz\right)e^{-st}dt = \int_{z=0}^{\infty}\left(\int_{t=z}^{\infty}f(z)g(t-z)e^{-st}dt\right)dz$$

$$= \int_{z=0}^{\infty}f(z)\left(\int_{t=z}^{\infty}g(t-z)e^{-st}t\right)dz.$$

In the integral

$$\int_{t=z}^{\infty}g(t-z)e^{-st}dt$$

we make the change of variables $u = t - z$. When $t = z$, then $u = 0$. When $t = \infty$, then $u = \infty$. Also, $dt = du$ and $e^{-st} = e^{-s(u+z)} = e^{-su}e^{-sz}$.

Thus

$$\int_{t=z}^{\infty} g(t-z)e^{-st}dt = \int_{u=0}^{\infty} g(u)e^{-su}e^{-sz}du = e^{-sz}\int_{u=0}^{\infty} g(u)e^{-su}du$$

so

$$\int_{z=0}^{\infty} f(z)\left(\int_{t=z}^{\infty} g(t-z)e^{-st}dt\right)dz$$

$$= \left(\int_{z=0}^{\infty} f(z)e^{-sz}dz\right)\left(\int_{u=0}^{\infty} g(u)e^{-su}du\right) = F(s)G(s).$$

Corollary:

With the notation of the theorem above

$$\mathcal{L}^{-1}(F(s)G(s)) = (f * g)(x).$$

As with the Fourier transform, it is the corollary that is useful in solving differential equations.

Some important results that we shall use in solving differential equations with the Laplace transform are the following theorem and its corollaries.

Theorem:

Suppose that the Laplace transform of $f(x)$ and $f'(x)$ exist and $f(x)$ is continuous. Then

$$(\mathcal{L}f')(s) = s(\mathcal{L}f)(s) - f(0).$$

Proof:

Integrating by parts with $u = e^{-sx}$ and $dv = f'$, we have

$$(\mathcal{L}f')(s) = \int_0^{\infty} e^{-sx}f'(x)dx = e^{-sx}f(x)\Big|_{x=0}^{\infty} + s\int_0^{\infty} e^{-sx}f(x)dx = s(\mathcal{L}f)(s) - f(0).$$

Corollary:

Suppose that the Laplace transform of $f(x)$, $f'(x)$, and $f''(x)$ exist and $f(x)$ and $f'(x)$ are continuous. Then

$$(\mathcal{L}f'')(s) = s^2(\mathcal{L}f)(s) - sf(0) - f'(0).$$

Proof:

By the theorem, we have

$$\begin{aligned}(\mathcal{L}f'')(s) &= s(\mathcal{L}f')(s) - f'(0) = s[s(\mathcal{L}f)(s) - f(0)] - f'(0)\\ &= s^2(\mathcal{L}f)(s) - sf(0) - f'(0).\end{aligned}$$

Continuing the idea of the above corollary, we have the following result.

Corollary:

Suppose the Laplace transform of $f(x)$, $f'(x)$, ..., $f^{(n)}(x)$ exist and $f(x), f'(x)$, ..., $f^{(n-1)}(x)$ are continuous. Then

$$\left(\mathcal{L}f^{(n)}\right)(s) = s^n(\mathcal{L}f)(s) - s^{n-1}f(0) - \cdots - sf^{(n-2)}(0) - f^{(n-1)}(0).$$

Table 12.2.1 gives some formulas for the Laplace transform. Several of these are proven in the exercises. In the table, $F(s)$ is the Laplace transform of $f(x)$.

Table 12.2.1 Some Properties and Formulas for the Laplace Transform

$\mathcal{L}(af(x) + bg(x)) = aF(s) + bG(s)$

$\mathcal{L}(xf(x)) = -\frac{d}{ds}F(s)$

$\mathcal{L}[x^n f(x)] = (-1)^n \frac{d^n}{ds^n}F(s)$

$\mathcal{L}[f(ax - b)] = \frac{1}{a}e^{-bs/a}F\left(\frac{s}{a}\right)$ $\quad a > 0,\ b \geq 0$

$\mathcal{L}[e^{ax}f(x)] = F(s - a)$

$\mathcal{L}\left[\int_0^x f(t)dt\right] = \frac{1}{s}F(s)$

$\mathcal{L}[(f * g)] = F(s)G(s)$

$\mathcal{L}\left[f^{(n)}(x)\right] = s^n F(s) - s^{n-1}f(0) - s^{n-2}f'(0) - \cdots - f^{(n-1)}(0)$

If $u(x) = \begin{cases} 1 & \text{if } x \geq 0 \\ 0 & \text{if } x < 0 \end{cases}$ then $\mathcal{L}[u(x)] = \frac{1}{s}$ and $\mathcal{L}[u(x-a)] = \frac{e^{-as}}{s}$ $\quad s > 0$, then

$\mathcal{L}[u(x - a)f(x - a)] = e^{-as}F(s)$

$\mathcal{L}[x] = \frac{1}{s^2}$ $\quad s > 0$

$\mathcal{L}[x^n] = \frac{n!}{s^{n+1}}$ $\quad s > 0$

$\mathcal{L}[e^{ax}] = \frac{1}{s-a}$

$\mathcal{L}[\sin(ax)] = \frac{a}{s^2+a^2}$ $\quad s > 0$

$\mathcal{L}[\cos(ax)] = \frac{s}{s^2+a^2}$ $\quad s > 0$

$\mathcal{L}[x\sin(a)] = \frac{2ax}{(s^2+a^2)^2}$ $\quad s > |a|$

$\mathcal{L}[x\cos(ax)] = \frac{s^2-a^2}{(s^2+a^2)^2}$ $\quad s > |a|$

$\mathcal{L}[f(ax)] = \frac{1}{a}F\left(\frac{s}{a}\right)$ $\quad a > 0$

$\mathcal{L}[e^{ax}x^n] = \frac{n!}{(s-a)^{n+1}}$ $\quad s > a$

$\mathcal{L}[e^{ax}\sin(bx)] = \frac{b}{(s-a)^2+b^2}$ $\quad s > a$

$\mathcal{L}[e^{ax}\cos(bx)] = \frac{s-a}{(s-a)^2+b^2}$ $\quad s > a$

$\mathcal{L}[\delta(s - a)] = e^{-as}$

$\mathcal{L}[(-x)^n f(x)] = F^{(n)}(s)$

EXERCISES

1. Show that
   a. $\mathcal{L}(xf(x)) = -\frac{d}{ds}F(s)$.
   b. $\mathcal{L}[\sin(ax)] = \frac{a}{s^2+a^2} \quad s > 0$.
   c. $\mathcal{L}[x\cos(ax)] = \frac{s^2-a^2}{(s^2+a^2)^2} \quad s > |a|$.
2. If $F(s) = \mathcal{L}\{f(t)\}$, show that if $F(s)$ exists for $s > b \geq 0$, then
   a. $\mathcal{L}\{f(at)\} = \frac{1}{a}F\left(\frac{s}{a}\right), \quad s > ab$.
   b. $\mathcal{L}\{e^{at}f(t)\} = F(s-a), \quad s > a+b$.
   c. $\mathcal{L}\left\{\frac{1}{a}e^{-bt/a}f\left(\frac{t}{a}\right)\right\} = F(as+b), \quad a > 0$.
3. Find the Laplace transform for the Dirac-δ function.
4. Find the Laplace transform of the functions below:
   a. $\int_0^t (t-\tau)\sin\tau d\tau$.
   b. $\int_0^t \cos(t-\tau)e^{\tau}d\tau$.
   c. $\int_0^t (t-\tau)\sin(4\tau)d\tau$.
5. Use the convolution theorem to find the inverse Laplace transform of the given functions:
   a. $\dfrac{1}{s^2(s^2+1)}$.
   b. $\dfrac{2}{s^2(s^2+4)}$.
   c. $\dfrac{s}{(s^2+4)^2}$.

12.3 SOLVING DIFFERENTIAL EQUATIONS USING THE LAPLACE TRANSFORM

The Laplace transform can be applied to solve both ordinary and partial differential equations.

We first demonstrate how the Laplace transform can be used to solve ordinary differential equations. The major steps of the process are as follows:

1. Apply the Laplace transform to the ordinary differential equation. We shall see that in the case where the coefficients are constants, this converts the ordinary differential equation to an algebraic equation. The point of the process is that the algebraic equation is easier to solve.
2. Solve the resulting algebraic equation.
3. Find the function whose Laplace transform is that of the solution found in Step (2). The result is the solution to the differential equation.

Like with solving differential equations with the Fourier transform, it is the third step, where we find the "inverse Laplace Transform" of a function that is the most difficult. There is a transform that accomplishes this, called the Mellin transform, but it is beyond the scope of this text.

We give two examples of how to use the Laplace transform to solve ordinary differential equations.

Example:
Solve the differential equation

$$y''(t) + 6y'(t) + 5y(t) = f(t), \quad y(0) = 1, \; y'(0) = 0$$

using the Laplace transform.

This example demonstrates how the Laplace transform converts an ordinary differential equation with constant coefficients into an algebraic equation and use of the convolution theorem.

Solution:
Letting $\mathcal{L}y(t) = Y(s)$, we have

$$\mathcal{L}y''(t) = s^2Y(s) - sy(0) - y'(0) = s^2Y(s) - s$$

$$\mathcal{L}y'(t) = sY(s) - y(0) = sY(s) - 1$$

and we let $F(s) = \mathcal{L}f(t)$.

Thus taking the Laplace transform of the equation yields

$$s^2Y(s) - s + 6[sY(s) - 1] + 5Y(s) = F(s)$$

or

$$(s^2 + 6s + 5)Y(s) - s - 6 = F(s).$$

So

$$Y(s) = \frac{s+6}{(s^2+6s+5)} + \frac{F(s)}{(s^2+6s+5)} = \frac{s+6}{(s+5)(s+1)} + \frac{F(s)}{(s+5)(s+1)}.$$

Taking the inverse Laplace transform gives

$$y(t) = \mathcal{L}^{-1}(Y(s)) = \mathcal{L}^{-1}\left(\frac{s+6}{(s+5)(s+1)} + \frac{F(s)}{(s+5)(s+1)}\right)$$

$$= \mathcal{L}^{-1}\left(\frac{s+6}{(s+5)(s+1)}\right) + \mathcal{L}^{-1}\left(\frac{F(s)}{(s+5)(s+1)}\right).$$

Using a CAS to decompose the first expression on the right into partial fractions yields

$$\frac{s+6}{(s+5)(s+1)} = \frac{5}{4(s+1)} - \frac{1}{4(s+5)}.$$

Now

$$\mathcal{L}^{-1}\left(\frac{s}{(s+5)(s+1)}\right) = \mathcal{L}^{-1}\left(\frac{5}{4(s+1)} - \frac{1}{4(s+5)}\right)$$

$$= \frac{5}{4}\mathcal{L}^{-1}\left(\frac{1}{s+1}\right) - \frac{1}{4}\mathcal{L}^{-1}\left(\frac{1}{s+5}\right) = \frac{5}{4}e^{-t} - \frac{1}{4}e^{-5t}.$$

We use the convolution theorem to get

$$\mathcal{L}^{-1}\left(\frac{F(s)}{(s+5)(s+1)}\right) = (f * g)(t)$$

where

$$g(t) = \mathcal{L}^{-1}\left(\frac{1}{(s+5)(s+1)}\right) = \mathcal{L}^{-1}\left(\frac{1}{4(s+1)} - \frac{1}{4(s+5)}\right) = \frac{1}{4}\left(e^{-t} - e^{-5t}\right).$$

So

$$(f * g)(t) = \int_0^t f(u)g(t-u)du = \int_0^t f(u)\left[\frac{1}{4}\left(e^{-(t-u)} - e^{-5(t-u)}\right)\right]du.$$

Finally,

$$y(t) = \frac{5}{4}e^{-t} - \frac{1}{4}e^{-5t} + \int_0^t f(u)\left[\frac{1}{4}\left(e^{-(t-u)} - e^{-5(t-u)}\right)\right]du.$$

The next example gives a linear equation where the coefficients are not constants. The main point of the example is to demonstrate a use of the relation

$$\mathcal{L}(ty') = -\frac{d}{ds}\mathcal{L}[y'].$$

Example:
Solve the equation

$$y''(t) + ty'(t) + y(t) = 0, \quad y(0) = 1, \; y'(0) = 0.$$

Solution:
Letting $\mathcal{L}y(t) = Y(s)$, we have

$$\mathcal{L}y''(t) = s^2Y(s) - sy(0) - y'(0) = s^2Y(s) - s$$

$$\mathcal{L}y'(t) = sY(s) - y(0) = sY(s) - 1.$$

Then

$$\mathcal{L}[ty'(t)] = -\frac{d}{ds}\mathcal{L}[y'] = -\frac{d}{ds}[sY(s) - 1] = -Y(s) - sY'(s).$$

Taking the Laplace transform of the differential equation thus yields

$$\left[s^2Y(s) - s\right] + \left[-Y(s) - sY'(s)\right] + \left[Y(s)\right] = 0$$

or

$$s^2Y(s) - sY'(s) - s = 0$$

or

$$Y'(s) - s\,Y(s) = 1.$$

In this case, taking the Laplace transform has reduced the order of the differential equation from two to one. The resulting differential equation is one that is familiar in a first differential equations course. It is solved by multiplying by an integrating factor, which in this case is $e^{-s^2/2}$, to give

$$e^{-s^2/2}Y'(s) - e^{-s^2/2}s\,Y(s) = e^{-s^2/2}.$$

But

$$\frac{d}{ds}\left(e^{-s^2/2}Y(s)\right) = e^{-s^2/2}Y'(s) - e^{-s^2/2}s\,Y(s)$$

so we have

$$\left(e^{-s^2/2}Y(s)\right)' = e^{-s^2/2}.$$

Then

$$e^{-s^2/2}Y(s) - C = \int_s^\infty \left(e^{-z^2/2}Y(z)\right)' dz = \int_s^\infty e^{-z^2/2}dz$$

where $C = \lim_{z\to\infty} e^{-z^2/2}Y(z) = 0$.

Integrating and simplifying gives

$$Y(s) = e^{s^2/2}\left[\int_s^\infty e^{-z^2/2}z\right].$$

Now,

$$Y(s) = \left[\int_s^\infty e^{-z^2/2}e^{s^2/2}dz\right] = \int_s^\infty e^{-(z^2-s^2)/2}dz.$$

We can find $y(t)$ by making the change of variables $u = z - s$. Then $du = dz$; $z^2 - s^2 = (z-s)(z+s) = u(u+2s) = u^2 + 2su$; if $z = s$, then $u = 0$; if $z = \infty$, then $u = \infty$.

So

$$Y(s) = \int_s^\infty e^{-(z^2-s^2)/2}dz = \int_0^\infty e^{-(u^2+2su)/2}du = \int_0^\infty e^{-su}e^{-u^2/2}du.$$

We recognize that the expression on the right is the Laplace transform of $e^{-t^2/2}$. Thus the solution to the differential equation is

$$y(t) = e^{-t^2/2}.$$

Example:

In this example we show how to find the Laplace transform of a function with a jump discontinuity. Let

$$u_c(t) = \begin{cases} 0 & t < c \\ 1 & t \geq c \end{cases}.$$

In Exercise 6 at the end of this section, we show $\mathcal{L}\{u_c(t)f(t-c)\} = e^{-cs}F(s)$. Let

$$f(t) = \begin{cases} t & 0 \leq t \leq 2 \\ t + (t-2)^3 & t > 2 \end{cases}.$$

Then $f(t) = t + u_c(t)(t-2)^3$. So

$$\mathcal{L}\{f(t)\} = \mathcal{L}\{t\} + \mathcal{L}\left\{u_c(t)(t-2)^3\right\} = \mathcal{L}\{t\} + e^{-2s}\mathcal{L}\{t^2\}.$$

Consider the initial value problem

$$y''(t) + 2y'(t) - 8y(t) = 3\delta_0(t-2); \quad y(0) = 2,\ y'(0) = -1.$$

Taking the Laplace transform, we have

$$\begin{aligned} &\left[s^2Y(s) - sy(0) - y'(0)\right] + 2[sY(s) - y(0)] - 8Y(s) \\ &= s^2Y(s) - 2s + 1 + 2sY(s) - 4 - 8Y(s) = 3e^{-2s} \end{aligned}$$

so

$$Y(s)\left(s^2 + 2s - 8\right) = 3e^{-2s} + 2s + 3.$$

Thus

$$Y(s) = e^{-2s}\frac{3}{(s+4)(s-2)} + \frac{2s+3}{(s+4)(s-2)}.$$

Using

$$\frac{3}{(s+4)(s-2)} = \frac{1}{2}\frac{1}{s-2} - \frac{1}{2}\frac{1}{s+4} \text{ and } \frac{2s+3}{(s+4)(s-2)} = \frac{5}{6}\frac{1}{(s+4)} + \frac{7}{6}\frac{1}{(s-2)}$$

we have

$$Y(s) = e^{-2s}Y_1(s) + Y_2(s)$$

where

$$Y_1(s) = \frac{1}{2}\frac{1}{(s-2)} - \frac{1}{2}\frac{1}{(s+4)} \text{ and } Y_2(s) = \frac{5}{6}\frac{1}{(s+4)} + \frac{7}{6}\frac{1}{(s-2)}.$$

So

$$y(t) = u_2(t)y_1(t-2) + y_2(t) = u_2(t)\left[\frac{1}{2}e^{2(t-2)} - \frac{1}{2}e^{-4(t-2)}\right] + \frac{5}{6}e^{-4t} + \frac{7}{6}e^{2t}.$$

EXERCISES

Use the Laplace transform to solve the following initial value problems:

1. $y''(x) + y(x) = 0, \quad y(0) = 2,\ y'(0) = 3.$
2. $y''(x) + 7y'(x) + 12y(x) = 0,\ y(0) = 3,\ y'(0) = -2.$
3. $y''(x) + 4y'(x) + 3y(x) = x, \quad y(0) = 0,\ y'(0) = 4.$
4. $y''(x) + 5y'(x) + 6y(x) = e^{-x}, \quad y(0) = y'(0) = 0.$
5. $y''(x) - 4y'(x) + 3y(x) = \sin x, \quad y(0) = 2,\ y'(0) = 0.$
6. For $c > 0$ define

$$u_c(t) = \begin{cases} 0 & t < c \\ 1 & t \geq c \end{cases}.$$

Show that $\mathcal{L}\{u_c(t)f(t-c)\} = e^{-cs}F(s)$. This is used in the next two exercises.

7. Use the Laplace transform to solve the initial value problem

$$y''(x) + y'(x) - 12y(x) = \delta_0(x-1), \quad y(0) = 0,\ y'(0) = 1.$$

8. A spring system is modeled by the differential equation

$$mx''(t) + cx'(t) + kx(t) = f(t)$$

where $m =$ mass of the spring, $c =$ damping factor, $k =$ spring constant, $f(t) =$ force, $x(t) =$ displacement from equilibrium. Suppose a spring is hit with a hammer at $t = 2$ that imparts a nearly instantaneous impulse of 1, so that $f(t) = \delta_0(t-2)$. Find $x(t)$ if $m = 1,\ c = 8$ and $k = 15$ using the Laplace transform.

12.4 SOLVING THE HEAT EQUATION USING THE LAPLACE TRANSFORM

As with the Fourier transform, we shall use the Laplace transform to solve partial differential equations. Consider two forms of the heat equation problem.

$$\frac{\partial u(x,t)}{\partial t} - \frac{\partial^2 u(x,t)}{\partial x^2} = 0, \quad -\infty < x < \infty,\ t > 0$$

$$u(x,0) = f(x), \quad u(0,t) = g(t)$$

and

$$\frac{\partial u(x,t)}{\partial t} - \frac{\partial^2 u(x,t)}{\partial x^2} = 0, \quad 0 < x < \infty, \ t > 0$$

$$u(x,0) = f(x), \quad u(0,t) = g(t)$$

where $g(t) = 0$ if $t < 0$.

Since $u(x,t)$ is a function of two variables, it might seem that we could apply the Laplace transform with respect to either variable. However, the Laplace transform is defined only for functions that are zero when the variable is negative. Thus we could apply the Laplace transform with respect to either variable in the second problem, but only with respect to t in the first problem.

Let $u(x,t)$ be a function for which $u(x,t) = 0$ if $x < 0$ or $t < 0$. Then we can take the Laplace transform of $u(x,t)$ with respect to either variable.

Let $\mathcal{L}_t$ denote the Laplace transform with respect to t and $\mathcal{L}_x$ denote the Laplace transform with respect to x. That is,

$$\mathcal{L}_t(u(x,t)) = U(x,s) = \int_0^\infty u(x,t)e^{-st}dt$$

$$\mathcal{L}_x(u(x,t)) = U(s,t) = \int_0^\infty u(x,t)e^{-sx}dx.$$

If we take the Laplace transform with respect to t, then

$$\mathcal{L}_t\left(\frac{\partial u(x,t)}{\partial t}\right) = \int_0^\infty \frac{\partial u(x,t)}{\partial t} e^{-st}dt = s\,U(x,s) - u(x,0) = s\,U(x,s) - f(x)$$

and

$$\mathcal{L}_t\left(\frac{\partial^2 u(x,t)}{\partial x^2}\right) = \int_0^\infty \frac{\partial^2 u(x,t)}{\partial x^2} e^{-st}dt = \frac{\partial^2}{\partial x^2}\int_0^\infty u(x,t)e^{-st}dt = \frac{\partial^2 U(x,s)}{\partial x^2}.$$

Analogous equations hold if one takes the Laplace transform with respect to x. The boundary condition transforms to

$$\mathcal{L}_t(u(0,t)) = \int_0^\infty g(t)e^{-st}dt = G(s).$$

Thus taking the Laplace transform with respect to t of the given equation yields

$$s\,U(x,s) - f(x) - \frac{\partial^2 U(x,s)}{\partial x^2} = 0 \quad U(0,s) = G(s).$$

In the next example, we solve such a problem for given functions $f(x)$ and $g(t)$. As one would expect because of the relationship between the Fourier transform and the Laplace transform given by Eq. (1) in Section 12.1, the process of solving partial differential equations with the two techniques is virtually identical. We recall that the major steps with the Fourier transform were as follows:

1. Convert the PDE to an ODE using the transform.
2. Solve the ODE.
3. Convert the solution back to the original space-time variables.

We reiterate that the Fourier transform when applied to the space (x) variable is valid for all problems, but to apply the Laplace transform to the x variable, the function must be zero when x is negative.

Example:
We solve the heat equation

$$\frac{\partial u(x,t)}{\partial t} - \frac{\partial^2 u(x,t)}{\partial x^2} = 0, \quad 0 < x < \infty, \ t > 0$$

$$u(x,0) = 1, \quad u(0,t) = \begin{cases} 1 & 0 < t < 2 \\ 0 & \text{otherwise} \end{cases}$$

using the Laplace transform.

Solution:
We apply the Laplace transform with respect to time to the equation. Since

$$\mathcal{L}\left(\frac{\partial u(x,t)}{\partial t}\right) = \int_0^\infty \frac{\partial u(x,t)}{\partial t} e^{-st} dt = s\,U(x,s) - u(x,0) = s\,U(x,s) - 1$$

and

$$\mathcal{L}\frac{\partial^2 u(x,t)}{\partial x^2} = \frac{\partial^2 U(x,s)}{\partial x^2}$$

we have

$$s\,U(x,s) - 1 - \frac{d^2 U(x,s)}{dx^2} = 0$$

or

$$\frac{d^2 U(x,s)}{dx^2} - s\,U(x,s) = 1$$

where the equation becomes an ordinary differential equation because we are assuming that s is fixed.

The boundary condition is

$$U(0,s) = \mathcal{L}[u(0,t)] = \int_0^2 e^{-st} dt = \frac{1}{s} - \frac{e^{-2s}}{s}.$$

The associated homogeneous equation is

$$\frac{d^2 U(x,s)}{dx^2} - sU(x,s) = 0,$$

for which the solution is

$$U_H(x,s) = c_1(s)e^{-\sqrt{s}x} + c_2(s)e^{\sqrt{s}x}. \tag{1}$$

It must be remembered that we are solving the equation for a fixed s, and the "constants" of integration depend on s.

To find a particular solution, we use variation of parameters as we did in the section on Green's functions.

Two linearly independent solutions to the homogeneous equation are

$$y_1(x) = e^{-\sqrt{s}x}, \quad y_2(x) = e^{\sqrt{s}x}.$$

So

$$y_1{}'(x) = -\sqrt{s}e^{-\sqrt{s}x}, \quad y_2{}'(x) = \sqrt{s}e^{\sqrt{s}x}$$

and

$$\begin{aligned}
U_p(x,s) &= \int_0^x \frac{y_1(t)y_2(x) - y_1(x)y_2(t)}{y_1(t)y_2{}'(t) - y_1{}'(t)\,y_2(t)} f(t)dt \\
&= \int_0^x \frac{e^{-\sqrt{st}}e^{\sqrt{sx}} - e^{-\sqrt{sx}}e^{\sqrt{st}}}{e^{-\sqrt{st}}\sqrt{s}e^{\sqrt{st}} - e^{-\sqrt{st}}\left(-\sqrt{s}\right)e^{\sqrt{st}}}(1)dt \\
&= \int_0^x \frac{e^{-\sqrt{st}}e^{\sqrt{sx}} - e^{-\sqrt{sx}}e^{\sqrt{st}}}{2\sqrt{s}}dt = \frac{1}{2\sqrt{s}}\int_0^x \left[e^{\sqrt{s}(x-t)} - e^{-\sqrt{s}(x-t)}\right]dt.
\end{aligned}$$

Now

$$\int_0^x \left[e^{\sqrt{s}(x-t)}\right]dt = -\frac{1}{\sqrt{s}} + \frac{1}{\sqrt{s}}e^{\sqrt{sx}}, \text{ and } \int_0^x \left[e^{-\sqrt{s}(x-t)}\right]dt = \frac{1}{\sqrt{s}} - \frac{1}{\sqrt{s}}e^{-\sqrt{sx}}$$

so

$$\begin{aligned}
y_p(x) &= \frac{1}{2\sqrt{s}}\left[\left(-\frac{1}{\sqrt{s}} + \frac{1}{\sqrt{s}}e^{\sqrt{sx}}\right) - \left(\frac{1}{\sqrt{s}} - \frac{1}{\sqrt{s}}e^{-\sqrt{sx}}\right)\right] \\
&= -\frac{1}{s} + \frac{1}{2s}e^{\sqrt{sx}} + \frac{1}{2s}e^{-\sqrt{sx}}.
\end{aligned}$$

We add the particular solution to the solution to the homogeneous equation given by Eq. (1) to get the general solution

$$\begin{aligned}
U(x,s) = U_H(x,s) + U_p(x,s) &= c_1(s)e^{-\sqrt{sx}} + c_2(s)e^{\sqrt{sx}} - \frac{1}{s} + \frac{1}{2s}e^{\sqrt{sx}} + \frac{1}{2s}e^{-\sqrt{sx}} \\
&= \left(c_1(s) + \frac{1}{2s}\right)e^{-\sqrt{sx}} + \left(c_2(s) + \frac{1}{2s}\right)e^{\sqrt{sx}} - \frac{1}{s} \\
&= C_1(s)e^{-\sqrt{sx}} + C_2(s)e^{\sqrt{sx}} - \frac{1}{s}.
\end{aligned}$$

$$C_1(s) = c_1(s) + \frac{1}{2s} \text{ and } C_2(s) = c_2(s) + \frac{1}{2s}.$$

We use the requirement that $U(x, s)$ is bounded as $s \to \infty$ to get $C_2(s) = 0$. Thus we have to this point

$$U(x, s) = C_1(s)e^{-\sqrt{s}x} + \frac{1}{s}.$$

We have the boundary condition

$$U(0, s) = \mathcal{L}[u(0, t)] = \int_0^2 e^{-st}\,dt = \frac{1}{s} - \frac{e^{-2s}}{s}.$$

Now

$$U(0, s) = C_1(s)e^{-\sqrt{s}0} + \frac{1}{s} = \frac{1}{s} - \frac{e^{-2s}}{s}$$

and thus

$$C_1(s) = -\frac{e^{-2s}}{s}.$$

Finally, we have

$$U(x, s) = -\frac{e^{-2s}}{s}\,e^{-\sqrt{s}x} + \frac{1}{s}.$$

To find $u(x, t)$ we must find the inverse Laplace transform of $U(x, s)$. Now

$$\mathcal{L}^{-1}\left(\frac{1}{s}\right) = 1$$

and we use the convolution theorem to find

$$\mathcal{L}^{-1}\left(-\frac{e^{-2s}}{s}\,e^{-\sqrt{s}x}\right).$$

We find

$$\mathcal{L}^{-1}\left(-\frac{e^{-2s}}{s}\right) \text{ and } \mathcal{L}^{-1}\left(e^{-s\sqrt{x}}\right).$$

We have

$$\mathcal{L}^{-1}\left(-\frac{e^{-2s}}{s}\right) = \begin{cases} 0 & \text{if } t < 2 \\ -1 & \text{if } t \geq 2 \end{cases}.$$

The inverse transform

$$\mathcal{L}^{-1}\left(e^{-\sqrt{s}x}\right)$$

often does not appear in tables. We show that

$$\text{if } f(t) = \frac{x}{\sqrt{4\pi t^3}}e^{-x^2/4t}, \quad t > 0$$

then

$$\mathcal{L}[f(t)] = e^{-x\sqrt{s}}.$$

In Section 1.5 we showed that

$$\int_0^\infty e^{-\left(a^2x^2+\frac{b^2}{x^2}\right)}dx = \frac{\sqrt{\pi}}{2a}e^{-2ab}.$$

Consider

$$\mathcal{L}[f(t)] = \int_0^\infty \frac{x}{\sqrt{4\pi t^3}}e^{-\frac{x^2}{4t}}e^{-st}dt.$$

Let $u = \frac{x}{2\sqrt{t}}$. Then

$$du = -\frac{x}{4\sqrt{t^3}}, \quad u^2 = \frac{x^2}{4t}, \quad t = \frac{x^2}{4u^2} \text{ and } st = \frac{sx^2}{4u^2}.$$

If $t = 0$, then $u = \infty$; and if $t = \infty$, then $u = 0$.

Thus

$$\int_0^\infty \frac{x}{\sqrt{4\pi t^3}}e^{-\frac{x^2}{4t}}e^{-st}dt = \frac{2}{\sqrt{\pi}}\int_\infty^0 e^{-\left(u^2+\frac{sx^2}{4u^2}\right)}(-du)$$

$$= \frac{2}{\sqrt{\pi}}\int_0^\infty e^{-\left(u^2+\frac{sx^2}{4u^2}\right)}du = \frac{2}{\sqrt{\pi}}\frac{\sqrt{\pi}}{2}e^{-\frac{2x\sqrt{s}}{2}} = e^{-x\sqrt{s}}.$$

Thus

$$\mathcal{L}\left[\frac{x}{\sqrt{4\pi t^3}}e^{-x^2/4t}\right] = e^{-x\sqrt{s}} \text{ and } \mathcal{L}^{-1}\left[e^{-x\sqrt{s}}\right] = \frac{x}{\sqrt{4\pi t^3}}e^{-\frac{x^2}{4t}}.$$

Finally,

$$u(x,t) = \mathcal{L}^{-1}\left[-\frac{e^{-2s}}{s}e^{-\sqrt{s}x} + \frac{1}{s}\right] = \mathcal{L}^{-1}\left[\frac{1}{s}\right] + \mathcal{L}^{-1}\left[-\frac{e^{-2s}}{s}e^{-\sqrt{s}x}\right]$$

$$= 1 - \left(\frac{x}{\sqrt{4\pi t^3}}e^{-\frac{x^2}{4t}}\right) * 1_{[2,\infty]}(x) = 1 - \int_2^\infty \frac{x}{\sqrt{4\pi(t-\tau)^3}}e^{-\frac{x^2}{(t-\tau)}}d\tau$$

where

$$1_{[2,\infty]}(x) = \begin{cases} 1 & \text{if } x \geq 2 \\ 0 & \text{otherwise} \end{cases}.$$

EXERCISES

1. Show that if $f(t) = \frac{1}{\sqrt{\pi t}} e^{-k^2/4t}$, $t > 0$ then $\mathcal{L}[f(t)] = \frac{1}{\sqrt{s}} e^{-k\sqrt{s}}$. Hint: Note that $\frac{d}{dk} \frac{1}{\sqrt{\pi t}} e^{-k^2/4t} = -\frac{k}{\sqrt{4\pi t^3}} e^{-k^2/4t}$.

2. Use the Laplace transform to solve the heat equation

$$\frac{\partial u(x,t)}{\partial t} - \frac{\partial^2 u(x,t)}{\partial x^2} = 0, \quad u(x,0) = 0, \ u(0,t) = 4; \quad \lim_{x\to\infty} u(x,t) = 0,$$
$$0 < x < \infty, \ t > 0.$$

3. Use the Laplace transform to solve the heat equation

$$\frac{\partial u(x,t)}{\partial t} - \frac{\partial^2 u(x,t)}{\partial x^2} = 0, \quad u(x,0) = 0, \ u\ (0,t) = \sin t; \quad \lim_{x\to\infty} u(x,t) = 0,$$
$$0 < x < \infty, \ t > 0.$$

4. Use the Laplace transform to solve the heat equation

$$\frac{\partial u(x,t)}{\partial t} - \frac{\partial^2 u(x,t)}{\partial x^2} = 0, \quad u(x,0) = 0, \ u(0,t) = f(t); \quad \lim_{x\to\infty} u(x,t) = 0,$$
$$0 < x < \infty, \ t > 0.$$

5.

   a. Use the Laplace transform to solve the heat equation

$$\frac{\partial u(x,t)}{\partial t} - \frac{\partial^2 u(x,t)}{\partial x^2} = 0, \quad u(x,0) = 0, \ u_x(0,t) = f(t); \quad \lim_{x\to\infty} u(x,t) = 0,$$
$$0 < x < \infty, \ t > 0.$$

   b. Show that for $t > 0$

$$u(0,t) = \int_0^t \frac{f(\tau)}{\sqrt{\pi(t-\tau)}} d\tau.$$

6. The flux at $x = 0$ is

$$-\frac{\partial}{\partial x} u(x,t)|_{x=0}.$$

Let

$$\frac{\partial u(x,t)}{\partial t} - \frac{\partial^2 u(x,t)}{\partial x^2} = 0, \quad u(x,0) = 0,\ u(0,t) = f(t); \quad \lim_{x\to\infty} u(x,t) = 0,$$
$0 < x < \infty,\ t > 0.$

We have shown that $U(x,s) = \widehat{f}(s)e^{-x\sqrt{s}}$ so that $\mathcal{L}\left[-\frac{\partial}{\partial x}u(x,t)\right] = -\frac{\partial}{\partial x}U(x,s) = \sqrt{s}\widehat{f}(s)e^{-x\sqrt{s}}$ so that $-\frac{\partial}{\partial x}U(x,s)|_{x=0} = \sqrt{s}\widehat{f}(s)$, and $-\frac{\partial}{\partial x}u(x,t)|_{x=0}$ $= \mathcal{L}^{-1}\left[\sqrt{s}\widehat{f}(s)\right] = \mathcal{L}^{-1}\left[\frac{1}{\sqrt{s}}s\widehat{f}(s)\right].$

Use

$$\mathcal{L}^{-1}\left[\frac{1}{\sqrt{s}}\right] = \frac{1}{\sqrt{\pi t}} \text{ and } \mathcal{L}^{-1}\left[s\widehat{f}(s)\right] = f'(t)$$

to find $\mathcal{L}^{-1}\left[\sqrt{s}\widehat{f}(s)\right]$.

12.5 THE WAVE EQUATION AND THE LAPLACE TRANSFORM

We apply the Laplace transform to solve the wave equation

$$\frac{\partial^2 u(x,t)}{\partial^2} = c^2\frac{\partial^2 u(x,t)}{\partial x^2}, \quad -\infty < x < \infty,\ t > 0,$$
$$u(x,0) = f(x), \quad \frac{\partial u(x,0)}{\partial t} = g(x). \tag{1}$$

The steps in the solution are as follows:

1. Take the Laplace transform with respect to t to change the partial differential equation to an ordinary differential equation.
2. Solve the ordinary differential equation using a variation of parameters.
3. Take the inverse Laplace transform of the ordinary differential equation to obtain the solution to the original partial differential equation.

Let $U(x,s)$ denote the Laplace transform of $u(x,t)$ with respect to t. That is,

$$U(x,s) = \mathcal{L}[u(x,t)] = \int_0^\infty e^{-st}u(x,t)dt.$$

Recall that

$$\mathcal{L}\left[\frac{\partial^2 u(x,t)}{\partial t^2}\right] = s^2U(x,s) - s\,u(x,0) - \frac{\partial u(x,0)}{\partial t}$$

so that taking the Laplace transform of Eq. (1) gives

$$s^2U(x,s) - s\,u(x,0) - \frac{\partial u(x,0)}{\partial t} = c^2\frac{\partial^2 U(x,s)}{\partial x^2}$$

or

$$s^2U(x,s) - sf(x) - g(x) = c^2\frac{\partial^2 U(x,s)}{\partial x^2}$$

so

$$-sf(x) - g(x) = c^2\frac{\partial^2 U(x,s)}{\partial x^2} - s^2U(x,s)$$

or

$$\frac{\partial^2 U(x,s)}{\partial x^2} - \frac{s^2}{c^2}U(x,s) = -\frac{s}{c^2}f(x) - \frac{1}{c^2}g(x). \tag{2}$$

We fix s and then regard Eq. (2) as an ordinary differential equation with respect to the variable x. Two linearly independent solutions of the associated homogeneous equation

$$\frac{d^2U(x,s)}{dx^2} - \frac{s^2}{c^2}U(x,s) = 0$$

are

$$y_1(x,s) = e^{\frac{-sx}{c}} \text{ and } y_2(x,s) = e^{\frac{sx}{c}}.$$

We find a particular solution to Eq. (2) using

$$y_p(x,s) = \int_0^x \left[\frac{y_1(\xi,s)y_2(x,s) - y_1(x,s)y_2(\xi,s)}{y_1(\xi,s)y_2{}'(\xi,s) - y_1{}'(\xi,s)y_2(\xi,s)}\right]h(\xi,s)d\xi$$

where

$$h(\xi,s) = -\frac{s}{c^2}f(\xi) - \frac{1}{c^2}g(\xi).$$

Now

$$y_1(\xi,s)y_2{}'(\xi,s) - y_1{}'(\xi,s)y_2(\xi,s) = e^{\frac{-s}{c}}e^{\frac{s\xi}{c}}\left(\frac{s}{c}\right) - e^{\frac{-s\xi}{c}}\left(-\frac{s}{c}\right)e^{\frac{s\xi}{c}} = \frac{2s}{c}$$

and

$$y_1(\xi,s)y_2(x,s) - y_1(x,s)y_2(\xi,s) = e^{\frac{-s(\xi-x)}{c}} - e^{\frac{-s(x-\xi)}{c}}.$$

Thus the general solution to Eq. (2) is

$$U(x,s) \equiv y(x,s)$$

$$= A(s)y_1(x,s) + B(s)y_2(x,s) + \int_0^x \frac{c}{2s}\left[e^{\frac{-s(\xi - x)}{c}} - e^{\frac{-s(x-\xi)}{c}}\right] h(\xi,s)d\xi. \tag{3}$$

Substituting

$$y_1(x,s) = e^{\frac{-sx}{c}} \text{ and } y_2(x,s) = e^{\frac{sx}{c}}$$

Eq. (3) may be written

$$U(x,s) \equiv y(x,s)$$

$$= A(s)e^{\frac{-sx}{c}} + B(s)e^{\frac{sx}{c}} + \frac{c}{2s}\left[\int_0^x e^{\frac{sx}{c}} e^{\frac{-s\xi}{c}}(h(\xi,s))d\xi - \int_0^x e^{\frac{-sx}{c}} e^{\frac{s\xi}{c}}(h(\xi,s))d\xi\right]$$

$$= e^{\frac{-sx}{c}}\left[A(s) - \frac{c}{2s}\int_0^x e^{\frac{s\xi}{c}}(h(\xi,s))d\xi\right] + e^{\frac{sx}{c}}\left[B(s) + \frac{c}{2s}\int_0^x e^{\frac{-s\xi}{c}}(h(\xi,s))d\xi\right].$$

If $\lim_{x\to -\infty} U(x,s)$ is finite, then we must have

$$\lim_{x\to -\infty} A(s) - \frac{c}{2s}\int_0^x e^{\frac{s\xi}{c}}(h(\xi,s))d\xi = 0$$

so

$$A(s) = \frac{c}{2s}\int_0^{-\infty} e^{\frac{s\xi}{c}}(h(\xi,s))d\xi = -\frac{c}{2s}\int_{-\infty}^0 e^{\frac{s\xi}{c}}(h(\xi,s))d\xi.$$

If $\lim_{x\to\infty} U(x,s)$ is finite, then we must have

$$\lim_{x\to\infty} B(s) + \frac{c}{2s}\int_0^x e^{\frac{-s\xi}{c}}(h(\xi,s))d\xi = 0$$

so

$$B(s) = -\frac{c}{2s}\int_0^{\infty} e^{\frac{-s\xi}{c}}(h(\xi,s))d\xi.$$

Thus

$$
\begin{aligned}
U(x,s) &\equiv y(x,s)\\
&= e^{\frac{-sx}{c}}\left[A(s) - \frac{c}{2s}\int_0^x e^{\frac{s\xi}{c}}(h(\xi,s))d\xi\right]\\
&\quad + e^{\frac{sx}{c}}\left[B(s) + \frac{c}{2s}\int_0^x e^{\frac{-s\xi}{c}}(h(\xi,s))d\xi\right]\\
&= e^{\frac{-sx}{c}}\left[-\frac{c}{2s}\int_{-\infty}^0 e^{\frac{s\xi}{c}}(h(\xi,s))d\xi - \frac{c}{2s}\int_0^x e^{\frac{s\xi}{c}}(h(\xi,s))d\xi\right]\\
&\quad + e^{\frac{sx}{c}}\left[-\frac{c}{2s}\int_0^\infty e^{\frac{-s\xi}{c}}(h(\xi,s))d\xi + \frac{c}{2s}\int_0^x e^{\frac{-s\xi}{c}}(h(\xi,s))d\xi\right]\\
&= -\frac{c}{2s}e^{\frac{-sx}{c}}\left[\int_{-\infty}^0 e^{\frac{s\xi}{c}}(h(\xi,s))d\xi + \int_0^x e^{\frac{s\xi}{c}}(h(\xi,s))d\xi\right]\\
&\quad - \frac{c}{2s}e^{\frac{sx}{c}}\left[\int_0^\infty e^{\frac{-s\xi}{c}}(h(\xi,s))d\xi - \int_0^x e^{\frac{-s\xi}{c}}(h(\xi,s))d\xi\right]\\
&= -\frac{c}{2s}\left\{e^{\frac{-sx}{c}}\left[\int_{-\infty}^x e^{\frac{s\xi}{c}}(h(\xi,s))d\xi\right] + e^{\frac{sx}{c}}\int_x^\infty e^{\frac{-s\xi}{c}}(h(\xi,s))d\xi\right\}\\
&= -\frac{c}{2s}\left[\int_{-\infty}^x e^{\frac{s(\xi-x)}{c}}h(\xi,s)d\xi + \int_x^\infty e^{\frac{s(x-\xi)}{c}}(h(\xi,s))d\xi\right]\\
&= -\frac{c}{2s}\int_{-\infty}^\infty e^{\frac{-s|x-\xi|}{c}}h(\xi,s)d\xi.
\end{aligned}
$$

Now

$$h(\xi,s) = -\frac{s}{c^2}f(\xi) - \frac{1}{c^2}g(\xi)$$

so

$$U(x,s) = -\frac{c}{2s}\int_{-\infty}^{\infty} e^{\frac{-s|x-\xi|}{c}} h(\xi,s)d\xi = -\frac{c}{2s}\int_{-\infty}^{\infty} e^{\frac{-s|x-\xi|}{c}}\left[-\frac{s}{c^2}f(\xi) - \frac{1}{c^2}g(\xi)\right]d\xi$$

$$= \frac{1}{2c}\int_{-\infty}^{\infty} e^{\frac{-s|x-\xi|}{c}} f(\xi)d\xi + \frac{1}{2c}\int_{-\infty}^{\infty}\frac{1}{s}e^{\frac{-s|x-\xi|}{c}} g(\xi)d\xi.$$

We now find $\mathcal{L}^{-1}[U(x,s)]$ We first note that

$$\mathcal{L}\left[\int_{-\infty}^{\infty} f(\xi)\delta\left(t-\frac{1}{c}|x-\xi|\right)d\xi\right] = \int_0^{\infty}\left[\int_{-\infty}^{\infty} f(\xi)\delta\left(t-\frac{1}{c}|x-\xi|\right)d\xi\right]e^{-st}dt.$$

Interchanging the order of integration, this is

$$\int_{-\infty}^{\infty}\left[\int_0^{\infty} e^{-st}\delta\left(t-\frac{1}{c}|x-\xi|\right)dt\right]f(\xi)d\xi = \int_{-\infty}^{\infty} e^{\frac{-s|x-\xi|}{c}} f(\xi)\xi.$$

Since

$$\mathcal{L}\left[\int_{-\infty}^{\infty} f(\xi)\delta\left(t-\frac{1}{c}|x-\xi|\right)d\xi\right] = \int_{-\infty}^{\infty} e^{\frac{-s|x-\xi|}{c}} f(\xi)d\xi$$

then

$$\mathcal{L}^{-1}\left[\int_{-\infty}^{\infty} e^{\frac{-s|x-\xi|}{c}} f(\xi)d\xi\right] = \int_{-\infty}^{\infty} f(\xi)\delta\left(t-\frac{1}{c}|x-\xi|\right)d\xi. \tag{5}$$

If $H(t)$ is the Heaviside function, then

$$\mathcal{L}\left[\int_{-\infty}^{\infty} g(\xi)H\left(t-\frac{1}{c}|x-\xi|\right)d\xi\right]$$

$$= \int_0^{\infty}\left[\int_{-\infty}^{\infty} g(\xi)H\left(t-\frac{1}{c}|x-\xi|\right)d\xi\right]e^{-st}dt.$$

Interchanging the order of integration, this is

$$\int_{-\infty}^{\infty}\left[\int_0^{\infty} e^{-st}H\left(t-\frac{1}{c}|x-\xi|\right)dt\right]g(\xi)\,d\xi. \tag{6}$$

Now

$$\int_0^{\infty} e^{-st} H\left(t - \frac{1}{c}|x-\xi|\right) dt = \int_{\frac{1}{c}|x-\xi|}^{\infty} e^{-st} dt = \frac{1}{s} e^{\frac{-s|x-\xi|}{c}}$$

so Eq. (6) is

$$\int_{-\infty}^{\infty} \frac{1}{s} e^{\frac{-s|x-\xi|}{c}} g(\xi)\, d\xi.$$

So, since

$$\mathcal{L}\left[\int_{-\infty}^{\infty} g(\xi) H\left(t - \frac{1}{c}|x-\xi|\right) d\xi\right] = \int_{-\infty}^{\infty} \frac{1}{s} e^{\frac{-s|x-\xi|}{c}} g(\xi)\, d\xi$$

then

$$\mathcal{L}^{-1}\left[\int_{-\infty}^{\infty} \frac{1}{s} e^{\frac{-s|x-\xi|}{c}} g(\xi)\, d\xi\right] = \int_{-\infty}^{\infty} g(\xi) H\left(t - \frac{1}{c}|x-\xi|\right) d\xi. \tag{7}$$

Thus

$$u(x,t) = \mathcal{L}^{-1}[U(x,s)] = \mathcal{L}^{-1}\left[\frac{1}{2c}\int_{-\infty}^{\infty} e^{\frac{-s|x-\xi|}{c}} f(\xi) d\xi + \frac{1}{2c}\int_{-\infty}^{\infty} \frac{1}{s} e^{\frac{-s|x-\xi|}{c}} g(\xi) d\xi\right]$$

$$= \frac{1}{2c}\left\{\mathcal{L}^{-1}\left[\int_{-\infty}^{\infty} e^{\frac{-s|x-\xi|}{c}} f(\xi) d\xi\right] + \mathcal{L}^{-1}\left[\int_{-\infty}^{\infty} \frac{1}{s} e^{\frac{-s|x-\xi|}{c}} g(\xi) d\xi\right]\right\}$$

$$= \frac{1}{2c}\left[\int_{-\infty}^{\infty} f(\xi)\delta\left(t - \frac{1}{c}|x-\xi|\right) d\xi + \int_{-\infty}^{\infty} g(\xi) H\left(t - \frac{1}{c}|x-\xi|\right) d\xi\right]. \tag{8}$$

Consider the first integral on the right-hand side of Eq. (8). We have

$$\frac{1}{2c}\int_{-\infty}^{\infty} f(\xi)\delta\left(t - \frac{1}{c}|x-\xi|\right) d\xi = \frac{1}{2}\int_{-\infty}^{\infty} f(\xi)\delta\left(t - \frac{1}{c}|x-\xi|\right) \frac{d\xi}{c}.$$

Let $z = \frac{\xi}{c}$. Then $dz = \frac{d\xi}{c}$, $\xi = cz$, $f(\xi) = f(cz)$ and $t - \frac{1}{c}|x-\xi| = t - \frac{1}{c}|x - cz|$. Now $t - \frac{1}{c}|x - cz| = 0$ if $x - cz = ct$ or $x - cz = -t$, which occurs when

$$z = \frac{x - ct}{c} \text{ or } z = \frac{x + ct}{c}.$$

Then

$$\frac{1}{2}\int_{-\infty}^{\infty} f(\xi)\delta\left(t-\frac{1}{c}|x-\xi|\right)\frac{d\xi}{c}=\frac{1}{2}\int_{-\infty}^{\infty} f(cz)\delta\left(t-\frac{1}{c}|x-cz|dz\right)$$

$$=\frac{1}{2}\left[f\left(c\left(\frac{x-ct}{c}\right)\right)+f\left(c\left(\frac{x+ct}{c}\right)\right)\right]=\frac{1}{2}[f(x-ct)+f(x+ct)].$$

Next, consider the second integral in the right-hand side of Eq. (8). Note that

$$H\left(t-\frac{1}{c}|x-\xi|\right)=\begin{cases}0 & \text{if } t-\frac{1}{c}|x-\xi|<0\\ 1 & \text{if } t-\frac{1}{c}|x-\xi|>0\end{cases}.$$

Now, $t-\frac{1}{c}|x-\xi|>0$ if and only if $|x-\xi|<ct$; that is, if and only if

$$-ct<\xi-x<ct \quad \text{or} \quad x-ct<\xi<x+ct.$$

So,

$$\frac{1}{2c}\int_{-\infty}^{\infty} g(\xi)H\left(t-\frac{1}{c}|x-\xi|\right)d\xi=\frac{1}{2c}\int_{x-ct}^{x+ct} g(\xi)d\xi.$$

Thus

$$u(x,t)=\frac{1}{2}[f(x+ct)+f(x-ct)]+\frac{1}{2c}\int_{x-ct}^{x+ct} g(\xi)d\xi$$

which is d'Alembert's formula.

EXERCISES

1. Use the Laplace transform to solve the wave equation

$$\frac{\partial^2 u(x,t)}{\partial t^2}=c^2\frac{\partial^2 u(x,t)}{\partial x^2}, \quad 0<x<\infty,\ t>0;\ u(x,0)=u_t(x,0)=0,\ u(0,t)$$
$$=f(t),\ \lim_{x\to\infty} u(x,t)=0.$$

2. Use the Laplace transform to solve the wave equation

$$\frac{\partial^2 u(x,t)}{\partial t^2}=c^2\frac{\partial^2 u(x,t)}{\partial x^2}, \quad 0<x<\pi,\ t>0;$$
$$u(x,0)=\sin x,\ u_t(x,0)=0,\ u(0,t)=u(\pi,t)=0.$$

3. Use d'Alembert's formula to solve the wave equation

$$\frac{\partial^2 u(x,t)}{\partial t^2} = c^2 \frac{\partial^2 u(x,t)}{\partial x^2}, \qquad -\infty < x < \infty,\ t > 0,\ \lim_{x\to\infty} u(x,t) = 0.$$

for the following initial conditions:

a. $u(x,0) = e^{-x^2}, \quad u_t(x,0) = \sin x.$
b. $u(x,0) = \frac{1}{1+x^2}, \quad u_t(x,0) = 0.$
c. $u(x,0) = \cos x, \quad u_t(x,0) = \sin x.$

Check the validity of your answer by substituting the solution into the wave equation.

CHAPTER

Solving PDEs With Green's Functions 13

13.1 SOLVING THE HEAT EQUATION USING GREEN'S FUNCTION

We construct Green's function for the heat equation using the Dirac-δ function, following the method of Section 3.1. The first form of the problem we consider is

$$\begin{aligned} Du_{xx}(x,t) &= u_t(x,t) \qquad -\infty < x < \infty, \quad t > 0 \\ u(x,0) &= \delta(x - x_0). \end{aligned} \tag{1}$$

We solve this problem using the Fourier transform. Recall that if the Fourier transform of $u(x,t)$ with respect to x is denoted by $U(k,t)$, then

$$U(k,t) = \frac{1}{\sqrt{2\pi}} \int_{-\infty}^{\infty} u(x,t) e^{-ikx} dx$$

$$u(x,t) = \frac{1}{\sqrt{2\pi}} \int_{-\infty}^{\infty} U(k,t) e^{ikx} dk$$

$$\frac{\partial^n u(x,t)}{\partial t^n} = \frac{1}{\sqrt{2\pi}} \int_{-\infty}^{\infty} \left[\frac{\partial^n U(k,t)}{\partial t^n} \right] e^{kx} dk$$

and

$$\frac{\partial^n u(x,t)}{\partial x^n} = \frac{1}{\sqrt{2\pi}} \int_{-\infty}^{\infty} U(k,t)(ik)^n e^{ikx} dk.$$

From Eq. (1), we have

$$0 = u_t(x,t) - Du_{xx}(x,t) = \frac{1}{\sqrt{2\pi}} \int_{-\infty}^{\infty} \left[\frac{\partial U(k,t)}{\partial t} + Dk^2 U(k,t) \right] e^{ikx} dk.$$

Thus, we need to solve the ordinary differential equation

$$\frac{dU(k,t)}{dt} + Dk^2 U(k,t) = 0. \tag{2}$$

Mathematical Physics with Partial Differential Equations. https://doi.org/10.1016/B978-0-12-814759-7.00013-2

The initial condition for Eq. (2) is

$$U(k,0) = \mathcal{F}(u(x,0)) = \mathcal{F}\left(\delta(x - x_0)\right)$$

$$= \frac{1}{\sqrt{2\pi}} \int_{-\infty}^{\infty} \delta(x - x_0) e^{-ikx} dx = \frac{1}{\sqrt{2\pi}} e^{-ikx_0}.$$

The solution to

$$\frac{dU(k,t)}{dt} + Dk^2 U(k,t) = 0, \quad U(k,0) = \frac{1}{\sqrt{2\pi}} e^{-ikx_0}$$

is

$$U(k,t) = \frac{1}{\sqrt{2\pi}} e^{-ikx_0} e^{-k^2 Dt}.$$

To find $u(x,t)$ we take the inverse Fourier transform of $U(k,t)$, which is

$$\frac{1}{\sqrt{2\pi}} \int_{-\infty}^{\infty} \frac{1}{\sqrt{2\pi}} e^{-ikx_0} e^{-k^2 Dt} e^{ikx} dk.$$

In Section 1.5, we showed that

$$u(x,t) = \frac{1}{2\pi} \int_{-\infty}^{\infty} e^{-ikx_0} e^{-k^2 Dt} e^{ikx} dk = \frac{1}{\sqrt{4\pi Dt}} \exp\left[\frac{(x - x_0)^2}{4Dt}\right].$$

Thus the Green's function is

$$G(x,t,s) = \frac{1}{\sqrt{4\pi Dt}} \exp\left[\frac{(x - s)^2}{4Dt}\right].$$

When we studied deriving Green's functions from the Dirac-δ function earlier, we gave a procedure to solve a differential equation with initial condition $u(x,0) = f(x)$, where $f(x)$ is a piecewise continuous function, once we knew the solution for the initial condition being a Dirac-δ function. We now review that procedure.

Suppose that $f(x)$ is a piecewise continuous function. Divide the x-axis into small subintervals of width Δx, and construct a step function

$$f^*(x) = \sum_k f(x_k) \psi_{\Delta_k}(x)$$

where $f(x_k)$ is the value of $f(x)$ in the center of the kth interval, and

$$\psi_{\Delta_k}(x) = \begin{cases} 1 & \text{if } x \text{ is in the } k\text{th interval} \\ 0 & \text{otherwise} \end{cases}.$$

By superposition and the fact that the effect of $f^*(x)$ is the same as $\sum_k f(x_k)\Delta x\delta(x - x_k)$, we have that $u(x,t)$ is approximated by

$$\sum_k f(x_k)\frac{1}{\sqrt{4\pi Dt}}\exp\left[\frac{(x - x_k)^2}{4Dt}\right]\Delta x.$$

Taking the limit as $\Delta x \to 0$, we have

$$u(x,t) = \int_{-\infty}^{\infty} f(s)\frac{1}{\sqrt{4\pi Dt}}\exp\left[\frac{(x - s)^2}{4Dt}\right]ds.$$

GREEN'S FUNCTION FOR THE NONHOMOGENEOUS HEAT EQUATION

Consider the equation

$$u_t(x,t) = Du_{xx}(x,t) + f(x,t) \quad -\infty < x < \infty, \quad t > 0, \quad u(x,0) = 0. \tag{3}$$

We find the Green's function for Eq. (3) by solving

$$u_t(x,t) - Du_{xx}(x,t) = \delta(x - y)\delta(t - s) \quad -\infty < x < \infty, \quad t > 0, \quad u(x,0) = 0. \tag{4}$$

Taking the Fourier transform of Eq. (4) with respect to x gives

$$\begin{aligned}\frac{dU(k,t)}{dt} &+ k^2DU(k,t)\\ &= \int_{-\infty}^{\infty} \delta(x - y)\delta(t - s)e^{-ikx}dx\\ &= \delta(t - s)\int_{-\infty}^{\infty} \delta(x - y)e^{-ikx}dx = \delta(t - s)e^{-iky}.\end{aligned}$$

The initial condition is

$$\mathcal{F}(u(x,0)) = U(k,0) = \mathcal{F}(0) = 0.$$

Thus, we seek to solve

$$\frac{dU(k,t)}{dt} + k^2DU(k,t) = \delta(t - s)e^{-iky}, \quad U(k,0) = 0. \tag{5}$$

One method of solving an equation such as Eq. (5) is to multiply by an integrating factor (see Edwards and Penny [2008] for example) which for this problem is e^{k^2Dt}. This gives

$$e^{k^2Dt}\frac{dU(k,t)}{dt} + e^{k^2Dt}k^2DU(k,t) = e^{k^2Dt}\delta(t - s)e^{-iky}.$$

The point of the integrating factor is

$$e^{k^2Dt}\frac{dU(k,t)}{dt} + e^{k^2Dt}k^2DU(k,t) = \frac{d}{dt}\left(e^{k^2Dt}U(k,t)\right)$$

so that

$$\frac{d}{dt}\left(e^{k^2Dt}U(k,t)\right) = e^{k^2Dt}\delta(t-s)e^{-iky}$$

and

$$e^{k^2Dt}U(k,t) = \int_0^t e^{k^2Dt}\delta(t-s)e^{-iky}dt = e^{k^2Ds}e^{-iky} \quad \text{for } s<t.$$

Thus

$$U(k,t) = \begin{cases} \exp(-k^2Dt)\exp(k^2Ds)\exp(-iky) & \text{for } s<t \\ 0 & \text{for } s>t \end{cases}.$$

To find the Green's function, we apply the inverse Fourier transform to $U(k,t)$. Thus, we have

$$G(x,t;s,y) = \frac{1}{\sqrt{2\pi}}\int_{-\infty}^{\infty} e^{ikx}e^{-k^2Dt}e^{k^2Ds}e^{-iky}dk$$

$$= \frac{1}{\sqrt{2\pi}}\int_{-\infty}^{-\infty} e^{-k^2D(t-s)-ik(y-x)}dk \quad \text{for } s<t.$$

Since

$$\frac{1}{2\pi}\int_{-\infty}^{\infty} e^{-ikx}e^{-\alpha tk^2}dk = \frac{1}{\sqrt{4\pi\alpha t}}e^{-x^2/4\alpha t}$$

we have

$$\frac{1}{2\pi}\int_{-\infty}^{\infty} e^{-k^2D(t-s)-ik(y-x)}dk = \frac{1}{\sqrt{4\pi D(t-s)}}\exp\left[\frac{-(y-x)^2}{4D(t-s)}\right]$$

and thus

$$G(x,t;s,y) = \begin{cases} \dfrac{1}{\sqrt{4\pi D(t-s)}}\exp\left[\dfrac{-(y-x)^2}{4D(t-s)}\right] & \text{for } s<t \\ 0 & \text{for } s>t \end{cases}.$$

By superposition, the solution to

$$u_t(x,t) = Du_{xx}(x,t) + f(x,t) \qquad -\infty<x<\infty,\ t>0,\ u(x,0)=0$$

is

$$u(x,t) = \int_{s=0}^{\infty}\int_{y=-\infty}^{\infty} G(x,t;s,y)f(y,s)dyds$$

$$= \int_{s=0}^{t}\int_{y=-\infty}^{\infty} \frac{1}{\sqrt{4\pi D(t-s)}}\exp\left[\frac{-(y-x)^2}{4D(t-s)}\right]f(y,s)dyds.$$

EXERCISES

In the following exercises it may be difficult to find the inverse Fourier transform that gives the solution. If that is the case, find the function that is to be inverted and give the solution as expressed by the Green's function.

1. Solve

$$u_t(x,t) = Du_{xx}(x,t) + e^{-|x|}\sin t, \quad t > 0, \quad -\infty < x < \infty; \quad u(x,0) = 0.$$

2. Solve

$$Du_{xx}(x,t) = u_t(x,t) \qquad -\infty < x < \infty, \quad t > 0; \; u(x,0) = e^{-|x|}.$$

3. Solve

$$Du_{xx}(x,t) = u_t(x,t) \qquad -\infty < x < \infty, \quad t > 0; \; u(x,0) = \frac{1}{1+x^2}.$$

4. Solve

$$u_t(x,t) = Du_{xx}(x,t) + te^{-t}, \quad t > 0, \quad -\infty < x < \infty; \; u(x,0) = 0.$$

13.2 THE METHOD OF IMAGES

The method of images is a technique for solving heat equation—type problems on a bounded interval or semiinfinite interval. It uses imaginary heat sources or sinks at points outside the interval to obtain the desired boundary conditions.

METHOD OF IMAGES FOR A SEMIINFINITE INTERVAL

Example:

Consider the heat equation

$$u_t(x,t) = Du_{xx}(x,t), \quad 0 < x < \infty, \quad 0 < t < \infty;$$

$$u(x,0) = \delta(x_0 - x), \quad u(0,t) = 0.$$

We present two methods to find the Green's function.

Method 1

In the first section of this chapter we saw that the Green's function for

$$u_t(x,t) = Du_{xx}(x,t), \quad -\infty < x < \infty, \ 0 < t < \infty;$$
$$u(x,0) = \delta(x_0 - x), \ u(0,t) = 0$$

is

$$G(x,t,\ x_0) = \frac{1}{\sqrt{4\pi Dt}} \exp\left[\frac{-(x-x_0)^2}{4Dt}\right].$$

The problem that we now consider is on a semiinfinite interval and has a boundary condition. We construct a second Green's function where an added Dirac-δ function balances the given Dirac-δ function at $x = 0$. That is, we construct the Green's function for

$$u_t(x,t) = Du_{xx}(x,t), \quad -\infty < x < \infty, \ 0 < t < \infty;$$
$$u(x,0) = -\delta(-x_0 - x), \quad u(0,t) = 0. \tag{1}$$

One way to visualize this is to consider the following example from electrostatics: To balance the effect at $x = 0$ of a positive unit charge placed at $x = x_0$, place a negative unit charge at $x = -x_0$.

The Green's function for Eq. (1) is

$$G_I(x,t,-x_0) = -\ G(x,t,\ -x_0) = -\exp\frac{\frac{[x-(-x_0)]^2}{4Dt}}{\sqrt{4\pi Dt}} = -\exp\frac{\frac{(x+x_0)^2}{4Dt}}{\sqrt{4\pi Dt}}.$$

Then

$$G(x,t,\ x_0) + G_I(x,t,-x_0) = \frac{1}{\sqrt{4\pi Dt}}\exp\frac{(x-x_0)^2}{4Dt} - \frac{1}{\sqrt{4\pi Dt}}\exp\frac{(x+x_0)^2}{4Dt}$$

solves

$$u_t(x,t) = Du_{xx}(x,t), \quad -\infty < x < \infty, \ 0 < t < \infty;$$
$$u(x,0) = \delta(x_0 - x) - \delta(-\ x_0 - x), \quad u(0,t) = 0.$$

Now consider the problem

$$u_t(x,t) = Du_{xx}(x,t), \quad 0 < x < \infty, \ 0 < t < \infty;$$
$$u(x,0) = f(x), \ u(0,t) = 0. \tag{2}$$

The solution to

$$u_t(x,t) = Du_{xx}(x,t), \quad -\infty < x < \infty, \ 0 < t < \infty;$$
$$u(x,0) = \tilde{f}(x), \ u(0,t) = 0$$

where

$$\widetilde{f}(x) = \begin{cases} f(x) & \text{if } x > 0 \\ 0 & \text{if } x \leq 0 \end{cases}$$

is given by

$$\begin{aligned} u(x,t) &= \int_{-\infty}^{\infty} [G(x,t,\ s) + G_I(x,t,-s)]\widetilde{f}(s)ds \\ &= \int_{0}^{\infty} [G(x,t,\ s) + G_I(x,t,-s)]\widetilde{f}(s)ds \\ &= \int_{0}^{\infty} [G(x,t,\ s) + G_I(x,t,-s)]f(s)ds \\ &= \int_{0}^{\infty} \left[\frac{1}{\sqrt{4\pi Dt}} \exp\frac{(x-s)^2}{4Dt} - \frac{1}{\sqrt{4\pi Dt}} \exp\frac{(x+s)^2}{4Dt} \right] f(s)ds. \end{aligned}$$

This is the same as the solution to

$$\begin{gathered} u_t(x,t) = Du_{xx}(x,t), \quad 0 < x < \infty, \ 0 < t < \infty; \\ u(x,0) = f(x), \quad u(0,t) = 0. \end{gathered} \tag{3}$$

Thus the Green's function for Eq. (3) is

$$\widetilde{G}(x,t,s) = \frac{1}{\sqrt{4\pi Dt}} \exp\frac{(x-s)^2}{4Dt} - \frac{1}{\sqrt{4\pi Dt}} \exp\frac{(x+s)^2}{4Dt}.$$

Method 2

In this method we use the fact that we know the solution to

$$\begin{gathered} u_t(x,t) = Du_{xx}(x,t), \quad -\infty < x < \infty, \ 0 < t < \infty; \\ u(x,0) = g(x) \end{gathered}$$

to construct the solution to

$$\begin{gathered} u_t(x,t) = Du_{xx}(x,t), \quad 0 < x < \infty, \ 0 < t < \infty; \\ u(x,0) = f(x), \ u(0,t) = 0. \end{gathered}$$

We are doing something similar to what we did in the first method. We are "balancing" the function $u(x,0) = f(x)$, $0 < x < \infty$, by extending the function to $-\infty < x < \infty$, so that it is an odd function. We let

$$g(x) = \begin{cases} f(x) & \text{if } x > 0 \\ -f(-x) & \text{if } x < 0 \end{cases}.$$

Then

$$\begin{aligned}
u(x,t) &= \int_{-\infty}^{\infty} \frac{1}{\sqrt{4\pi Dt}} \exp\left[\frac{(x-s)^2}{4Dt}\right] g(s)ds \\
&= \int_{-\infty}^{0} \frac{1}{\sqrt{4\pi Dt}} \exp\left[\frac{(x-s)^2}{4Dt}\right] g(s)ds + \int_{0}^{\infty} \frac{1}{\sqrt{4\pi Dt}} \exp\left[\frac{(x-s)^2}{4Dt}\right] g(s)ds \\
&= -\int_{-\infty}^{0} \frac{1}{\sqrt{4\pi Dt}} \exp\left[\frac{(x-s)^2}{4Dt}\right] f(-s)ds + \int_{0}^{\infty} \frac{1}{\sqrt{4\pi Dt}} \exp\left[\frac{(x-s)^2}{4Dt}\right] f(s)ds.
\end{aligned}$$

In

$$-\int_{-\infty}^{0} \frac{1}{\sqrt{4\pi Dt}} \exp\left[\frac{(x-s)^2}{4Dt}\right] f(-s)ds$$

we make the change of variables $w=-s$. Then $ds=-dw$; if $s=-\infty$, then $w=\infty$; if $s=0$, then $w=0$, so

$$\begin{aligned}
-\int_{-\infty}^{0} \frac{1}{\sqrt{4\pi Dt}} \exp\left[\frac{(x-s)^2}{4Dt}\right] f(-s)ds &= -\int_{\infty}^{0} \frac{1}{\sqrt{4\pi Dt}} \exp\left[\frac{(x-w)^2}{4Dt}\right] f(w)(-dw) \\
&= -\int_{0}^{\infty} \frac{1}{\sqrt{4\pi Dt}} \exp\left[\frac{(x-w)^2}{4Dt}\right] f(w)dw.
\end{aligned}$$

Thus, we have

$$\begin{aligned}
u(x,t) &= -\int_{-\infty}^{0} \frac{1}{\sqrt{4\pi Dt}} \exp\left[\frac{(x-s)^2}{4Dt}\right] f(-s)ds + \int_{0}^{\infty} \frac{1}{\sqrt{4\pi Dt}} \exp\left[\frac{(x-s)^2}{4Dt}\right] f(s)ds \\
&= -\int_{0}^{\infty} \frac{1}{\sqrt{4\pi Dt}} \exp\left[\frac{(x-w)^2}{4Dt}\right] f(w)dw + \int_{0}^{\infty} \frac{1}{\sqrt{4\pi Dt}} \exp\left[\frac{(x-s)^2}{4Dt}\right] f(s)ds \\
&= \int_{0}^{\infty} \left[\frac{1}{\sqrt{4\pi Dt}} \exp\left[\frac{(x-s)^2}{4Dt}\right] - \frac{1}{\sqrt{4\pi Dt}} \exp\left[\frac{(x+s)^2}{4Dt}\right]\right] f(s)ds.
\end{aligned}$$

which is what we concluded in Method 1.

We can use a modification of the ideas in Method 2 to solve the heat equation with a Neumann boundary condition at $x=0$. Now we want to solve

$$u_t(x,t) = Du_{xx}(x,t), \quad 0 < x < \infty, \quad 0 < t < \infty;$$

$$u(x,0) = f(x), \quad u_x(0,t) = 0.$$

The difference in the approach to the problems is that now we want to balance the derivative of the function $u(x,0) = f(x)$, $0 < x < \infty$, by extending the function $f(x)$ to $-\infty < x < \infty$. We do this by extending $f(x)$ to be an even function. For if

$$f(-x) = f(x) \text{ then } f'(-x) = -f'(-x) = -f'(x).$$

Accordingly, we define

$$g(x)=\begin{cases} f(x) & \text{if } x>0 \\ f(-x) & \text{if } x<0 \end{cases}.$$

If we repeat the ideas of Method 2, we need to evaluate

$$\begin{aligned} u(x,t) &= \int_{-\infty}^{\infty} \frac{1}{\sqrt{4\pi Dt}} \exp\left[\frac{(x-s)^2}{4Dt}\right] g(s)ds \\ &= \int_{-\infty}^{0} \frac{1}{\sqrt{4\pi Dt}} \exp\left[\frac{(x-s)^2}{4Dt}\right] g(s)ds + \int_{0}^{\infty} \frac{1}{\sqrt{4\pi Dt}} \exp\left[\frac{(x-s)^2}{4Dt}\right] g(s)ds \\ &= -\int_{-\infty}^{0} \frac{1}{\sqrt{4\pi Dt}} \exp\left[\frac{(x-s)^2}{4Dt}\right] f(-s)ds + \int_{0}^{\infty} \frac{1}{\sqrt{4\pi Dt}} \exp\left[\frac{(x-s)^2}{4Dt}\right] f(s)ds. \end{aligned}$$

We again make the change of variables $w=-s$ in

$$\int_{-\infty}^{0} \frac{1}{\sqrt{4\pi Dt}} \exp\left[\frac{(x-s)^2}{4Dt}\right] f(-s)ds$$

to obtain

$$\begin{aligned} \int_{-\infty}^{0} \frac{1}{\sqrt{4\pi Dt}} \exp\frac{(x-s)^2}{4Dt} f(-s)ds &= \int_{\infty}^{0} \frac{1}{\sqrt{4\pi Dt}} \exp\frac{(x+w)^2}{4Dt} f(w)(-dw) \\ &= \int_{0}^{\infty} \frac{1}{\sqrt{4\pi Dt}} \exp\frac{(x+w)^2}{4Dt} f(w)dw \end{aligned}$$

so that

$$\begin{aligned} u(x,t) &= \int_{-\infty}^{0} \frac{1}{\sqrt{4\pi Dt}} \exp\frac{(x-s)^2}{4Dt} f(-s)ds + \int_{0}^{\infty} \frac{1}{\sqrt{4\pi Dt}} \exp\frac{(x-s)^2}{4Dt} f(s)ds \\ &= \int_{0}^{\infty} \frac{1}{\sqrt{4\pi Dt}} \exp\frac{(x+w)^2}{4Dt} f(w)dw + \int_{0}^{\infty} \frac{1}{\sqrt{4\pi Dt}} \exp\frac{(x-s)^2}{4Dt} f(s)ds \\ &= \frac{1}{\sqrt{4\pi Dt}} \int_{0}^{\infty} \left[\exp\frac{(x+s)^2}{4Dt} + \exp\frac{(x-s)^2}{4Dt}\right] f(s)ds. \end{aligned}$$

Thus, for the Neumann boundary condition we have

$$G(x,t,x_0) = \frac{1}{\sqrt{4\pi Dt}} \left[\exp\frac{(x+s)^2}{4Dt} + \exp\frac{(x-s)^2}{4Dt}\right].$$

METHOD OF IMAGES FOR A BOUNDED INTERVAL

Now we consider the heat equation

$$u_t(x,t) = Du_{xx}(x,t), \quad -L \le x < L,$$
$$0 < t < \infty; \quad u(L,t) = u(-L,t) = 0, \quad u(x,0) = \delta(x_0 - x).$$

We want to do something similar to the semiinfinite case, but now the "balancing" is more involved. Our approach will be similar to Method 1 for the semiinfinite interval. We again use an example from electrostatics to develop some intuition.

Consider a charge q to be placed at $x = 0$ on the interval $[-L, L]$. We want to place charges $-q$ or q outside the interval so that the potential at $x = -L$ and $x = L$ will be zero.

We proceed in steps:

Step 1. We make the potential at $x = L$ be 0 by balancing the charge q at $x = 0$ with a charge of $-q$ at $x = 2L$ as shown in Fig. 13.2.1A.

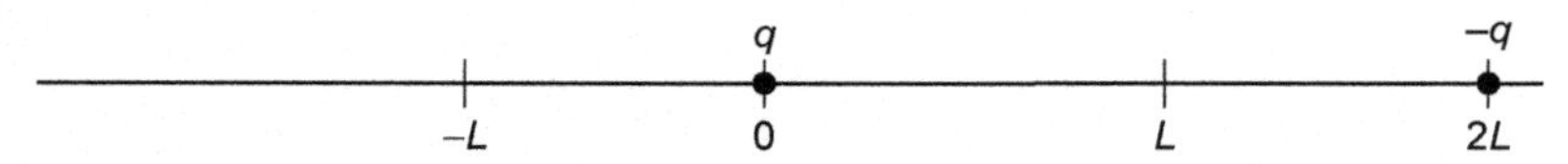

FIGURE 13.2.1A

Step 2. We make the potential at $x = -L$ be 0 by adding two charges; one to balance the charge at $x = 0$ and another to balance the charge at $x = 2L$. To balance the charge q at $x = 0$ we add a charge of $-q$ at $x = -2L$. To balance the charge $-q$ at $2L$ we add a charge q at $x = -4L$. We now have the charges as shown in Fig. 13.2.1B.

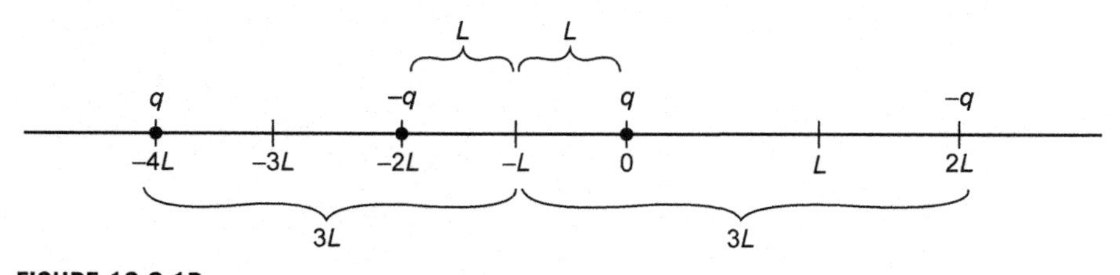

FIGURE 13.2.1B

Step 3. We return to make the potential at $x = L$ be 0. To do this, we must balance the charges we added in Step 2. In Fig. 13.2.1C we show only the charges we added in Step 2.

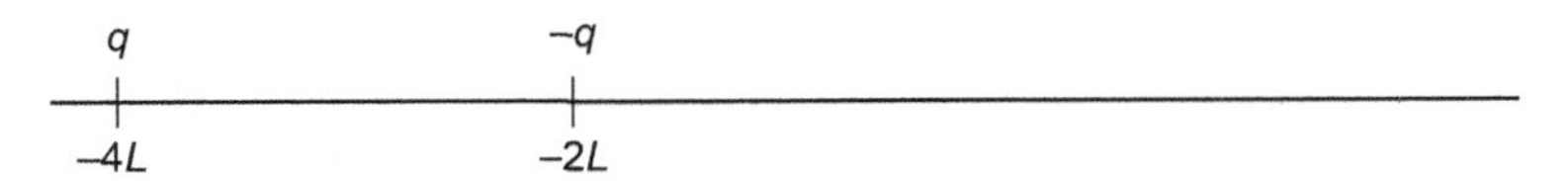

FIGURE 13.2.1C

To balance the charge of $-q$ at $x = -2L$ we add the charge of q at $x = 4L$. To balance the charge of q at $x = -4L$ we add the charge of $-q$ at $x = 6L$. See Fig. 13.2.1D.

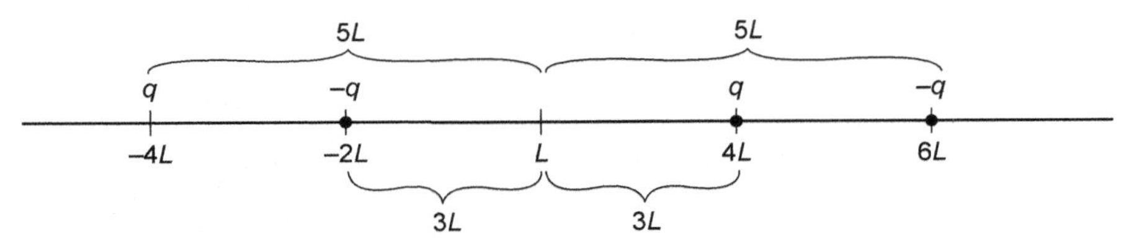

FIGURE 13.2.1D

The charges to this point are shown in Fig. 13.2.1E.

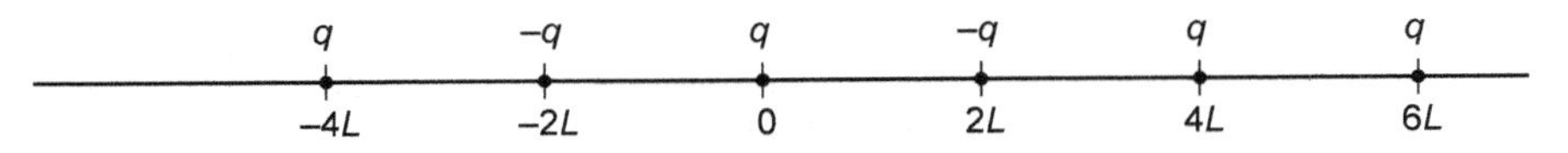

FIGURE 13.2.1E

We can perhaps see a pattern emerging that we are adding charges $-q$ at $x = \pm 2L, \pm 6L, \pm 10L, \ldots$ and charges q at $x = \pm 4L, \pm 8L, \pm 12L, \ldots$.

We can now find Green's function for

$$u_t(x,t) = Du_{xx}(x,t), \quad -L \le x < L, \quad 0 < t < \infty;$$
$$u(L,t) = u(-L,t) = 0, \quad u(x,0) = \delta(0-x).$$

(Conceptually, this is the problem we just solved.) We get

$$G(x,t,0) = \frac{1}{\sqrt{4\pi Dt}}\left[\exp\left(-\frac{x^2}{4Dt}\right) - \exp\left(-\frac{(x-2L)^2}{4Dt}\right) - \exp\left(-\frac{(x+2L)^2}{4Dt}\right)\right.$$
$$\left. + \exp\left(-\frac{(x-4L)^2}{4Dt}\right) + \exp\left(-\frac{(x+4L)^2}{4Dt}\right) - \cdots\right].$$

For the last example of using the method of images, we demonstrate how to balance a charge of q at $x_0 \in (-L, L)$ so that the potential at $x = L$ and $x = -L$ is zero. Consider Fig. 13.2.2A.

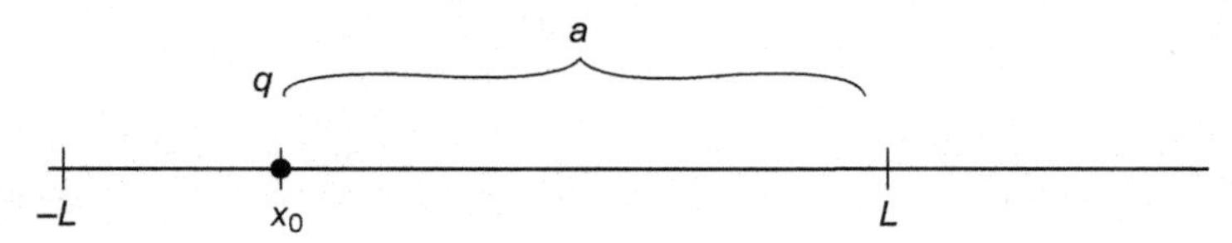

FIGURE 13.2.2A

We let

$$a = L - x_0 = \text{the distance from } x_0 \text{ to } L.$$

We follow the same ideas as in the previous example.

We first make the potential at $x = L$ be zero by putting a charge of $-q$ at $x = L + a$ (see Fig. 13.2.2B).

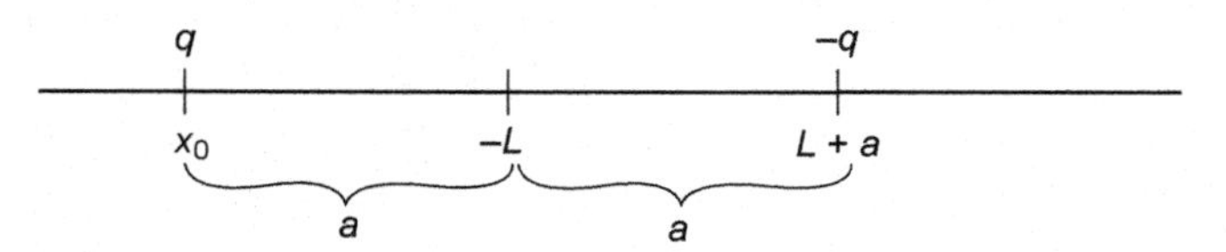

FIGURE 13.2.2B

Figs. 13.2.2C–2F illustrate how we proceed. In Fig. 13.2.2C we show how the charges after the first step are configured with respect to $-L$. In Fig. 13.2.2D we show how to add charges so that the potential at $x = -L$ will be zero.

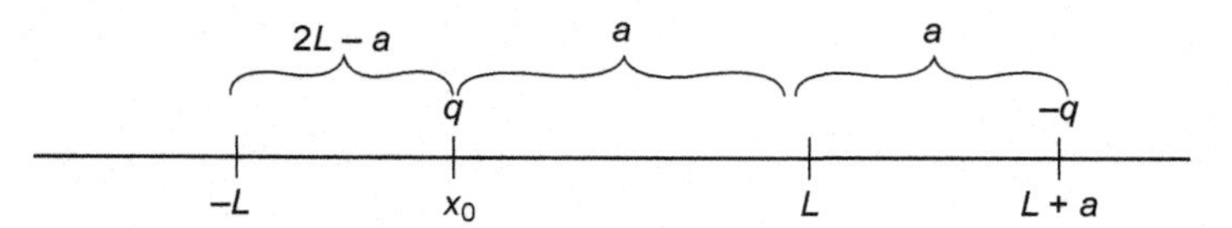

FIGURE 13.2.2C

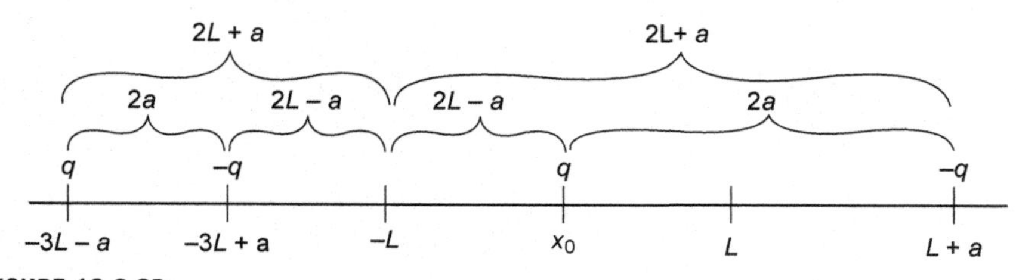

FIGURE 13.2.2D

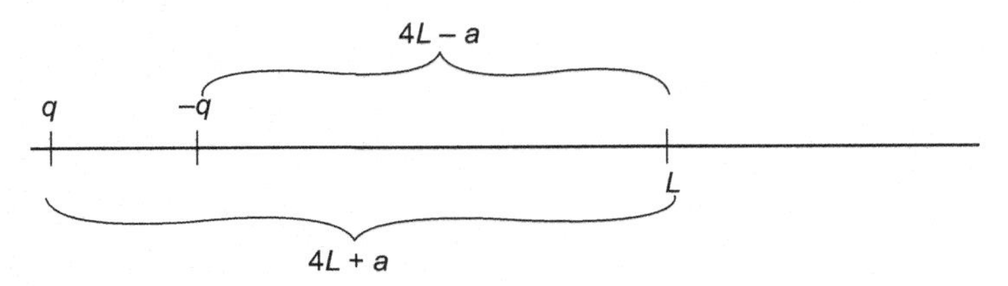

FIGURE 13.2.2E

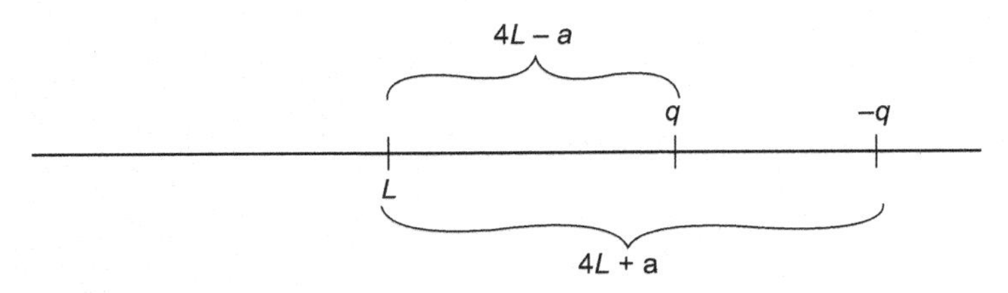

FIGURE 13.2.2F

In Fig. 13.2.2E we show the distances the new charges are from L, and in Fig. 13.2.2F we show how to place additional charges so that the potential at $x = L$ will be zero.

Finally, Fig. 13.2.2G shows the positioning of the charges so far.

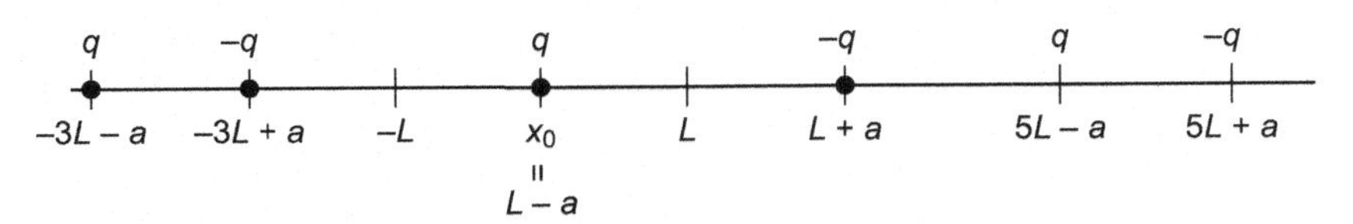

FIGURE 13.2.2G

We now determine a pattern for the placement of the charges.

We have placed charges q at

$$x = -3L - a,\ L - a, \quad \text{and} \quad 5L - a.$$

Since $a = L - x_0$, these points are

$$\begin{aligned} x &= -3L - (L - x_0), \quad L - (L - x_0), \text{ and } 5L - (L - x_0) \\ &= -4L + x_0, \quad x_0, \quad 4L + x_0. \end{aligned}$$

It appears that it is correct to place a charge q at $x = 4nL + x_0$ where n is an integer.

We have placed charges $-q$ at

$$\begin{aligned} x &= -3L+a, \quad L+a, \quad \text{and } 5L+a \\ &= -3L+(L-x_0), \quad L+(L-x_0), \quad \text{and } 5L+(L-x_0) \\ &= -2L-x_0, \quad 2L-x_0, \quad 6L-x_0. \end{aligned}$$

It appears that a charge of $-q$ should be placed at $x = 4\,nL + 2L - x_0$ where n is an integer.

We can now find the Green's function for

$$\begin{aligned} &u_t(x,t) = Du_{xx}(x,t), \quad -L \le x < L, \quad 0 < t < \infty; \\ &u(L,t) = u(-L,t) = 0, \quad u(x,0) = c\delta(x_0 - x). \end{aligned}$$

Following the analysis for the electrostatic potential, we get

$$\begin{aligned} G(x,t,x_0) = \frac{c}{\sqrt{4\pi Dt}} \Bigg\{ &\sum_{n=-\infty}^{\infty} \left[\exp\left[-(x-x_0+4nL)^2 \Big/ 4Dt \right] \right] \\ &- \sum_{n=-\infty}^{\infty} \left[\exp\left[-(x-x_0+2L+4nL)^2 \Big/ 4Dt \right] \right] \Bigg\}. \end{aligned}$$

Thus the solution to

$$\begin{aligned} &u_t(x,t) = Du_{xx}(x,t), \quad -L < x < L, \quad 0 < t < \infty; \\ &u(L,t) = u(-L,t) = 0, \quad u(x,0) = f(x) \end{aligned}$$

is

$$\begin{aligned} u(x,t) = \frac{c}{\sqrt{4\pi Dt}} \sum_{n=-\infty}^{\infty} \Bigg\{ \frac{1}{2L} \int_{-L}^{L} f(u) \Bigg\{ &\left[\exp\left[-(x-x_0+4nL)^2 \Big/ 4Dt \right] \right] \\ &- \left[\exp\left[-(x-x_0+2L+4nL)^2 \Big/ 4Dt \right] \right] \Bigg\} du \Bigg\}. \end{aligned}$$

We have found the Green's function for several forms of the heat equation. There are a few factors that can complicate a particular form of the heat equation, including boundary terms, initial conditions, and existence of a heat source or sink. When complicating factors are present, it is sometimes advantageous to separate the problem into pieces, each of which contains one of the complicating factors. The solutions to each of the pieces are added together to give the solution to the original problem. We give two examples of this method.

Example:
Consider

$$u_t(x,t) = Du_{xx}(x,t) + f(x,t), \quad -\infty < x < \infty, \quad 0 < t < \infty; \quad u(x,0) = g(x).$$

Suppose that $v(x,t)$ solves

$$v_t(x,t) = Dv_{xx}(x,t) + f(x,t), \quad -\infty < x < \infty, \quad 0 < t < \infty; \quad v(x,0) = 0$$

and $w(x,t)$ solves

$$w_t(x,t) = Dw_{xx}(x,t) \quad -\infty < x < \infty, \quad 0 < t < \infty, \quad w(x,0) = g(x).$$

(Note that we have found $v(x,t)$ and $w(x,t)$ earlier.)
Let $u(x,t) = v(x,t) + w(x,t)$. Then

$$\begin{aligned} u_t(x,t) &= v_t(x,t) + w_t(x,t) = Dv_{xx}(x,t) + f(x,t) + Dw_{xx}(x,t) \\ &= D(v_{xx}(x,t) + w_{xx}(x,t)) + f(x,t) \\ &= Du_{xx}(x,t) + f(x,t), \quad -\infty < x < \infty, \quad 0 < t < \infty; \end{aligned}$$

$$u(x,0) = v(x,0) + w(x,0) = 0 + g(x) = g(x).$$

Example:
Suppose

$$\begin{aligned} &u_t(x,t) = Du_{xx}(x,t) + f(x,t), \quad 0 < x < \infty, \quad 0 < t < \infty; \\ &u(x,0) = g(x), \quad u(0,t) = h(t). \end{aligned}$$

Suppose $p(x,t)$ solves

$$p_t(x,t) = Dp_{xx}(x,t) + f(x,t), \ 0 < x < \infty, \ 0 < t < \infty; \ p(x,0) = 0, \ p(0,t) = 0,$$

and $q(x,t)$ solves

$$q_t(x,t) = Dq_{xx}(x,t), \quad 0 < x < \infty, \quad 0 < t < \infty; \quad q(x,0) = g(x), \ q(0,t) = 0$$

and $r(x,t)$ solves

$$r_t(x,t) = Dr_{xx}(x,t), \quad 0 < x < \infty, \quad 0 < t < \infty; \quad r(x,0) = 0, \ r(0,t) = h(t).$$

We have already found the solutions for $p(x,t)$, $q(x,t)$, and $r(x,t)$. Let

$$u(x,t) = p(x,t) + q(x,t) + r(x,t).$$

Then

$$\begin{aligned} u_t(x,t) &= p_t(x,t) + q_t(x,t) + r_t(x,t) \\ &= Dp_{xx}(x,t) + f(x,t) + Dq_{xx}(x,t) + Dr_{xx}(x,t) \\ &= D(p_{xx}(x,t) + q_{xx}(x,t) + r_{xx}(x,t)) + f(x,t) \\ &= Du_{xx}(x,t) + f(x,t), \quad 0 < x < \infty; \end{aligned}$$

$$u(x,0) = p(x,0) + q(x,0) + r(x,0) = 0 + g(x) + 0 = g(x)$$

$$u(0,t) = p(0,t) + q(0,t) + r(0,t) = 0 + 0 + h(t) = h(t).$$

EXERCISES

1. Solve the heat equation

$$u_t(x,t) - Du_{xx}(x,t) = 0, \quad -1 < x < 1, \ t > 0; \ u(-1,t) = u(1,t) = 0,$$
$$u(x,0) = f(x)$$

using the method of images.

2. Solve

$$u_t(x,t) - Du_{xx}(x,t) = 0, \quad 0 < x < \infty, \quad t > 0; \ u(x,0) = 0,$$
$$u_x(0,t) = \cos t \lim_{x\to\infty} u(x,t) = 0.$$

If you do this using the Laplace transform,

$$\mathcal{L}^{-1}\left\{\frac{e^{-Dx\sqrt{s}}}{\sqrt{s}}\right\} = \frac{1}{\sqrt{\pi t}}e^{-\frac{x^2D^2}{4t}}$$

is helpful.

3. Solve

$$u_t(x,t) - Du_{xx}(x,t) = 0,$$
$$-L < x < L, \ t > 0; \ u(x,0) = \sin x, \ u(L,t) = u(-L,t) = 0.$$

4. Solve

$$u_t(x,t) = Du_{xx}(x,t) + e^{-t}, \quad 0 < x < \infty, \quad 0 < t < \infty;$$
$$u(x,0) = \sin x, \quad u(0,t) = \frac{1}{1+t}.$$

13.3 GREEN'S FUNCTION FOR THE WAVE EQUATION

Consider the wave equation in one dimension. Let $G(x,t;x_0,t_0)$ denote the deflection of a string initially at rest when a unit force is applied at the point x_0 at the time t_0. Then $G(x,t;x_0,t_0)$ satisfies the equation

$$\frac{\partial^2 G(x,t;x_0,t_0)}{\partial t^2} = c^2\frac{\partial^2 G(x,t;x_0,t_0)}{\partial x^2} + \delta(x-x_0)\delta(t-t_0), \quad t \geq t_0;$$

$$G(x,t;x_0,t_0) = 0 \text{ for } t < t_0.$$

Applying the Fourier transform with respect to x gives

$$\frac{\partial^2 \widehat{G}(k,t;x_0,t_0)}{\partial t^2} = c^2(ik)^2\widehat{G}(k,t;x_0,t_0) + \frac{1}{\sqrt{2\pi}}e^{-ikx_0}\delta(t-t_0) \qquad (1)$$

where

$$\widehat{G}(k,t;x_0,t_0) = \frac{1}{\sqrt{2\pi}}\int_{-\infty}^{\infty} G(x,t;x_0,t_0)e^{-ikx}dx.$$

Also,

$$\widehat{G}(k,t;x_0,t_0) = 0 \text{ for } t < t_0. \qquad (2)$$

Solving for (1) and (2) gives

$$\widehat{G}(k,t;x_0,t_0) = \begin{cases} 0 & \text{for } t < t_0 \\ ae^{ikct} + be^{-ikct} & \text{for } t > t_0 \end{cases} \qquad (3)$$

where

$$a = a(k,x_0,t_0) \text{ and } b = b(k,x_0,t_0).$$

For $\varepsilon > 0$ we have

$$\int_{t_0-\varepsilon}^{t_0+\varepsilon} \frac{\partial^2 \widehat{G}(k,t;x_0,t_0)}{\partial t^2}dt + \int_{t_0-\varepsilon}^{t_0+\varepsilon} c^2k^2\widehat{G}(k,t;x_0,t_0)dt$$
$$= \frac{1}{\sqrt{2\pi}}e^{-ikx_0}\int_{t_0-\varepsilon}^{t_0+\varepsilon}\delta(t-t_0)dt. \qquad (4)$$

Since G and $\widehat{G}$ are continuous,

$$\lim_{\varepsilon\downarrow 0}\int_{t_0-\varepsilon}^{t_0+\varepsilon} c^2k^2\widehat{G}(k,t;x_0,t_0)dt = 0.$$

Also

$$\int_{t_0-\varepsilon}^{t_0+\varepsilon} \frac{\partial^2 \widehat{G}(k,t;x_0,t_0)}{\partial t^2}dt = \frac{\partial \widehat{G}(k,t+\varepsilon;x_0,t_0)}{\partial t} - \frac{\partial \widehat{G}(k,t-\varepsilon;x_0,t_0)}{\partial t}$$

and

$$\frac{1}{\sqrt{2\pi}}e^{-ikx_0}\int_{t_0-\varepsilon}^{t_0+\varepsilon}\delta(t-t_0)dt = \frac{1}{\sqrt{2\pi}}e^{-ikx_0}.$$

Thus from Eq. (4), we have

$$\lim_{\varepsilon \downarrow 0}\left[\frac{\partial \widehat{G}(k,t+\varepsilon;x_0,t_0)}{\partial t}-\frac{\partial \widehat{G}(k,t-\varepsilon;x_0,t_0)}{\partial t}\right]=\frac{1}{\sqrt{2\pi}}e^{-ikx_0}. \tag{5}$$

We use these conditions to solve for $a = a(k,x_0,t_0)$ and $b = b(k,x_0,t_0)$ in Eq. (3). Since $\widehat{G}(k,t;x_0,t_0)$ is continuous in t,

$$\lim_{t \uparrow t_0} \widehat{G}(k,t;x_0,t_0)=\lim_{t \downarrow t_0} \widehat{G}(k,t;x_0,t_0).$$

Then

$$ae^{ikct_0}+be^{-ikct_0}=0$$

so

$$b=-ae^{2ikct_0}.$$

Differentiating Eq. (3) gives

$$\frac{\partial \widehat{G}(k,t;x_0,t_0)}{\partial t}=\begin{cases}0 & \text{for } t<t_0 \\ ikcae^{ickt}-ikcbe^{-ickt} & \text{for } t>t_0\end{cases}. \tag{6}$$

Substituting $b = ae^{2ikct_0}$ into Eq. (6) gives

$$ikc\left[ae^{ickt}-\left(-ae^{2ikct_0}\right)e^{-ickt}\right]=ikca\left(e^{ickt}+e^{2ikct_0-ikct}\right).$$

As $t \to t_0$

$$ikca\left(e^{ickt}+e^{2ikct_0-ikct}\right) \to 2ikcae^{ikct_0}.$$

Eq. (5) says that the jump condition on the derivative of $\widehat{G}$ is $\frac{1}{\sqrt{2\pi}}e^{-ikx_0}$, so by Eq. (6), we have

$$2ikcae^{ikct_0}=\frac{1}{\sqrt{2\pi}}e^{-ikx_0},$$

so

$$a=\frac{e^{-ikx_0}e^{-ikct_0}}{\sqrt{2\pi}2ikc}$$

and thus

$$b=-ae^{2ikct_0}=-\frac{e^{-ikx_0}e^{-ikct_0}}{\sqrt{2\pi}2ikc}e^{2ikct_0}=-\frac{e^{-ikx_0}e^{ikct_0}}{\sqrt{2\pi}2ikc}.$$

Thus

$$\widehat{G}(k, t; x_0, {}_0) = \begin{cases} 0 & \text{for } \ t < t_0 \\ ae^{ikct} + be^{-ikct} & \text{for } \ t > t_0 \end{cases}$$

$$= \begin{cases} 0 & \text{for } \ t < t_0 \\ \left(\dfrac{e^{-ikx_0} e^{-ikct_0}}{\sqrt{2\pi} 2ikc} \right) e^{ikct} + \left(- \dfrac{e^{-ikx_0} e^{ikct_0}}{\sqrt{2\pi} 2ikc} \right) e^{-ikct} & \text{for } \ t > t_0 \end{cases}$$

$$= \begin{cases} 0 & \text{for } \ t < t_0 \\ e^{-ikx_0} \left[\dfrac{e^{ikc(t-t_0)} - e^{-ikc(t-t_0)}}{\sqrt{2\pi} 2ikc} \right] & \text{for } \ t > t_0 \end{cases}. \tag{7}$$

Now

$$e^{i\theta} - e^{-i\theta} = (\cos\,\theta + i\,\sin\,\theta) - (\cos\,\theta - i\,\sin\,\theta) = 2i\,\sin\,\theta$$

so

$$\sin\,\theta = \frac{e^{i\theta} - e^{-i\theta}}{2i}.$$

Thus, Eq. (7) can be expressed

$$\widehat{G}(k, t; x_0, t_0) = \begin{cases} 0 & \text{for } \ t < t_0 \\ \dfrac{e^{-ikx_0}}{\sqrt{2\pi} kc} \sin[kc(t - t_0)] & \text{for } \ t > t_0 \end{cases}.$$

We now show that taking the inverse Fourier transform of $\widehat{G}(k, t; x_0, t_0)$ gives

$$G(x, t; x_0, t_0) = \begin{cases} \dfrac{1}{2} & \text{if } \ |x - x_0| < c(t - t_0) \\ 0 & \text{if } \ |x - x_0| > c(t - t_0) \end{cases}.$$

We note that $|x - x_0| < c(t - t_0)$ if and only if $x_0 - x \, \varepsilon \, (-c(t - t_0), c(t - t_0))$ if and only if $x_0 \, \varepsilon \, (x - c(t - t_0), x + c(t - t_0))$.

To prove the claim, we compute for $t > t_0$

$$\frac{1}{\sqrt{2\pi}}\int_{-\infty}^{\infty} e^{ikx}\frac{e^{-ikx_0}}{kc}\sin[kc(t-t_0)]dk$$
$$= \frac{1}{\sqrt{2\pi}}\frac{1}{2}\int_{-\infty}^{\infty} e^{ikx}e^{-ikx_0}\frac{e^{ikc(t-t_0)} - e^{-ikc(t-t_0)}}{ikc}dk \tag{8}$$
$$= \frac{1}{\sqrt{2\pi}}\frac{1}{2c}\int_{-\infty}^{\infty} e^{-ikx_0}\left(\int_{x-c(t-t_0)}^{x+c(t-t_0)} e^{ik\xi}d\xi\right) dk.$$

Reversing the limits of integration, we have

$$\frac{1}{\sqrt{2\pi}}\frac{1}{2c}\int_{-\infty}^{\infty} e^{-ikx_0}\left(\int_{x-c(t-t_0)}^{x+c(t-t_0)} e^{ik\xi}d\xi\right) dk$$
$$= \frac{1}{\sqrt{2\pi}}\frac{1}{2c}\int_{x-c(t-t_0)}^{x+c(t-t_0)}\left(\int_{-\infty}^{\infty} e^{-ikx_0}e^{ik\xi}dk\right) d\xi.$$

Now

$$\frac{1}{\sqrt{2\pi}}\int_{-\infty}^{\infty} e^{-ikx_0}e^{ik\xi}dk = \mathcal{F}^{-1}\left(e^{-ikx_0}\right) = \mathcal{F}^{-1}(\mathcal{F}(\delta(x_0))) = \delta(x_0).$$

Thus

$$\frac{1}{\sqrt{2\pi}}\frac{1}{2c}\int_{-\infty}^{\infty} e^{-ikx_0}\left(\int_{x-c(t-t_0)}^{x+c(t-t_0)} e^{ik\xi}d\xi\right) dk$$
$$= \frac{1}{\sqrt{2\pi}}\frac{1}{2c}\int_{x-c(t-t_0)}^{x+c(t-t_0)}\left(\int_{-\infty}^{\infty} e^{-ikx_0}e^{ik\xi}dk\right) d\xi$$
$$= \frac{1}{2c}\int_{x-c(t-t_0)}^{x+c(t-t_0)}(\delta(x_0))\, d\xi$$
$$= \begin{cases} \dfrac{1}{2c} & \text{if } x_0\ \varepsilon\ (x-c(t-t_0),\ x+\ c(t-t_0)) \\ 0 & \text{otherwise} \end{cases}.$$

We note that $x_0\, \varepsilon (x - c(t - t_0), x + c(t - t_0))$ if and only if $x_0 - x\ \varepsilon\ (-c(t - t_0), c(t - t_0))$ if and only if $|x - x_0| < c(t - t_0)$.

Thus the solution to

$$\frac{\partial^2 u(x,t)}{\partial t^2} = c^2 \frac{\partial^2 u(x,t)}{\partial x^2} + f(x,t), \quad -\infty < x < \infty, \ t > 0;$$

$$u(x,0) = 0, \quad \frac{\partial u(x,0)}{\partial t} = 0$$

is

$$u(x,t) = \int_{x_0=-\infty}^{\infty} \int_{t_0=0}^{\infty} G(x,t;x_0,t_0) f(x_0,t_0) dt_0 dx_0 = \frac{1}{2c} \iint_{\Omega} f(x_0,t_0) dt_0 dx_0$$

where Ω is the region $\left\{ (x_0,t_0) \mid 0 < t_0 < t - \frac{1}{c}|x - x_0| \right\}$.

Example:
Solve

$$\frac{\partial^2 u(x,t)}{\partial t^2} = c^2 \frac{\partial^2 u(x,t)}{\partial x^2}, \quad -\infty < x < \infty, \quad t > 0; \quad u(x,0) = 0, \quad \frac{\partial u(x,0)}{\partial t} = g(x).$$

Solution:
In this problem there is no forcing term, but there is an initial velocity. We show that this is equivalent to having a forcing term that acts only at $t = 0$. We replace the given equation by

$$\frac{\partial^2 u(x,t)}{\partial t^2} = c^2 \frac{\partial^2 u(x,t)}{\partial x^2} + g(x)\delta(t-0); \quad u(x,0) = 0, \quad \frac{\partial u(x,0)}{\partial t} = 0.$$

If this replacement is valid, then the solution will be

$$\begin{aligned} u(x,t) &= \int_{x_0=-\infty}^{\infty} \int_{t_0=0}^{\infty} G(x,t;x_0,t_0) g(x_0) \delta(t_0 - 0) dt_0 dx_0 \\ &= \int_{x_0=-\infty}^{\infty} G(x,t;x_0,t_0) g(x_0) dx_0. \end{aligned}$$

We now demonstrate that this is the case.
Suppose $G(x,t;x_0,t_0)$ satisfies

$$\frac{\partial^2 G(x,t;x_0,t_0)}{\partial t^2} = c^2 \frac{\partial^2 G(x,t;x_0,t_0)}{\partial x^2} + \delta(x - x_0)\delta(t - t_0).$$

We show

$$u(x,t) = \int_{-\infty}^{\infty} G(x,t;x_0,t_0) g(x_0) dx_0$$

satisfies

$$\frac{\partial^2 u(x,t)}{\partial t^2} = c^2 \frac{\partial^2 u(x,t)}{\partial x^2}, \quad u(x,0) = 0, \quad \frac{\partial u(x,0)}{\partial t} = g(x).$$

We have

$$\frac{\partial^2 u(x,t)}{\partial t^2} - c^2\frac{\partial^2 u(x,t)}{\partial x^2} = \int_{-\infty}^{\infty}\left[\frac{\partial^2 G(x,t;x_0,0)}{\partial t^2} - c^2\frac{\partial^2 G(x,t;x_0,0)}{\partial x^2}\right]g(x_0)dx_0$$

$$= \delta(t-0)\int_{-\infty}^{\infty}\delta(x-x_0)g(x_0)dx_0 = \delta(t-0)g(x) = 0 \quad \text{if } t>0.$$

Also

$$u(x,0) = \int_{-\infty}^{\infty} G(x,0;x_0,0)g(x_0)dx_0.$$

Now

$$G(x,t;x_0,t_0) = \begin{cases} \dfrac{1}{2c} & \text{if } |x-x_0| < c(t-t_0) \\ 0 & \text{if } |x-x_0| > c(t-t_0) \end{cases}$$

so

$$G(x,0;x_0,0) = 0 = u(x,0).$$

Now

$$G(x,t+\Delta t;x_0,0) = \begin{cases} \dfrac{1}{2c} & \text{if } |x-x_0| < c(t+\Delta t) \\ 0 & \text{otherwise} \end{cases}$$

so

$$G(x,t+\Delta t;x_0,0) - G(x,t;x_0,0) = 0 \text{ unless } |x-x_0| < c\Delta t;$$

that is, unless

$$-c\Delta t < x_0 - x < c\Delta t \text{ or } x - c\Delta t < x_0 < x + c\Delta t.$$

So

$$\frac{u(x,\Delta t)-u(x,0)}{\Delta t} = \int_{-\infty}^{\infty}\left[\frac{G(x,t+\Delta t;x_0,0)-G(x,t;x_0,0)}{\Delta t}\right]g(x_0)dx_0$$

$$= \frac{1}{\Delta t}\int_{x-c\Delta t}^{x+c\Delta t}\frac{1}{2c}g(x_0)dx_0 \approx \frac{1}{2c\Delta t}g(x)2c\Delta t \to g(x) \quad \text{as } \Delta t\to 0.$$

Thus,

$$\frac{\partial u(x,0)}{\partial t} = g(x).$$

Example:
Show that

$$\frac{\partial^2 u(x,t)}{\partial t^2} = c^2 \frac{\partial^2 u(x,t)}{\partial x^2} \qquad -\infty < x < \infty, \quad t > 0, \quad u(x,0) = f(x), \quad \frac{\partial u(x,0)}{\partial t} = 0$$

has the solution

$$u(x,t) = -\int_{-\infty}^{\infty} \frac{\partial G(x,t;x_0,0)}{\partial t_0} f(x_0) dx_0 = \frac{f(x+ct) + f(x-ct)}{2}.$$

It is easy to check that

$$u(x,t) = \frac{f(x+ct) + f(x-ct)}{2}$$

satisfies the equation.

To show that the integral form of the solution is valid, we express $G(x,t;x_0,t_0)$ in terms of the Heaviside function. Recall that the Heaviside function, $H(x)$, is defined by

$$H(x) = \begin{cases} 0 & \text{if } x < 0 \\ 1 & \text{if } x > 0 \end{cases}.$$

Also, the derivative of the Heaviside function is the Dirac-δ function. We have

$$G(x,t;x_0,t_0) = \begin{cases} \dfrac{1}{2c} & \text{if } |x - x_0| < c(t - t_0) \\ 0 & \text{if } |x - x_0| > c(t - t_0) \end{cases}.$$

Note that $H(\alpha) - H(\beta) = 1$ if and only if $\alpha > 0$ and $\beta < 0$.
Now $-b < a < b$ if and only if $|a| < b$, so

$$|x - x_0| < c(t - t_0)$$

if and only if

$$-c(t - t_0) < x - x_0 < c(t - t_0)$$

which is true if and only if

$$(x - x_0) + c(t - t_0) > 0 \text{ and } (x - x_0) - c(t - t_0) < 0.$$

Thus

$$H((x - x_0) + c(t - t_0)) - H((x - x_0) - c(t - t_0)) = 1$$

if and only if

$$|x - x_0| < c(t - t_0).$$

So we have

$$G(x,t;x_0,t_0) = \frac{1}{2c}[H((x-x_0)+ c(t-t_0)) - H((x-x_0)- c(t-t_0))].$$

Keeping in mind that the derivative of the Heaviside function is the Dirac-δ function, we have

$$\begin{aligned}\frac{\partial G(x,t;x_0,t_0)}{\partial t_0} &= \frac{1}{2c}[\{\delta[(x-x_0)+ c(t-t_0)](-c) - \delta[(x-x_0)- c(t-t_0)](c)\}] \\ &= \frac{c}{2c}[\{-\delta[(x-x_0)+ c(t-t_0)] - \delta[(x-x_0)- c(t-t_0)]\}] \\ &= -\frac{1}{2}\{\delta[(x-x_0)+ c(t-t_0)] - \delta[(x-x_0)- c(t-t_0)]\}.\end{aligned}$$

Thus,

$$\frac{\partial G(x,t;x_0,0)}{\partial t_0} = -\frac{1}{2}\{\delta[(x-x_0)+ ct] - \delta[(x-x_0)- ct]\}$$

and so

$$\begin{aligned}&= -\int_{-\infty}^{\infty}\frac{\partial G(x,t;x_0,0)}{\partial t_0}f(x_0)dx_0 \\ &= \frac{1}{2}\left\{\int_{-\infty}^{\infty}\delta[(x-x_0)+ ct]f(x_0)dx_0 + \int_{-\infty}^{\infty}\delta[(x-x_0)- ct]f(x_0)dx_0\right\}.\end{aligned}$$

Now

$$(x-x_0)+ ct = 0 \text{ if } x_0 = x+ct \text{ and } (x-x_0)- ct = 0 \text{ if } x_0 = x-ct$$

so

$$\begin{aligned}&\frac{1}{2}\left\{\int_{-\infty}^{\infty}\delta[(x-x_0)+ ct]f(x_0)dx_0 + \int_{-\infty}^{\infty}\delta[(x-x_0)- ct]f(x_0)dx_0\right\} \\ &\quad = \frac{f(x+ct)+f(x-ct)}{2}.\end{aligned}$$

Combining the examples above, we have the solution to

$$\frac{\partial^2 u(x,t)}{\partial t^2} = c^2\frac{\partial^2 u(x,t)}{\partial x^2} + Q(x,t), \quad -\infty < x < \infty, \ t > 0;$$

$$u(x,0) = f(x), \quad \frac{\partial u(x,0)}{\partial t} = g(x)$$

is

$$u(x,t)=\frac{1}{2c}\iint_{\Omega}Q(x_0,t_0)dx_0dt_0+\frac{f(x+ct)+f(x-ct)}{2}+\frac{1}{2c}\int_{x-ct}^{x+ct}g(x_0)dx_0$$

where Ω is the region $\left\{(x_0,t_0)\mid 0<t_0<t-\frac{1}{c}|x-x_0|\right\}$.

EXERCISES

1. Solve

$$\frac{\partial^2 u(x,t)}{\partial t^2}=c^2\frac{\partial^2 u(x,t)}{\partial x^2},\quad -\infty<x<\infty,\ t>0;$$
$$u(x,0)=0,\quad \frac{\partial u(x,0)}{\partial t}=e^{-x}.$$

2. Solve

$$\frac{\partial^2 u(x,t)}{\partial t^2}=c^2\frac{\partial^2 u(x,t)}{\partial x^2}+e^{-t}\sin x,\quad -\infty<x<\infty,\ t>0;$$
$$u(x,0)=\frac{1}{1+x^2},\quad \frac{\partial u(x,0)}{\partial t}=\cos x.$$

3. Solve

$$\frac{\partial^2 u(x,t)}{\partial t^2}=c^2\frac{\partial^2 u(x,t)}{\partial x^2},\quad -\infty<x<\infty,\ t>0;\ u(x,0)=e^{-x^2},\quad \frac{\partial u(x,0)}{\partial t}=0.$$

13.4 GREEN'S FUNCTION AND POISSON'S EQUATION

In this section we find the solution to Laplace's equation and Poisson's equation in $\mathbb{R}^2$ using Green's function.

A boundary value problem for Laplace's equation on a domain D is

$$\Delta u=0 \text{ on } D;\quad u=f \text{ on } \partial D.$$

Poisson's equation is the nonhomogeneous form of Laplace's equation. A boundary value problem for Poisson's equation is

$$\Delta u=-g \text{ on } D;\quad u=f \text{ on } \partial D.$$

The pattern that we follow is conceptually the same as what one does in ordinary differential equation to find the solution to a nonhomogeneous problem. Namely, we find the solution to the homogeneous problem and a particular solution. One then gets the general solution by adding the two. To determine the solution to Poisson's equation, we find the superposition of u_f and u_g where u_f is the solution to Laplace's equation

$$\Delta u_f = 0 \text{ on } D; \quad u_f = f \text{ on } \partial D$$

and u_g is the solution to

$$\Delta u_g = -g \text{ on } D; \quad u_g = 0 \text{ on } \partial D.$$

We consider the problem on the disk

$$\Delta u(r,\theta) = 0, \quad u(R,\theta) = f(\theta).$$

We have previously noted (in Exercise 8, Section 4.3 and at the end of Section 10.5) that the solution is given by the Poisson integral formula

$$u(r,\theta) = \frac{1}{2\pi}\int_0^R f(\tau)\frac{R^2 - r^2}{R^2 - 2rR\cos(\theta-\tau) + r^2}\,d\tau.$$

The second part of the problem is a boundary value problem with homogeneous boundary conditions, and we find the solution using Green's function.

We find Green's function for Poisson's equation inside a circle of radius R centered at the origin. We want to find $G(\widehat{x}, \widehat{x}_0)$ so that

$$\Delta G(\widehat{x}, \widehat{x}_0) = \delta(\widehat{x} - \widehat{x}_0)$$

inside the circle and satisfies the boundary condition

$$G(\widehat{x}, \widehat{x}_0) = 0 \text{ if } |\widehat{x}| = R.$$

To achieve the boundary condition we use the method of images. We follow the ideas of the previous section and hypothesize that we balance a charge q within the circle at point $\widehat{x}_0$ by a charge $-q$ at point outside the circle at point $\widehat{x}_0^*$. We situate the point $\widehat{x}_0^*$ so that $\widehat{x}_0$ and $\widehat{x}_0^*$ lie on the same radial line from the origin. See Fig. 13.4.1.

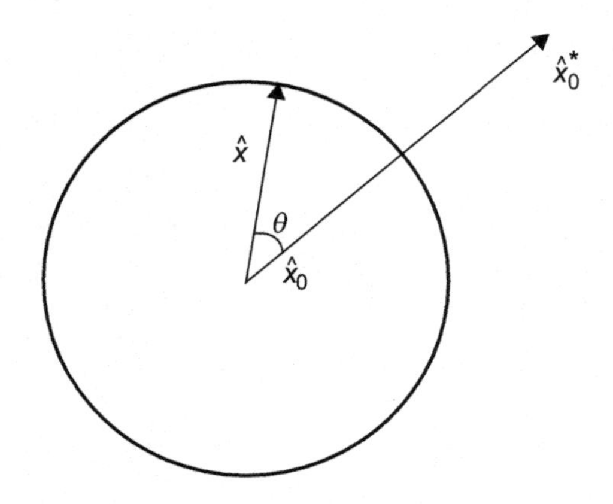

FIGURE 13.4.1

We must determine the distance from the origin to $\widehat{x}_0^*$ so that

$$G(\widehat{x},\ \widehat{x}_0) = 0 \text{ if } |\widehat{x}| = R.$$

We have

$$\begin{aligned} G(\widehat{x},\ \widehat{x}_0) &= \frac{1}{2\pi}\ln|\widehat{x}-\widehat{x}_0| - \frac{1}{2\pi}\ln\left|\widehat{x}-\widehat{x}_0^*\right| + C \\ &= \frac{1}{4\pi}\ln\frac{|\widehat{x}-\widehat{x}_0|^2}{\left|\widehat{x}-\widehat{x}_0^*\right|^2} - \frac{1}{4\pi}\ln e^{-4\pi C} \\ &= \frac{1}{4\pi}\ln\frac{|\widehat{x}-\widehat{x}_0|^2}{e^{-4\pi C}\left|\widehat{x}-\widehat{x}_0^*\right|^2} \\ &= \frac{1}{4\pi}\ln\frac{|\widehat{x}-\widehat{x}_0|^2}{D\left|\widehat{x}-\widehat{x}_0^*\right|^2} \quad \text{where } D = e^{-4\pi C}. \end{aligned}$$

For $|\widehat{x}| = R$, we want $G(\widehat{x},\ \widehat{x}_0) = 0$,which will be true if and only if

$$\frac{|\widehat{x}-\widehat{x}_0|^2}{D\left|\widehat{x}-\widehat{x}_0^*\right|^2} = 1;$$

that is, if and only if

$$|\widehat{x}-\widehat{x}_0|^2 = D\left|\widehat{x}-\widehat{x}_0^*\right|^2.$$

Since $\widehat{x}_0$ and $\widehat{x}_0^*$ are on the same radial line from the origin, then $\widehat{x}_0^* = \alpha\widehat{x}_0$. We must find D and α. Now

$$\begin{aligned} |\widehat{x}-\widehat{x}_0|^2 &= \langle\widehat{x}-\widehat{x}_0, \widehat{x}-\widehat{x}_0\rangle = \langle\widehat{x},\widehat{x}\rangle + \langle\widehat{x}_0,\widehat{x}_0\rangle - 2\langle\widehat{x},\widehat{x}_0\rangle \\ &= |\widehat{x}|^2 + |\widehat{x}_0|^2 - 2|\widehat{x}||\widehat{x}_0|\cos\theta \end{aligned}$$

where θ is the angle between $\widehat{x}$ and $\widehat{x}_0$. Likewise,

$$\left|\widehat{x}-\widehat{x}_0^*\right|^2 = |\widehat{x}|^2 + \left|\widehat{x}_0^*\right|^2 - 2|\widehat{x}|\left|\widehat{x}_0^*\right|\cos\theta.$$

If $\widehat{x}$ is on the boundary of the circle, then $|\widehat{x}| = R$. We let $|\widehat{x}_0| = r$, so that $\left|\widehat{x}_0^*\right| = \alpha r$. Thus, we have

$$R^2 + r^2 - 2rR\cos\theta = D\left(R^2 + \alpha^2 r^2 - 2\alpha rR\cos\theta\right) \tag{1}$$

and Eq. (1) must hold for all angles θ. Letting $\theta = \pi/2$ we have

$$R^2 + r^2 = D\left(R^2 + \alpha^2 r^2\right).$$

Then

$$R^2 + r^2 - 2rR\cos\theta = D(R^2 + \alpha^2 r^2) - 2rR\cos\theta = D(R^2 + \alpha^2 r^2 - 2\alpha rR\cos\theta)$$
$$= D(R^2 + \alpha^2 r^2) - D2\alpha rR\cos\theta$$

so

$$2rR\cos\theta = D2\alpha rR\cos\theta \text{ and thus } \alpha D = 1.$$

One can use the quadratic formula applied to

$$R^2 + r^2 = D(R^2 + \alpha^2 r^2) = \frac{1}{\alpha}(R^2 + \alpha^2 r^2) \quad \text{or} \quad \alpha^2 r^2 - \alpha(R^2 + r^2) + R^2 = 0$$

to get

$$\alpha = \frac{R^2}{r^2} \text{ and thus } D = \frac{1}{\alpha} = \frac{r^2}{R^2}.$$

Finally, we have

$$G(\widehat{x}, \widehat{x}_0) = \frac{1}{4\pi}\ln\frac{|\widehat{x} - \widehat{x}_0|^2}{D|\widehat{x} - \widehat{x}_0^*|^2} = \frac{1}{4\pi}\ln\frac{R^2}{r^2}\frac{|\widehat{x} - \widehat{x}_0|^2}{|\widehat{x} - \widehat{x}_0^*|^2} = \frac{1}{2\pi}\ln\frac{R}{r}\frac{|\widehat{x} - \widehat{x}_0|}{|\widehat{x} - \widehat{x}_0^*|}$$
$$= \frac{1}{2\pi}\ln\frac{R}{|\widehat{x}_0|}\frac{|\widehat{x} - \widehat{x}_0|}{|\widehat{x} - \widehat{x}_0^*|}$$

where

$$\widehat{x}_0^* = \frac{R^2}{|\widehat{x}_0|^2}\widehat{x}_0.$$

We thus have the following result:

Theorem:

The solution to

$$\Delta u_g = -g \text{ on } D; \quad u_g = 0 \text{ on } \partial D$$

where D is a disc of radius R is

$$u(P) = \int_{|Q|=R}\frac{1}{2\pi}\ln\frac{R}{|Q|}\frac{|P - Q|}{|P - Q'|}g(Q)dS_Q$$

where

$$Q' = \frac{R^2}{|Q|^2}Q.$$

We thus have the following result:
Theorem:
The solution to

$$\Delta u = -g \text{ on } D; \quad u = f \text{ on } \partial D$$

where D is the disc of radius R is given by

$$u(r,\theta) = \frac{1}{2\pi}\int_{-\pi}^{\pi} f(\theta)\frac{R^2 - r^2}{R^2 - 2Rr\cos(\theta-\theta_0) + r^2}d\theta + \int_{|Q|=R}\frac{1}{2\pi}\ln\frac{R}{|Q|}\frac{|P-Q|}{|P-Q'|}g(Q)dS_Q$$

where $P = (r,\theta)$.

We note that

$$\int_{|Q|=R}\frac{1}{2\pi}\ln\frac{R}{|Q|}\frac{|P-Q|}{|P-Q'|}g(Q)dS_Q = \frac{1}{2\pi}\int_{|Q|=R}\left(\ln\frac{1}{|P-Q|} - \ln\frac{1}{|P-Q'|}\right)g(Q)dS_Q.$$

We note also that if $P = (r_0,\theta_0)$ and $Q = (r,\theta)$ then $|P-Q| = r^2 + r_0^2 - 2rr_0\cos(\theta-\theta_0)$ and $Q' = \left(\frac{R^2}{r},\theta\right)$ so that $|P-Q'| = \frac{R^4}{r^2} + r_0^2 - 2\frac{R^2}{r}\cos(\theta-\theta_0)$. Thus, $u(r,\theta)$ can be expressed as

$$u(r,\theta) = \frac{1}{2\pi}\int_{-\pi}^{\pi} f(\theta)\frac{R^2 - r^2}{R^2 - 2Rr\cos(\theta-\theta_0) + r^2}d\theta + \frac{1}{2\pi}\int_{-\pi}^{\pi}\left[\ln\left(\frac{1}{r^2 + r_0^2 - 2rr_0\cos(\theta-\theta_0)}\right) - \ln\left(\frac{1}{\frac{R^4}{r^2} + r_0^2 - 2\frac{R^2}{r}\cos(\theta-\theta_0)}\right)\right]g(R\theta)d\theta.$$

EXERCISES

1. Let D be the disc of radius.
2. Find an expression for the solution to $\Delta u = -g$ on $D, u = f$ on ∂D for the following functions:
   a. $f = r \sin \theta,\ g = \cos \theta.$
   b. $f = 3,\ g = \sin^2 \theta.$
   c. $f = \dfrac{\cos \theta}{1 + r^2},\ g = 1.$

Appendix 1

COMPUTING THE LAPLACIAN WITH THE CHAIN RULE

Here we compute the Laplacian in polar and spherical coordinates using the chain rule. The computations are tedious and are presented only because this is the approach of many texts.

Suppose that we are in the (x,y,z) coordinate system, and we want to convert to the (a,b,c) coordinate system where a, b, and c are functions of x, y, and z. The chain rule says that

$$\frac{\partial f}{\partial x}=\frac{\partial f}{\partial a}\cdot\frac{\partial a}{\partial x}+\frac{\partial f}{\partial b}\cdot\frac{\partial b}{\partial x}+\frac{\partial f}{\partial c}\cdot\frac{\partial c}{\partial x}. \tag{1}$$

Note that $\frac{\partial f}{\partial x}$ is expressed in terms of x, y, and z, and the right-hand side of Eq. (7) is in terms of a, b, and c. Also $\frac{\partial f}{\partial y}$ and $\frac{\partial f}{\partial z}$ are expressed in a similar manner. We shall apply the idea of Eq. (1) several times to find

$$\frac{\partial^2 f}{\partial x^2}+\frac{\partial^2 f}{\partial y^2}+\frac{\partial^2 f}{\partial z^2}$$

in cylindrical and spherical coordinates.

CYLINDRICAL COORDINATES

Cylindrical coordinates are related to rectangular coordinates by

$$x = r\cos\theta, \quad y = r\sin\theta, \quad z = z,$$

or

$$r^2 = x^2+y^2, \quad \theta = \tan^{-1}\left(\frac{y}{x}\right), \quad z = z.$$

Thus $r=\left(x^2+y^2\right)^{\frac{1}{2}}$, so

$$\frac{\partial r}{\partial x}=x\left(x^2+y^2\right)^{-\frac{1}{2}}=\frac{x}{r}=\frac{r\cos\theta}{r}=\cos\theta;$$

$$\frac{\partial\theta}{\partial x}=\frac{1}{1+\left(\frac{y}{x}\right)^2}\cdot y\left(\frac{-1}{x^2}\right)=\frac{-y}{x^2+y^2}=-\frac{r\sin\theta}{r^2}=-\frac{\sin\theta}{r};$$

$$\frac{\partial z}{\partial x} = 0.$$

Similarly,

$$\frac{\partial r}{\partial y} = \sin\theta, \quad \frac{\partial \theta}{\partial y} = \frac{\cos\theta}{r}, \quad \frac{\partial z}{\partial y} = 0,$$

and

$$\frac{\partial r}{\partial z} = 0, \quad \frac{\partial \theta}{\partial z} = 0, \quad \frac{\partial z}{\partial z} = 1.$$

So

$$\frac{\partial f}{\partial x} = \frac{\partial f}{\partial r}\cdot\frac{\partial r}{\partial x} + \frac{\partial f}{\partial \theta}\cdot\frac{\partial \theta}{\partial x} + \frac{\partial f}{\partial z}\cdot\frac{\partial z}{\partial x} = \frac{\partial f}{\partial r}\cos\theta + \frac{\partial f}{\partial \theta}\left(-\frac{\sin\theta}{r}\right).$$

Also

$$\frac{\partial f}{\partial y} = \frac{\partial f}{\partial r}\sin\theta + \frac{\partial f}{\partial \theta}\frac{\cos\theta}{r}$$

$$\frac{\partial f}{\partial z} = \frac{\partial f}{\partial z}.$$

We next compute $\frac{\partial^2 f}{\partial x^2}$. We have

$$\frac{\partial^2 f}{\partial x^2} = \frac{\partial}{\partial x}\left(\frac{\partial f}{\partial x}\right) = \frac{\partial}{\partial x}\left(\frac{\partial f}{\partial r}\cos\theta\right) - \frac{\partial}{\partial x}\left(\frac{\partial f}{\partial \theta}\frac{\sin\theta}{r}\right).$$

Now

$$\frac{\partial}{\partial x}\left(\frac{\partial f}{\partial r}\cos\theta\right) = \frac{\partial}{\partial r}\left(\frac{\partial f}{\partial r}\cos\theta\right)\frac{\partial r}{\partial x} + \frac{\partial}{\partial \theta}\left(\frac{\partial f}{\partial r}\cos\theta\right)\frac{\partial \theta}{\partial x} + \frac{\partial}{\partial z}\left(\frac{\partial f}{\partial r}\cos\theta\right)\frac{\partial z}{\partial x}.$$

Consider

$$\frac{\partial}{\partial r}\left(\frac{\partial f}{\partial r}\cos\theta\right) = \frac{\partial^2 f}{\partial r^2}\cos\theta + \frac{\partial f}{\partial r}\cdot\frac{\partial}{\partial r}\cos\theta = \frac{\partial^2 f}{\partial r^2}\cos\theta.$$

So

$$\frac{\partial}{\partial r}\left(\frac{\partial f}{\partial r}\cos\theta\right)\frac{\partial r}{\partial x} = \frac{\partial^2 f}{\partial r^2}\cos^2\theta.$$

Now

$$\frac{\partial}{\partial \theta}\left(\frac{\partial f}{\partial r}\cos\theta\right) = \frac{\partial^2 f}{\partial r\partial \theta}\cos\theta - \frac{\partial f}{\partial r}\sin\theta$$

so

$$\frac{\partial}{\partial \theta}\left(\frac{\partial f}{\partial r}\cos\theta\right)\frac{\partial \theta}{\partial x} = \left(\frac{\partial^2 f}{\partial r \partial \theta}\cos\theta - \frac{\partial f}{\partial r}\sin\theta\right)\left(-\frac{\sin\theta}{r}\right)$$
$$= -\frac{\partial^2 f}{\partial r \partial \theta}\frac{\cos\theta\sin\theta}{r} + \frac{\partial f}{\partial r}\frac{\sin^2\theta}{r}.$$

Also

$$\frac{\partial z}{\partial x} = 0.$$

Thus

$$\frac{\partial}{\partial x}\left(\frac{\partial f}{\partial r}\cos\theta\right) = \frac{\partial}{\partial r}\left(\frac{\partial f}{\partial r}\cos\theta\right)\frac{\partial r}{\partial x} + \frac{\partial}{\partial \theta}\left(\frac{\partial f}{\partial r}\cos\theta\right)\frac{\partial \theta}{\partial x} + \frac{\partial}{\partial z}\left(\frac{\partial f}{\partial r}\cos\theta\right)\frac{\partial z}{\partial x}$$
$$= \frac{\partial}{\partial r}\left(\frac{\partial f}{\partial r}\cos\theta\right)\frac{\partial r}{\partial x} + \frac{\partial}{\partial \theta}\left(\frac{\partial f}{\partial r}\cos\theta\right)\frac{\partial \theta}{\partial x}$$
$$= \left[\frac{\partial^2 f}{\partial r^2}\cos^2\theta - \frac{\partial^2 f}{\partial r \partial \theta}\frac{\cos\theta\sin\theta}{r} + \frac{\partial f}{\partial r}\frac{\sin^2\theta}{r}\right]. \tag{2}$$

Similarly,

$$\frac{\partial}{\partial x}\left(\frac{\partial f}{\partial \theta}\frac{\sin\theta}{r}\right) = \frac{\partial}{\partial r}\left(\frac{\partial f}{\partial \theta}\frac{\sin\theta}{r}\right)\frac{\partial r}{\partial x} + \frac{\partial}{\partial \theta}\left(\frac{\partial f}{\partial \theta}\frac{\sin\theta}{r}\right)\frac{\partial \theta}{\partial x} + \frac{\partial}{\partial z}\left(\frac{\partial f}{\partial \theta}\frac{\sin\theta}{r}\right)\frac{\partial z}{\partial x}$$
$$= \frac{\partial}{\partial r}\left(\frac{\partial f}{\partial \theta}\frac{\sin\theta}{r}\right)\frac{\partial r}{\partial x} + \frac{\partial}{\partial \theta}\left(\frac{\partial f}{\partial \theta}\frac{\sin\theta}{r}\right)\frac{\partial \theta}{\partial x}.$$

We have

$$\frac{\partial}{\partial r}\left(\frac{\partial f}{\partial \theta}\frac{\sin\theta}{r}\right) = \frac{\partial^2 f}{\partial r \partial \theta}\cdot\frac{\sin\theta}{r} - \frac{\partial f}{\partial \theta}\frac{\sin\theta}{r^2},$$

so

$$\frac{\partial}{\partial r}\left(\frac{\partial f}{\partial \theta}\frac{\sin\theta}{r}\right)\frac{\partial r}{\partial x} = \frac{\partial^2 f}{\partial r \partial \theta}\cdot\frac{\sin\theta\cos\theta}{r} - \frac{\partial f}{\partial}\frac{\sin\theta\cos\theta}{r^2}.$$

Likewise,

$$\frac{\partial}{\partial \theta}\left(\frac{\partial f}{\partial \theta}\frac{\sin\theta}{r}\right) = \frac{\partial^2 f}{\partial \theta^2}\cdot\frac{\sin\theta}{r} + \frac{\partial f}{\partial \theta}\cdot\frac{\cos\theta}{r},$$

so

$$\frac{\partial}{\partial \theta}\left(\frac{\partial f}{\partial \theta}\frac{\sin\theta}{r}\right)\frac{\partial \theta}{\partial x} = \left(\frac{\partial^2 f}{\partial \theta^2}\cdot\frac{\sin\theta}{r} + \frac{\partial f}{\partial \theta}\cdot\frac{\cos\theta}{r}\right)\left(-\frac{\sin\theta}{r}\right)$$

$$= -\frac{\partial^2 f}{\partial \theta^2}\frac{\sin^2\theta}{r^2} - \frac{\partial f}{\partial \theta}\frac{\sin\theta\cos\theta}{r^2}.$$

Thus

$$\frac{\partial}{\partial x}\left(\frac{\partial f}{\partial \theta}\frac{\sin\theta}{r}\right) = \frac{\partial^2 f}{\partial r\partial\theta}\cdot\frac{\sin\theta\cos\theta}{r} - \frac{\partial f}{\partial \theta}\frac{\sin\theta\cos\theta}{r^2} - \frac{\partial^2 f}{\partial \theta^2}\frac{\sin^2\theta}{r^2} - \frac{\partial f}{\partial \theta}\frac{\sin\theta\cos\theta}{r^2}. \tag{3}$$

To obtain $\frac{\partial^2 f}{\partial x^2}$, we subtract expression (3) from expression (2) to get

$$\begin{aligned}\frac{\partial^2 f}{\partial x^2} &= \frac{\partial}{\partial x}\left(\frac{\partial f}{\partial r}\cos\theta\right) - \frac{\partial}{\partial x}\left(\frac{\partial f}{\partial \theta}\frac{\sin\theta}{r}\right)\\ &= \left[\frac{\partial^2 f}{\partial r^2}\cos^2\theta - \frac{\partial^2 f}{\partial r\partial\theta}\frac{\cos\theta\sin\theta}{r} + \frac{\partial f}{\partial r}\frac{\sin^2\theta}{r}\right]\\ &\quad - \left[\frac{\partial^2 f}{\partial r\partial\theta}\cdot\frac{\sin\theta\cos\theta}{r} - \frac{\partial f}{\partial \theta}\frac{\sin\theta\cos\theta}{r^2} - \frac{\partial^2 f}{\partial \theta^2}\frac{\sin^2\theta}{r^2} - \frac{\partial f}{\partial \theta}\frac{\sin\theta\cos\theta}{r^2}\right]\\ &= \frac{\partial^2 f}{\partial r^2}\cos^2\theta - 2\frac{\partial^2 f}{\partial r\partial\theta}\frac{\cos\theta\sin\theta}{r} + \frac{\partial^2 f}{\partial \theta^2}\frac{\sin^2\theta}{r^2}\\ &\quad + 2\frac{\partial f}{\partial \theta}\frac{\sin\theta\cos\theta}{r^2} + \frac{\partial f}{\partial r}\frac{\sin^2\theta}{r}.\end{aligned} \tag{4}$$

Likewise, we must calculate $\frac{\partial^2 f}{\partial y^2}$.
We have

$$\frac{\partial f}{\partial y} = \frac{\partial f}{\partial r}\sin + \frac{\partial f}{\partial \theta}\frac{\cos\theta}{r},$$

so

$$\frac{\partial^2 f}{\partial y^2} = \frac{\partial}{\partial y}\left(\frac{\partial f}{\partial r}\sin\theta\right) + \frac{\partial}{\partial y}\left(\frac{\partial f}{\partial \theta}\frac{\cos\theta}{r}\right).$$

Now

$$\frac{\partial}{\partial y}\left(\frac{\partial f}{\partial r}\sin\theta\right) = \frac{\partial}{\partial r}\left(\frac{\partial f}{\partial r}\sin\theta\right)\frac{\partial r}{\partial y} + \frac{\partial}{\partial \theta}\left(\frac{\partial f}{\partial r}\sin\theta\right)\frac{\partial \theta}{\partial y}.$$

We have

$$\frac{\partial}{\partial r}\left(\frac{\partial f}{\partial r}\sin\theta\right)\frac{\partial r}{\partial y}=\frac{\partial^2 f}{\partial r^2}\sin\theta\cdot\sin\theta,$$

and

$$\frac{\partial}{\partial\theta}\left(\frac{\partial f}{\partial r}\sin\theta\right)\frac{\partial\theta}{\partial y}=\left[\frac{\partial^2 f}{\partial\theta\partial r}\sin\theta+\frac{\partial f}{\partial r}\cos\theta\right]\frac{\cos\theta}{r}.$$

So

$$\frac{\partial}{\partial y}\left(\frac{\partial f}{\partial r}\sin\theta\right)=\frac{\partial^2 f}{\partial r^2}\sin^2\theta+\frac{\partial^2 f}{\partial\theta\partial r}\frac{\sin\theta\cos\theta}{r}+\frac{\partial f}{\partial r}\frac{\cos^2\theta}{r}. \tag{5}$$

Also

$$\frac{\partial}{\partial y}\left(\frac{\partial f}{\partial\theta}\frac{\cos\theta}{r}\right)=\frac{\partial}{\partial r}\left(\frac{\partial f}{\partial\theta}\frac{\cos\theta}{r}\right)\frac{\partial r}{\partial y}+\frac{\partial}{\partial\theta}\left(\frac{\partial f}{\partial\theta}\frac{\cos\theta}{r}\right)\frac{\partial\theta}{\partial y}.$$

Now

$$\begin{aligned}\frac{\partial}{\partial r}\left(\frac{\partial f}{\partial\theta}\frac{\cos\theta}{r}\right)\frac{\partial r}{\partial y}&=\left[\frac{\partial^2 f}{\partial r\partial\theta}\frac{\cos\theta}{r}-\frac{\partial f}{\partial\theta}\frac{\cos\theta}{r^2}\right]\sin\theta\\&=\frac{\partial^2 f}{\partial r\partial\theta}\frac{\cos\theta\sin\theta}{r}-\frac{\partial f}{\partial\theta}\frac{\cos\theta\sin\theta}{r^2},\end{aligned} \tag{6}$$

and

$$\begin{aligned}\frac{\partial}{\partial\theta}\left(\frac{\partial f}{\partial\theta}\frac{\cos\theta}{r}\right)\frac{\partial\theta}{\partial y}&=\left[\frac{\partial^2 f}{\partial\theta^2}\frac{\cos\theta}{r}-\frac{\partial f}{\partial\theta}\frac{\sin\theta}{r}\right]\frac{\cos\theta}{r}\\&=\frac{\partial^2 f}{\partial\theta^2}\frac{\cos^2\theta}{r^2}-\frac{\partial f}{\partial\theta}\frac{\cos\theta\sin\theta}{r^2}.\end{aligned} \tag{7}$$

Adding (6) and (7) gives

$$\begin{aligned}\frac{\partial}{\partial y}\left(\frac{\partial f}{\partial\theta}\frac{\cos\theta}{r}\right)&=\left[\frac{\partial^2 f}{\partial r\partial\theta}\frac{\cos\theta\sin\theta}{r}-\frac{\partial f}{\partial\theta}\frac{\cos\theta\sin\theta}{r^2}\right]\\&\quad+\left[\frac{\partial^2 f}{\partial\theta^2}\frac{\cos^2\theta}{r^2}-\frac{\partial f}{\partial\theta}\frac{\cos\theta\sin\theta}{r^2}\right]\\&=\frac{\partial^2 f}{\partial r\partial\theta}\frac{\cos\theta\sin\theta}{r}-2\frac{\partial f}{\partial\theta}\frac{\cos\theta\sin\theta}{r^2}+\frac{\partial^2 f}{\partial\theta^2}\frac{\cos^2\theta}{r^2}.\end{aligned} \tag{8}$$

Adding (5) and (8) yields

$$\frac{\partial^2 f}{\partial y^2} = \frac{\partial}{\partial y}\left(\frac{\partial f}{\partial r}\sin\theta\right) + \frac{\partial}{\partial y}\left(\frac{\partial f}{\partial \theta}\frac{\cos\theta}{r}\right)$$

$$= \left[\frac{\partial^2 f}{\partial r^2}\sin^2\theta + \frac{\partial^2 f}{\partial\theta\partial r}\frac{\sin\theta\cos\theta}{r} + \frac{\partial f}{\partial r}\frac{\cos^2\theta}{r}\right]$$

$$+ \left[\frac{\partial^2 f}{\partial r\partial\theta}\frac{\cos\theta\sin\theta}{r} - 2\frac{\partial f}{\partial\theta}\frac{\cos\theta\sin\theta}{r^2} + \frac{\partial^2 f}{\partial\theta^2}\frac{\cos^2\theta}{r^2}\right].$$

Adding the expressions for $\frac{\partial^2 f}{\partial x^2}$ and $\frac{\partial^2 f}{\partial y^2}$ gives

$$\frac{\partial^2 f}{\partial x^2} + \frac{\partial^2 f}{\partial y^2} = \left\{\frac{\partial^2 f}{\partial r^2}\cos^2\theta - 2\frac{\partial^2 f}{\partial r\partial\theta}\frac{\cos\theta\sin\theta}{r} + \frac{\partial^2 f}{\partial\theta^2}\frac{\sin^2\theta}{r^2}\right.$$

$$\left. + 2\frac{\partial f}{\partial\theta}\frac{\sin\theta\cos\theta}{r^2} + \frac{\partial f}{\partial r}\frac{\sin^2\theta}{r}\right\}$$

$$+ \left\{\left[\frac{\partial^2 f}{\partial r^2}\sin^2\theta + \frac{\partial^2 f}{\partial\theta\partial r}\frac{\sin\theta\cos\theta}{r} + \frac{\partial f}{\partial r}\frac{\cos^2\theta}{r}\right]\right.$$

$$\left. + \left[\frac{\partial^2 f}{\partial r\partial\theta}\frac{\cos\theta\sin\theta}{r} - 2\frac{\partial f}{\partial\theta}\frac{\cos\theta\sin\theta}{r^2} + \frac{\partial^2 f}{\partial\theta^2}\frac{\cos^2\theta}{r^2}\right]\right\}$$

$$= \frac{\partial^2 f}{\partial r^2}\cos^2\theta + \frac{\partial^2 f}{\partial r^2}\sin^2\theta + \frac{\partial f}{\partial r}\frac{\sin^2\theta}{r}$$

$$+ \frac{\partial f}{\partial r}\frac{\cos^2\theta}{r}\frac{\partial^2 f}{\partial\theta^2}\frac{\sin^2\theta}{r^2} + \frac{\partial^2 f}{\partial\theta^2}\frac{\cos^2\theta}{r^2}$$

$$= \frac{\partial^2 f}{\partial r^2} + \frac{1}{r}\frac{\partial f}{\partial r} + \frac{1}{r^2}\frac{\partial^2 f}{\partial\theta^2}.$$

Thus the Laplacian in cylindrical coordinates is given by

$$\Delta f = \frac{\partial^2 f}{\partial r^2} + \frac{1}{r}\frac{\partial f}{\partial r} + \frac{1}{r^2}\frac{\partial^2 f}{\partial\theta^2} + \frac{\partial^2 f}{\partial z^2}. \tag{9}$$

Appendix 2

THE LAPLACIAN IN SPHERICAL COORDINATES

To compute the Laplacian in spherical coordinates, we use the variables r, θ, ϕ, where θ and ϕ are as shown in Fig. A.1 and $r^2 = x^2 + y^2 + z^2$. We have

$$x = r \sin\theta \cos\phi;$$

$$y = r \sin\theta \sin\phi;$$

$$z = r \cos\theta.$$

We shall adapt some of our computations from cylindrical coordinates. In doing so we must be careful, because the standard representation of r in spherical coordinates is not the same r as in cylindrical coordinates. If we take ρ in spherical coordinates to be

$$\rho^2 = x^2 + y^2,$$

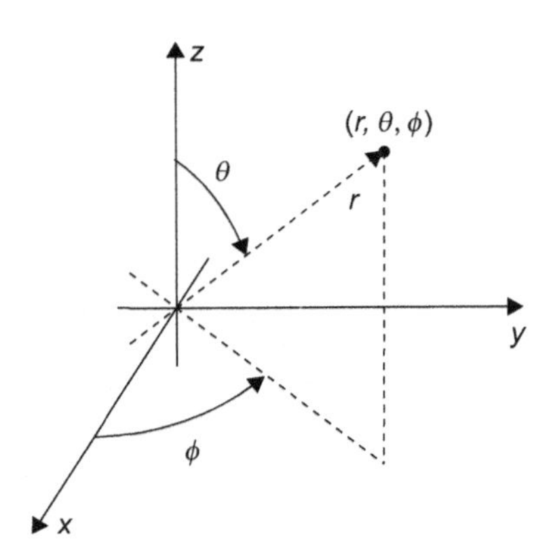

FIGURE A.1

then ρ corresponds to r in cylindrical coordinates. Also ϕ in spherical coordinates was θ in cylindrical coordinates. We shall also use the subscript notation for partial derivatives.

Adapting what we found in cylindrical coordinates to spherical coordinate notation, and using u to stand for the function instead of f, we have

$$u_{xx} + u_{yy} = u_{\rho\rho} + \frac{1}{\rho}u_\rho + \frac{1}{\rho^2}u_{\phi\phi}. \tag{10}$$

We need to convert from the variable ρ to the standard spherical coordinates of r, θ, and ϕ.

A key observation is that z and ρ in spherical coordinates are obtained from r and θ by the same functions that give x and y from ρ and ϕ. Namely,

$$x = \rho \cos\phi \quad \text{and} \quad y = \rho \sin\phi,$$

and

$$z = r\cos\theta \quad \text{and} \quad \rho = r\sin\theta,$$

The last relationship follows from

$$\begin{aligned}\rho^2 = x^2 + y^2 &= (r\sin\theta\cos\phi)^2 + (r\sin\theta\sin\phi)^2 \\ &= (r\sin\theta)^2\left(\cos^2\phi + \sin^2\phi\right) = (r\sin\theta)^2\end{aligned}$$

so

$$\rho = r\sin\theta.$$

This correspondence means that since we have in cylindrical coordinates

$$u_{xx} + u_{yy} = u_{\rho\rho} + \frac{1}{\rho}u_\rho + \frac{1}{\rho^2}u_{\phi\phi},$$

we have in spherical coordinates

$$u_{zz} + u_{\rho\rho} = u_{rr} + \frac{1}{r}u_r + \frac{1}{r^2}u_{\theta\theta}. \tag{11}$$

We want to compute $u_{xx} + u_{yy} + u_{zz}$. Adding Eqs. (10) and (11) gives

$$u_{xx} + u_{yy} + u_{zz} + u_{\rho\rho} = u_{\rho\rho} + \frac{1}{\rho}u_\rho + \frac{1}{\rho^2}u_{\phi\phi} + u_{rr} + \frac{1}{r}u_r + \frac{1}{r^2}u_{\theta\theta}$$

so

$$u_{xx} + u_{yy} + u_{zz} = \frac{1}{\rho}u_\rho + \frac{1}{\rho^2}u_{\phi\phi} + u_{rr} + \frac{1}{r}u_r + \frac{1}{r^2}u_{\theta\theta}. \tag{12}$$

To convert Eq. (12) to the desired form, we must eliminate the variable ρ. If we can express $u\rho$ in terms of r, θ, and ϕ, the rest will be easy. The chain rule gives

$$\frac{\partial u}{\partial \rho} = \frac{\partial u}{\partial r} \cdot \frac{\partial r}{\partial \rho} + \frac{\partial u}{\partial \theta} \cdot \frac{\partial \theta}{\partial \rho} + \frac{\partial u}{\partial \phi} \cdot \frac{\partial \phi}{\partial \rho}.$$

Now $r = \left(\rho^2 + z^2\right)^{\frac{1}{2}}$, $\quad \theta = \tan^{-1}\frac{\rho}{z}$, $\quad \phi = \phi$, so

$$\frac{\partial r}{\partial \rho} = \frac{\rho}{\left(\rho^2 + z^2\right)^{\frac{1}{2}}} = \frac{\rho}{r};$$

$$\frac{\partial \theta}{\partial \rho} = \frac{1}{1 + \left(\frac{\rho}{z}\right)^2} \cdot \frac{1}{z} = \frac{z}{z^2 + \rho^2} = \frac{z}{r^2} = \frac{r \cos \theta}{r^2} = \frac{\cos \theta}{r};$$

$$\frac{\partial \phi}{\partial \rho} = 0.$$

Thus

$$\frac{\partial u}{\partial \rho} = \frac{\partial u}{\partial r} \cdot \frac{\rho}{r} + \frac{\partial u}{\partial \theta} \cdot \frac{\cos \theta}{r} + \frac{\partial u}{\partial \phi} \cdot 0.$$

Or, converting to subscript notation,

$$u_\rho = u_r \frac{\rho}{r} + u_\theta \frac{\cos \theta}{r}.$$

Substituting into Eq. (12) gives

$$\begin{aligned} u_{xx} + u_{yy} + u_{zz} &= \frac{1}{\rho} u_\rho + \frac{1}{\rho^2} u_{\phi\phi} + u_{rr} + \frac{1}{r} u_r + \frac{1}{r^2} u_{\theta\theta} \\ &= \frac{1}{\rho}\left(u_r \frac{\rho}{r} + u_\theta \frac{\cos \theta}{r}\right) + \frac{1}{\rho^2} u_{\phi\phi} + u_{rr} + \frac{1}{r} u_r + \frac{1}{r^2} u_{\theta\theta} \\ &= \frac{u_r}{r} + u_\theta \frac{1}{\rho} \frac{\cos \theta}{r} + \frac{1}{\rho^2} u_{\phi\phi} + u_{rr} + \frac{1}{r} u_r + \frac{1}{r^2} u_{\theta\theta} \\ &= \frac{2}{r} u_r + u_\theta \frac{1}{\rho} \frac{\cos \theta}{r} + \frac{1}{\rho^2} u_{\phi\phi} + u_{rr} + \frac{1}{r^2} u_{\theta\theta}. \end{aligned}$$

Substituting $r \sin \theta$ for ρ gives

$$u_{xx} + u_{yy} + u_{zz} = \frac{2}{r} u_r + u_\theta \frac{1}{r \sin \theta} \frac{\cos \theta}{r} + \frac{1}{(r \sin \theta)^2} u_{\phi\phi} + \frac{1}{r^2} u_{\theta\theta} + u_{rr},$$

which may be rearranged to yield

$$\Delta u = u_{rr} + \frac{2}{r} u_r + \frac{1}{r^2}\left(u_{\theta\theta} + \cot \theta u_\theta + \frac{1}{\sin^2 \theta} u_{\phi\phi}\right).$$

We note that other sources may express this formula in a slightly different way.

Appendix 3

SOME OTHER COORDINATE SYSTEMS

ELLIPTIC CYLINDRICAL COORDINATES

In elliptic cylindrical coordinates the transformations are

$$x = a \cosh u \sin v$$
$$y = a \sinh u \sin v$$
$$z = z$$
$$0 \leq u < \infty, \quad 0 \leq v < 2\pi, \quad -\infty < z < \infty$$

where a is a positive constant.

We take $u_1 = u, \quad u_2 = v, \quad u_3 = z$. We have

$$\widehat{r} = x\widehat{i} + y\widehat{j} + z\widehat{k} = a \cosh u \cos v\widehat{i} + a \sinh u \sin v\widehat{j} + z\widehat{k},$$

so

$$\frac{\partial \widehat{r}}{\partial u_1} = \frac{\partial \widehat{r}}{\partial u} = a \sinh u \cos v\widehat{i} + a \cosh u \sin v\widehat{j}$$
$$\frac{\partial \widehat{r}}{\partial u_2} = \frac{\partial \widehat{r}}{\partial v} = -a \cosh u \sin v\widehat{i} + a \sinh u \cos v\widehat{j}$$
$$\frac{\partial \widehat{r}}{\partial u_3} = \frac{\partial \widehat{r}}{\partial z} = \widehat{k}.$$

1. We demonstrate that the system is orthogonal. We have

$$\left\langle \frac{\partial \widehat{r}}{\partial u}, \frac{\partial \widehat{r}}{\partial v} \right\rangle = -a^2 \sinh u \cos v \cosh u \sin v + a^2 \cosh u \sin v \sinh u \cos v$$
$$= 0.$$

Clearly,

$$\left\langle \frac{\partial \widehat{r}}{\partial u}, \frac{\partial \widehat{r}}{\partial z} \right\rangle = 0 \quad \text{and} \quad \left\langle \frac{\partial \widehat{r}}{\partial v}, \frac{\partial \widehat{r}}{\partial z} \right\rangle = 0.$$

2. We compute the scaling factors. We have

$$h_1 = \left[(a \sinh u \cos v)^2 + (a \cosh u \sin v)^2\right]^{1/2}$$
$$= a\left[\sinh^2 u \cos^2 v + \cosh^2 u \sin^2 v\right]^{1/2}$$
$$= a\left[\sinh^2 u \cos^2 v + \left(1 + \sinh^2 u\right)\sin^2 v\right]^{1/2}$$

since $\cosh^2 u = 1 + \sinh^2 u$.

Then

$$a\left[\sinh^2 u \cos^2 v + \left(1 + \sinh^2 u\right) \sin^2 v\right]^{1/2}$$
$$= a\left[\sinh^2 u\left(\cos^2 v + \sin^2 v\right) + \sin^2 v\right]^{1/2}$$
$$= a\left[\sinh^2 u + \sin^2 v\right]^{1/2}.$$

Thus

$$h_1 = h_u = a\left[\sinh^2 u + \sin^2 v\right]^{1/2}.$$

Also

$$h_2 = h_v = \left[(-a \cosh u \sin v)^2 + (a \sinh u \cos v)^2\right]^{1/2}$$
$$= a\left[\cosh^2 u \sin^2 v + \sinh^2 u \cos^2 v\right]^{1/2}$$
$$= a\left[\sinh^2 u + \sin^2 v\right]^{1/2}.$$

Finally, $h_3 = h_z = 1$.

3. The orthonormal basis $\{\widehat{e}_1, \widehat{e}_2, \widehat{e}_3\}$ is

$$\widehat{e}_1 = \widehat{e}_u = \frac{\sinh u \cos v}{[\sinh^2 u + \sin^2 v]^{1/2}}\widehat{i} + \frac{\cosh u \sin v}{[\sinh^2 u + \sin^2 v]^{1/2}}\widehat{j}$$
$$\widehat{e}_2 = \widehat{e}_v = \frac{-\cosh u \sin v}{[\sinh^2 u + \sin^2 v]^{1/2}}\widehat{i} + \frac{\sinh u \cos v}{[\sinh^2 u + \sin^2 v]^{1/2}}\widehat{j}$$
$$\widehat{e}_3 = \widehat{e}_k = \widehat{k}.$$

4. We have

$$dV = h_1 h_2 h_3 du dv dz = a^2\left[\sinh^2 u + \sin^2 v\right] du dv dz.$$

5. Now,

$$\nabla f = \frac{1}{h_1}\widehat{e}_1\frac{\partial f}{\partial u_1} + \frac{1}{h_2}\widehat{e}_2\frac{\partial f}{\partial u_2} + \frac{1}{h_3}\widehat{e}_3\frac{\partial f}{\partial u_3}$$
$$= \frac{1}{a\left[\sinh^2 u + \sin^2 v\right]^{1/2}}\frac{\partial f}{\partial u}\widehat{e}_u + \frac{1}{a\left[\sinh^2 u + \sin^2 v\right]^{1/2}}\frac{\partial f}{\partial v}\widehat{e}_v + \frac{\partial f}{\partial z}\widehat{e}_z.$$

6. Next

$$\nabla^2 f = \frac{1}{h_1 h_2 h_3}\left[\frac{\partial}{\partial u_1}\left(\frac{h_2 h_3}{h_1}\frac{\partial f}{\partial u_1}\right) + \frac{\partial}{\partial u_2}\left(\frac{h_1 h_3}{h_2}\frac{\partial f}{\partial u_2}\right) + \frac{\partial}{\partial u_3}\left(\frac{h_1 h_2}{h_3}\frac{\partial f}{\partial u_3}\right)\right]$$

$$= \frac{1}{a^2\left[\sinh^2 u + \sin^2 v\right]}\left[\frac{\partial}{\partial u}\frac{\left[\sinh^2 u + \sin^2 v\right]^{1/2}}{\left[\sinh^2 u + \sin^2 v\right]^{1/2}}\frac{\partial f}{\partial u}\right.$$

$$\left.+\frac{\partial}{\partial v}\frac{\left[\sinh^2 u + \sin^2 v\right]^{1/2}}{\left[\sinh^2 u + \sin^2 v\right]^{1/2}}\frac{\partial f}{\partial v} + \frac{\partial}{\partial z}a^2\left[\sinh^2 u + \sin^2 v\right]\frac{\partial f}{\partial z}\right]$$

$$= \frac{1}{a^2\left[\sinh^2 u + \sin^2 v\right]}\left[\frac{\partial^2 f}{\partial u^2} + \frac{\partial^2 f}{\partial v^2} + a^2\left(\sinh^2 u + \sin^2 v\right)\frac{\partial^2 f}{\partial z^2}\right].$$

7. Finally,

$$\nabla \times \vec{F} = \frac{1}{h_1 h_2 h_3}\begin{vmatrix} h_1 \widehat{e}_1 & h_2 \widehat{e}_2 & h_3 \widehat{e}_3 \\ \frac{\partial}{\partial u_1} & \frac{\partial}{\partial u_2} & \frac{\partial}{\partial u_3} \\ h_1 F_1 & h_2 F_2 & h_3 F_3 \end{vmatrix}$$

$$= \frac{1}{a^2\left[\sinh^2 u + \sin^2 v\right]}\begin{vmatrix} a\left[\sinh^2 u + \sin^2 v\right]^{1/2}\widehat{e}_u & a\left[\sinh^2 u + \sin^2 v\right]^{1/2}\widehat{e}_v & \widehat{e}_z \\ \frac{\partial}{\partial u} & \frac{\partial}{\partial v} & \frac{\partial}{\partial z} \\ a\left[\sinh^2 u + \sin^2 v\right]^{1/2}F_1 & a\left[\sinh^2 + \sin^2 v\right]^{1/2}F_2 & F_3 \end{vmatrix}$$

$$= \frac{1}{a^2\left[\sinh^2 u + \sin^2 v\right]}\left\{\left[\frac{\partial F_3}{\partial v} - a\left[\sinh^2 u + \sin^2 v\right]^{1/2}\frac{\partial F_2}{\partial z}\right]a\left[\sinh^2 u + \sin^2 v\right]^{1/2}\widehat{e}_u\right.$$

$$-\left[\frac{\partial F_3}{\partial u} - a\left[\sinh^2 u + \sin^2 v\right]^{1/2}\frac{\partial F_1}{\partial z}\right]a\left[\sinh^2 u + \sin^2 v\right]^{1/2}\widehat{e}_v$$

$$+\left[a\left[\sinh^2 u + \sin^2 v\right]^{1/2}\frac{\partial F_2}{\partial u} + a\left[\sinh^2 u + \sin^2 v\right]^{-1/2}\sinh u \cosh u\, F_2\right.$$

$$\left.\left. -a\left[\sinh^2 u + \sin^2 v\right]^{1/2}\frac{\partial F_1}{\partial v} - a\left[\sinh^2 u + \sin^2 v\right]^{-1/2}\sin v \cos v\right]\widehat{e}_z\right\}.$$

We present some additional examples.

1. Parabolic cylindrical coordinates (ξ, η, z) whose transformation equations are

$$x = \xi\eta$$
$$y = \frac{1}{2}\left(\eta^2 - \xi^2\right)$$
$$z = z.$$

The ranges of the variables are $-\infty < \xi < \infty$, $0 \le \eta < \infty$, and $-\infty < z < \infty$. It can be shown that

$$h_\xi = h_v = \sqrt{\eta^2 + \xi^2}, \quad h_z = 1;$$
$$dV = \left(\eta^2 + \xi^2\right) d\eta\, d\xi\, dz;$$
$$\Delta f = \nabla^2 f = \frac{1}{\eta^2 + \xi^2}\left(\frac{\partial^2 f}{\partial \eta^2} + \frac{\partial^2 f}{\partial \eta^2}\right) + \frac{\partial^2 f}{\partial z^2}.$$

One example where these coordinates could be used is to describe an electric field around a semiinfinite conducting plate.

2. Parabolic coordinates (ξ, η, φ) whose transformation equations are

$$x = \xi\eta \cos\varphi$$
$$y = \xi\eta \sin\varphi$$
$$z = \frac{1}{2}\left(\eta^2 - \xi^2\right).$$

The ranges of the variables are $0 < \xi < \infty$, $0 \le \eta < \infty$, are $0 \le \varphi < 2\pi$. It can be shown that

$$h_\xi = h_\eta = \sqrt{\eta^2 + \xi^2}, \quad h_z = \xi\eta;$$
$$dV = \xi\eta\left(\eta^2 + \xi^2\right) d\xi\, d\eta\, dz;$$
$$\Delta f = \nabla^2 f = \frac{1}{\eta^2 + \xi^2}\left[\frac{1}{\xi}\frac{\partial}{\partial \xi}\left(\xi\frac{\partial f}{\partial \xi}\right) + \frac{1}{\eta}\frac{\partial}{\partial \eta}\left(\eta\frac{\partial f}{\partial \eta}\right)\right] + \frac{1}{\xi^2\eta^2}\frac{\partial^2 f}{\partial \varphi^2}.$$

3. Bipolar coordinates (ξ, η, z) whose transformation equations are

$$x = \frac{a \sinh\eta}{\cosh\eta - \cos\xi}$$
$$y = \frac{a \sin\xi}{\cosh\eta - \cos\xi}$$
$$z = z.$$

where $0 \leq \xi < 2\pi$, $-\infty < \eta < \infty$ and $-\infty < z < \infty$. It can be shown that

$$h_\xi = h_\eta = \frac{a}{\cosh \eta - \cos \xi}, \quad h_z = 1;$$

$$dV = \frac{a^2}{(\cosh \eta - \cos \xi)^2} d\xi \, d\eta \, dz;$$

$$\Delta f = \nabla^2 f = \frac{(\cosh \eta - \cos \xi)^2}{a^2} \left(\frac{\partial^2 f}{\partial \eta^2} + \frac{\partial^2 f}{\partial \xi^2} \right) + \frac{\partial^2 f}{\partial z^2}.$$

An example where these coordinates could be used is in describing the electric field around two parallel cylindrical cylinders.

4. Prolate spheroidal coordinates (u, v, ϕ) whose transformation equations are

$$x = a \sinh u \sin v \cos \phi$$

$$y = a \sinh u \sin v \sin \phi$$

$$z = a \cosh u \cos v.$$

The ranges of the variables are $0 \leq u < \infty$, $0 \leq v \leq \pi$, are $0 \leq \phi < 2\pi$. It can be shown that

$$h_u = h_v = a\left[\sinh^2 u + \sin^2 v\right]^{1/2}, \quad h_\phi = a \sinh u \sin v;$$

$$dV = a^3 \left(\sinh^2 u + \sin^2 v\right) \sinh u \sin v;$$

$$\Delta f = \nabla^2 f$$

$$= \frac{1}{a^2(\sinh^2 u + \sin^2 v)} \left[\frac{\partial^2 f}{\partial u^2} + \frac{\partial^2 f}{\partial v^2} + \frac{1}{\tanh u} \frac{\partial f}{\partial u} + \frac{1}{\tan v} \frac{\partial f}{\partial v} \right]$$

$$+ \frac{1}{{}^2\sinh^2 u \sin^2 v} \frac{\partial^2 f}{\partial \phi^2}.$$

Prolate spheroidal coordinates is a coordinate system that results from rotating an ellipse about the axis on which the foci are located. Oblate spheroidal coordinates is a coordinate system that results from rotating an ellipse about the axis that separates the foci. Both coordinates are sometimes used to solve partial differential equations when the boundary conditions are defined on particular shapes.

Prolate spheroidal coordinates could be used to describe the electric field generated by two electrode tips.

5. Oblate spheroidal coordinates (u, v, ϕ) whose transformation equations are

$$x = a \cosh u \cos v \cos \phi$$

$$y = a \cosh u \cos v \sin \phi$$

$$z = a \sinh u \sin v.$$

The ranges of the variables are $0 \le u < \infty$, $-\frac{\pi}{2} \le v \le \frac{\pi}{2}$, are $0 \le \phi < 2\pi$. It can be shown that

$$h_u = h_v = a\left[\sinh^2 u + \sin^2 v\right]^{1/2}, \quad h_\phi = \cosh u \cos v;$$

$$dV = a^3 \cosh u \cos v\left(\sinh^2 u + \sin^2 v\right) du\, dv\, d\phi;$$

$$\begin{aligned}\Delta f &= \nabla^2 f \\ &= \frac{1}{a^2(\sinh^2 u + \sin^2 v)\cosh u \cos v}\left[\frac{\partial}{\partial u}\left(a \cosh u \cos v \frac{\partial f}{\partial u}\right)\right. \\ &\quad \left. + \frac{\partial}{\partial v}\left(a \cosh u \cos v \frac{\partial f}{\partial v}\right) + \frac{a^2(\sinh^2 u + \sin^2 v)}{a \cosh u \cos v}\frac{\partial^2 f}{\partial \phi^2}\right].\end{aligned}$$

Oblate coordinates can be used in diffusion problems such as the scattering of sound through a circular hole or flow of liquid through a hole.

Bibliography

Apostol, T.M., 1974. Mathematical Analysis, second ed. Addison −Wesley Publishing Company, Reading MA.

Arfken, G., 1970. Mathematical Methods for Physicists, second ed. Academic Press, New York.

Boas, M.L., 1966. Mathematical Methods in the Physical Sciences. John Wiley & Sons, Inc., New York.

Boyce, W., DiPrima, R., 2008. Elementary Differential Equations and Boundary Value Problems. John Wiley & Sons, Inc., New York.

Brown, J.W., Churchill, R.V., 2008. Fourier Series and Boundary Value Problems. McGraw-Hill Book Company, New York.

Courant, R., Hilbert, D., 1989. Methods of Mathematical Physics, Vol. 1, p. c1937. New York.

Edwards, C., Penny, D., 1994. Elementary Differential Equations with Applications, third ed. Prentice Hall, Englewood Cliffs, NJ.

Edwards, C., Penny, D., 2008. Elementary Differential Equations.

Gelbaum, B., Olmsted, J., 1964. Counterexamples in Analysis. Holden Day, San Francisco.

Growers, T. (Ed.), 2008. The Princeton Companion to Mathematics. Princeton University Press, Princeton, NJ.

Hoskins, R.F., 1979. Generalized Functions. Halsted Press, New York.

Kirkwood, J.R., 1995. An Introduction to Analysis, second ed. Waveland Press, Inc., Prospect Heights, IL.

Kreysig, E., 1967. Advanced Engineering Mathematics, second ed. John Wiley & Sons, Inc., New York.

Marsden, J.E., Tromba, A.J., 1988. Vector Calculus, third ed. W.H. Freeman and Company, New York.

McQuarrie, D., 2003. Mathematical Methods for Scientists and Engineers. University Science Books, Sausalito, CA.

Morse, P.M., Feshbach, H., 1953. Methods of Theoretical Physics. McGraw-Hill Book Company, New York.

Park, D., 1964. Introduction to the Quantum Theory. McGraw-Hill Book Company, New York.

Pinsky, M.A., 1998. Partial Differential Equations and Boundary − Value Problems with Applications, third ed. Waveland Press, Inc, Prospect Heights, IL.

Rogawski, J., 2008. Calculus. W.H. Freeman and Company, New York.

Rudin, W., 1964. Principles of Mathematical Analysis. McGraw-Hill Book Company, New York.

Rudin, W., 1973. Functional Analysis. McGraw-Hill Book Company, New York.

Stakgold, I., 1967a−68. Boundary Problems in Mathematical Physics. Macmillan, New York.

Stakgold, I., 1967b−68. Green's Functions and Boundary Value Problems. Macmillan, New York.

Schey, H.M., 1973. Div, Grad, Curl and All that; an Informal Text on Vector Calculus. Norton, New York.

Weinberger, H.F., 1965. A First Course in Partial Differential Equations. John Wiley and Sons, New York.

Index

'*Note*: Page numbers followed by "f" indicate figures and "t" indicate tables.'

S

T

U

V

W